KdV '95

PROCEEDINGS OF THE INTERNATIONAL SYMPOSIUM
HELD IN AMSTERDAM, THE NETHERLANDS, 1995

Table of Contents

Acta Applicandae Mathematicae **39:** 1–2, 1995.

Preface

The now famous paper "*On the Change of Form of Long Waves Advancing in a Rectangular Canal and on a New Type of Long Stationary Waves*" by D. J. Korteweg and G. de Vries appeared in the *Philosophical Magazine* a century ago. For the board of the Dutch Association for Mathematical Physics, this centennial is a welcome opportunity to commemorate the event with an international symposium on those developments in mathematics and physics that have their roots in what is now known as the Korteweg–de Vries equation. The circumstances that Korteweg held the chair of mathematics and mechanics at the University of Amsterdam from 1881 until 1918 and that de Vries wrote his thesis on the above-mentioned long waves under the supervision of Korteweg are motives to organize this symposium in Amsterdam.

The new type of long stationary waves were called cnoidal waves (in analogy with sinusoidal waves) by Korteweg and de Vries. In the particular case that the modulus of the Jacobian elliptic function equals one, the wave takes the form of the "rounded, smooth and well defined heap of water", discovered by Scott Russell in 1834 and coined much later in 1965 by Zabusky and Kruskal as the *soliton.*

Meanwhile, it has become commonplace to state that the rediscovery of the soliton by Zabusky and Kruskal precursed a breakthrough in nonlinear analysis and mechanics. Besides a better understanding of nonlinear wave phenomena this also brought new important developments in the theory of completely integrable systems.

The board of the Dutch Association of Mathematical Physics and the organizing scientific committee are pleased with the response of many mathematicians and physicists to the invitation to participate in the symposium; in particular we thank those participants who have contributed to the symposium by giving a lecture or presenting a poster.

A very fascinating feature of the theory of solitons lies in the application of many branches of mathematics, such as analysis, differential geometry, algebraic geometry, topology and infinite-dimensional algebraic structures. Apart from this, one should also mention a multitude of applications in physics, for instance in statistical mechanics, point mechanics, hydrodynamics, optics and field theory. These proceedings reflect all these aspects and are in line with the aims and scope that the scientific committee had in mind. Besides this, attention has also been paid to the interesting history of solitons.

The symposium has been held under the auspices of the Royal Netherlands Academy of Arts and Sciences, KNAW.

Nowadays, a symposium is not possible without external help and funding; we express our gratitude to the following organizations and institutions for their generous support, financial or otherwise: the Royal Netherlands Academy of Arts and Sciences; University of Amsterdam; Municipality of the City of Amsterdam; Foundation for Fundamental Research of Matter, FOM; Foundation for Mathematics "Mathematical Centre", SMC; Foundation "Physics"; Centre for Mathematics and Informatics, CWI; Committee "Nonlinear Systems"; Netherlands Organization for Scientific Research; Johan Enschedé BV, Haarlem; IBM Nederland BV, Amsterdam; Kluwer Academic Publishers, Dordrecht; OCE van der Grinten NV, Venlo; PTT Research, Leidschendam; Shell Nederland BV, Rotterdam.

To give PhD students the opportunity to get the optimal profit from the symposium activities, it was considered appropriate to organize during the week preceding the symposium, an introductory course on integrable systems connected with the KdV equation and its relatives. The lectures of this course will be published by the Centre for Mathematics and Informatics in the CWI Syllabus series.

We are indebted to the following organizations who made this course possible by granting financial support: Erasmus Inter University Cooperation Programme, ICP 94; Research School JM Burgers Centre, Delft; Research School MRI, Utrecht; Research School Thomas Stieltjes, Leiden; Dutch Association for Mathematical Physics, FOM/SMC.

Further, it is our duty to acknowledge the secretarial support of Mrs M. I. van der Kooij, University of Twente, who conducted the secretariat of the symposium, Mrs R. Koopmans, Eindhoven University, who assisted the treasurer, and Mrs Ph. Zijlstra, Universty of Amsterdam, who was engaged with the administration of the PhD course.

Finally, we highly appreciate that these proceedings have been published as a special issue of *Acta Applicandae Mathematica*. It is our pleasure to thank Kluwer Academic Publishers for this, and to thanks them for their generous financial support.

E. M. DE JAGER
President Scientific Committee KdV'95

Part I: Invited Plenary Lectures

Acta Applicandae Mathematicae **39:** 5–37, 1995.

Integrability, Computation and Applications

M. J. ABLOWITZ, S. CHAKRAVARTY
Program in Applied Mathematics, University of Colorado, Boulder, CO 80309, U.S.A.

and

B. M. HERBST
Department of Applied Mathematics, The University of the Orange Free State, Bloemfontein 9300, South Africa

(Received: 13 January 1995)

Abstract. The study of integrable systems and the notion of integrability has been re-energized with the discovery that infinite-dimensional systems such as the Korteweg–de Vries equation are integrable. In this paper, the following novel aspects of integrability are described: (i) solutions of Darboux, Brioschi, Halphen-type systems and their relationships to monodromy problems and automorphic functions, (ii) computational chaos in integrable systems, (iii) we explain why we believe that homoclinic structures and homoclinic chaos associated with nonlinear integrable wave problems, will be observed in appropriate laboratory experiments.

Mathematics Subject Classification (1991): 58F07.

Key words: integrability, KdV equation, nonlinear integrable wave functions, monodromy problems, automorphic functions.

1. Introduction

The study of 'integrable' or exactly solvable systems has a long history. Classical mathematicians such as Euler, Lagrange, Liouville, Riemann, Poincaré, Painlevé amongst many others, investigated nonlinear systems which could either be integrated more or less explicitly or possessed special analytic structures in the complex plane. Perhaps surprisingly, there is still no single adequate definition of 'integrability'. Certainly explicit integration of nonlinear systems in the real domain should be considered as integrable, as should cases where suitable transformations exist to allow an elementary solution, such as occurs in the Hamiltonian case with action-angle variables (often called integrability in the Liouville sense). Less clear is the still developing notion of integrability in the complex plane. For example, if the general solution of an ODE has appropriate analytic properties, such as (i) having poles as the only movable singular points (a movable singular point is one whose location depends on the initial conditions; a fixed singular point is fixed by the coefficients in the equation) or more generally (ii) being everywhere single-valued in its domain of existence, then we consider the equation to be integrable in the complex plane. Many examples of nonlinear

ODEs have shown that when (i) or (ii) are satisfied then the equation falls into a class in which the general solution can be either obtained by explicit integration or can be linearized via an associated Riemann–Hilbert (RH) factorization problem.

The investigation of integrable systems has been an interesting and active field in recent years due to the fact that numerous physically interesting infinite-dimensional systems have been linearized via the method of the Inverse Scattering Transform (IST) and large classes of explicit solutions, such as soliton solutions, have been obtained. The best known example is the Korteweg–de Vries (KdV) equation,

$$u_t + 6uu_x + u_{xxx} = 0, \tag{1}$$

which these proceedings have been named after. The KdV equation arises in many physical problems. It is the canonical equation describing weakly dispersive and weakly nonlinear wave phenomena arising in shallow water waves (this was the original application of Korteweg and de Vries), internal waves in fluids, plasma waves, lattice dynamics, etc.

The IST method yields the general solution to the Cauchy problem on the infinite line corresponding to rapidly decaying initial conditions (cf. [1–3]), a special case of this is the general N-soliton solution. The IST method uses direct and inverse scattering analysis to obtain a matrix RH factorization problem which leads to a Gel'fand–Levitan–Marchenko integral equation as the linearization of KdV. The KdV equation has been investigated in a variety of contexts and we note that there is one other case in which the general initial-value solution is known. This is the periodic boundary-value problem where the solution of the KdV equation can be expressed in terms of Riemann theta functions of arbitrary genus (cf. [1, 2]).

The IST is an effective tool to obtain linearizations and solutions to many other well known nonlinear wave equations in $(1 + 1)$ dimensions, e.g., the nonlinear Schrödinger equation (NLS), sine-Gordon equation, three wave interaction $(1 + 1)$ equations, etc. Moreover, it is significant that there are many $(2 + 1)$-dimensional equations which have solutions via the IST method. The best known are the Kadomtsev–Petviashvili (KP), Davey–Stewartson (DS) and three wave interaction $(2 + 1)$ equations. We note that the IST method needs to be generalized appropriately in order to handle $(2 + 1)$ problems; it was found that nonlocal RH problems and DBAR problems played essential roles (cf. [3]).

In this paper, we shall discuss two issues related to integrability. In the first part, we shall focus on specific differential equations which have novel properties. These equations are obtained from a well known integrable system in four dimensions, the so-called self-dual Yang–Mills (SDYM) system. The other topic we shall discuss involves the computation of integrable systems in the vicinity of what are called homoclinic manifolds. In these regions, the solutions of the nonlinear equation are very sensitive. This allows numerical approximations to the equations to induce a chaotic response. In fact we have observed computational chaos due to (i) truncation effects and (ii) very small errors even on the order of roundoff error. Our investigations and analysis also strongly suggest that analogous experimental investigations (e.g., in water waves and/or nonlinear optics) will be possible and that homoclinic manifolds and related chaotic phenomena will be observable experimentally.

In higher dimensions the SDYM system is one of the few (perhaps the only(?)) known physically important four-dimensional system to which the IST method applies. Significantly, the SDYM system turns out to encompass many of the integrable systems we already know. The self-dual Yang–Mills equations (SDYM) arise in particle physics and relativity (see [3] for many references on SDYM).

Specification of a self-dual system requires the specification of (i) the Lie algebra in which the field variables reside and (ii) the specification of which independent variables will be chosen to keep (or equivalently which variables to ignore). Reductions of the SDYM equations to the well-known soliton equations in $(1+1)$ dimensions such as KdV, NLS, sine-Gordon, N-wave and respective hierarchies, etc., all follow by considering two independent variables with the field variables lying in a finite-dimensional Lie algebra (cf. [3]). However, the well-known soliton equations in $(2+1)$ dimensions, such as KP, DS, N-wave and respective hierarchies follow from SDYM by assuming two independent variables with the field variables lying in a suitable infinite-dimensional algebra [4].

Important and interesting $(0+1)$-dimensional equations (ODEs) are also obtained as reductions of SDYM. By reducing to one independent variable with the field variables lying in a finite-dimensional Lie algebra, it can be shown that the free spinning Euler top, the 'heavy' Kovalevskya top [5], and the six irreducible Painlevé transcendents [6] are special cases. Moreover, a novel system which we call the Darboux–Brioschi–Halphen (DBH) top [5, 7], and is related to self-dual Bianchi IX cosmologies, is also obtained from the SDYM equations with one independent variable, but in this case we take the field variables to lie in an infinite-dimensional algebra of volume preserving diffeomorphisms. The DBH system as the interesting analytic property that it is single-valued everywhere in its domain of existence, but there is a natural boundary beyond which the solution cannot be analytically continued.

The solution of the DBH top has been discussed earlier and there are important connections with modular functions (cf. [7]). However, we have only recently obtained the associated linear pair (by reduction of the associated SDYM linear pair) and now have obtained the solution by investigating the relevant monodromy problem [8]. In this paper, we shall show that the linear problem leads to a monodromy problem whose monodromy *is not* fixed ('nonisomonodromy') in time. Nevertheless, we can 'factorize' the problem into isomonodromy and nonisomonodromy parts in such a way so as to be able to solve the DBH system.

We have also been able to solve interesting generalizations of the DBH system, of fifth order (we call this the DBH-5 system), which relate to nondiagonal self-dual Bianchi IX cosmologies. Once again there is an associated linear pair which yields solutions. However, it turns out that in this case the solution is not always single-valued, and in fact depending on the initial values, sometimes the solution is densely branched. In the latter case, the equation is not of Painlevé type despite the fact that there is an associated linear pair and that the DBH-5 system is obtained by reduction (with an infinite-dimensional gauge algebra) from the SDYM equations.

In our computational studies, we have been carrying out moderate to long-time numerical calculations associated with integrable nonlinear wave equations with periodic boundary conditions. It is well known that many of these equations have solutions which are linearly unstable and have interesting associated mathematical and physical phenomena. These unstable solutions are in a sense close to homoclinic manifolds. This fact has serious consequences which are reflected in numerical calculations (and we believe physical experments). Prototypical examples are the focusing nonlinear Schrödinger, sine-Gordon and modified KdV equations. (We note in passing that the (real-valued) KdV equation, the defocusing NLS and the sinh-Gordon equations are not amongst this class of equations.) We have observed that corresponding to certain initial/boundary conditions, these equations admit a special class of solutions which we refer to as homoclinic orbits. Alternatively, one can view these solutions geometrically as being part of an infinite-dimensional phase space consisting of an infinite-dimensional torus which has associated homoclinic manifolds ('pinched tori' [10]). In analogy with low-dimensional dynamics, due to perturbations, crossing of these homoclinic regions can drive a chaotic flow. In the numerical problem, the perturbations are due to either truncation or roundoff effects.

We have shown [11, 12, 14] that in the *true* nonlinear wave equation, small perturbations around a particular solution, say of order epsilon, can translate into close proximity to the homoclinic manifolds with a 'distance' of *any* order of epsilon and possibly even BEYOND all orders in epsilon. The choice of these initial conditions is straightforward, depending only on (a) the number of unstable modes that are contained in the initial data, and (b) the frequency of the initial perturbation. In order to demonstrate this situation mathematically, we have found that it is useful to employ the underlying IST formalism. In particular, we have found that homoclinic orbits correspond to suitable double points in the spectra of the associated scattering problem. The 'distance' from these orbits (manifolds) correspond to the order of epsilon in the asymptotic expansion around the double points. We have also found that exhibiting the 'sides' of the homoclinic orbits is straightforward, depending on the choice of phase in the initial conditions.

With this information in hand, we can understand why numerical experiments exhibit chaotic dynamics either due to truncation or roundoff effects. Truncation effects become important when the distance to the homoclinic manifolds are less than the order of truncation. On the other hand, roundoff effects are critical when the distance to the homoclinic manifold is less than the order of roundoff. We have observed these effects in many numerical experiments.

As mentioned earlier, we believe that observations of homoclinic orbits and associated chaotic dynamics should be possible in laboratory experiments. For example, the NLS model is well known to describe weakly nonlinear slow modulations of a wave train in water waves or nonlinear optics. In the physical problem, homoclinic orbits should be 'seen' by observing *different* wave structures on either 'side' of the orbit; since the homoclinic orbit is an unstable wave mode, one does *not* observe it directly. The different 'sides' depend on the phase of the perturbation of the initial conditions, or alternatively, how the double point in the eigenvalue spectrum, splits in the complex plane. We shall discuss this more concretely later in this paper.

On the other hand, we know that the distance to the homoclinic manifold depends on the number of unstable modes in the initial conditions and the frequency of the perturbation to the true solution. With this information we can create data which is arbitrarily close to the homoclinic manifold. Physically this means that the magnitude of the higher-order terms in the perturbation expansion, from which we obtained the NLS equation at leading order, would be large enough so that these (small) terms are, in fact, larger than the distance to the homoclinic orbit. The perturbations should therefore be capable of driving a chaotic response via crossing of homoclinic manifolds. Naturally, the precise conditions for which chaos will be observed will depend on laboratory considerations; e.g., the size of the amplitude of the true periodic wave, the number of unstable modes that can be generated; and other small perturbations which we have omitted from our discussions, such as dissipation. Nevertheless, we believe that the ability to observe homoclinic structures in wave problems is fundamental, and important. It is well understood theoretically and consequently we make the prediction that in the experiments such as we have sketched here they should be observed.

2. Integrability, PDEs and ODEs

2.1. Reduction of self-dual Yang–Mills equations

In this section we give a brief overview of the SDYM equations in a complex Euclidean space $\mathbb{C}^4$, and outline the method of symmetry reductions to integrable systems in lower dimensions. We only outline the reduction procedure in general terms. By necessity, we leave out most of the technical details which can be found in the references cited in the text.

2.1.1. *SDYM in* $\mathbb{C}^4$

We introduce the complex coordinates, $y, \bar{y}, z, \bar{z}$ in $\mathbb{C}^4$ and the differential operators D_1, A_1, D_2 and A_2 defined by

$$D_1 \equiv \partial_y + \lambda\partial_{\bar{z}}, \qquad D_2 \equiv \partial_z - \lambda\partial_{\bar{y}},$$

$$A_1 \equiv A_y + \lambda A_{\bar{z}}, \qquad A_2 \equiv A_z - \lambda A_{\bar{y}}, \tag{2}$$

where $\partial_\alpha \equiv \partial/\partial_\alpha$, $\alpha \in \{y, \bar{y}, z, \bar{z}\}$, $\lambda \in \mathbb{C}P1$ is a complex parameter. $A_\alpha = A_\alpha(x) \in g$, $x \in \mathbb{C}^4$, where g is the Lie algebra corresponding to some fixed gauge group G.

The A_α's are called the Yang–Mills gauge potentials and are defined up to the gauge freedom,

$$A_\alpha \longrightarrow A'_\alpha = fA_\alpha f^{-1} + \partial_\alpha f f^{-1}$$

for $f \in G$. Both A_α and A'_α satisfy the SDYM equations which can be expressed as

$$[D_1 - A_1, D_2 - A_2] = 0, \tag{3}$$

which are the compatibility conditions for the linear system

$$(D_1 - A_1)\psi = (D_2 - A_2)\psi, \tag{4}$$

where $\psi(x, \lambda) \in G$. Equation (3) is quadratic in λ, so by setting each of its coefficients to zero, yields a system of three nonlinear PDEs in $\mathbb{C}^4$ given by

$$F_{yz} = F_{\bar{y}z} = F_{y\bar{y}} + F_{z\bar{z}} = 0 \tag{5}$$

Here $F \equiv \partial_\alpha A_\beta - \partial_\beta A_\alpha - [A_\alpha, A_\beta]$ and $[\,,\,]$ denotes the Lie bracket associated with the Lie algebra g. The linear system (3) is the Lax pair for the SDYM equations given by (5). The above formulation of the SDYM equations has a natural geometric structure: Equation (2) is the zero curvature condition for a connection (with components A_1, A_2) associated with holomorphic sections $\psi(x, \lambda) \in G$ of the Yang–Mills vector bundle. The bundle is restricted to a family of two-planes, called the anti-self-dual (ASD) planes, passing through each $x \in \mathbb{C}^4$ and parametrized by $\lambda \in \mathbb{C}P1$; the vector fields D_1, D_2 are tangent to each ASD plane. The connection is defined by the linear system (4) by requiring that the sections $\psi(x, \lambda)$ be covariantly constant on each ASD plane. The zero curvature condition (2) (equivalently, (5)) guarantees that (4) is integrable on the ASD surfaces and that it admits solutions defined globally over each ASD plane. The space of such solutions considered over the family of two-planes determine a holomorphic vector bundle. Therefore, this geometric construction establishes a one-to-one correspondence between holomorphic vector bundles and solutions of the SDYM equations in $\mathbb{C}^4$. This geometric formulation was given by Ward [15].

Conversely, given such a vector bundle, it is possible to obtain the solutions to (3) that are holomorphic in appropriate neighborhoods of λ (e.g., $\lambda = 0$ and $\lambda = \infty$) via a Riemann–Hilbert factorization problem parametrized by $x \in \mathbb{C}^4$.

2.1.2. *The Reduction Scheme*

The main relationship between the SDYM equations and integrable systems arises from the fact that an enormous number of integrable equations can be obtained from the SDYM equations by imposing appropriate symmetry conditions. This is a very large program of study and a systematic analysis is still lacking. Nevertheless, significant work during the past 15 years or so indicates that most of the well-known equations solvable by IST are embedded in the SDYM equations. The essential steps of the reduction procedure are outlined below:

(i) Coordinate symmetries are imposed such that the SDYM equations are invariant under a subgroup of the conformal group (the full symmetry group of the SDYM equations). In a suitable gauge, the SDYM connection A_α is invariant on the orbits of the symmetry group. Introducing a new set of coordinates $\{x^{K'}, y^K\}$ on $\mathbb{C}^4$, it is then possible to reduce the SDYM equations to a lower-dimensional system of PDEs which are expressed only in terms of the y^K's. Here $x^{K'}, K' = 1, 2, \ldots, 4 - m$, are the coordinates on each orbit, while $y^K, K = 1, \ldots, m$, are the coordinates that label the orbits. For example, when the symmetry group consists of $4 - m$ translations, the SDYM potentials are simply given by $A_\alpha(x) = A_\alpha(y^K)$. Then in (5) all the derivatives involving the ignorable coordinates vanish so that the resulting SDYM equations are reduced to a system of PDEs in the independent variables y^K, $K = 1, \ldots, m$.

(ii) A choice of the gauge group G is necessary such that the gauge potential A_α lie in the corresponding Lie algebra g. For a finite-dimensional gauge group, the elements of g (and hence the A_α's) can be represented by finite rank matrices. Most $(1+1)$ integrable equations, e.g., the KdV, NLS and the N-wave equations and many ODEs including the Euler–Arnold and Kovalevskya 'tops', arise from such a choice of gauge group. On the other hand, interesting reductions are also obtained when the A_α's belong to specific representations of *infinite*-dimensional Lie algebras. Examples include the self-dual Einstein equation, the KP, DS and the $(2+1)$ N-wave equations. A special ODE reduction of the self-dual Einstein equation, called the DBH system will be discussed in the next section.

(iii) The SDYM equations reduced via steps (i) and (ii) can be further simplified by the residual gauge freedom and by imposing certain autonomy conditions on the dependent variables. After imposing the coordinate symmetries in step (i), the remaining freedom in the gauge transformation, $A_\alpha \to A'_\alpha = fA_\alpha f^{-1} + \partial_\alpha f f^{-1}$, $f(x) \in G$, is essentially restricted to $f(x) = f(y^K)$. Nevertheless, by making judicious choices for f (that takes some of the A_α's into irreducible ('normal') forms), certain components of the reduced SDYM system can often be trivially solved. The autonomy condition is applied to certain arbitrary functions arising as 'constants' of integration in the reduction process. It requires that if a function is constant with respect to at least one of the independent variables, then it should be set to a constant. Thus this condition guarantees that the symmetry reduced SDYM equations will be an autonomous system.

The following additional remarks are also relevant to the reduction scheme described above.

Remarks. 1. Many equations which are symmetry reductions follow directly from the reduced system of SDYM equations obtained after applying steps (i)–(iii) and most importantly, without any further assumptions. It is a remarkable feature that a large class of integrable equations are naturally embedded in the SDYM equations in four-dimensions. However, sometimes it is necessary to make specific algebraic choices for the SDYM potentials in order to identify the reduced equations with a particular integrable system. The KdV equation is an example of a natural embedding, whereas the N-wave system is derived from the SDYM equations with additional assumptions on the A_α's (i.e. certain components are closen to be skew-symmetric matrices).

2. Another ingredient is to impose an appropriate reality condition on the reduced system in order to derive the soliton-type integrable equations which are usually defined on the real domain. The equations derived from the reduction of SDYM in $\mathbb{C}^4$ are complexified versions of these soliton-type equations. However, it is indeed possible to choose real slices of $\mathbb{C}^4$ with a metric of appropriate signature, and impose suitable reality conditions on the connection components in order to obtain integrable equations defined on the real domain. We will not discuss the question of the reality condition in this article, since our (primary) objective here is to discuss issues related to integrability in the complex domain. Interested readers are referred to [3] for many references on SDYM.

2.1.3. *Reduction to KdV*

We illustrate the reduction scheme outlined in Subsection 2.1.2 with an example. Here we will derive the KdV equation as a symmetry reduction of the SDYM equations. We point out only the key features of the derivation, leaving aside the technical details which can be found in [4, 17]. We begin by imposing the translational symmetries (step (i)) that the A_α's be independent of the variables $x^{K'} = (y - \bar{y}, z) \in \mathbb{C}^4$. Then we redefine the remaining coordinates as $y^K = (y - \bar{y} \equiv x,\ z \equiv t)$ such that each component of the gauge potential is now of the form $A_\alpha = A_\alpha(x, t)$. Here it is convenient to work with the Lax pair, (4), rather than the SDYM equations (5) themselves. Under these assumptions, we have $D_1 = \partial_x$, $D_2 = \partial_t - \lambda\partial_x$, and (4) becomes

$$\partial_x\psi = (A_y + \lambda A_{\bar{x}})\psi, \qquad (\partial_t - \lambda\partial_x)\psi = (A_z - \lambda A_{\bar{y}})\psi.$$

Substituting the first equation into the second, we obtain a $(1+1)$ linear system for the reduced form of the Lax pair

$$\partial_x\psi = (A_y + \lambda A_{\bar{z}})\psi, \qquad \partial_t\psi = (A_z - \lambda(A_{\bar{y}} + A_y) + \lambda^2 A_{\bar{z}})\psi. \tag{6}$$

Next we assume that the A_α's are elements of the Lie algebra $\mathrm{sl}(2,\mathbb{C})$ (step (ii)) and represent them by the (2×2) trace-free matrices, $A_y \equiv U$, $A_{\bar{z}} \equiv A$, $A_{\bar{y}} \equiv B$, $A_z \equiv V$. The integrability condition of (6) yields the following reduced SDYM equations,

$$\partial_t U - \partial_x V + [U, V] = 0, \tag{7}$$

$$\partial_x(U + B) = [U, B] + [A, V], \tag{8}$$

$$\partial_x A = [A, B]. \tag{9}$$

These equations are simplified considerably by step (iii) of the reduction scheme. It can be shown [4] that there exists a gauge choice in which A, B satisfies $[A, B] = 0$, so that from (9), $\partial_x A = 0$. It follows immediately from the autonomy condition that A must be a constant $\mathrm{sl}(2,\mathbb{C})$ matrix. The remaining gauge freedom allows A to be determined up to conjugation, $A \to fAf^{-1}$ by a constant $\mathrm{sl}(2,\mathbb{C})$ matrix f. This can be used to put A in either of the normal forms for the (2×2) trace-free matrices

$$\text{(i)} \quad A = \begin{pmatrix} 0 & 0 \\ 1 & 0 \end{pmatrix} \quad \text{or} \quad \begin{pmatrix} 0 & 1 \\ 0 & 0 \end{pmatrix},$$

$$\text{(ii)} \quad A = \begin{pmatrix} 1 & 0 \\ 0 & -1 \end{pmatrix}.$$

Cases (i) and (ii) lead to the KdV and NLS equations, respectively. We consider only case (i) here; the treatment of case (ii) is very similar. The residual gauge freedom (that preserves the normal form of A given above as well as leaves the condition $[A, B] = 0$ invariant) is used to choose U to be off-diagonal. Therefore, we have the following matrices,

$$A = \begin{pmatrix} 0 & 0 \\ 1 & 0 \end{pmatrix}, \qquad U = \begin{pmatrix} 0 & \alpha \\ u & 0 \end{pmatrix},$$

$$B = \beta\begin{pmatrix} 0 & 0 \\ 1 & 0 \end{pmatrix}, \qquad V = \begin{pmatrix} v_1 & v_2 \\ v_3 & -v_1 \end{pmatrix},$$

which must satisfy the system given by (7), (8); (9) is identically satisfied in this gauge. From the (1,2) element of (8), it follows (after using the autonomy condition) that α must be a constant. We set $\alpha = 1$, since the case $\alpha = 0$ does not lead to any nontrivial equation. The remaining equations in (8) and (7) can now be solved systematically to determine β and the elements of V in terms of u and its x-derivatives

$$\beta = \tfrac{1}{2}u, \qquad V = \tfrac{1}{4}\begin{pmatrix} \partial_x u & -2u \\ \partial_{xx}u - 2u^2 & -\partial_x u \end{pmatrix}.$$

Now it follows (from the (2,1) element of Equation (7)) that $u(x,t)$ must satisfy the KdV equation

$$4\partial_x u = \partial_{xxx}u - 6u\partial_x u.$$

2.2. The Darboux–Brioschi–Halphen (DBH) system

The symmetry reductions of the SDYM equations gives rise to ODEs in the case where the gauge potentials are restricted to depend on one independent variable. For example, the classical ODEs mentioned in the introduction arise in this way where the underlying gauge group is finite-dimensional. Here we will discuss an interesting system of three coupled ODEs arising from the SDYM equations but in this case the gauge field variables are associated with an infinite-dimensional Lie algebra of volume preserving diffeomorphisms on a 3-sphere, i.e., g = sdiff($\mathbf{S}^3$). This system possesses several remarkable properties which are dramatically different from the other classical ODEs. Although they have been found recently as a SDYM reduction [5], the origin of this system dates back to the last century! It was found by Darboux in his study of orthogonal systems of curvilinear coordinates in $\mathbb{R}^N$. Brioschi and Halphen solved this system in 1881 and we will call it the DBH system. It is interesting that it has appeared only recently in problems in mathematical physics. It was found as the self-dual Einstein equations [19] corresponding to an Euclidian Bianchi-type-IX metric in diagonal form and more recently, it was discovered by Dubrovin [18] as a special similarity solution of the Witten–Dijkgraaf–Verlinde equations in topological field theory. We begin with a brief outline of its derivation from the SDYM equations.

Let the gauge potentials depend on only one coordinate $y+\bar{y} \equiv t$ by imposing three translational symmetries on the SDYM equations. Then fix the gauge by setting $A_y + A_{\bar{y}} = 0$ and introduce the variables

$$A_1 = (A_z + A_{\bar{z}})/2, \quad A_2 = i(A_z - A_{\bar{z}})/2 \quad \text{and} \quad A_3 = i(A_y - A_{\bar{y}})/2,$$

where $A_l = A_l(t)$, $l = 1, 2, 3$. In terms of these new variables, the SDYM equations (5) are expressed as

$$\frac{\mathrm{d}A_l}{\mathrm{d}t} = [A_j, A_k], \quad l \neq j \neq k \text{ and cyclic.} \tag{10}$$

where the [,] is the Lie bracket associated with the left-invariant su(2)-generators X_l of the algebra sdiff($\mathbf{S}^3$) satisfying $[X_l, X_j] = X_k$, $l \neq j \neq k$, and cyclic. The vector fields X_l's are the infinitesimal generators for the left SU(2) action on $\mathbf{S}^3$ leaving the standard metric and the volume form on $\mathbf{S}^3$ invariant. The gauge potentials are chosen as follows: Let O: SU(2) $\to$ SO(3) be a map defined by $O(g) \equiv O_{ij}(g) \in \mathrm{SO}(3)$, $i, j = 1, 2, 3$, $g\sigma_i g^{-1} = \sum_j O_{ij}\sigma_j$, $g \in \mathrm{SU}(2)$ and σ_i's are the standard Pauli matrices. Then the A_i's assume the special form

$$A_i(t) = \sum_j O_{ij}(g)\omega_j(t)X_j.$$

Substituting this in (10) and using the property $2iX_k(g) = g\sigma_k$ for the vector fields (which follows from the definition of left action), yields the DBH system for the ω_i's

$$\partial_t\omega_i = \omega_j\omega_k - \omega_i(\omega_j + \omega_k), \quad i \neq j \neq k, \text{ and cyclic.} \tag{11}$$

The important features of the DBH system which distinguish it from the classical ODEs are as follows:

(i) Its solutions are single-valued in their domain of existence in the complex plane; but they cannot be analytically continued beyond a dense set of essential singularities which form a natural boundary. In contrast, the solutions to classical ODEs which are integrable in the complex plane are single-valued meromorphic functions in the complex plane with simple poles as the only movable singularities.

(ii) The DBH system can be explicitly integrated in terms of the Schwartz triangle function $S(0, 0, 0; t)$ which is the unique inverse of a conformal map from the upper half S-plane to an interior region F in the t-plane bounded by 3 circular arcs intersecting at 0 angles. This fundamental triangle F 'tiles' the interior of a circle C orthogonal to (each side of) F uniformly by a fractional linear transformation $t \to \gamma(t)$, $\gamma \in \Gamma$; Γ being a discrete subgroup of the modular group PSL$(2, \mathbb{Z})$. This uniformization (or 'tiling') is restricted only to the interior of C and cannot be continued beyond. The Schwartz function is invariant under the transformation $\gamma(t)$: $S(\gamma(t)) = S(t)$ and is the fundamental modular function for the subgroup Γ. $S(t)$ is single-valued within all of C but has no analytic continuation beyond C which forms its natural boundary. The solution to (11) is given by

$$\omega_i = -\tfrac{1}{2}\partial_t \log E_i, \tag{12}$$

where the E_i's are defined by

$$E_1 = \frac{\partial_t S}{S}, \qquad E_2 = \frac{\partial_t S}{S-1}, \qquad E_3 = \frac{\partial_t S}{S(S-1)}, \tag{13}$$

are the (regular) modular forms of weight 2 associated with Γ. The ω_i's also inherit the same natural boundary C whose center and radius can be prescribed in terms of their initial values. Using (12), (13) the following transformation property can be obtained for the ω_i's: If $\omega_i(t)$ is a solution of (11), then

$$\widetilde{\omega}_i(t) = \frac{\omega_i(\phi(t))}{(ct+d)^2} + \frac{c}{ct+d},$$

is also a solution, where

$$\phi = \begin{pmatrix} a & b \\ c & d \end{pmatrix} \in \mathrm{PSL}(2,\mathbb{C}), \quad \phi(t) = \frac{at+b}{ct+d}.$$

This can be also verified directly from (11) which is invariant under this transformation. Equation (11) reduces to a single third-order ODE studied by Chazy in 1909 [20],

$$\partial_{ttt}y = 2\partial_t y \partial_{tt} y - 3(\partial_t y)^2, \tag{14}$$

in terms of the variable $y(t) = -2(\omega_1 + \omega_2 + \omega_3)$. Its solution can be obtained by introducing a new variable s such that $t(s) = z_2/z_1, y(t(s)) = 6\partial_t \log z_1$, where z_1 and z_2 are two linearly independent solutions of certain hypergeometric equations. The inverse function $s(t)$ is a different Schwartz triangle function corresponding to the fundamental domain F' with angles $0, \pi/2$, and $\pi/3$. In this case, the transformation group is the full modular group $\mathrm{PSL}(2,\mathbb{Z})$ and the solution $y(t)$ can be expressed as

$$y(t) = \tfrac{1}{2}\partial_t \log\left[\frac{(\partial_t s)^6}{s^4(s-1)^3}\right] = \tfrac{1}{2}\partial_t \log\Delta = \pi i E_2.$$

Here Δ is the unique modular 'cusp' form of weight 12, and E_2 is the normalized Eisenstein series of weight 2 associated with the modular group $\mathrm{PSL}(2,\mathbb{Z})$. It should be clear from above that the solutions of DBH system are quite different from the classical ODEs whose solutions are (a) singly or doubly periodic in the complex plane and are given in terms of the Jacobi elliptic functions (for, e.g., classical 'tops'), (b) expressible in terms of Riemann theta functions of finite genus (associated with periodic solutions of the soliton-type equations), or (c) one of the Painlevé equations.

(iii) Unlike the classical ODEs, the DBH system does not possess any polynomial or rational first integrals, so that it is not algebraically integrable. However, it does admit two transcendental first integrals A and B [21] which can be expressed in terms of complete elliptic integrals. We do not include here the rather complicated formulas for A and B but note that (11) can be expressed in terms of A and B by the (vector product) formula

$$\partial_t \omega_i = M(\omega_i)(\nabla A \times \nabla B)_i,$$

where $(\nabla f)_i \equiv \partial f/\partial\omega_i$ and the integrating factor $M(\omega_i) = (\omega_1 - \omega_2)(\omega_2 - \omega_3)(\omega_3 - \omega_1)$. The equation given above is an example of a Nambu–Hamilton equation of motion [22]. Such systems are natural generalizations of the canonical Hamiltonian dynamical systems with $2N$ degrees of freedom to those with $3N$ degrees of freedom. The above formula indicates that the phase space (excluding the planes $\omega_i - \omega_j = 0$, $i \neq j$, of the DBH system) admits a nondegenerate Nambu-bracket structure in analogy to the Poisson-bracket structure of the usual Hamiltonian dynamics.

2.2.1. *Lax Pair for the DBH System*

We consider now the Lax pair of the DBH system and discuss how to recover its solution from the linear problem. It will be shown that the Lax pair actually leads to a monodromy problem associated with a special case of the Painlevé VI equations whose solutions are hypergeometric functions. It is then possible to express the solutions of the DBH system in terms of these special P-VI solutions. In this sense, the Lax pair provides an effective 'linearization' of the DBH system. Starting from the linear system (4) for the SDYM equations, it is possible to show that the DBH system (11) is obtained (after a trivial rescaling of the independent variable t) as the integrability conditions for the following pair of ODEs,

$$L\psi \equiv \left(P\frac{\mathrm{d}}{\mathrm{d}\lambda} + \ell\right)\psi = \mu\psi,$$

$$M\psi \equiv \left(\frac{\mathrm{d}}{\mathrm{d}\lambda} + \ell_1 + f_1\frac{\mathrm{d}}{\mathrm{d}\lambda}\right)\psi = \nu\psi \tag{15}$$

for $\psi(\lambda, t) \in \mathbb{C}^2$ provided the function ν satisfies the auxiliary condition

$$\frac{\mathrm{d}\nu}{\mathrm{d}\lambda} = \frac{A}{P}\mu, \tag{16}$$

μ being a complex constant. P, ℓ, ℓ_1, f_1 and A are defined below in Equations (18)–(20). Alternatively, (11) also follows from the commutator relation

$$[L, M] = A(\lambda, t)L \tag{17}$$

in terms of the linear differential operators L and M. The relevant quantities in (15), (16) (and (17)) are given as follows: First consider the $\mathrm{sl}(2, \mathbb{C})$ matrices,

$$X = \omega_1 X_1 - \omega_2 X_2, \qquad Y = \omega_1 X_1 + \omega_2 X_2, \qquad Z = \omega_3 X_3,$$

where

$$X_1 = i\sigma_1, \qquad X_2 = i\sigma_2, \qquad X_3 = \sigma_3,$$

σ_i's being the standard Pauli matrices, and $\omega_i = \omega_i(t)$ are the DBH variables. Then, we define

$$\ell = \lambda^2 X + 2\lambda Z + Y, \qquad \ell_1 = Z + \lambda X. \tag{18}$$

Next we have the polynomials (in λ) with the coefficients depending on the ω_i's,

$$P = \alpha(\lambda^4 + 2r\lambda^2 + 1), \qquad f_1 = \alpha\lambda^3 + \beta\lambda,$$
$$A = (\omega_1 + \omega_2 + 2\omega_3) - \alpha\lambda^2, \tag{19}$$

where

$$\begin{aligned} \alpha &= \omega_1 - \omega_2, \qquad \beta = \omega_1 + \omega_2 - 2\omega_3, \\ r &= \beta/\alpha = (s+1)/(s-1), \qquad s = (\omega_1 - \omega_3)/(\omega_2 - \omega_3). \end{aligned} \tag{20}$$

It is a remarkable fact that the original infinite-dimensional Lax pair (recall that the SDYM potentials $A_i \in \mathrm{sdiff}(\mathbf{S}^3)$) can indeed be reduced to a pair of complex vector-valued ODEs. However, the derivation of (15), (16) from (4) is nontrivial and involves technicalities that are outside the scope of the present article. Therefore, they will be presented elsewhere. Equation (15) can be simplified further by eliminating the variables μ and ν via a simple rescaling $\psi = f\psi'$ for some scalar function $f(\lambda, t)$. Then ψ' satisfies

$$\frac{\mathrm{d}\psi'}{\mathrm{d}\lambda} = \frac{\ell}{P}\psi', \tag{21a}$$

$$\frac{\mathrm{d}\psi'}{\mathrm{d}t} + \ell_1\psi' + f_1\frac{\mathrm{d}\psi'}{\mathrm{d}\lambda} = 0, \tag{21b}$$

with the additional equations for f: $\partial_\lambda f = \mu f/P$ and $\partial_t f = \nu f - f_1\partial_\lambda f$. The compatibility condition for the existence of such a function f given by $\partial_{\lambda t} f = \partial_{t\lambda} f$ requires that

$$\partial_t P + f_1\partial_\lambda P - P\partial_\lambda f_1 + AP = 0. \tag{22}$$

Equations (21), (22) provide an equivalent formulation of the DBH Lax pair (15); the DBH system is obtained from the compatibility condition of (21a) and (21b) together with the additional condition given by (22). Note that since P is a polynomial in λ with distinct roots, (21a) is a Fuchsian ODE with regular singular points $\{\lambda_i\}$ and whose fundamental solution ψ' is a multivalued function, analytic in $\lambda \in \mathbb{C}P1 - \{\lambda_i\}$.

2.2.2. *The Monodromy Problem Associated with the DBH System*

In 1857, Riemann formulated the classical problem (the so-called Riemann problem) of constructing a fundamental set of solutions with regular singularities from a prescribed monodromy data. He also studied the deformation of the solution with respect to the singular points when the monodromy remains invariant. The

Riemann problem was subsequently studied extensively by others. Plemelj [23] proved the existence of solutions (under certain assumptions) corresponding to an nth order Fuchsian equation with m regular singular points (for arbitrary m and n), by reducing the Riemann problem to Fredholm-type integral equations. This was also investigated independently by Birkhoff [24] who also generalized the problem to the case of irregular singular points. The monodromy preserving deformation (isomonodromy) problem of such ODEs (with regular as well as irregular singularities) was investigated by Schlesinger, Fuchs, Garnier and more recently by Jimbo *et al.* [25]. For a class of second-order ODEs or (2×2) matrix Fuchsian equations, it was found that the integrability conditions for the associated deformation equations led to the six classical Painlevé transcendents. We show that (21a) belongs to this class of ODEs, namely, the compatibility conditions for (21a) and (21b) is precisely the isomonodromy conditions associated with the Fuchsian ODE (21a) and leads to a Painlevé VI equation with special parameter values. We rewrite (21a) as

$$\frac{\mathrm{d}\psi'}{\mathrm{d}\lambda} = \sum_{i=1}^{4} \left(\frac{\ell_i}{\lambda - \lambda_i} \right) \psi'. \tag{23}$$

Here the singular points $\lambda = \lambda_i$, $i = 1, \ldots, 4$, are the distinct roots of

$$P' = P/\alpha = \lambda^4 + 2r\lambda^2 + 1 = 0,$$

$$\ell_i = \mathrm{Res}(l/P)|_{\lambda=\lambda_i} \in \mathrm{sl}(2,\mathbb{C}), \quad \sum \ell_i = 0$$

(since $\lambda = \infty$ is a regular point for (23)).

The isomonodromy condition for (23) is given by the Schlesinger equations,

$$\mathrm{d}\ell_i = \sum [\ell_i, \ell_j]\, \mathrm{d}(\lambda_i - \lambda_j) + [\mathrm{d}g\, g^{-1}, \ell_i],$$

where $g(t) = \psi'(\lambda = \infty, t)$. It follows from the asymptotic expansion of (21b) near $\lambda = \infty$ that $g(t)$ satisfies $\mathrm{d}g\, g^{-1} = Z\, \mathrm{d}t$.

It can be verified from the definitions in (18)–(20) that the elements of ℓ_i are given only by the ratios $\widehat{\omega}_i \equiv \omega_i/\alpha$ and that the λ_i's are parametrized by the variable s, i.e., $\lambda_i = \lambda_i(s(t))$. Substituting the expressions for the ℓ_i's in the Schlessinger equations (isomonodromy equations) yields

$$\begin{aligned}
\frac{\mathrm{d}\widehat{\omega}_1}{\mathrm{d}s} &= 2\widehat{\omega}_2\widehat{\omega}_3 + (1-r)\widehat{\omega}_1, \\
\frac{\mathrm{d}\widehat{\omega}_2}{\mathrm{d}s} &= 2\widehat{\omega}_1\widehat{\omega}_3 - (1+r)\widehat{\omega}_2, \\
\frac{\mathrm{d}\widehat{\omega}_3}{\mathrm{d}s} &= 2\widehat{\omega}_1\widehat{\omega}_2 - 2r\widehat{\omega}_3,
\end{aligned} \tag{24}$$

together with

$$\alpha = \omega_1 - \omega_2 = \frac{1}{4s}\frac{ds}{dt}. \tag{25}$$

It can also be verified in a straightforward manner that the compatibility of Equations (21a) and 21b) give rise to the *same* set of equations as in (24), (25). Furthermore, the Schlessinger equations corresponding to equations of the form (23) with four regular singular points is shown in [25] to yield the Painlevé VI (P-VI) equations, parametrized by the eigenvalues of the ℓ_i's. Therefore, (24) corresponds to a Painlevé VI equation with special values of parameters (the eigenvalues of the ℓ_i's in (23) are $\pm\frac{1}{4}$) in the independent variables s. Note that there is, in fact, only one independent equation in (24) because of the algebraic relations among the $\widehat{\omega}_i$'s, namely, $\widehat{\omega}_1 - \widehat{\omega}_2 = 1, \widehat{\omega}_1 + \widehat{\omega}_2 - 2\widehat{\omega}_3 = r = (s+1)/(s-1)$. This is a Riccati equation for one of the dependent variables, e.g.,

$$\frac{d\widehat{\omega}_1}{ds} = \frac{\widehat{\omega}_1^2}{2s} - \frac{\widehat{\omega}_1}{s-1} + \frac{1}{2s(s-1)}, \tag{26}$$

which is equivalent to a second-order linear ODE. The solution of (26) leads to hypergeometric functions. It is well known that hypergeometric functions are special solutions of P-VI. Equation (26) plays a crucial role in the 'linearization' procedure of the DBH system which is now described.

From (25) and the solutions of (26), it is possible to express the residual variables ω_i's in terms of s and ds/dt. But in order to obtain the ω_i's explicitly, one still needs to know $s = s(t)$. Also, (24), (25) constitutes only that part of the DBH system, which corresponds to an isomonodromy deformation problem. To obtain the full DBH system, we also need to consider the auxiliary Equation (22) which yields the following equation,

$$\frac{d}{dt}\ln\alpha = -4\omega_3 \tag{27}$$

in addition to (25). Note that (27) with (24) and (25) are equivalent to the full DBH system (11). From the definition of s in (20) and Equations (25), (27) one can express the ω_i's in terms of s and its time derivatives as follows,

$$\begin{aligned}
\omega_1 &= -\frac{1}{4}\frac{d}{dt}\ln\left(\frac{1}{4s(s-1)}\frac{ds}{dt}\right), \\
\omega_2 &= -\frac{1}{4}\frac{d}{dt}\ln\left(\frac{1}{4(s-1)}\frac{ds}{dt}\right), \\
\omega_3 &= -\frac{1}{4}\frac{d}{dt}\ln\left(\frac{1}{4s}\frac{ds}{dt}\right).
\end{aligned} \tag{28}$$

Note that these are the expressions for the ω_i's which are in terms of the modular function S, given earlier (12), (13), where $s(t)$ is related to S by a Möbius transformation (and recalling the rescaling of time). Using (27) and (25) and interchanging roles of the independent variables t and s one can solve for the quantity $\widehat{\omega}_3$ as,

$$\widehat{\omega}_3 = 1 + \frac{st''}{t'},$$

where $'$ means $\mathrm{d}/\mathrm{d}s$. Substituting this, using the relationship for $\widehat{\omega}_1, \widehat{\omega}_2$, and $\widehat{\omega}_3$, into the Riccati equation (26), one obtains the Schwartzian equation for $t = t(s)$,

$$\left(\frac{t''}{t'}\right)' - \frac{1}{2}\left(\frac{t''}{t'}\right)^2 = \frac{1}{2}\left[\frac{1}{s^2} + \frac{1}{(s-1)^2} - \frac{1}{s(s-1)}\right]$$

which can be solved by transforming

$$t(s) = \frac{z_2}{z_1}, \tag{29}$$

where z_1 and z_2 are two linearly independent solutions of the second-order linear equation

$$z'' + v(s)z = 0, \qquad v(s) = \frac{1}{4}\left[\frac{1}{s^2} + \frac{1}{(s-1)^2} - \frac{1}{s(s-1)}\right]. \tag{30}$$

Finally, inverting (29) to obtain $s(t)$, one can express the solution of the DBH system $\omega_i(s(t))$ via (28).

2.2.3. *Remarks*

There are three remarks to be made which are important when considering the monodromy problem and the associated linearization of the DBH system.

1. The compatibility of the Lax pair (15) introduced earlier with $\mu \neq 0, \nu \neq 0$ is *not* an isomorphic deformation problem. This can be seen by expressing the L-equation in (15) as

$$\frac{\mathrm{d}\psi}{\mathrm{d}\lambda} = \left(\frac{\mu I - \ell}{P}\right)\psi = \left(\sum_{i=1}^{4}\frac{R_i}{\lambda - \lambda_i}\right)\psi$$

and noting that here the residue matrices satisfy $\mathrm{Tr}(R_i) = \mathrm{Res}(2\mu/P)|_{\lambda=\lambda_i}$. When $\mu \neq 0$, the eigenvalues of R_i and hence the monodromy is not constant in time. Nevertheless, we have shown that it is possible to 'factorize' this non-isomonodromic problem into two separate parts – the isomonodromic part is given by Equations (21a) and (21b), whereas (22) reflects the departure from isomonodromy. A careful analysis of the direct problem [9] shows that the time evolution equations for the eigenvalues of R_i are equivalent to (22) which leads to the transformation $s = s(t)$ between the deformation parameter s of the isomonodromy problem and the original independent variable t.

2. In more general terms, the inverse function $s(t)$ obtained from (29), corresponding to more general potentials $v(s)$, may not be single-valued in the complex t-plane. Although the Schwartzian equation can always be solved for $t(s)$ via (29) for arbitrary potentials $v(s)$, the resulting function $s = s(t)$ can be densely branched. Single valuedness in this sense, i.e. integrability in the complex t-plane, is achieved only if $v(s)$ has the form [27]

$$v(s) = \frac{1}{4}\left[\frac{1-\alpha_1^2}{s^2} + \frac{1-\alpha_2^2}{(s-1)^2} - \frac{1+\alpha_3^2-\alpha_1^2-\alpha_2^2}{s(s-1)}\right],$$

where $\alpha_i = 1/n_i$, $n_i \in \mathbb{Z}^+$, or $\alpha_i = 0$, $i = 1, 2, 3$. The latter case (i.e. $\alpha_i = 0$) corresponds to (30), hence the corresponding function $s(t)$ is single-valued. Consequently, the solutions $w_i(s(t))$ to the DBH system are also single-valued functions. An example where the solutions are not always single-valued (even though the underlying system is 'linearizable' in the same way as the DBH system) is illustrated by the next remark.

3. We now describe a novel fifth-order system which is a generalization of the DBH system (4). It also arises as an ODE reduction of the self-dual Einstein equations associated with a (nondiagonal) Euclidean metric and can be derived as a SDYM reduction in the same manner as the DBH equations via (10). The connection components are given by

$$A_i(t) = \sum_{j=1}^{3} O_{i,j}(g)T_j,$$

where the T_j's are the sdiff$(\mathbf{S}^3)$ vector fields defined as

$$T_1 = \omega_1 x_1 + \theta x_2, \qquad T_2 = \phi x_1 + \omega_2 x_2, \qquad T_3 = \omega_3 x_3,$$

where ω_i, θ, and ϕ are functions of t. Substituting these in (10) yields a system of five ODEs for $\omega_i(t)$, $\theta(t)$, and $\phi(t)$ (see [7]),

$$\begin{aligned}
\frac{d\omega_1}{dt} &= \omega_2\omega_3 - \omega_1(\omega_2+\omega_3) + \phi^2,\\
\frac{d\omega_2}{dt} &= \omega_1\omega_3 - \omega_2(\omega_1+\omega_3) + \theta^2,\\
\frac{d\omega_3}{dt} &= \omega_1\omega_2 - \omega_3(\omega_1+\omega_2) - \phi\theta,\\
\frac{d\phi}{dt} &= (\theta-\phi)\omega_1 - (\theta+\phi)\omega_3,\\
\frac{d\theta}{dt} &= -(\theta-\phi)\omega_2 - (\theta+\phi)\omega_3.
\end{aligned} \tag{31}$$

When $\theta = \phi = 0$, (31) reduces to the DBH system. This system can also be solved (in terms of linear equations) in the same way as the DBH system and it is possible to express the dependent variables in terms of a function $s(t)$ and its time derivatives. However, it turns out that $s(t)$ is not single-valued in this case.

We outline the key features of the linearization procedure, space considerations require that we omit most of the details. The Lax pair for (31), also given by (15), (16) (but with different choices for the $\mathrm{sl}(2,\mathbb{C})$ matrices and polynomials) can be separated into isomonodromic and nonisomonodromic parts. The isomonodromic deformation problem yields four equations (the Schlessinger equations) involving the ratios of the dependent variables. They also admit two first integrals arising from the (constant) eigenvalues of the corresponding residue matrices ℓ_i. Using the first integrals, it is possible to reduce these equations by quadrature to a single Riccati equation with respect to the deformation parameter $s(t)$ which in this case is defined by $s(t) = (\omega + R)/(\omega - R)$, where $\omega = \omega_1 = \omega_2 - 2\omega_3$, and $R^2 = (\omega_1 + \omega_2)^2 + (\phi + \theta)^2$. The rest of the system in (31) is recovered from the nonisomonodromic part which, together with the Riccati equation, provides a Schwartzian equation for $t(s)$. Finally, we arrive at a system where the dependent variables can be expressed in terms of s and its time derivatives in the following form:

$$\omega = \omega_1 + \omega_2 - 2\omega_3 = \tfrac{1}{4}\frac{\mathrm{d}s}{\mathrm{d}t}\left(\frac{s+1}{s(s-1)}\right),$$

$$\omega_3 = -\tfrac{1}{4}\frac{\mathrm{d}}{\mathrm{d}t}\ln\left(\frac{1}{4s}\frac{\mathrm{d}s}{\mathrm{d}t}\right),$$

$$(\omega_1 - \omega_2) \pm i(\phi + \theta) = \frac{1}{4s}\frac{\mathrm{d}s}{\mathrm{d}t}\exp(\pm i\Omega),$$

$$\Omega - \Omega_0 = ia\ln\left(\frac{1+\sqrt{s}}{1-\sqrt{s}}\right),$$

$$\theta - \phi = \left(\frac{ia}{2}\frac{\mathrm{d}s}{\mathrm{d}t}\right)\left(\frac{1}{\sqrt{s}(s-1)}\right),$$

where the constants a and Ω_0 are related to the eigenvalues of ℓ_i, as mentioned above. In addition, $t(s)$ satisfies

$$\left(\frac{t''}{t'}\right)' - \frac{1}{2}\left(\frac{t''}{t'}\right)^2 = \frac{1}{2}\left[\frac{1}{s^2} + \frac{1-a^2}{(s-1)^2} - \frac{1-a^2}{s(s-1)}\right] \equiv 2v(s),$$

which can be solved in the same way as in (29). As mentioned above, here the parameter a is a first integral given explicitly by

$$a^2 = \frac{(\theta - \phi)^2}{(\theta + \phi)^2 + \omega_1 - \omega_2)^2 - (\omega_1 + \omega_2 - 2\omega_3)^2},$$

and is prescribed by the initial values of ω_i, θ, and ϕ. Therefore, a in general, is not zero or the reciprocal of a positive integer which is necessary for the inverse function $s(t)$ to be single-valued, as we mentioned in Remark 2. Consequently, solutions of (31) are in general not single-valued functions in the complex t-plane, although they can essentially be described by a linear second-order equation similar to (30). The system (31) provides a concrete example where linearization does not imply integrability in the sense of single-valuedness in the complex plane! Note that $\theta = \phi$ implies $a = 0$, and in this case the solutions to the resulting fourth-order system are single-valued. (In fact, by suitable transformations of the dependent variables this fourth-order system can be transformed to the DBH system.) But when $a \neq 0$ the solution to (31) is densely branched in the complex domain whenever the initial conditions are such that the value of a differs from being a reciprocal of an integer. Conversely, when a is suitably 'quantized', $a = 1/n$, $n \in \mathbb{Z}$, then the solution of the generalized DBH system is single-valued, and we consider it 'integrable in the complex plane'.

3. Homoclinic Manifolds and Water Waves

Although Stokes published his famous work *On the Theory of Oscillatory Waves* in 1847 [30], it was only during the 1960s that Benjamin [29] realized that the periodic Stokes wave on deep water is, in fact, unstable (see also Whitham [31]). About the same time, Zabusky and Kruskal studied the Korteweg–de Vries (KdV) equation [32], a pioneering investigation that paved the way for the development of soliton theory. In a suitable asymptotic limit, the KdV equation also describes weakly nonlinear water waves but in *shallow* water. Periodic waves of the KdV equation are stable and, in fact, Benjamin's instabilities are not observed in the KdV equation. Actually, the asymptotic equation governing the one-dimensional slow modulation of periodic wave-trains (such as the Stokes wave) in *deep* water is the nonlinear Schrödinger (NLS) equation. It is the NLS equation where one indeed encounters instabilities similar to those discovered by Benjamin.

Benjamin observed the instability through a linearized analysis of the Euler equations, accordingly the analysis is only valid for the short period of time while the linearized assumption remains valid. The linearized stability analysis does not give qualitative information about the long time evolution of the instabilities. It is perhaps reasonable to conjecture that the end-state is one of complete breakdown with energy equally distributed between all Fourier modes, i.e. thermalization (see, e.g., [35]). However, the remarkable laboratory experiments by Lake *et al.* [34] indicated that thermalization may not occur. They found that the initial instabilities saturate and that the initial state is approximately recovered at some stage of the evolution. These experiments dealt with a small number of unstable modes and relatively short time observations. Since the nonlinear Schrödinger equation is a useful model for water waves on deep water, they also turned to numerical studies of the (focusing) NLS equation

$$iu_t + u_{xx} + 2|u|^2 u = 0, \tag{32}$$

with periodic boundary conditions, $u(x+L,t) = u(x,t)$. Using up to five unstable modes, their numerical experiments did not indicate any thermalization – the energy remained confined to a relatively small number of Fourier modes. They concluded that the initial-wave profile should be recovered after a sufficiently long time.

From a mathematical point of view it is now known that the instabilities of the periodic waves in the NLS analogous to those of Benjamin and Feir, are closely related to the so-called homoclinic manifolds associated with the NLS equation. We next proceed to explain how this may shed new light on the Benjamin–Feir instabilities and the subsequent time evolution of these instabilities.

3.1. Benjamin–Feir-type instabilities in the NLS equation

Our goal is to develop a deeper understanding of the stability of the Stokes water wave. In the NLS model of water waves it is represented in its simplest form by the solution

$$\tilde{u}(x,t) = a \exp(2i|a|^2 t), \tag{33}$$

where a is any complex constant. In order to investigate its stability, we follow a well-known procedure and add small 'side-bands' to the Stokes wave

$$u(x,t) = \tilde{u}(x,t)(1 + \varepsilon(x,t)), \tag{34}$$

where $|\varepsilon| \ll 1$ and

$$\varepsilon(x,t) = \hat{\varepsilon}_{-n}(t) \exp(-i\mu_n x) + \hat{\varepsilon}_n(t) \exp(i\mu_n x) \tag{35}$$

and $\mu_n = 2\pi n/L$. Substituting (34) into (32) and keeping linear terms in ε yield,

$$\varepsilon_t = i\varepsilon_{xx} + 2i|a|^2(\varepsilon + \varepsilon^*).$$

Making use of (35), it follows that the growth rate σ_n of the nth mode, $\hat{\varepsilon}_n(t) = \hat{\varepsilon}_n(0) \exp(\sigma_n t)$, is given by

$$\sigma_{n\pm} = \pm\mu_n \sqrt{4|a|^2 - \mu_n^2}\,. \tag{36}$$

Hence, the side bands are unstable provided that

$$0 < \mu_n^2 < 4|a|^2. \tag{37}$$

The number of unstable modes (which also determines the complexity of the homoclinic structure) is given by the largest integer N, satisfying

$$0 < N < |a|L/\pi.$$

This instability is the analogue (for the NLS model) of the Benjamin–Feir instability of the Stokes water wave described above.

It is possible to associate a specific initial value with the unstable case, namely (see [11])

$$u(x,0) = a + \varepsilon_0 \exp(i\phi)\cos(\mu_n x + \xi), \tag{38}$$

where $0 < |\varepsilon_0| \ll 1$, $\tan(\phi) = \sigma_{n+}/\mu_n^2$ and ξ is an arbitrary real constant. Since the NLS equation admits the symmetry $(x,t,u) \to (x+\xi,t,u)$, it is sufficient to consider the initial condition

$$u(x,0) = a + \varepsilon_0 \exp(i\phi)\cos(\mu_n x). \tag{39}$$

The significance of the phase ϕ will become clear in a moment.

In order to explain how these instabilities are related to homoclinic manifolds, we recall (see, e.g., [11]) the following exact (homoclinic) solution of the NLS equation,

$$\begin{aligned} u(x,t) &= a\exp(2ia^2t)\times \\ &\times \frac{1 + 2\cos(px)\exp(\Omega t + 2i\phi + \gamma) + A_{12}\exp(2\Omega t + 4i\phi + 2\gamma)}{1 + 2\cos(px)\exp(\Omega t + \gamma) + A_{12}\exp(2\Omega t + 2\gamma)}, \end{aligned} \tag{40}$$

where

$$\Omega = \pm p\sqrt{4a^2 - p^2},$$

$$p = 2a\sin\phi \quad \text{and} \quad A_{12} = \frac{1}{\cos^2\phi}.$$

The periodic boundary conditions are satisfied by imposing $p = 2\pi n/L$.

Thus, for any p satisfying the condition (which is identical to (37))

$$0 < p^2 < 4a^2, \tag{41}$$

the solution (40) represents an orbit homoclinic to the Stokes wave, $u_p(x,t) = a\exp(2ia^2t)$. More specifically, assuming $\Omega > 0$ it is easily verified that $u \to u_p(x,t)$ as $t \to -\infty$ and $u \to u_p(x,t)\exp(4i\phi)$ as $t \to \infty$. Alternatively, it follows that $|u(x,t)| \to |a|$ as $t \to \pm\infty$. In an attempt to follow the homoclinic orbit *numerically*, we choose $L = 2\sqrt{2}\pi$, $a = 0.5$, and γ large negative. Making use of the spatial translational invariance of the NLS equation, an approximate (e.g., for γ large negative) initial value for a homoclinic orbit is obtained from (40),

$$u(x,0) = \tfrac{1}{2}(1 + \varepsilon_0 \exp(i\phi)\cos(\mu x)), \tag{42}$$

where ε_0 is small, $\mu = 2\pi/L$, and $\sin\phi = \mu$, i.e. $\phi = \pi/4$. (Again note that the initial value for the homoclinic orbit (42) is identical to the initial value (38) obtained from the linearized stability analysis.) A typical numerical solution using $\varepsilon_0 = 10^{-5}$ is shown in Figure 1. Note that the numerical solution stays *close* to, but it is not *on* the homoclinic orbit – instead of returning to a constant modulus, the solution displays quasi-periodic behavior. This is not unexpected since the initial value (42) ignores second-order terms in ε_0 and almost any small perturbation moves the solution away from the homoclinic obit. This also means that one does not expect to observe homoclinic orbits directly in laboratory experiments. However, it is possible to observe homoclinic orbits indirectly numerically by adjusting the phase in the initial value (42). More specifically, Figures 2(a) and (b) show the result for $\phi = \pi/4 \pm 0.01$, respectively, and with $\varepsilon_0 = 0.1$. These figures show that the homoclinic orbit forms a separatrix between two different wave patterns – *we believe that the different sides of the homoclinic orbit should be observable in laboratory experiments*. We remind the reader that similar situations occurs in classical mechanics, for example the pendulum problem. One does not 'observe' the separatrix (homoclinic) solution when the pendulum is in the pure inverted position. Rather, the two 'sides' of the separatrix are observed as the 'oscillating' or 'rotating' states.

The homoclinic orbit (40) is the simplest homoclinic structure (we call this a homoclinic manifold of order one) associated with the focusing NLS equation. More complicated structures (e.g., homoclinic manifolds of order N) can also be

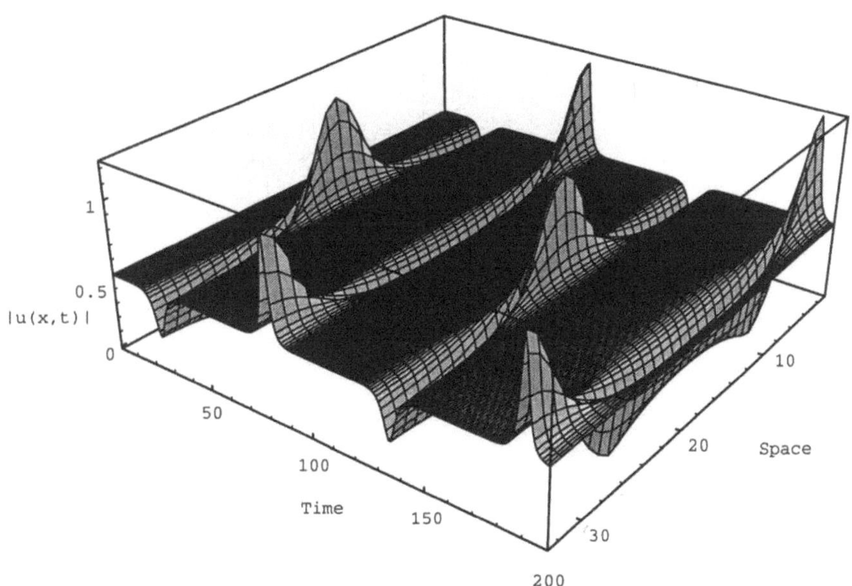

Fig. 1. The homoclinic orbit.

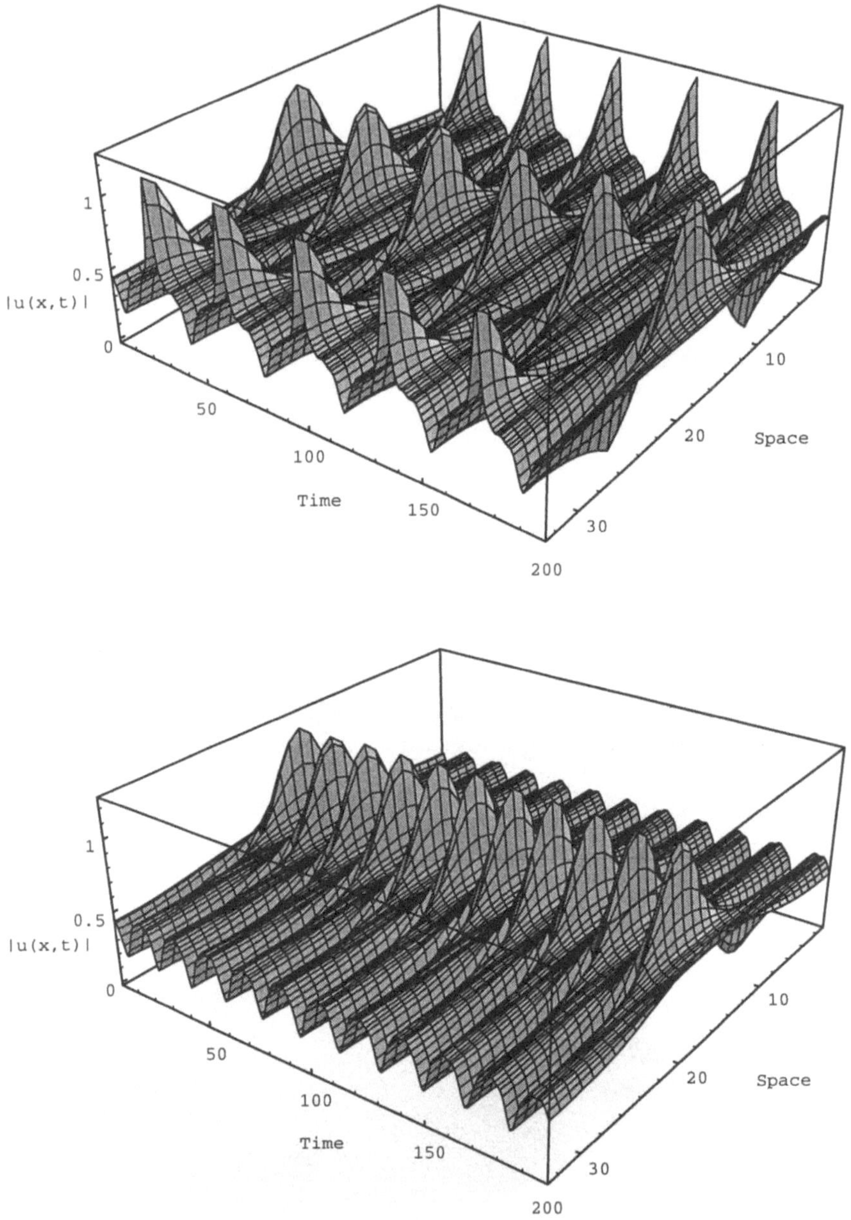

Fig. 2. (a) Outside and (b) inside the homoclinic orbit.

obtained, see, for example, [11]. We now turn our attention to some of the consequences of the presence of homoclinic structures on numerical solutions of the NLS equation. We are particularly interested in the performance of *nonintegrable* vs *integrable* discretizations.

3.2. Numerical Schemes

In order to compare the behavior of integrable and nonintegrable discretizations of the NLS equation in the vicinity of homoclinic orbits, we consider two finite difference schemes. The first one is the nonintegrable, standard second-order scheme,

$$i\dot{U}_j + (U_{j-1} - 2U_j + U_{j+1})/h^2 + 2|U_j|^2 U_j = 0, \tag{43}$$

where $h = L/N$ is the grid spacing; periodic boundary conditions are used, $U_{j+N} = U_j$. This scheme is not integrable and the solution may become highly irregular. For example, consider the initial values

$$u(x,0) = a\Big(1 + \tfrac{1}{10}\cos(\mu x)\Big), \tag{44}$$

with $a = \frac{1}{2}$, $L = 4\sqrt{2}\pi$ and $\mu = 2\pi/L$, i.e. a comparison with the instability condition (37) reveals two unstable side bands. A more detailed investigation shows that this initial value is close, but not on a homoclinic orbit (see, for example, [11]). For the time integration the Runge–Kutta–Merson routine, D02BBF, in the NAG library is used with the relative error specified as 10^{-10}. The results using $N = 32$ are shown in Figure 3. One observes the quick and complete breakdown of the solution which has been related to homoclinic crossings (see [33]). This breakdown in the spatial structure may be viewed as thermalization. However, here it is the result of an inadequate spatial resolution and the instability disappears as the mesh is further refined.

The following numerical scheme was carefully designed to be an *integrable* discretization of the NLS equation (see [28]),

$$i\dot{U}_j + (U_{j-1} - 2U_j + U_{j+1})/h^2 + |U_j|^2(U_{j-1} + U_{j+1}) = 0. \tag{45}$$

Note that it is also of second-order accuracy and that the differences with the standard finite difference scheme (43) appear at fourth order in h and only in the *nonlinear* term. However, despite these apparently superficial differences, the two schemes behave very differently. Using the same initial values (44), the Runge–Kutta–Merson scheme and $N = 32$ as before, the solution is shown in Figure 4. There is no indication of any instability and the spatial structure remains smooth, i.e. all the energy are confined to a small number of low modes. The (quasi) periodicity is analogous to the recurrence observed experimentally by Lake *et al.* [34].

Figure 4 shows that the integrable discretization approximates the solutions of the continuous problem at moderate levels of discretization – in the limit as the mesh is further refined, the integrable discretization captures all the features of the continuous problem. However, if the complexity of the homoclinic structure is increased by increasing the number of unstable modes allowed by (37), the integrable discretization may also break down, as we discuss next.

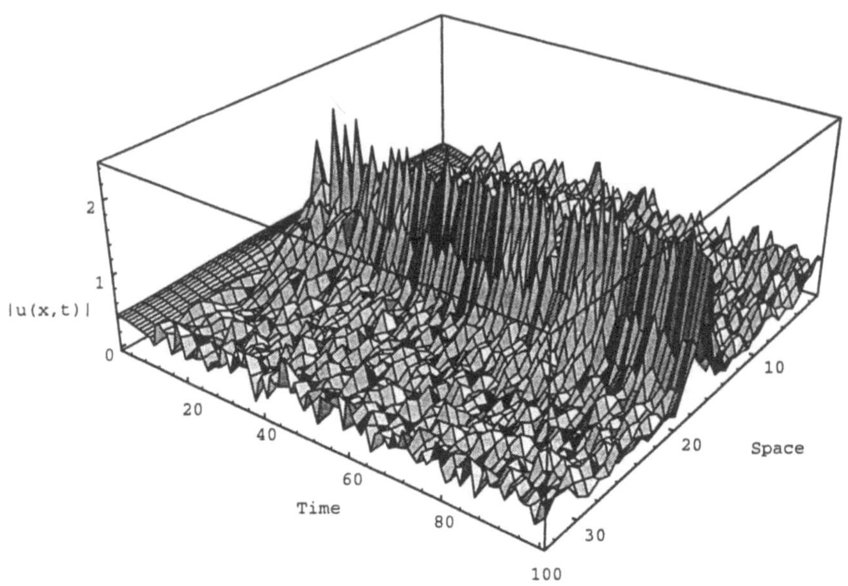

Fig. 3. The finite difference solution.

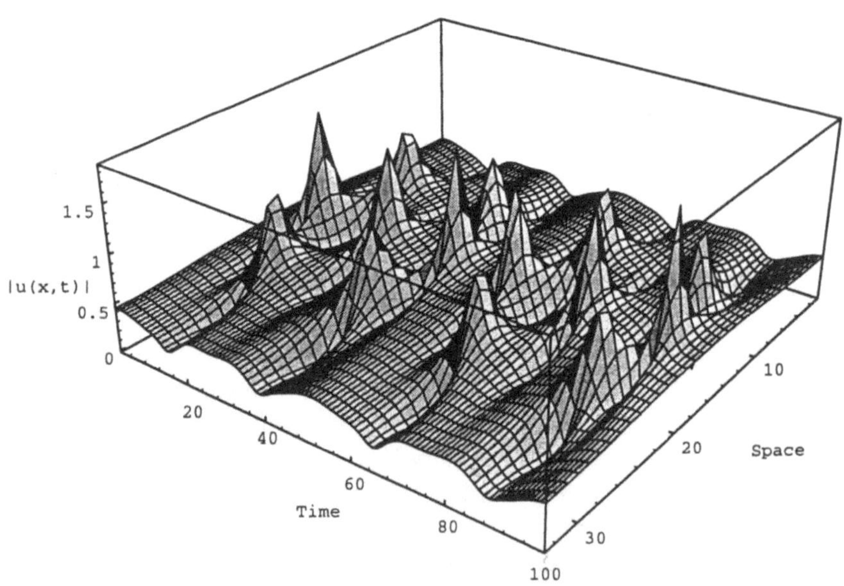

Fig. 4. The integrable discretization.

3.3. More Unstable Modes

If we increase the number of unstable modes, we find that the solution may become sensitively dependent on small perturbations. In order to illustrate the sensitivity, we return to the initial value (44) and increase the number of unstable modes to 7 by choosing $a = \frac{5}{2}$ and $L = 2\sqrt{2}\pi$. Consider the integrable discretization (45) as well as the mathematically equivalent one,

$$i\dot{U}_j + (U_{j-1} - 2U_j + U_{j+1})/h^2 + |U_j|^2 U_{j-1} + |U_j|^2 U_{j+1} = 0. \tag{46}$$

The results integrating over 100 time units are shown in Figure 5. In this figure lighter shades of gray indicate larger values of the modulus. This figure, shows that despite their mathematical equivalence, completely different numerical solutions are obtained after a sufficiently long time (about 50 time units)! Calling the two solutions $u_j^{(1)}(t)$ and $u_j^{(2)}(t)$, respectively, Figure 6 shows a plot of $\log_{10} e(t)$ vs t, where $e(t)$ given by

$$e(t) = \sum_{j=0}^{N} \log_{10}\left(|u_j^{(1)}(t) - u_j^{(2)}(t)|\right)/(N+1),$$

is the normalized difference between the two solutions as a function of t. Note that the difference becomes $\mathrm{O}(1)$ after about 50 time units.

The two schemes are mathematically equivalent – the only difference computationally is roundoff error (the calculations were done in double precision). Thus, tiny differences due to roundoff error are rapidly amplified to the extent that they have a marked effect on the solution. This sensitive behavior on small perturbations is an indication of an unpredictable, even chaotic, time evolution. However, it should be observed that the spatial structure does not break down completely. Figure 7 shows the Fourier decomposition of the spatial structure (only the lowest 64 modes are shown to facilitate comparison with Figure 8(b)); the energy remains confined to approximately the first 20 modes.

Next we solve the same problem as before (i.e. seven unstable modes), using the Fourier pseudospectral method with $N = 128$ modes and a fourth-order symplectic time integrator (fourth-order split-step) with a time step of 0.001. Since we are dealing with only seven unstable modes in the problem, one may expect the $N = 128$ Fourier modes of the spectral scheme to be large enough to resolve the spatial structure. The solution is shown in Figure 8(a). Note that the solution breaks down reasonably quickly and bears little resemblance to the solutions of Figure 5 after a short period of time – about 10 time units. Its Fourier decomposition is shown in Figure 8(b). As in the case of Figure 7, apart from a few brief excursions to higher modes, the energy is limited to a relatively small number of Fourier modes.

The sensitivity observed in the calculations shown in Figures 5–8 can be explained in terms of the underlying homoclinic manifold (see [12]). The homoclinic manifold is represented by double eigenvalues of the associated linear

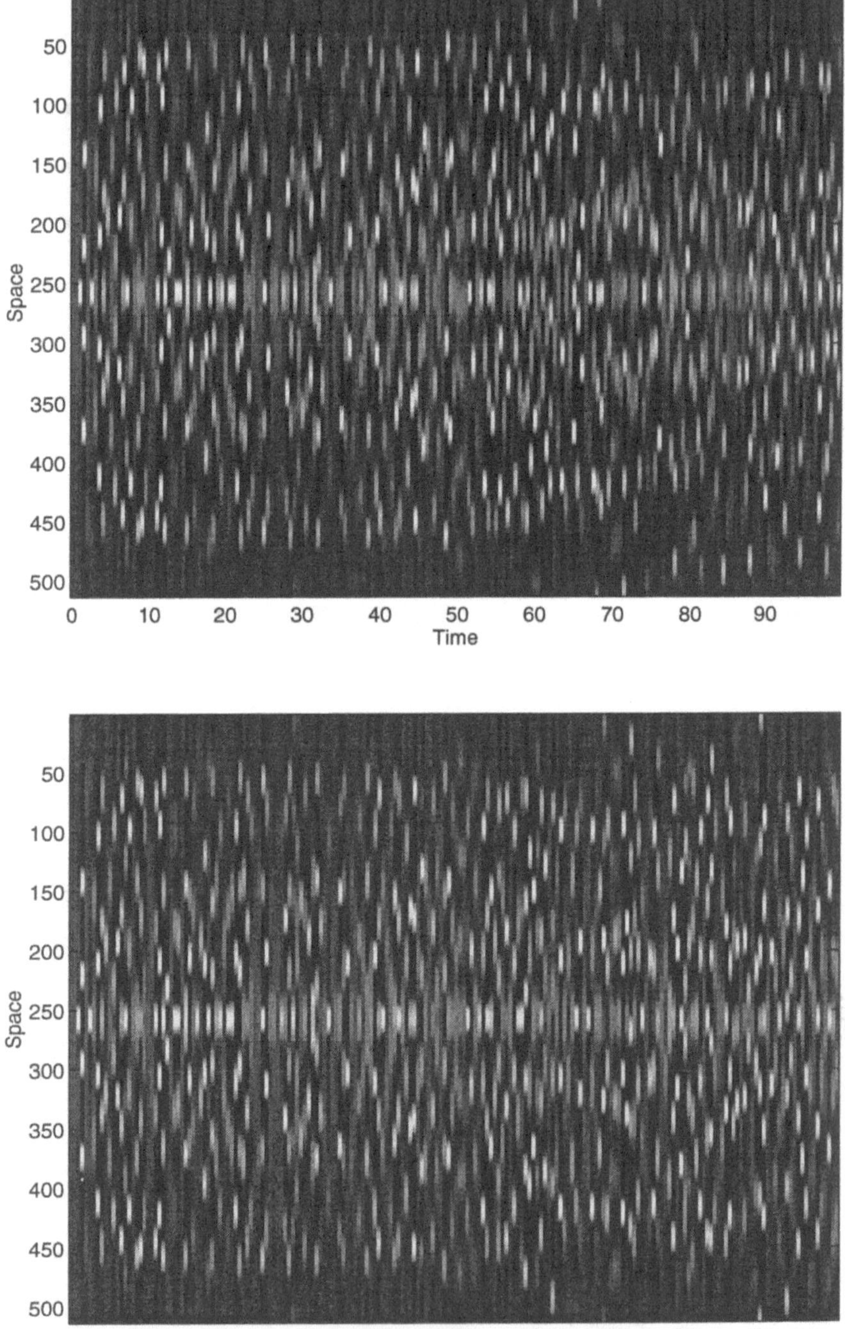

Fig. 5. Two mathematically equivalent forms of the integrable discretization.

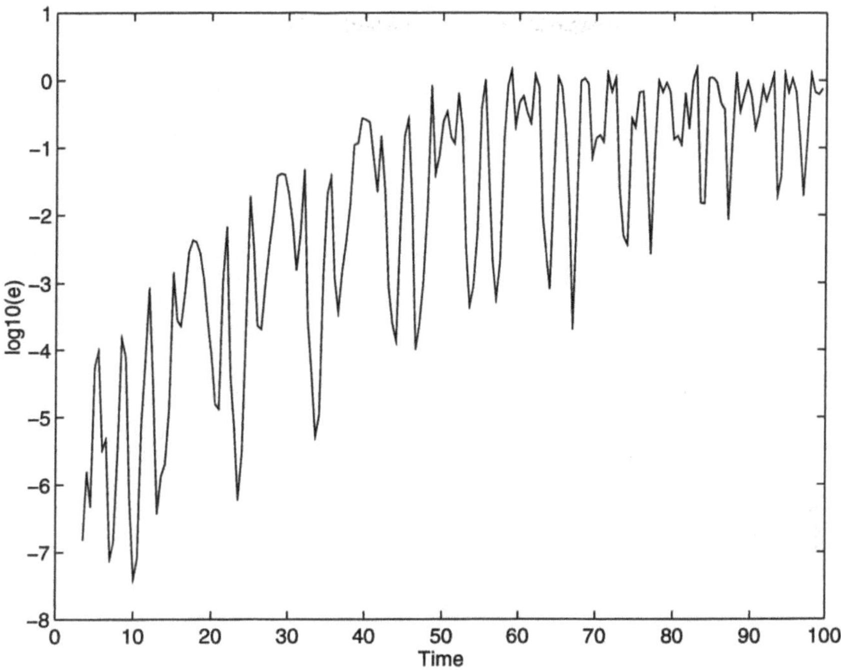

Fig. 6. The difference between the two solutions shown in Figure 5.

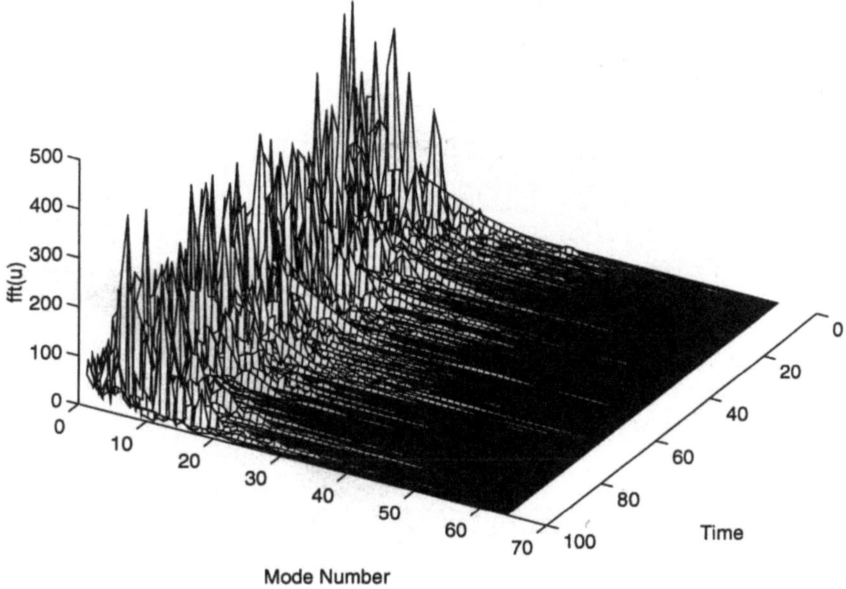

Fig. 7. The Fourier decomposition of the solution of Figure 5 (top).

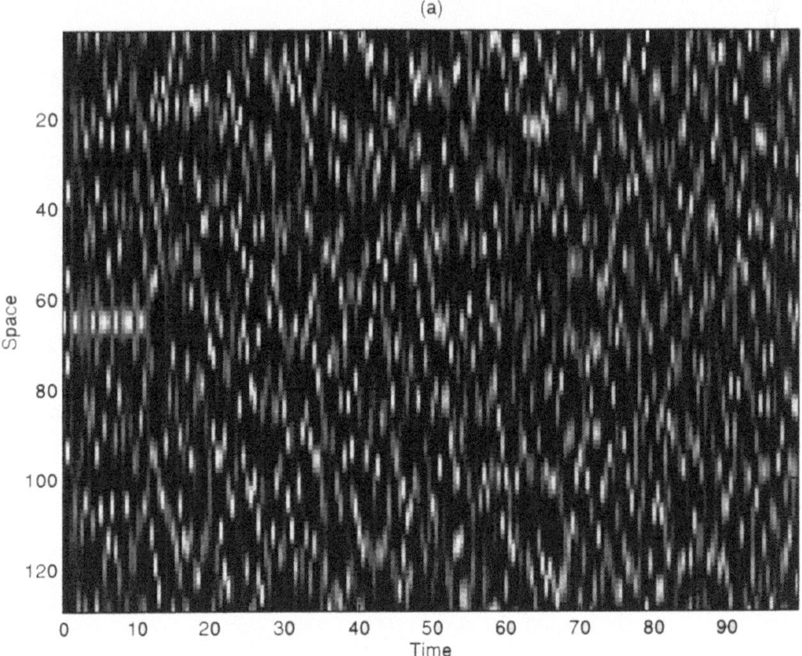

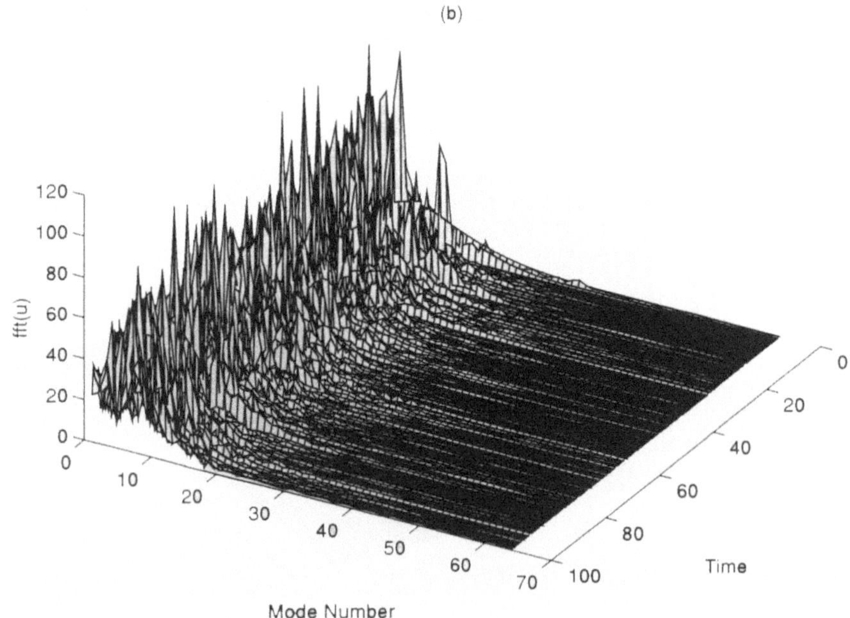

Fig. 8. (a) The Fourier spectral solution and (b) its Fourier decomposition.

spectral problem; each unstable mode represents one double point. If one moves away from the homoclinic manifold, the double eigenvalues are split into gaps and the distance from the homoclinic manifold is given by the 'size' (distance between the eigenvalues) of the gaps. If initial values are chosen on the homoclinic manifold, one expects that small perturbations will cause frequent homoclinic crossings, i.e. the solution will be very sensitive to small perturbations. However, the initial values used in the numerical experiments above (see [12, 14] for a more complete description), are not on the homoclinic manifold. In fact, a perturbation analysis (see [12]) shows that the first double point (unstable mode) is split $O(\varepsilon_0)$, the second one $O(\varepsilon_0^2)$, etc., with the last unstable mode being split $O(\varepsilon_0^n)$ (assuming n unstable modes). Thus, when the numerical errors (including roundoff error) are of $O(\varepsilon_0^n)$, one expects homoclinic crossings and sensitive dependence on the perturbations.

Finally, we turn to a brief discussion of the potential significance of the results obtained from the NLS model for the long-time behavior of the Stokes wave in deep water.

3.4. Conclusions

Our numerical experiments with the NLS equation indicate that the long-time behavior of the Stokes wave depends critically on the number of unstable modes, i.e. the amplitude of the Stokes wave. For a small number of unstable modes (with a perturbation that has the fundamental frequency; this is the most stable situation) our calculations show that the instability saturates and the initial wave is eventually (approximately) recovered. Moreover, our numerical and analytical calculations demonstrate that this situation is fairly robust with respect to small perturbations, i.e. the distance from the homoclinic manifold is larger than the numerical perturbations. However, if the number of unstable modes is increased, the homoclinic manifold becomes more complex, fills a larger part of phase space and the distance from it may become of the same order of magnitude as the numerical perturbations. In this case, the time evolution becomes irregular and no recurrence is observed.

Thermalization and a complete break-down of the spatial structure is only observed if the spatial resolution of the numerical scheme is inadequate, i.e. if the numerical perturbations are large enough.

It is well known that the NLS equation is a good, although idealized model of the actual physical water waves in deep water. However, laboratory experiments of water waves may be interpreted as introducing small perturbations to the NLS equation. For a small number of unstable modes, the NLS equation has been shown to be robust to the small, conservative perturbations introduced by the numerical schemes. One might therefore expect (but it is no less remarkable) to observe the recurrence associated with the NLS equation, also in the laboratory, as did Lake *et al.* [34]. Moreover, for one unstable mode the homoclinic orbit, more specifically, the two sides of the homoclinic should be observable. If the number of unstable modes are increased, the sensitivity of the NLS solutions prevents any recurrence in numerical simulations – likewise we do *not* expect to observe recurrence in laboratory experiments for a large number of unstable modes. Based on the above arguments, we expect to see a more or less random 'sea of humps' as in Figure 8(a). In the future we will return to the NLS equation and investigate the effect of higher order corrections to the NLS equation on the solutions described above.

Acknowledgements

This work was partially supported by AFOSR grant F49620–94–0120, ONR grant N00014–94–0194 and NSF grants DMS–9024528, DMS–9404265.

References

1. Ablowitz, M. J. and Segur, H.: *Solitons and the Inverse Scattering Transform*, SIAM, Philadelphia, 1981.
2. Novikov, S. P., Manakov, S. V., Pitaevskii, L. P., and Zakharov, V. E.: *Theory of Solitons. The Inverse Scattering Method*, Plenum, New York, 1984.
3. Ablowitz, M. J. and Clarkson, P. A.: *Solitons, Nonlinear Evolution Equations and Inverse Scattering*, Cambridge University Press, Cambridge, 1991.
4. Ablowitz, M. J., Chakravarty, S., and Takhtajan, L. A.: *Comm. Math. Phys.* **158** (1993), 289–314.
5. Chakravarty, S., Ablowitz, M. J., and Clarkson, P. A.: *Phys. Rev. Lett.* **65** (1990), 1085–1087.
6. Mason, L. J. and Woodhouse, N. M. J.: *Nonlinearity* **6** (1993), 569–581.
7. Ablowitz, M. J., Chakravarty, S., and Takhtajan, L. A.: Integrable systems, self-dual Yang–Mills equations and connections with modular forms, in S. Xiao (ed.), *Proc. of Nonlinear Problems in Engineering and Science*, Science Press, Beijing, China, 1992.
8. Takhtajan, L. A.: Modular forms as τ-functions for certain integrable reductions of the Yang–Mills equations, in O. Babelon *et al.* (eds.), *Proc. Verdier Memorial Conference on Integrable Systems*, Birkhauser, Berlin, 1993.
9. Chakravarty, S. and Ablowitz, M. J.: to be published, 1995.
10. Ercolani, N., Forest, M. G., and McLaughlin, D. W.: *Physica* **D43** (1990), 349–384.
11. Ablowitz, M. J. and Herbst, B. M.: *SIAM J. Appl. Math.* **50** (1990), 339–251.
12. Ablowitz, M. J., Schober, C. M., and Herbst, B. M.: *Phys. Rev. Lett.* **71** (1993), 2683–2686.
13. Ablowitz, M. J., Herbst, B. M., and Schober, C. M.: On the numerical solution of the sine-Gordon equation, I. Integrable discretizations and homoclinic manifolds, PAM Report 214, University of Colorado, Boulder, 1994.
14. Ablowitz, M. J. and Schober, C. M.: *Contemp. Math.* **172** (1994), 253–268.
15. Ward, R.S.: *Phys. Lett. A* **61** (1977), 81–82.
    See also, R. S. Ward and R. O. Wells: *Twistor Geometry and Field Theory*, Cambridge University Press, Cambridge, 1990.
16. Chakravarty, S. and Ablowitz, M. J.: On reductions of self-dual Yang–Mills equations, in P. Winternitz and D. Levi (eds.), *Proc. NATO Advanced Research Workshop*, Plenum Press, New York, 1990.
17. Mason, D. and Sparling, G. A. J.: *J. Geom. Phys.* **8** (1991), 263–271.
18. Dubrovin, B.: Private communication.
19. Gibbons, G. W. and Pope, C. N.: *Commun. Math. Phys.* **66** (1979), 267–290.
20. Chazy, J.: *C.R. Acad. Sci. Paris* **149** (1909), 563–565.
21. Chakravarty, S.: To be published.
22. Takhtajan, L. A.: *Commun. Math. Phys.* **160** (1994), 295–315.

23. Plemelj, J.: *Problems in the Sense of Riemann and Klein*, Interscience Publishers, New York, 1964.
24. Birkhoff, G. D.: *Collected Mathematical Papers*, Vol. 1, Dover, New York, 1968.
25. Jimbo, M., Miwa, T., and Ueno, K.: *Physica* **D2** (1981), 306–352 and the references therein.
26. Jimbo, M. and Miwa, T.: *Physica* **D2** (1981), 407–448.
27. Nehari, Z.: *Conformal Mapping*, McGraw-Hill, New York, 1952; (Reprinted by Dover, New York, 1975).
28. Ablowitz, M. J. and Ladik, J. F.: *Stud. Appl. Math.* **55** (1976), 213–229.
29. Benjamin, T. B.: *Proc. Roy. Soc. A* **299** (1967), 59–75.
30. Stokes, G. G.: *Camb. Trans.* **8** (1847), 441–473.
31. Whitham, G. B.: *Linear and Nonlinear Waves*, Wiley, New York, 1974.
32. Zabusky, N. J. and Kruskal, M. D.: *Phys. Rev. Lett.* **15** (1965), 240–243.
33. Schober, C. M. and McLaughlin, D. W.: *Physica* **D57** (1992), 447–465.
34. Lake, B. M., Yuen, H. C., Rungaldier, H., and Ferguson, W. E.: *J. Fluid. Mech.* **83** (1977), 49–74.
35. Hasselmann, K.: Discussion, *Proc. Royal Soc. London (A)*, **299** (1967), 67.

Acta Applicandae Mathematicae **39:** 39–67, 1995.

Applications of KdV

D. G. CRIGHTON
Department of Applied Mathematics and Theoretical Physics, University of Cambridge, Silver Street, Cambridge CB3 9EW, U.K.

(Received: 6 February 1995)

Mathematics Subject Classification (1991): 58F07.

Key words: gravity waves, internal solitons, nonlinear accoustics, plasma physics.

1. Introduction

The discovery of the remarkable interaction properties of solitary wave solutions to KdV by Zabusky and Kruskal [62] and the invention by Gardner, Greene, Kruskal and Miura [9] of the Inverse Spectral Transform for the solution of the Cauchy problem for KdV stand as two of the most far-reaching breakthroughs in the development of modern nonlinear mathematical science. Of all the completely integrable systems discovered since 1967, KdV certainly remains the most fully understood and arguably the most important for applications to macroscopic phenomena and processes.

Here we shall survey some of these applications, with emphasis on several cases where there is extensive experimental confirmation of the predictions of KdV theory. Not surprisingly, many of these applications are in fluid mechanics – not surprisingly, because fluid mechanics continues to maintain its seminal role in physics as the discipline in which many key nonlinear structures were first discovered (shocks, bifurcations, solitons, deterministic chaos, hypercomplex systems, ...) and in which the underlying theories receive their most convincing experimental validation. As the prototypical integrable nonlinear system, KdV has also had enormous indirect impact on many parts of theoretical physics, pure mathematics, and the areas in between. Vast areas of mathematics, including ordinary differential equations, algebraic geometry, Lie group theory, differential geometry and asymptotics have been opened up 'on the back', as it were, of the solving of KdV, and brought to bear on issues in quantum field theory, string and conformal field theory, quantum gravity and classical general relatively, to say nothing of the myriad applications in concrete settings of other famous integrable systems including nonlinear Schrödinger (NLS) and sine-Gordon (SG). These latter applications range from condensed matter and semiconductor physics through nonlinear optics and laser physics, hydrodynamics, meteorology and plasma physics to protein systems and neurophysiology. Robin Bullough (private communication) has devised a 'map' of considerable complexity, showing the astonishingly rich range of interconnections between mathematical and physical structures in integrable nonlinear systems, while the applications receive many treatments, though not in recent survey form, in the very many monographs and collections of reviews on soliton systems that now exist, and in the large number of volumes of Proceedings of Advanced Research Workshops and Advanced Study Institutes organized under the NATO Special Programme *Chaos, Order and Patterns: Aspects of Nonlinearity*, to say nothing of the large number of articles that have appeared not only in the long-established scientific journals, but also in the dozen or so journals devoted to nonlinear science that have sprung up in the early 1990's.

It is tempting to include with KdV its one-dimensional modified form MKdV and its two-dimensional counterpart KP, but to do so would be to divert attention from the remarkable success of KdV theory, to which we turn in a moment. Amusingly, the first application of KdV theory was actually made some years or even decades before the publication of KdV in 1895, as pointed out by Bullough and Caudrey ([4, p. 374]). John Scott and Russell, in his book *The Wave of Translation in the Oceans of Water, Air and Ether*, published posthumously in 1882, applied the KdV formula for the propagation speed of a soliton, discovered experimentally by Russell himself, to two problems. First, given the velocity of sound, the depth of the Earth's atmosphere was calculated to be five miles on the basis that sound is evidently carried by waves free of distortion and therefore by Scott Russell's solitary waves. Remarkably, this prediction is correct numerically, or would be if the matter in the atmosphere were distributed with uniform density. The same argument was used, second, with the velocity of light as the propagation speed for a distortion-free signal, to infer the radius of the universe as 5×10^{17} miles. Bullough and Caudrey point out that this estimate is wrong by at least five orders of magnitude, and in any event is incorrect not least in using a value of g which Russell arbitrary reduced by a factor 10^{-5}. Largely free of contention, however, are Russell's own experiments on solitary waves in water channels in which, as is now well known, he not only correctly extracted the propagation velocity from measurements, and convincingly showed that solitary waves of depression are impossible and that an arbitrary initial elevation would break up into a finite number of solitary waves, but also that the interaction between solitary waves had the particle-like property which was not picked up for more than a further century.

It is appropriate to turn, for our first example, to modern careful experiments in the water-wave problem that started the whole subject off.

2. Surface Gravity Waves

Weakly nonlinear surface waves on uniform water of shallow depth provided the first recorded sighting of a soliton, the first formulation, by Boussinesq and by

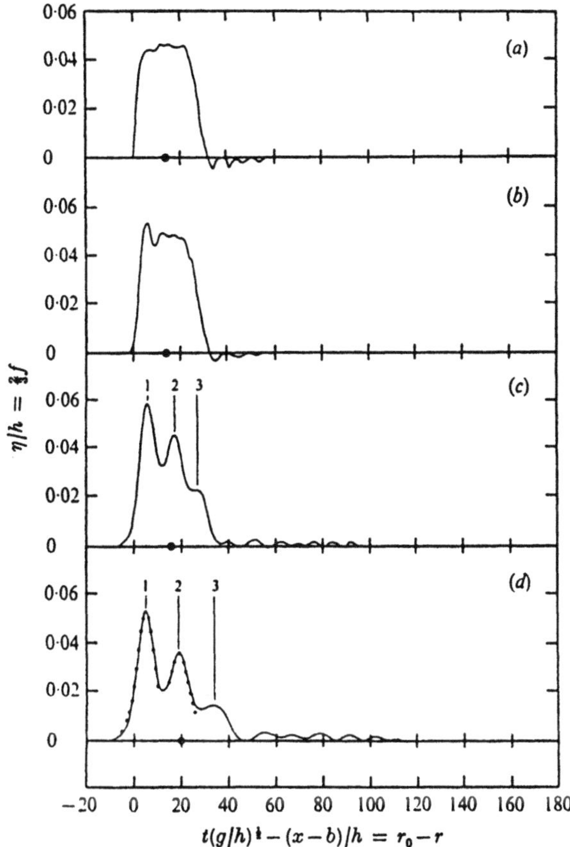

Fig. 1. Emergence of KdV solitons of elevation in water-channel experiments. The plots correspond to measuring stations (a) $x/h = 0$, (b) $x/h = 20$, (c) $x/h = 180$, (d) $x/h = 400$. Plotted on the abscissa is retarded time, so that a soliton should, in the absence of dissipation, shift to the left as time increases. (From Hammack and Segur [13], with permission.)

Korteweg and de Vries, of model equations to describe it, and the first example through KdV of an integrable system solvable by IST. These waves also find, in the work of Hammack and Segur [12–14], one of the most complete and convincing confirmations of IST theory for KdV. This work comprised experiments in a tank of length 31.6 m and breadth 39.4 cm, with waves of typical length-scale $10h$ and amplitude $0.05h$ generated in water of depth $h = 5$ cm or 10 cm by the controlled and recorded action of a rectangular wavemaker at one end. Measurements were taken for $h = 5$ cm at $x/h = 0, 20, 180, 400$ and for $h = 10$ cm at $x/h = 0, 50, 100, 150, 200$; in each case the first two stations correspond to linear propagation, the last two to ranges equivalent to the long time-scale for separation out of nonlinear KdV features. The initial wave profile was measured in each case and used to determine the number of solitons expected ($N = 0, 3, 4$ typically).

First, the prediction of N was very satisfactorily checked, and then confirmation that the emerging features (see Figure 1) were KdV solitons was made by measuring the amplitude of each feature and plotting points, as the dots in Figure 1, on the profile of an isolated KdV soliton with that amplitude. Second, an attempt was made to predict the amplitude of the leading soliton at the most remote measuring station. For this it was necessary to determine the largest Schrödinger eigenvalue of the initial waveform, and then to use results due to Keulegan [26] for the viscous decay, mainly in side-wall and bottom boundary layers, of a solitary wave of the corresponding amplitude. The viscous corrections are large enough to prevent any of the soliton velocities exceeding $(gh)^{1/2}$, which they should all do in lossless theory (the corresponding excess over the linear speed was detected, and found to be in good agreement with theory, in the propagation of pressure pulses in a tube of bubbly liquid: see Section 4), but if they are taken into account, the decaying amplitude and speed of the surface wave solitons can be reasonably well predicted.

But especially valuable in the work of Hammack and Segur is the study of the radiation corresponding to the continuous part of the spectral problem, particularly in the case of an initial wave of depression, which has no discrete eigenvalues and leads only to decaying oscillations. Although for realistic initial wave profiles there are no exact solutions of KdV involving radiation, the asymptotic evolution as $t \to +\infty$ is well understood in overall structure (see [1, Section 1.7c]).

(i) For $x \gg t^{1/3}$ the solution $\eta(x,t)$ is exponentially small.

(ii) For $|x| \lesssim t^{1/3}$ the solution is self-similar, and given by the solution $\eta = (x + x_0)/6t$ of the nondispersive simple wave equation $\eta_t + 6\eta\eta_x = 0$.

(iii) The transition away from (ii) takes place in a thin 'collisionless shock' at $x \sim -t^{1/3}(\ln t)^{p+2/3}$ for some p, $0 \leqslant p \ll 1$.

(iv) Beyond the collisionless shock, for $(-x) \gg t^{1/3}(\ln t)^{p+2/3}$, the solution breaks into a set of decaying wave packets, each packet having many oscillations at a fixed wavenumber and travelling at the (linearized) group velocity for that wavenumber. The nodes separating adjacent wave packets are defined by the zeros of the reflexion coefficient of the spectral problem for $\eta(x,0)$.

Figure 2, from Hammack and Segur [13], shows that this structure is indeed borne out in experiment; and in the paper just quoted, the prediction of the leading wave (the simple wave ramp) is brought even more closely into agreement with experiment if account is taken of viscous effects. This leading wave is completely different, in form and strength, from that predicted by the solution of the linearized KdV with the same initial data (see Hammack and Segur [13]).

An interesting and important issue also settled by Hammack and Segur [13] is as to whether the presence of extraneous fine-scale waves, often present in practice and not correctly described by KdV, might invalidate the KdV predictions. That it would not was shown by driving the wavemaker with the same

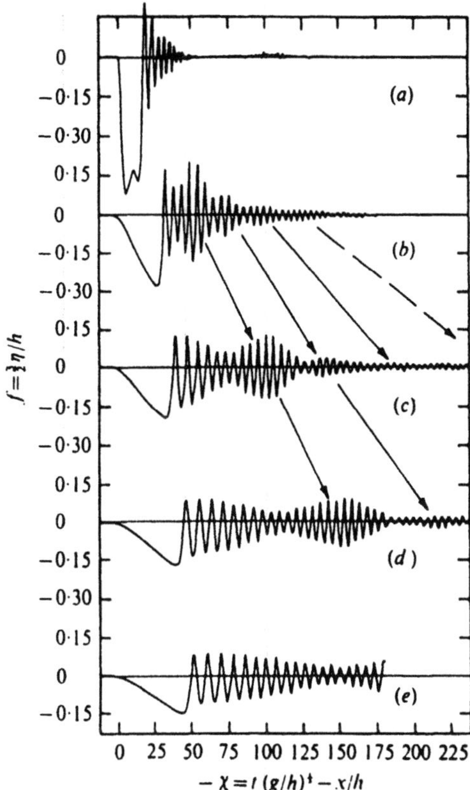

Fig. 2. Dispersive radiation produced by an initial depression of a water-channel surface. The plots correspond to measuring stations (a) $x/h = 0$, (b) $x/h = 50$, (c) $x/h = 100$, (d) $x/h = 150$, (e) $x/h = 200$. Plotted on the abscissa is retarded time. The arrows indicate wave packet trajectories based on the average wavenumber of each packet and the linear dispersion relation. Note the simple wave ramp at the left. (From Hammack and Segur [13], with permission.)

mean motion in each of three experiments, but with different types of high-frequency fluctuation superimposed. The fine-scale fluctuations were found to disperse rapidly, and at $x/h = 400$ precisely the same (four) solitons were found, identical to within experimental accuracy.

Segur and Hammack [55] used the same water channel to study internal waves at the interface between two layers of fluid of different densities. (The dispersion relation has two branches, in general distinct, one representing surface gravity waves, one internal waves localized at the density discontinuity.) If $kH \ll 1$ where H is the total depth, then in terms of a coordinate X translating with the nondispersive long-internal-wave speed and a long time variable τ proportional to the square root of the (small) density difference, the interface elevation η satisfies, after scaling, KdV in the form

$$\eta_\tau + 6(h_1 - h_2)\eta\eta_x + \eta_{xxx} = 0. \qquad (2.1)$$

Solitons were generated by the operation of the wavemaker, and the interface displacement η measured. Close agreement was again found between measured wave profiles and those of individual KdV solitons, and agreement in other aspects was much improved by the inclusion of viscous effects at the interface, as well as in the side-wall and bottom boundary layers, through an extension of the Keulegan theory [27]. Very similar conclusions – resoundingly in favour of KdV – were reached by Koop and Butler [31], who also provided theoretical extensions of KdV to include higher-order nonlinearity.

If $h_1 = h_2$, the coefficient of the quadratic term in (2.1) vanishes, and then MKdV is the appropriate evolution equation, but for a still longer time-scale with τ now proportional to the density difference itself. The corresponding propagation length for the emergence of MKdV solitons may then be beyond the range of the Segur and Hammack facility.

Returning to KdV internal solitons, the theory for these has proved crucial in understanding some extraordinary observations of very large amplitude waves in the Andaman Sea, and we now turn to Osborne's work on this topic.

3. Internal Solitons in the Ocean

The internal solitons studied in the laboratory by Segur and Hammack [55] have been seen at full scale in the seas in the Far East, with amplitudes dangerously large in terms of human activities. In the Andaman Sea, between Sumatra and Thailand, severe sub-surface currents were experienced by oil drilling rigs, one of which, in a depth of 1100 m, was spun around through 90 degrees and translated more than 30 m as a large-amplitude internal soliton propagated along the thermocline beneath it. Satellite photographs confirmed, through surface wave activity of much lower amplitude, and generated by an internal-surface wave resonance mechanism, the presence of these solitons in very distinctive groupings, and Osborne and Burch [49] made extensive *in situ* temperature and velocity measurements in the Andaman Sea over a four-day period in 1976. The analysis of their work lends strong support to the KdV theory of internal solitons (rather than the competing weakly-nonlinear models, namely the intermediate long-wave equation for intermediate depth and the Benjamin–Ono equation for large depth). Similar observations of large-amplitude internal solitons were made in the Sulu Sea by Apel *et al.* [2]. The longevity of these waves is attested to by the fact that the research vessel followed an internal soliton, of wavelength 1700 m and amplitude 100 m, for more than two days as it propagated at about 8 km/h. Since then many studies have shown the ubiquitous presence of oceanic internal solitons where there are topographic features to excite them via tidal action, as, for example, in the Strait of Messina, the Strait of Gibraltar, and the Gulf of California, but not, for example, in the Gulf of Mexico.

The situation in the Andaman Sea is reviewed by Osborne [48]. The internal waves can be thought of, in first approximation, as localized on the thermocline at a depth of several hundred metres, and at such depths the vertical excursions during the passage of an internal soliton are large, of the order of 100 m, with much smaller surface waves of amplitude 1–2 m. Nevertheless, the surface waves are locked by the resonant coupling to the internal wave, are similarly one-dimensional, with coherent crests extending over 100 km or more, and for each soliton are confined to bands about 1 km wide normal to the crests. These surface wave 'rips' are very prominent in satellite photographs, in which each band in a rather regularly spaced sequence of six or more appears as a dark striation separating bright bands of wave-free activity between the rips. Successive sequences are separated by 12.4 h, the local tidal period. Topographical obstructions to the tidal flow in different locations generate such packets of solitons and rips travelling in different directions, but satellite photographs indicate that each packet is able to maintain its identity over hundreds of kilometres despite oblique interaction with other soliton packets.

All this, together with the shapes of the pulses measured in the displacements of isotherms (or isopycnals) and the fact that generally the pulses are rank-ordered by amplitude and correspond to a depression of the thermocline, suggests an underlying soliton mechanism, and Osborne [48] compares the merits of KdV, the intermediate depth equation of Kubota, Ko and Dobbs [32], and the Benjamin–Ono equation [3, 47]. The typical pulse widths are of order 1–3 km and the depth between 300 m and 1100 m, so the temptation is to assume the depth small compared with the pulse width and expect that (of these three models, all integrable) KdV will provide the best approximation. That is confirmed by the analysis of Osborne and his collaborators, with extensive measurements over the full water depth and analysis for a two-layer model and for continuous density profiles. The measurements show that throughout most of the depth, and certainly for a considerable range on either side of the thermocline, the wave-induced particle velocities are independent of depth, as required in the approximations leading to KdV.

For the analysis the streamfunction is written as $\psi(x,z,t) = \phi(z)\eta(x,t)$, where the vertical mode shape $\phi(z)$ is determined by a differential eigenvalue problem and $\eta(x,t)$, which to the same approximation represents the displacement of an isopycnal, satisfies KdV

$$\eta_t + c_0\eta_x + \alpha\eta\eta_x + \beta\eta_{xxx} = 0. \tag{3.1}$$

With a two-layer model (upper and lower layer thicknesses and densities h_1, ρ_1, and h_2, ρ_2) the coefficients are simply

$$c_0 = \left[\left(\frac{\rho_2 - \rho_1}{\rho_2 + \rho_1}\right)\left(\frac{2gh_1h_2}{h_1 + h_2}\right)\right]^{1/2},$$

$$\alpha = -\frac{3c_0(h_2 - h_1)}{2h_1h_2}, \qquad \beta = \frac{c_0h_1h_2}{6}, \tag{3.2}$$

for $\rho_2 - \rho_1 \ll \rho_1, \rho_2$. Since $h_2 > h_1$ here, a soliton is, as observed, one in which the thermocline is depressed. In the detailed application of KdV theory, Osborne and his collaborators computed the coefficients in (3.1) by numerically solving the eigenvalue problem using measured mean density profiles and performing the necessary integrals involving the eigenfunction (only the lowest mode was considered, other measurements showing that more than 90% of the total energy was carried by this mode) and the mean state.

A time series of recorded data in the Andaman Sea was then used as the scattering potential for the Schrödinger problem, and the number and magnitudes of the discrete eigenvalues were calculated numerically, along with the reflexion coefficient as a function of scattering frequency (or wavenumber). In some cases, as many as 12 bound states were predicted, and the first seven of these gave soliton magnitudes in good agreement with observed maxima in the time series. Attention was also paid to the continuous spectrum, which was generally dominated by a single lobe in the frequency range for which KdV is applicable, corresponding to a single radiative wave packet in the time series (cf. Section 2).

Some anomalies remain, for example, in the observation that in some similar experiments involving internal waves in the ocean, the solitons were not always ordered according to amplitude. This may simply reflect the fact that a sufficiently large time had not elapsed to quality for the asymptotic limit for the particular complicated evolution in question, or it may have to do with the fact that the infinite-line spectral problem is not strictly appropriate to these problems where there is periodic tidal excitation. This would be settled if, as promised by Osborne, the periodic-initial-value problem for KdV were solved by numerical implementation of the periodic spectral transform. Whatever the outcome, it is clear that these large-amplitude waves (large in relation to human marine activities) are actually sufficiently weakly nonlinear and of sufficiently long wavelength that they should be described by KdV, and that KdV theory does indeed give an excellent description of all their essential features.

4. Nonlinear Acoustics of Bubbly Liquids

No account of applications of KdV should omit reference to theoretical and experimental work on the propagation of nonlinear acoustic waves in liquids with small volume concentrations of gas bubbles. Such a suspension has remarkable acoustic properties, even for very small volume concentrations of gas bubbles. First, for example, for air bubbles in water at standard pressure and temperature, the low-frequency sound speed c_0 is of the order of 40 m s^{-1} for a concentration $\alpha_0 = 0.1$, not simply well below the pure water sound speed $c_l = 1500$ m s^{-1} but also well below that of the pure gas phase (340 $\mathrm{m\,s^{-1}}$). This can be understood by noting that in the expression $c_0^2 =$ (bulk modulus)/(density) the bubbles endow the suspension simultaneously with the low modulus of the gas phase and the high density of the liquid phase. These low values of c_0 imply that nonlinear effects are much more important for bubbly liquids, even with very small bubble concentrations, than for monophase liquids.

Second, the bubbly liquid exhibits strong dispersion. For a lossless suspension of identical-size bubbles, the acoustic phase speed $c_p(\omega)$ is reduced, parabolically, below c_0 for low frequencies, and reaches zero at the bubble monopole resonance frequency ω_0. There is then a forbidden band $\omega_0 < \omega < \omega_1$ in which $c_p(\omega)$ is purely imaginary and all motion is purely reactive, and then $c_p(\omega)$ becomes real again, decreasing from $+\infty$ to essentially c_l as ω increases from ω_1 to ∞. If losses are included, there is a small real part of $c_p(\omega)$ in the forbidden band, and a high value of $\mathrm{Im}\, c_p(\omega)$ there, peaking at resonance. These features have been substantially confirmed by experiment. See, for example, the review by van Wijngaarden [58], in which, in his Figure 3, we see measured phase speeds as low as 500 $\mathrm{m\,s^{-1}}$ and as high as 2800 $\mathrm{m\,s^{-1}}$ in experiments at different frequencies on the *same* bubbly mixture with α_0 very small, $\alpha_0 = 2 \times 10^{-4}$.

The parabolic form of $c_p(\omega)$ for $\omega/\omega_0 \ll 1$ and the presence of the usual convective derivatives in the mass and momentum conservation equations indicate that low-frequency pressure pulses of moderate amplitude should satisfy KdV, or (Burgers) BKdV if dissipative effects are significant. Two groups have done impressive analytical and experimental studies for such cases. That led by van Wijngaarden was the first to carry out experiments to verify qualitatively the predictions of KdV theory for soliton formation and of BKdV theory for the formation of shocks with either monotonic or oscillatory structure depending upon the strength of dissipation relative to dispersion. Later, van Wijngaarden and his student Roelofsen carried out experiments in which a triangular compression pulse was delivered by a piston at one end of a tube of bubbly mixture. For such initial data the Schrödinger spectral problem for KdV can be solved explicitly in terms of Airy functions, and the number of solitons expected can be predicted in closed form as a function of the duration and amplitude of the initial wave and the properties of the medium. Gratifying agreement was obtained between the experiment and theory, as is explained in more detail in van Wijngaarden's contribution to these Proceedings. This also amounts to an impressive confirmation of the theory of bubbly liquids, modelled as two co-existing continua. There is a spectacular difference between prediction or observation for a pure liquid or gas phase, where the triangular pulse simply lengthens as $t^{1/2}$, remaining triangular of constant area, and with a leading diffusion or relaxation-controlled shock with amplitude decreasing as $t^{-1/2}$; and the prediction or observation of the response of a liquid with a minute gas bubble concentration, where a finite number of rank-ordered solitons may be produced, with no shock-like features under appropriate circumstances.

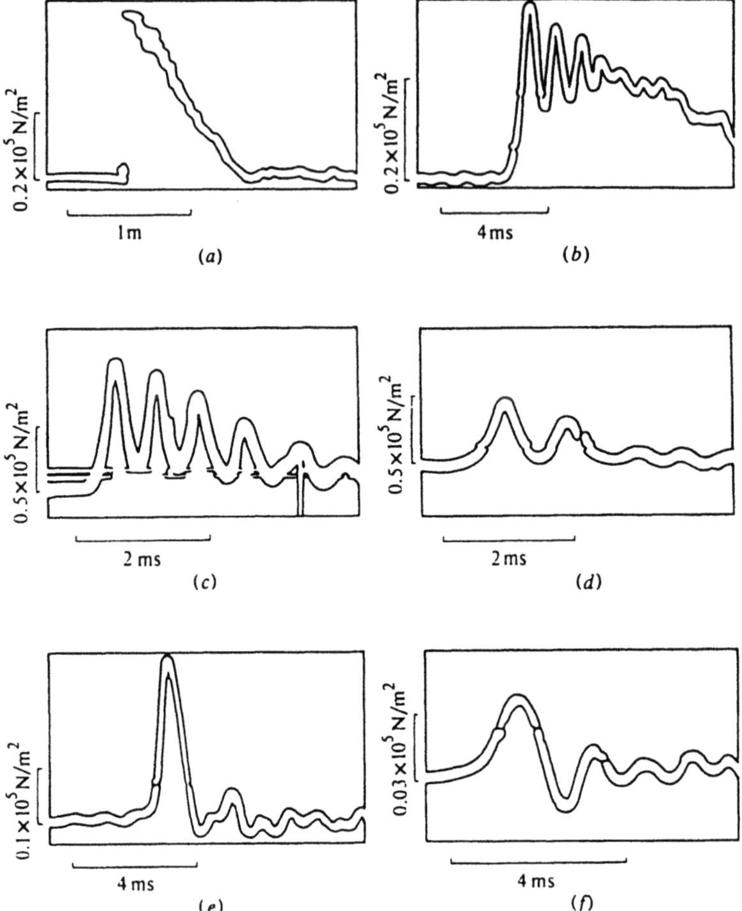

Fig. 3. Experimental observations of oscillatory shock and soliton features in propagation of pressure pulses in liquid with CO_2 bubbles. (From Kuznetsov *et al.* [33], with permission.)

Similar theoretical and experimental studies were also conducted in Novosibirsk (see Kuznetsov *et al.* [33] and Nakoryakov *et al.* [40]). A BKdV equation for the pressure $p(\xi, \tau)$,

$$\frac{\partial p}{\partial \tau} + p\frac{\partial p}{\partial \xi} - \frac{1}{\text{Re}}\frac{\partial^2 p}{\partial \xi^2} + \frac{1}{\sigma^2}\frac{\partial^3 p}{\partial \xi^3} = 0 \tag{4.1}$$

was derived, in scaled variables, with Re a Reynolds number based on characteristic velocity and length-scales u_0, ℓ, respectively, of the initial signal and an effective viscosity coefficient incorporating viscothermal diffusion and acoustic radiation damping, and with σ a dispersion parameter,

$$\sigma = \ell_0 u_0^{1/2} \beta^{-1/2}, \qquad \beta = R_0^2 c_0 / 6\alpha_0(1 - \alpha_0), \tag{4.2}$$

for a mixture of bubbles all of the same radius R_0, and with c_0 the low-frequency sound speed in the mixture. Low-amplitude wave packets were studied, and also larger-amplitude compression pulses produced by breaking a diaphragm separating a tube of bubbly liquid (CO_2 or He bubbles with $\alpha_0 \sim 0.01$) from a high-pressure chamber. Calculations were made for initial profiles of Gaussian form, and for triangular profiles corresponding to the diaphragm rupture.

For sufficiently small σ/Re, solitons were produced whose number agreed with spectral theory prediction. Reasonable agreement was obtained for the soliton propagation velocity (and, incidentally, for the propagation speed c_0 of low-amplitude wave packets) as a function of the medium parameters and soliton amplitude, for the dependence of soliton width on amplitude, and for the soliton shape, compared with KdV prediction. See Figure 3 above (Figure 10 of Kuznetsov *et al.* [33]) for oscillograms showing the emergence of 5, 2, 1 or 0 solitons in appropriate cases, together with dispersive radiation. The damping of solitons and wave packets was also studied, with agreement between theory and experiment which is astonishingly good given all the idealizations of the underlying model.

5. Voidage Slugs in Fluidized Beds

Multiphase media with dynamical microstructure will evidently display frequency dispersion, as is to be expected in a bubbly liquid where each micro-element has its own internal resonance, often lightly damped. It is less obvious that dispersion with only small dissipation will arise in the relaxation processes that take place in a gas or liquid phase with an included rigid particulate phase, such as one finds in dusty gas flows, in particulate matter sedimenting under gravity, or in fluidized beds as widely used in chemical processing. In a fluidized bed, gas or liquid is forced upwards under pressure in a vertical tube containing a fine bed of particles. When the fluidizing velocity exceeds a threshold, the bed expands – homogeneously one hopes – the particles being held at rest, on average, against gravity through their interactions with the fluid and with each other. But the homogeneously-expanded state is almost invariably instable, the instability leading in wide tubes to the production of large bubbles of particle-free fluid rising through the bed and creating a state visually similar to the boiling of a liquid. In sufficiently narrow tubes, a one-dimensional state persists, the instability here leading to 'slugging', the production of horizontal bands or slugs of particle-lean fluid propagating upward, with a region below each slug of increased particle concentration.

This instability is undesirable in technological applications, leading, for example, to highly uneven performance, excessive temperatures and fatigue damage. Much effort has been put into the modelling of fluidized beds (this is still a matter of considerable controversy) and into understanding the linear instability, although experimental study of the linearized stages is almost impossible. Weakly nonlinear analyses have been conducted by numerous authors, and Kluwick [28] was perhaps the first to show that in appropriate circumstances a fluidized bed or a sedimenting suspension would support KdV solitons. (In fact there is a change of sign of the nonlinear coefficient for a particular value $\phi_0 = n/(n+2)$ of the background voidage – the fraction of unit volume of mixture occupied by the gas phase – with n an empirically-determined constant in an assumed power law expression for the drag coefficient on a particle as a function of the local voidage ϕ, and Kluwick allowed for this, retaining quadratic and cubic nonlinearities to give an equation of Gardner type (KdV–MKdV).) However, weak nonlinearity is not necessarily stabilizing, and if the particle-phase interaction pressure is not sufficiently large, the essential linear instability will cause growth of the KdV solitons through what looks, for long waves, like negative diffusion. The question then is "what role does KdV have in the overall evolution, and what is the ultimate fate of any solitons that may be produced at the KdV stage of events?"

Harris and Crighton [15] sought to answer this question – and thereby in effect to show much more generally how the nongeneric integrable systems can be embedded, for limited ranges of time and space, in a much richer space-time evolution – by systematically analyzing one of several popular models for fluidized beds using asymptotic techniques model can, for a gas-fluidized bed, be reduced to the nondimensional system

$$-\frac{\partial \phi}{\partial t} + \frac{\partial}{\partial x}[(1-\phi)v] = 0, \tag{5.1}$$

$$\begin{aligned}(1-\phi)&\left(\frac{\partial v}{\partial t} + v\frac{\partial v}{\partial x}\right)\\ &= \frac{(1-\phi)}{F^2}\left(\frac{\phi_0}{\phi}\right)^{n+1}\left(1-\frac{v}{\phi_0}\right) - \frac{(1-\phi)}{F^2} - p_s'(\phi)\frac{\partial \phi}{\partial x} + \frac{1}{R}\frac{\partial^2 v}{\partial x^2},\end{aligned} \tag{5.2}$$

expressing conservation of the mass and momentum, respectively, of the particle phase which has voidage $\phi(x,t)$ and upward velocity $v(x,t)$. Terms on the right of (5.2) correspond, in order, to the drag force between particles and gas, to gravitational forces, to the interparticle pressure field, and to viscous effects in the particulate phase (this being treated as a Newtonian fluid with an isotropic interparticle pressure field and a deviatoric stress with a particulate phase viscosity). The parameters in (5.2) are the uniform background voidage ϕ_0, the empirical Richardson–Zaki index n, a Froude number F and a particle-phase Reynolds number R. Laboratory experiments in narrow tubes suggest $n = \mathrm{O}(1)$, $R = \mathrm{O}(1)$, $\phi_0 = \mathrm{O}(1)$, and $F \ll 1$, so Harris and Crighton seek asymptotics for $F \to 0$ for fixed values of all other parameters.

Bypassing the linear instability and going straight to the weakly nonlinear stage via

$$\begin{aligned} v &= F^2 v_1 + F^3 v_2 + \cdots, \\ \phi &= \phi_0 + F^2\phi_1 + F^3\phi_2 + \cdots, \end{aligned} \tag{5.3}$$

with $X = x - (n+1)(1-\phi_0)t$, $\tau = F^2 t$, we get

$$\frac{\partial \phi_1}{\partial \tau} + \beta\phi_1 \frac{\partial \phi_1}{\partial X} + \gamma \frac{\partial^3 \phi_1}{\partial X^3} = -F\delta_0 \frac{\partial^2 \phi_1}{\partial X^2} + \mathrm{O}(F^2), \tag{5.4}$$

$$\beta = -(n+1)\left(n+2-\frac{n}{\phi_0}\right), \qquad \gamma = (n+1)\frac{\phi_0}{R}. \tag{5.5}$$

The left side contains cubic dispersion from what in the momentum equation for v was a diffusive term and predicts, for $\phi_0 < n/(n+2)$, the formation of KdV solitons of voidage greater than ϕ_0, rising vertically. The condition on ϕ_0 is usually satisfied, since typically $0.4 < \phi_0 < 0.5$ and $3 < n < 4$.

On the right of (5.4) the perturbing term is diffusive if $\delta_0 < 0$, which corresponds to a sufficiently strong particle-pressure $p_s'(0)$, and then perturbation theory [1, 24, 25, 29, 30] for nearly integrable systems shows that the KdV solitons simply decay adiabatically, with amplitude $\mathrm{O}(T^{-1/2})$, $T = F\tau$. If $\delta_0 > 0$ the linear elements of (5.4) represent the essential instability of the fluidized bed, and the solitons grow. (The problem is then ill-posed, but the high-wavenumber growth is actually controlled by a fourth derivative term (cf. Appendix A of Harris and Crighton [15], and Hayakawa *et al.* [16]).) Then we seek an expansion

$$\phi = \phi_0 + F^2\phi_1(X,\tau,T) + F^3\phi_2(X,\tau,T) + \cdots, \tag{5.6}$$

in which $\tau = F^2 t$, $T = F^3 t$, and ϕ_1 is an adiabatically changing soliton,

$$\phi_1 = \frac{12\gamma}{\beta}\kappa^2(T)\,\mathrm{sech}^2\left[\kappa(T)(X - X_0(T) - \xi)\right] \tag{5.7}$$

and where, as usual, it is important to uniformize the phase by writing

$$\xi = \frac{4\gamma}{F}\int \kappa^2(T)\,\mathrm{d}T. \tag{5.8}$$

Inspection of the problem for ϕ_2 gives the growth law

$$\kappa_T = \frac{8}{15}\delta_0\kappa^3, \qquad \kappa(T) = \left(\frac{T_0}{T_0 - T}\right)^{1/2}, \qquad T_0 = \frac{15}{16\delta_0}, \tag{5.9}$$

so that the soliton amplitude has a finite-time singularity. One can then complete the solution for ϕ_2 and by examining its asymptotics below the soliton we deduce the existence of the now-familiar 'shelf', in this context an extensive region of negative voidage. As in the first successful application of perturbation theory to KdV solitons, the amplitude change mandated by (5.9) is actually what one gets by substituting the slowly-varying soliton expression in the 'energy' balance equation obtained by multiplying (5.4) by ϕ_1 and integrating over all X. But the resulting soliton variation cannot simultaneously satisfy conservation of particles, expressed by integrating (5.4) itself over all X. In fact, particles must continually be lost from the soliton and they find their way into the shelf below, forming the particle-rich regions mentioned earlier. Although the amplitude of the shelf is formally small (as indeed is that of the soliton), it is proportional to $\kappa(T)$ and has the finite-time singularity too. This singularity suggests that the KdV solitons will amplify beyond the weakly nonlinear limit ($\phi - \phi_0 = \mathrm{O}(F^2)$), and possibly even to $\mathrm{O}(1)$.

Before that happens, however, other perturbations become comparable with, and then dominate over, the negative diffusion in (5.4), and the growth law changes. For $T_0 - T = F\widehat{T}, \widehat{T} = \mathrm{O}(1)$, various new quadratically nonlinear terms are comparable with diffusion, and the growth law – for the voidage fluctuation, which has now grown from $\mathrm{O}(F^2)$ to $\mathrm{O}(F)$, but is still given by a KdV soliton – is of the form

$$\kappa_{\widehat{T}} = b\kappa^3 + d\kappa^5, \tag{5.10}$$

and when the κ^5 term dominates

$$\kappa \sim \left[4d(\widehat{T}_0 - \widehat{T})\right]^{-1/4}, \tag{5.11}$$

so that the soliton continues to grow, but less rapidly. When $T_0 - T = \mathrm{O}(F^2)$ all perturbations are comparable and we must leave the KdV soliton in favour of a solution of the fully nonlinear system. However, we are permitted, through the matching backward in time, to seek a slowly-varying travelling wave solution of the full system, the slow variation being determined, once again, by examination of corrections to the leading order ($\mathrm{O}(1)$) voidage solitary wave.

Analysis of the travelling solitary wave is given by Harris and Crighton [15] along with a determination of the amplitude variation, and a proof that in the long-time limit the amplitude and propagation speed tend to *constant* values, dependent on the background voidage and related to each other in a certain functional way. Going backward in time, the expressions for the amplitude and velocity and for the solitary wave shape match asymptotically those for the explosively growing KdV soliton.

Computed solitary wave profiles are not much different from those of sech^2 solitons. Harris and Crighton [15] give estimates appropriate to typical small-scale laboratory experiments for the times and distances over which one may expect to see the full nonlinear development or merely the emergence of the weak solitons. For example, with glass beads of 50 μm diameter and a typical disturbance length-scale of 1 cm, the KdV soliton stage corresponds to upward propagation over about 0.5 m, and the fully developed with experiments has yet been made, but the theory provides a clear suggestion as to how the large-scale orderly structures initially form from general disturbances, and how they amplify to the large-amplitude state associated with voidage bands. It also provides the theory for the particle-rich shelf below each voidage band, and gives estimates for the length- and time-scales associated with solitons, solitary waves and shelves.

According to Harris and Crighton [15], other competing continuum models for fluidized beds lead to similar results; at some stage in the spatio-temporal development they all contain a KdV equation with unstable perturbation. This leads to the emergence of nascent voidage bands as solitons, though the details of their subsequent amplification and equilibration will differ with the model.

We have discussed this topic at some length because it is perhaps the first in which all stages of development in a complicated nonintegrable system can be at least partially analyzed and asymptotically related to each other. As part of this, the integrable KdV enters at one stage, and its solitons are subject to different types of perturbation as they grow. There is no need – indeed no possibility short of writing the whole system, very artificially, as a perturbed KdV system – to try to represent the system uniformly as perturbed KdV.

From this it is clear that although integrable, and indeed nearly-integrable, systems are a set of measure zero, a vast number of much more complicated systems generally far from integrable may contain an integrable element at some stage. Asymptotic methods tell us how to handle the complicated scenario, as above. Soliton large-scale order will emerge at the integrable stage, though whether it will persist, as here, and grow to large amplitude and a change of form, depends on the nonintegrable ingredients.

6. Magma Flow and Conduit Waves

KdV theory comes into play in several areas of geophysics. In conduit flows, buoyant fluid introduced below a layer of fluid of greater viscosity rises through a conduit which it creates, with buoyancy and viscous shear stress in balance for steady flow in a conduit of uniform area. If the supply rate varies, axisymmetric bulges propagate upward as conduit waves. Helfrich and Whitehead [17] offer simple elegant theory and laboratory experiments on large-amplitude conduit waves, and emphasize that they transport with them a substantial blob of recirculating fluid. It is thought that such essentially diffusion-free conduit-wave transport of magma from the Earth's interior to the surface could contribute to the formation of hot spots and of volcanic island chains.

If the conduit has area A and carries a volume flux Q, then mass conservation gives

$$\frac{\partial A}{\partial t} + \frac{\partial Q}{\partial z} = 0, \tag{6.1}$$

while the momentum equation for a local Poiseuille flow and a small-slope approximation give

$$Q = A^2\left[1 + \frac{\partial}{\partial z}\left(\frac{1}{A}\frac{\partial Q}{\partial z}\right)\right], \tag{6.2}$$

where all quantities are dimensionless. In the weakly-nonlinear limit we replace Q by A^2 on the right of (6.2), substitute in (6.1) and drop certain derivatives on a long-waves basis. This immediately gives KdV for $A(z,t)$.

Olson and Christensen [46] studied solitary wave solutions of the full system (6.1), (6.2), showing, for instance, that the relation between propagation speed c and amplitude A_{m} is

$$c = (2A_{\mathrm{m}}^2 \ln A_{\mathrm{m}} - A_{\mathrm{m}}^2 + 1)/(A_{\mathrm{m}} - 1)^2, \tag{6.3}$$

which reduces to the KdV result for $(A_{\mathrm{m}} - 1)$ small, and that for $A_{\mathrm{m}} \gg 1$ the wave profile is Gaussian,

$$A(\xi) = A_{\mathrm{m}} \exp(-\xi^2/2c). \tag{6.4}$$

Helfrich and Whitehead tested these predictions, and those of KdV theory, against experiments in which controlled excitation of solitary waves was possible, and in which provision was made for marking – with dye – the fluid particles in a given solitary wave. They found that the predictions of the full nonlinear theory were generally well borne out in the experiments, which involved waves with A (nondimensionalized on the uniform area of the conduit with steady flow) as large as 20. The overtaking of one solitary wave by another was observed to take place essentially elastically, although for reasons as yet unexplained, the larger wave increased in amplitude by about 5% on average, and its speed decreased by 4%, while the smaller wave was unchanged in amplitude and speed to within experimental accuracy. Numerical solutions of (6.1) and (6.2) gave an over-prediction, by as much as 40%, of the phase shifts experienced in interaction. Clearly mass is not conserved by the two solitary waves in such interactions, and the excess must presumably be supplied unsteadily from the reservoir at the foot of the conduit.

As mentioned earlier, these conduit solitary waves transport matter with them, and for large amplitude waves the transported volume asymptotes to the whole volume excess

$$\int_{-\infty}^{+\infty} (A - A_0)(\xi)\, \mathrm{d}\xi$$

carried by a given solitary wave. This geologically significant transport property persists, though to a lesser degree, in the KdV limit. When a larger wave overtakes a smaller, matter from the smaller is retained in the larger afterward, while the smaller contains only fluid particles that were present in the original larger, as was shown very convincingly in the Helfrich and Whitehead [17] experiments.

Most of these studies, and indeed of the interest, refer to amplitudes far beyond the reach of the KdV limit, but KdV properties dominate the experiments, the way they are conducted, and the inferences drawn from them. Another geophysical process within the same general category is referred to as 'compaction-driven flow', where buoyant interstitial melt (corresponding to the intrusive conduit fluid) is forced through a deformable porous crystalline matrix representing the Earth's mantle, and corresponding to the viscous fluid exterior to a conduit. The functional relation corresponding to (6.2) is different for compaction-driven flows, but again KdV emerges in the weakly-nonlinear limit, with solitary wave solutions of the fully nonlinear system [51]. These solitary wave solutions have been called *magmons*, although numerical evidence is that the result of a collision between two magmons is not confined simply to phase shifts. The magmons are also quite different from the conduit waves in that they do not transport matter, the fluid particles simply experiencing a finite displacement in the passage of a magmon [54]. By contrast, conduit solitary waves are rather efficient in transporting matter right through the mantle with negligible diffusion; Whitehead and Helfrich [60] estimate that the occurrence of a single solitary wave (of the geologically plausible scales) every 500 million years would double the magma flux to the surface over what would be produced by steady flow in the same conduit!

7. Jupiter

Perhaps the boldest application of KdV to date has been to the understanding of the Great Red Spot (GRS) and other features in the Jovian atmosphere, seen in cloud patterns, such as the South Equatorial Disturbance, the Dark South Tropical Streak, the Hollow and the White Ovals and, in particular also, the South Tropical Disturbance (STD). The soliton theory is due to Maxworthy and Redekopp [38] and to Maxworthy, Redekopp and Weidman [39]. Although now seen to be inadequate quantitatively and even qualitatively, this work was highly influential ("a noticeable milestone in the history of contemporary ideas of the GRS nature" – Nezlin [43]) and contained many of the essential ingredients for current understanding in what has become, with the recent advent of high-quality spacecraft images, a topic of intense research interest. The reader is referred to Marcus [36], Nezlin [43], and a number of articles in the Focus Issue of *Chaos* (1994, Vol. 4, No. 2) coordinated by Nezlin, for accounts of the current position.

In rapidly rotating planetary atmospheres, the dominant linear wave modes are highly dispersive westward-propagating Rossby waves. The long waves have highest velocity and are weakly dispersive, with KdV-type cubic dispersion. Now in the 'shallow-water equations' for a thin atmosphere with a free upper surface, and with the β-plane approximation $f = f_0 + \beta y$ (y northward) for the local Coriolis parameter, a KdV nonlinearity $\beta h h_x$ arises (see [43, Equation (16)]) which has the effect of changing the phase velocity of long Rossby waves from $V_R = \beta r_R^2$ in the linear regime (r_R is the Rossby deformation radius, $r_R = (gH_0)^{1/2}/f_0$ in obvious symbols) to $V = V_R(1+h)$, $h = \delta H/H_0$, for a nonlinear solitary Rossby wave. Such a solitary wave can preserve itself over long times only if it is free of resonance (phase velocity matching) with any linear waves, so we must have $h > 0$, which means that the vortical structures producing the free surface elevation must be *anticyclonic*. By contrast, disturbances with cyclonic vorticity disperse by linear Rossby wave radiation. This anticyclonic circulation is completely in accord with almost all observations of large-scale organized structures in giant planets (Jupiter, Saturn, Neptune).

This further suggests that in a weakly nonlinear approximation, the vortical structure should be described by KdV, as first suggested in [38]. There the streamfunction for the quasi-geostrophic equations was expanded in the form of a mean horizontally-sheared east-west zonal shear flow plus a perturbation

$$\Psi = \int^{y} U(y)\,\mathrm{d}y + \varepsilon\psi(x, y, z, t) \tag{7.1}$$

and ψ taken as a single mode

$$\psi = A_n(\xi, \tau)\phi_n(y)\cos n\pi z + \text{integral harmonics}, \tag{7.2}$$

with scaled variables $\xi = \varepsilon(x - C_n^{(0)}t)$, $\tau = \varepsilon^3 t$ and with rigid top and bottom walls ($z = \pm 1$). Then A_n was found to satisfy MKdV for $n \neq 0$ (i.e. for baroclinic waves),

$$A_{n\tau} + \alpha_n A_n^2 A_{nx} + \beta_n A_{nxxx} = 0, \tag{7.3}$$

with coefficients defined through integrals of the basic state and the mode function $\phi_n(y)$ (itself the solution of an eigenvalue problem). The scalings $\xi = \varepsilon^{1/2}(x - C_0^{(0)}t)$, $\tau = \varepsilon^{3/2}t$ are needed for barotropic disturbances ($n = 0$), and A_0 satisfies standard KdV.

Maxworthy and Redekopp [38] calculated the coefficients for the simple shear flow $U(y) = S\tanh y$ ($S = \pm 1$ for anticyclonic and cyclonic basic shear) and displayed contours of constant ψ in a horizontal plane $z =$ const in coordinates travelling (westward) with a soliton solution of KdV or MKdV. They noted that such solitons have the forms, respectively,

$$A_0 = \operatorname{sgn}(\alpha_0\beta_0)\operatorname{sech}^2 X, \qquad A_n = \pm\operatorname{sech} X, \tag{7.4}$$

and so, at $z = 0$ say, one can have both an E-soliton of elevation and a D-soliton of depression for MKdV, corresponding to $\pm$ for A_n, but, for the same propagation speed only either an E-soliton or a D-soliton for KdV.

Of course, the features of a baroclinic D-soliton where $\cos n\pi z > 0$ become those of an E-soliton where $\cos n\pi z < 0$, and vice-versa. Typical streamline patterns for E- and D-solitons are shown in Figures 4a and 4b, together with, in Figure 4c, a suggested combination of D- and E-solitons on a jet profile modelling one of the east–west zones or belts on giant planets. This suggested structure was later confirmed by Maxworthy, Redekopp and Weidman ([39, Figure 16]), where analysis was carried out for a Bickley jet profile $U(y) = \operatorname{sech}^2 y$ and the MKdV equation. The morphology of Figure 4c is strongly reminiscent of

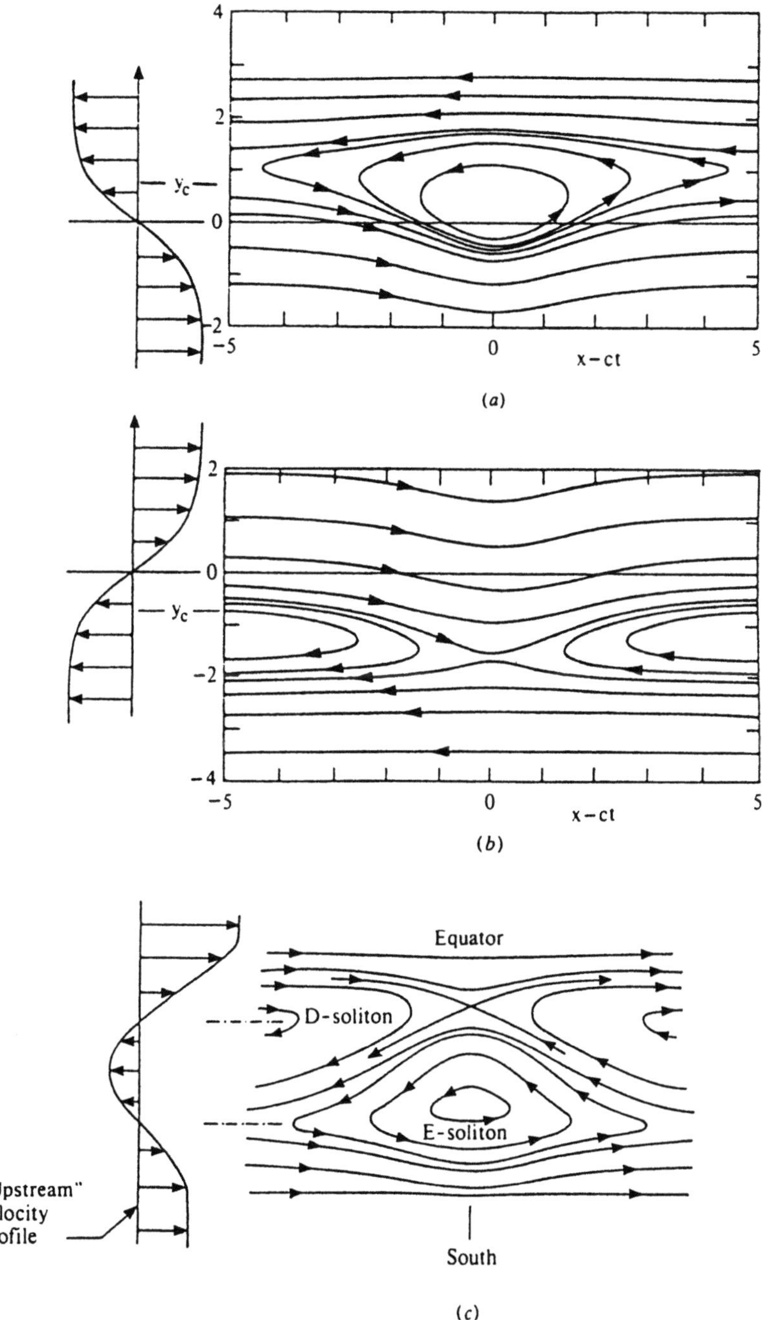

Fig. 4. (a) An E-soliton on an anticyclonic shear in the southern hemisphere; (b) A D-soliton on a cyclonic shear in the southern hemisphere; (c) A combined D–E soliton on an asymmetric jetlike profile. (Figure 4 from [38], with permission.)

the structure of the GRS and the Hollow immediately northward of it. Numerous other isolated Jovian features also have the typical structure of an isolated E- or D-soliton, the White Oval, for example, and the Dark South Tropical Streak, respectively.

A particularly striking aspect of these large-scale structures lies in their interaction properties, the most famous and well-documented of which [50] concerns that which occurred nine times between the GRS-plus-Hollow combination and the South Tropical Disturbance between 1901 and 1938. Peek's description of this has to be set alongside Scott Russell's vivid account of his first sighting of an individual soliton:

> "During the six weeks which would have been required for the p end of the Disturbance to pass from one end of the Hollow to the other... there was no sign whatever of any encroachment upon the region; instead, within a few days of its arrival at the f end of the Hollow, a facsimile of the p end of the Disturbance was seen... to be forming at the other end of the Hollow. ... The new development... proved to be a true p end of the Disturbance which drew away from the Red Spot at approximately the same rate (at which it had approached).... Thus its passage through... the Red Spot, which would have taken 3 months at its normal rate of progress, must have been accomplished in a matter of fourteen days."

Armed with KdV theory, it is impossible to refrain from speculating that Peek was describing the identity preservation of solitons and their generation of an abrupt phase shift on interaction. The phase shift in this case is equivalent to increasing the propagation speed of the STD from 7 $\mathrm{m\,s^{-1}}$ to 42 $\mathrm{m\,s^{-1}}$ as it traverses the GRS. Maxworthy, Redekopp and Weidman [39] analyzed the observations of this repeated interaction at length, modelling the GRS as an E-soliton and the STD as a double-D-soliton (while mentioning other possibilities for modelling the STD, namely a breather solution of MKdV, or a dispersive wave packet associated with the continuous spectrum). They calculated phase shifts not for their favoured MKdV system, but for Rossby waves in a homogeneous atmosphere. If the basic perturbation is the sum of two isolated solitary waves,

$$\psi = A_1(x,t)\phi_1(y) + A_2(x,t)\phi_2(y),$$

then the mode shapes $\phi_i(y)$ satisfy a barotropic Rayleigh equation, and the amplitudes satisfy coupled KdV equations,

$$A_{1t} + C_1 A_{1x} - 2r_1 A_1 A_{1x} - s_1 A_{1xxx} = \lambda_1 A_1 A_{2x} + \nu_1 A_2 A_{1x}, \tag{7.5}$$

$$A_{2t} + C_2 A_{2x} - 2r_2 A_2 A_{2x} - s_2 A_{2xxx} = \lambda_2 A_2 A_{1x} + \nu_2 A_1 A_{2x}. \tag{7.6}$$

The coefficients were evaluated for the periodic shear flow $U(y) = S\,\mathrm{sn}(y, m)$, and the interactions calculated by the asymptotic method of Oikawa and Yajima [45]. Direct comparison with the observations was not possible, however, because with the barotropic KdV solitons a D-soliton modelling the STD can approach an E-soliton modelling the GRS only from the east, and not from the west as on Jupiter. Nevertheless, the similarities between the various possible D–D, D–E and E–E interaction flow patterns and the observations are highly suggestive of the notion that although a realistic description cannot be attained at the level originally proposed by Maxworthy and Redekopp [38] (indeed, their rigid lid boundary condition precludes precisely the free surface variations that lead to the asymmetry between the anticyclonic and cyclonic solitary waves and thereby the exclusion, in agreement with observations, of the latter), there is nevertheless the crucial ingredient, within the more complete description, of an underlying integrable KdV model.

Finally, one might ask how large solitary Rossby waves are generated and maintained against dissipation. After all, the GRS has been observed now for more than 300 years, which exceeds by three orders of magnitude the lifetime of a Rossby wave packet of the same scale under linear dispersion mechanisms. Although it is possible that such waves are fed from some lower-atmosphere convective energy flux, it is more likely that they are formed from large-scale instability of the shear across the edges of the planetary belts and zones. Under the influence of linear amplification mechanisms, waves amplify to a weakly nonlinear stage, at which dispersion and nonlinearity enter in KdV or MKdV fashion. Integrability at this stage for the anticyclonic case implies the emergence of localized orderly soliton structures which may (cf. [38, p. 266]) be able to extract further energy from the shear and grow to larger amplitudes in which we have quasi-steady large-amplitude solitary Rossby waves with shear amplification and dissipation in balance. (Compare the corresponding evolution of large-amplitude solitary waves of voidage in fluidized beds, described in Section 5 above.) This would be consistent with the observations by Nezlin [43] of a laboratory model of Jovian flows where, in particular, he showed also how the number of such large features is a decreasing function of the velocity shear, only one feature, akin to GRS, being produced – and then maintained apparently indefinitely – for sufficiently strong shear.

Whatever the final theoretical framework might involve, it is clear that the astonishingly beautiful and well-ordered features of planetary atmospheres, and the attendant smaller-scale Lee waves, Kármán vortex streets, and hydraulic jumps – have been illuminated in an equally astonishingly beautiful way by KdV theory.

8. Plasma Physics

No catalogue of applications of KdV should fail to mention plasma physics. It was indeed in the context of hydromagnetic plasma waves [10] that the KdV equation was again encountered for the first time since 1895, and again in 1966 in the work of Washimi and Taniuti [59] on ion-acoustic waves in a cold plasma.

Ion-acoustic waves provided an early confirmation of the soliton aspects of KdV theory in the experiments of Hershkowitz, Romesser and Montgomery [18], the generation and interaction of ion-acoustic solitons having been demonstrated already by Ikezi, Taylor and Baker [23]; and Hershkowitz and Romesser [19] made observations of cylindrical ion-acoustic solitons before the cylindrical KdV has either been derived [37] or shown to be integrable [5].

Confining attention to one-dimensional ion-acoustic waves in a plasma with cold ions, negligible ion pressure and an isothermal perfect gas law for the electron field, one has, in the absence of any current, the set of four equations

$$\frac{\partial n_{\mathrm{i}}}{\partial t}+\frac{\partial}{\partial x}\left(n_{\mathrm{i}} v_{\mathrm{i}}\right)=0, \tag{8.1}$$

$$\frac{\partial v_{\mathrm{i}}}{\partial t}+v_{\mathrm{i}} \frac{\partial v_{\mathrm{i}}}{\partial x}=E, \tag{8.2}$$

$$n_{\mathrm{e}} E+\frac{\partial n_{\mathrm{e}}}{\partial x}=0, \tag{8.3}$$

$$\frac{\partial E}{\partial x}=n_{\mathrm{i}}-n_{\mathrm{e}}, \tag{8.4}$$

for the ion velocity v_{i} and number density n_{i}, the electric field E and the electron number density n_{e}. All quantities are nondimensional, the ambient state has $n_{\mathrm{i}}=n_{\mathrm{e}}=1$, $v_{\mathrm{i}}=E=0$, and ions and electrons are assumed to carry the same magnitude of charge. Equations (8.1)–(8.4) are, respectively, the expressions of mass and momentum conservation for the ions, a force balance equation for the electrons, whose inertia in a low-frequency wave is negligible, and Poisson's equation for the electric field given the charges.

The long-wave dispersion relation from (8.1)–(8.4) is

$$\omega^{2}/k^{2}=1-k^{2}+\mathrm{O}(k^{4}), \tag{8.5}$$

and therefore when this dispersion (valid for $\omega \ll \Omega_{\mathrm{p_i}}$ where $\Omega_{\mathrm{p_i}}$ is the ion-plasma frequency, and closely analogous to the dispersion in a bubbly liquid for $\omega \ll \omega_0$ (cf. Section 4)) is coupled with the convective derivatives in (8.1) and (8.2) in an appropriate weakly-nonlinear weakly-dispersive limit it is evident that fluctuations n_{i}' in the ion density must satisfy KdV (there being no symmetry under which the quadratic term would vanish).

Hershkowitz, Romesser and Montgomery [18] generated solitons from various square-wave input voltages in a narrow column of cold plasma, and were able to predict accurately the number of solitons produced in any given case from the well-known solutions to the Schrödinger equation for square-well potentials. The calculation also gave the soliton amplitudes in reasonable agreement with experiment, and the experiment further gave good agreement with theory on the speed-amplitude-width relation for KdV solitons. Ikezi [21] reported on similar experiments with one half-cycle of a cosine-squared wave parameters were reasonably predicted, the number of solitons observed was systematically greater than predicted, for reasons which seem not to have been explained. Ikezi [21] also showed the results of interaction between unequal solitons travelling in the same direction, and between equal solitons travelling in opposite directions. In the former case identity preservation was observed, though with very strong amplitude loss in both waves because of large damping experienced in the long interaction period; in the latter case, modelled by a Boussinesq equation, interaction was rapid and close to that expected theoretically. A further interesting point of these experiments was an observation of FPU recurrence. A sinusoidal input signal was transformed to one dominated by two solitons per period at 5 cm from input and back effectively to the original sinusoidal at 9 cm (excitation frequency 0.35 MHz, ion-plasma resonance frequency $\Omega_{\mathrm{p_i}} = 4\pi$ MHz). Theory of the recurrence for the two-solitons-per-period case was developed by Ikezi, Taylor and Baker [23] and Tappert and Judice [57].

For the plane wave case, experimental observations of KdV solitons were also made in the context of the Trivelpiece–Gould plasma wave by Ikezi *et al.* [22] and Saeki [52]. Ion-acoustic plasma waves provided the first experimental evidence of cylindrical KdV solitons [19]. The plasma was contained within a circular cylinder, excitation being provided by the application of a bias voltage across a surrounding cylindrical plasma sheath. Break-up of the input signal was observed, with soliton-like pulses emerging in both the ingoing and outgoing (from the axis) waves for compression input, the soliton velocities exceeding the linear ion-acoustic low-frequency speed. The solitons had a pronounced asymmetry, trailing a wake generally familiar for two-dimensional linear waves, and a scaling law asserting constancy of the product of soliton width and square root of amplitude, and following from an adiabatically varying KdV soliton approximate solution to the cylindrical KdV, was found to be tolerably well satisfied. Application of rarefaction excitation produced no evidence of soliton-like structures.

Since the time of these early experiments, Cylindrical KdV has been proved integrable [5] and much of the apparatus of integrable systems has been at least partially established for it (e.g., Bäcklund transformations, together with a proof that BTs exist only for Cylindrical KdV out of the whole class of Generalized KdV equations [44]). Consequently it should be possible now to produce key relations for experimental test, though to the author's knowledge this has not been attempted. The experiments in plasma are difficult and costly at the scales that would allow detailed confirmation of CKdV theory. The two-wave nonlinear equation for weakly-nonlinear cylindrical waves (the counterpart to the Boussinesq equation for plane waves) is probably not integrable, so observation should be restricted to one-way waves in the period prior to any reflexion, and a preferable configuration maintaining the assumptions underlying CKdV for longer would involve excitation of a large annulus of plasma at some small inner radius with initially purely diverging waves. Experiments to test CKdV for cylindrically diverging waves on the surface of shallow water seem just as difficult to execute.

9. Electrical Transmission Lines

In the early days of soliton mathematics and physics, there were numerous simulations of physical systems through experiments on long transmission lines, and analytical and numerical solutions of discrete and continuum models for them. Such lines are relatively cheap and easy to construct, and versatile, in that they often lead to equations of the form

$$u_t + \alpha u^m u_x + \beta u_{xxx} = \delta u_{xx}, \tag{9.1}$$

where the relative importance of dissipation and dispersion is controllable, where the nonlinearity index m can be controlled by suitable choice of nonlinear elements, and where there is also the possibility of spatial variation of the coefficients. The nonlinear elements are typically variable-capacitance diodes, or saturating ferromagnetic inductances.

Lonngren [35] gives a review of numerous physical experiments on lines which may be as long as 20 m with as many as 1000 nominally identical sections. The typical nonlinear wave equation for the line voltage V, in the continuum limit and nondimensional variables, is

$$\frac{\partial^4 V}{\partial x^2 \partial t^2} + \frac{1}{LC}\frac{\partial^2 V}{\partial x^2} - \frac{1}{C}\frac{\partial^2 Q(V)}{\partial t^2} = 0, \tag{9.2}$$

where L, C are linear inductance capacitance, and the charge $Q(V)$ held by a nonlinear capacitance is assumed to have the form $Q(V) = C_0 V - C_N V^2$ as $V \to 0$. The linear dispersion relation is

$$\omega^2/k^2 = (LCk^2 + LC_0)^{-1} \tag{9.3}$$

and we readily find KdV,

$$\frac{\partial u}{\partial \tau} + \frac{C_N}{C_0} u \frac{\partial u}{\partial \xi} + \frac{C}{2C_0}\frac{\partial^3 u}{\partial \xi^3} = 0 \tag{9.4}$$

in appropriate variables for right-running waves. In principle it would be possible to go to higher nonlinearity and a higher leading-order dispersive term by appropriate choice of linear and nonlinear circuit elements.

Experiments on a line for which (9.4) should hold confirm quite reasonably the main predictions, namely the propagation of linear dispersive waves of either polarity in the linear limit; the emergence in strong waves of a definite number of compression solitons ($V > 0$) and a decaying oscillatory tail together conserving $\int_{-\infty}^{+\infty} V(t)\,dt$ along the line, with merely decaying radiation for rarefaction input signals; the amplitude–speed–width relation for KdV solitons; and identity preservation in both overtaking and head-on collisions. The generation of shocks with oscillations on the high-V side was also found for input signals corresponding to shock-type boundary conditions $u(\xi = -\infty) = u_1$, $u(\xi = +\infty) = u_2 > u_1$, also in accord with KdV theory. Fissioning of a KdV soliton into two solitons was demonstrated by Stewart and Corones [56] on a line comprising three segments, the first and third of uniform but different linear properties, with a spatial variation of dispersion coefficient in the middle section. FPU recurrence of an initial sinusoid, again for the two-solitons-per-period case as in Section 8, was observed by Hirota and Suzuki [20].

In all these cases dissipation is weak compared with dispersion. Gorshkov *et al.* [11] report similar experiments where MKdV is the basic equation and, in one limit, where the modified Burgers equation is appropriate. Among the distinctive features of solutions of that equation is the production of sonic shocks in finite time and their maintenance as sonic shocks thereafter, with an unusual shock profile; this is in agreement with asymptotic theory, for small diffusivity, for the (nonintegrable) modified Burgers equation.

Concluding his review, Lonngren [35] suggested that

> "With the rapid development of microcircuit technology, it is not inconceivable that 'pocket size' soliton (transmission line) experiments could be eventually built and sold to serious soliton students."

That this suggestion has not been taken up is perhaps simply a reflexion of the fact soliton theory has rapidly acquired an accepted place in so many areas of science and that numerical simulation of soliton processes has becomes so easy.

10. Epilogue

There are countless other contexts in which KdV arises, of which just a few can receive a brief note here.

First, although in the water wave context the first application was to the profile of long waves, KdV also turns up in modulation problems where one immediately thinks instead of NLS (see, e.g., [42, pp. 40–48]). For the modulation on long scales of a carrier wave we do have NLS,

$$ia_T + \tfrac{1}{2}\,\omega_0'' a_{\xi\xi} + \beta a|a|^2 = 0, \tag{10.1}$$

for the amplitude $a(\xi, T)$, where ω_0'' refers to the dispersion relation at the carrier wavenumber k_0 and β is defined in terms of k_0 and the medium properties. However, if $\beta\omega_0'' > 0$ the monochromatic wave given by $|a| = A = A_0$, $\phi = \arg a = \beta A_0^2 T$ is unstable to long-wave perturbations (the celebrated Benjamin–Feir instability), but if $\beta\omega_0'' < 0$ the perturbation $\tilde{A}$, where $|a| = A = A_0 + \tilde{A}$, evolves under KdV, with solitons and radiation. The solitons have $\tilde{A} < 0$ and represent a local reduction in intensity of the otherwise uniform intensity, and are called *dark solitons*.

Second, although the issue of energy transport over long ranges in the protein α-helix is dominated by Davydov solitons (see [6] and [34] for a recent review) on a discrete lattice, with the Zakharov equations of plasma physics and NLS in the continuum limit, KdV nevertheless arises in this field too. In the Yomosa [61] model, the emphasis is different, and is on the large-amplitude dynamics of the peptide groups in the nonlinear hydrogen-bonded spines of the α-helix, rather than on the interaction, as in Davydov theory, between a molecular mode and a low-frequency intermolecular acoustic mode. In the Yomosa lattice model we have a polypeptide chain with the nth peptide group undergoing a displacement u_n, with $r_n = u_{n+1} - u_n$ the extension of the nth peptide bond, and with an assumed potential $V_n(r_n) = Ar_n^2 - Br_n^3$ for the nth hydrogen bond. The equation of motion of the nth group is

$$m\ddot{r}_n = 2A(r_{n+1} + r_{n-1} - 2r_n) - 3B(r_{n+1}^2 + r_{n-1}^2 - 2r_n^2), \tag{10.2}$$

and in the continuum limit $n\ell \to x$, $r_n \to r(x,t)$ with right-propagating waves we get KdV,

$$u_\tau - 6uu_\xi + u_{\xi\xi\xi} = 0 \tag{10.3}$$

with

$$\xi = x/\ell - (2A/m)^{1/2}t, \quad \tau = (2A/m)^{1/2}t/24, \quad u = (6B/A)r. \tag{10.4}$$

The KdV solitons are supersonic, relative to the sound velocity $V_0 = (2A/m)^{1/2}\ell$. In his review Lomdahl [34] says that such supersonic lattice KdV solitons represent a 'reasonable alternative' to the Davydov and Takeno models for long-range coherent transport of biological energy.

In the field of solid mechanics, a number of authors, beginning with Nariboli and Sedov [41], have given analysis leading to KdV for the propagation of longitudinal waves in a nonlinearly elastic medium, but no successful confirmation of the KdV predictions has come from experiments. Indeed, Samsonov and Sokurinskaya [53] showed that for a thin circular rod of radius R, the weakly-nonlinear wave equation is not Boussinesq,

$$v_{tt} - c_0^2 v_{xx} = \frac{1}{2}\left[\frac{\beta}{\rho}v^2 + \nu^2 R^2 v_{tt}\right]_{xx}, \tag{10.5}$$

which leads to KdV under the one-wave restriction, but rather, through coupling of longitudinal and transverse waves, a double-dispersion equation

$$v_{tt} - c_0^2 v_{xx} = \frac{1}{2}\left[\frac{\beta}{\rho}v^2 + \nu^2 R^2 (v_{tt} - c_1^2 v_{xx})\right]_{xx}. \tag{10.6}$$

In these, v is the longitudinal strain, ρ and ν the density and Poisson's ratio, β a nonlinearity parameter and c_0, c_1 the speeds of linear P and S waves. Equation (10.6) has sech^2 solitary wave solutions, which coincide with those of KdV in the appropriate limit, though (10.6) appears generally nonintegrable. Experiments were carried out by Dreiden *et al.* [8] on the excitation of intense longitudinal waves in a polystyrene rod, for which all the numerous restrictions leading to (10.6) were reasonably well satisfied. These were the first in which such solitary waves had successfully been identified in solid matter, and confirmed the solitary wave features predicted by (10.6) to the exclusion of the earlier predictions of KdV.

In many of these applications, the initial stimulus came from the recognition that some complicated partial differential or difference system would reduce to KdV in an appropriate limit, suggesting an association between observed long-lived large-scale orderly structures and KdV solitons. Currently, the focus is much more on large-amplitude phenomena, well beyond the range of KdV theory and exemplified by the *magmons* of compaction-driven or conduit flows in geophysics (cf. Section 6), the *planetons* (large-amplitude Rossby waves (cf. Section 7)) introduced by Dewar [7], and perhaps, to coin a new term, the *voidons* (large-amplitude voidage slugs (cf. Section 5)) in fluidized beds. This shift of focus is reflected in the early dates, in the history of soliton systems, associated with the many discoveries of the application of KdV theory to diverse physical phenomena. It must be seen as a tribute to Korteweg and de Vries that their celebrated equation has had such profound effects, not simply on its own terms as a rich differential equation, nor simply for the great range of phenomena that it describes in some limited part of parameter space, but also for the impetus it has given to the analysis of more complex and general fully nonlinear systems.

The material for a review such as this is now to be found in literally thousands of papers in journals covering many disparate branches of science. I apologize to all those many researchers whose work has inevitably been overlooked, or omitted, or inadequately described in the above, and I hope that at least some feel has been given for the importance – impossible to exaggerate – of KdV in applications.

References

1. Ablowitz, M. J. and Segur, H.: *Solitons and the Inverse Scattering Transform*, SIAM, Philadelphia, 1981.
2. Apel, J. R., Holbrook, J. R., Liu, A. K., and Tsai, J. J.: *J. Phys. Oceanogr.* **15** (1985), 1625.
3. Benjamin, T. B.: *J. Fluid Mech.* **25** (1967), 559.
4. Bullough, R. K. and Caudery, P. J. (eds): *Solitons*, Springer-Verlag, Berlin, 1980.
5. Calogero, F. and Degasperis, A.: *Lett. Nuovo Cimento* **23** (1978), 150.
6. Davydov, A. S.: *Solitons in Molecular Systems*, D. Reidel, Dordrecht, 1985.
7. Dewar, W. K.: *Geophys. Astrophys. Fluid Dynamics* **51** (1990), 53.
8. Dreiden, G. V., Ostrovskii, Y. I., Samsonov, A. M., Semenova, I. V., and Sokurinskaya, E. V.: *Sov. Phys. Tech. Phys.* **33**(10) (1988), 1237.
9. Gardner, C. S., Greene, J. M., Kruskal, M. D., and Miura, R. M.: *Phys. Rev. Lett.* **19** (1967), 1095.
10. Gardner, C. S. and Morikawa, G. K.: Courant Inst. Math. Sci. Res. Rep. NYO-9082, New York University, 1960.
11. Gorshkov, K. A., Ostrovsky, L. A., and Pelinovsky, E. N.: *Proc. IEEE* **62** (1974), 1511.

12. Hammack, J. L. and Segur, H.: *J. Fluid Mech.* **65** (1974), 289.
13. Hammack, J. L. and Segur. H.: *J. Fluid Mech.* **84** (1978), 337.
14. Hammack, J. L. and Segur. H.: *J. Fluid Mech.* **84** (1978), 359.
15. Harris, S. E. and Crighton, D. G.: *J. Fluid Mech.* **266** (1994), 243.
16. Hayakawa, H., Komatsu, T. S., and Tsuzuki, T.: *Physica* **A 204** (1994), 277.
17. Helfrich, K. R. and Whitehead, J. A.: *Geophys. Astrophys. Fluid Dynamics* **51** (1990), 35.
18. Hershkowitz, N., Romesser, T., and Montgomery, D.: *Phys. Rev. Lett.* **29** (1972), 1586.
19. Hershkowitz, N. and Romesser, T.: *Phys. Rev. Lett.* **32** (1974), 581.
20. Hirota, R. and Suzuki, K.: *J. Phys. Soc. Japan* **28** (1970), 1366.
21. Ikezi, H.: Experiments on solitons in plasmas, in K. Lonngren and A. C. Scott (eds), *Solitons in Action*, Academic Press, New York, 1978, 153.
22. Ikezi, H., Barrett, P. J., White, R. B., and Wong, A. Y.: *Phys. Fluids* **14** (1971), 1997.
23. Ikezi, H., Taylor, R. J., and Baker, D. R.: *Phys. Rev. Lett.* **25** (1970), 11.
24. Karpman, V. I. and Maslov, E. M.: *Sov. Phys. JETP* **46**(2) (1977), 281.
25. Karpman, V. I. and Maslov, E. M.: *Sov. Phys. JETP* **48**(2) (1978), 252.
26. Keulegan, G. H.: *J. Res. N.B.S.* **40** (1948), 487.
27. Keulegan, G. H.: *J. Res. N.B.S.* **51** (1953), 133.
28. Kluwick, A.: *Z. Angew. Math. Mech.* **63** (1983), 161.
29. Knickerbocker, C. J. and Newell, A. C.: *J. Fluid Mech.* **98** (1980), 803.
30. Knickerbocker, C. J. and Newell, A. C.: *J. Fluid Mech.* **153** (1985), 1.
31. Koop, C. G. and Butler, G.: *J. Fluid Mech.* **112** (1981), 225.
32. Kubota, T., Ko, D. R. S., and Dobbs, L. D.: *AIAA J. Hydronautics* **12** (1978), 157.
33. Kuznetsov, V. V., Nakoryakov, V. E., Pokusaev, B. G., and Shreiber, I. R.: *J. Fluid Mech.* **85** (1978), 85.
34. Lomdahl, P. S.: Chapter 9, in A. P. Fordy (ed.), *Soliton Theory: A Survey of Results*, Manchester Univ. Press, 1990.
35. Lonngren, K.: Observations of solitons on nonlinear dispersive transmission waves, in: K. Lonngren and A. C. Scott (eds), *Solitons in Action*, Academic Press, New York, 1978, 127.
36. Marcus, P. S.: *Ann. Rev. Astron. Astrophys.* **31** (1993), 523.
37. Maxon, S. and Viecelli, J.: *Phys. Fluids* **17** (1974), 1614.
38. Maxworthy, T. and Redekopp, L. G.: *Icarus* **29** (1976), 261.
39. Maxworthy, T., Redekopp, L. G., and Weidman, P. D.: *Icarus* **33** (1978), 388.
40. Nakoryakov, V. E., Pokusaev, B. G., and Shreiber, I. R.: *Wave Propagation in Gas-Liquid Media*, 2nd edn, CRC Press, Boca Raton, 1993.
41. Nariboli, G. A. and Sedov, A.: *J. Math. Anal. Appl.* **32** (1970), 661.
42. Newell, A. C.: *Solitons in Mathematics and Physics*, SIAM, Philadephia, 1985.
43. Nezlin, M. V.: *Chaos* **4**(2) (1994), 187.
44. Nimmo, J. J. C. and Crighton, D. G.: *Phys. Lett.* **82A** (1981), 211.
45. Oikawa, M. and Yajima, N.: *J. Phys. Soc. Japan* **34** (1973), 1093.
46. Olson, M. and Christensen, U.: *J. Geophys. Res.* **91**(B) (1986), 6367.
47. Ono, H.: *J. Phys. Soc. Japan* **39** (1975), 1082.
48. Osborne, A. R.: Chapter 7, in A. P. Fordy (ed.), *Soliton Theory: A Survey of Results*, Manchester Univ. Press, 1990.
49. Osborne, A. R. and Burch. T. L.: *Science* **208** (1980), 451.
50. Peek, B. M.: *The Planet Jupiter*, Faber and Faber, London, 1958.
51. Richter, F. M. and McKenzie, D. P.: *J. Geol.* **92** (1984), 729.
52. Saeki, K.: *J. Phys. Soc. Japan* **35** (1973), 251.
53. Samsonov, A. M. and Sokurinskaya, E. V.: *Sov. Phys. Tech. Phys.* **33**(8) (1988), 989.
54. Scott, D. R. and Stevenson, D. J.: *Geophys. Res. Letter.* **11** (1986), 1161.
55. Segur, H. and Hammack, J. L.: *J. Fluid Mech.* **118** (1982), 285.
56. Stewart, R. W. and Corones, J.: *Rocky Mountain J. Math.* **8** (1978), 227.
57. Tappert, F. D. and Judice, C. N.: *Phys. Rev. Lett.* **29** (1972), 1308.
58. van Wijngaarden, L.: *Ann. Rev. Fluid Mech.* **4** (1972), 369.
59. Washimi, H. and Taniuti, T.: *Phys. Rev. Lett.* **17** (1966), 996.

60. Whitehead, J. A. and Helfrich, K. R.: *Nature* **336** (1988), 59.
61. Yomosa, S.: *J. Phys. Soc. Japan* **53** (1984), 3692.
62. Zabusky, N. J. and Kruskal, M. D.: *Phys. Rev. Lett.* **15** (1965), 240.

Acta Applicandae Mathematicae **39**: 69–84, 1995.

Instructive History of the Quantum Inverse Scattering Method

L. FADDEEV*
Steklov Mathematical Institute, St. Petersburg Branch, Fontanka 27, St. Petersburg, Russia

(Received: 2 November 1994)

Mathematics Subject Classification (1991): 58F07.

Key words: classical inverse scattering method, nonlinear Schrödinger equation, soliton theory.

The classical inverse scattering method was invented during the course of investigation of the KdV equation [1] (100 years of which we celebrate). It was described in a short and famous research letter by Gardner, Green, Kruskal, and Miura (GGKM) [2] in 1967. Its quantum version, which is ten years younger, was devised mostly in Leningrad (now St. Petersburg). In what follows, I shall underline some highlights and lessons of this latter development.

I shall specially stress the fusion of different apparently unrelated directions in one-dimensional mathematical physics (term coined by E. Lieb and D. Mattis [3]) which occurs thereby, making it an exemplary showcase of interrelation of ideas and methods of theoretical and mathematical physics.

First I shall briefly describe the evolution of the classical inverse scattering method underlying the aspects relevant to its quantum generalization (1971–1974). Then I shall treat its quasiclassical quantization (1973–1977), which in particular led to the proper particle-like interpretation of solitons. Finally, the exact quantization (1978–1983) will be discussed. An unexpected connection with already-existing methods of one-dimensional physics, namely:

1) Bethe Ansatz in the theory of magnetic chains developed by Bethe [4], Hulthén [5], Gaudin [6], De-Cloizeaux and Pearson [7], C. N. Yang and C. P. Yang [8], and for Bose gas by Lieb and Liniger [9];
2) transfer matrix method in two-dimensional lattice statistical mechanics, worked out by Onsager, Lieb, and Baxter [10–12];
3) factorized scattering in quantum N-body systems of one-dimensional particle of McGuire [13], Berezin *et al.* [14], C. N. Yang [15], Brezin and Zinn-Justin [16];

will be emphasised.

* The preparation of this report was partly supported by the ISF grant R2H000.

The story begins with the paper [2], in which it was shown that the reflection coefficient $r(k)$ of a one-dimensional Schrödinger equation

$$\psi'' + k^2\psi = v(x)\psi, \quad -\infty < x < \infty,$$

with potential $v(x)$ vanishing at infinity so that $r(k)$ is defined by the asymptotics of $\psi(x,k)$

$$\psi(x,k) = \begin{cases} e^{ikx} + r(k)e^{-ikx}, & x \to +\infty, \\ t(k)e^{ikx}, & x \to -\infty, \end{cases}$$

evolves as

$$r(k) \longrightarrow e^{ik^3 t} r(k),$$

if the potential develops according to the Korteweg–de Vries equation

$$v_t = 6vv_x + v_{xxx}.$$

If the nonlinear term above was absent, the evolution of $r(k)$ would be like that for the Fourier transform of $v(x)$. However, the nonlinear transform

$$v(x) \longrightarrow r(k)$$

defined implicitly through the solution of a (linear!) Schrödinger equation serves as the same for the fully nonlinear equation. Moreover, the inverse transform

$$r(k) \longrightarrow v(x)$$

is a content of the so-called inverse scattering problem which has had a rich history. Inspired by Heisenberg's general ideas, it was discussed by many people at the end of the 40s and at the beginning of the 50s until it was solved by Gelfand and Levitan [17], Krein [18] and Marchenko [19].

In 1969, P. Lax [20] interpreted rather explicit formulae of GGKM in an elegant form. Let L be a Schrödinger operator

$$L = -\frac{d^2}{dx^2} + v(x).$$

Then the KdV equation can be written as

$$\frac{dL}{dt} = [L, M],$$

where M is a differential operator of third order, coefficients of which are constructed in terms of v and its x derivatives. The Lax equation shows that the KdV evolution is a similarity transformation of L

$$L \longrightarrow U(t)LU^{-1}(t)$$

and thus describes a spectral conserving deformation of L.

Up to that time, the trick of GGKM looked like an isolated example of a particular ingenuous method applicable to just one solvable nonlinear evolution equation. All this changed by the third article – that by Zakharov and Shabat [21], showing in 1970 that one more nonlinear equation, namely the so-called nonlinear Schrödinger equation (NSE)

$$i\Psi_t = \Psi_{xx} + g|\Psi|^2\Psi$$

allows for the inverse scattering transform, with the role of L played by the the 2×2 first-order system

$$L = \frac{1}{i}\begin{pmatrix} 1 & 0 \\ 0 & -1 \end{pmatrix}\frac{\mathrm{d}}{\mathrm{d}x} + \begin{pmatrix} 0 & g\Psi^* \\ \Psi & 0 \end{pmatrix}.$$

This broke the isolation of the KdV example and led to the impressive development of the method in 70s which gradually attracted a lot of developers and followers and produced an elaborate industry of the construction of exactly solvable models.

It is instructive to stress that the pioneers of the method belonged to the community of the applied mathematical physicists dealing mostly with the problems of the classical many-body systems, including plasma physics, hydrodynamics, etc. They were very far from quantum physics and its mathematical methods. This is illustrated by the fact that the authors of the inverse scattering method did not use the formalism of quantum scattering theory for several years after the letter [2] in their further investigation of the KdV equation.

I learned the IST method at the beginning of 1971 from V. Zakharov. It excited me immediately, partly because the mathematically rigorous investigation of the variant of an inverse scattering problem suitable for the KdV equation was a part of my candidate thesis [22], so that I knew a lot of formulae in this domain.

From a discussion with Zakharov the idea came to look upon the KdV equation as a Hamiltonian system (with an infinite number of degrees of freedom) which is completely integrable. Indeed, as was already clear, the KdV equation is generated by the usual formulae of Hamiltonian mechanics if one introduces the Poisson brackets

$$\{v(x), v(y)\} = \delta'(x-y)$$

for 'coordinates' $v(x)$ in the phase space and take the functional

$$H = \int_{-\infty}^{\infty} \left(v^3(x) + \tfrac{1}{2}v_x^2(x)\right)\mathrm{d}x$$

as a Hamiltonian. Further, the solution in terms of new coordinates $r(k)$ shows that half of the initial data, namely $|r(k)|$ conserves with time. This is a characteristic property of completely integrable case. We calculated the Poisson brackets in terms of variables $r(k)$ and found the expression for the Hamiltonian in terms of it. In this way the complete construction of the angle-action variables for the KdV equation was achieved [23].

The idea of integrability was not foreign to the GGKM group. Indeed, earlier work of Kruskal and Zabusky was motivated by a famous numerical experiment of Fermi, Pasta and Ulam on one of the first electronic computers. Having at their disposal a new calculation facility and looking for an interesting application of it, they decided to check the equipartition of an energy hypothesis for a Hamiltonian system with many degrees of freedom. They have chosen as such a chain of nonlinearly interacting oscillators with the interaction seeming generic enough. However, the system under investigation showed unexpected behavior and equipartition was not achieved in a reasonable time. So the system has shown itself rather as completely integrable. Later it was realized that indeed the Fermi–Pasta–Ulam experiment could be well approximated by an integrable one. But this was after my joint paper with Zakharov. The GGKM group also certainly worked in this direction and Gardner has independently shown the complete integrability of the KdV equation. However, he did not use the formalism of scattering theory and did not produce the explicit angle-action variables.

One striking feature of the KdV equation is the existence of the solitary wave solutions (term from hydrodynamics) which were dubbed 'solitons' by Kruskal and Zabusky, evidently following the fashion of particle physics. However, the particle interpretation of solitons was obscure in the hydrodynamic setting. I believe that it was the angle-action formulation of KdV which established the particle-like feature of solitons on a firm basis.

Indeed, the formula for the Hamiltonian H and momentum P

$$P = \int v^2 \, dx = \sum^{N} \lambda_i + \int k^2 \rho(k) \, dk;$$

$$H = \sum^{N} \lambda_i^{3/2} + \int_0^{\infty} k^3 \rho(k) \, dk,$$

found by Zakharov and me looks exactly as that for the system of noninteracting particles with momenta λ_i and energies $\varepsilon_i = \lambda_i^{3/2}$ and wave-like excitations with the same dispersion law $\varepsilon(k) = k^{3/2}$. Quantization of these last excitations also leads to particles. In the terms of the spectral theory of operator L, parameters $(-\lambda_i)$ are negative eigenvalues and

$$\rho(k) = \frac{k}{2\pi} \ln \left(1 - |r(k)|^2\right)$$

is a spectral density; they constitute a complete set of action variables. The appearance of eigenvalues having a nonperturbative effect, the particles corresponding to solitons looked exactly like candidates to circumvent the perturbation paradigm still reigning at that time in the Quantum Field Theory, namely, to go beyond the rule: one field-one particle.

Two circumstances blackened this new tantalizing opportunity. First, the KdV equation looks irrelevant from the quantum point of view. Second, and much more serious was the fact that corresponding space-time was two-dimensional. The first obstacle was very soon circumvented with the appearance of more relevant examples, but the second remained until now. So for a long time the development of this subject had only a model character, far from the real applications in particle physics. Mathematical interest prevailing, this was not a decisive obstruction for me to continue to pursue the problem of quantization of solitons.

The nonlinear Schrödinger equation for which the inverse scattering formalism was also available, is a much more suitable model. Moreover, it has a new feature – the soliton has two degrees of freedom rather than one (it corresponded to the complex eigenvalue of the corresponding L-operator which is not selfadjoint for the negative sign of the coupling constant g). The momentum-energy relations [24]

$$P = \sum p_i + \int p\rho(p)\,\mathrm{d}p;$$

$$H = \sum \frac{1}{2M_i} p_i^2 + \int p^2 \rho(p)\,\mathrm{d}p,$$

look exactly like those in nonrelativistic mechanics. And due to the fact that the mass variable M is conjugate to the angle ϕ, $0 \leqslant \phi \leqslant 2\pi$, its quasiclassical quantization leads to the discrete spectrum of masses $M = 1, 2, \ldots$. The particle interpretation of a soliton obtained new support in this picture, but for complete satisfaction, one needed a relativistic example. This was readily produced by the famous sine-Gordon equation:

$$\phi_{tt} - \phi_{xx} + \frac{m^2}{\gamma} \sin \gamma\phi = 0.$$

I was given a hint that this equation could be integrable by John Klander at the beginning of 1972 when relating my KdV paper with Zakharov at Princeton University. Two years of work began with Zakharov and, accomplished with new student Leon Takhtajan [26], led at the end of 1973 to the construction of the appropriate angle action variables. Now the energy-momentum relations looked manifestly Lorentz covariant [26, 27]

$$P = \sum p_i + \sum k_l + \int k\rho(k)\,\mathrm{d}k;$$

$$H = \sum \sqrt{p_i^2 + M^2} + \sum \sqrt{k_l^2 + \mu^2} + \int \sqrt{k^2 + m^2}\rho(k)\,\mathrm{d}k,$$

where the first sum corresponds to solitons without internal degrees of freedom with mass $M = 8m/\gamma$; the second sum is over solitons with internal degrees of freedom (picturesquely called breathers), μ being the dynamical variable $\mu = (16m/\gamma)\sin\theta$, where the action variable θ is canonically conjugate to angle ϕ, $0 \leqslant \phi \leqslant 2\pi/\gamma$; finally the third term is a contribution of wave-like solutions. Quasiclassical quantization immediately leads to particle interpretation of solitons with breathers giving a rich mass spectrum

$$m_n = \frac{16m}{\gamma}\sin\frac{\gamma n}{16}, \quad n = 1, 2, \ldots, \left[\frac{\pi}{2\gamma}\right].$$

The short announcement of this result (with the unneeded change $n \to n+1$ due to the uncritical treating of quasiclassical formalism) was published in [26].

I was quite excited by this development and an article coauthored with Takhtajan with the rather pretentious title "The relativistic one-dimensional model, generating several particles" was sent to *Physics Letters* at the end of 1973. However, due either to the Iron Curtain or to the nonattentiveness of the editor, it never appeared. And then in 1974 the idea of soliton-particle correspondence suddenly acquired prominence, being proposed by many authors who had different points of departure [28–31].

It is necessary to mention that prior to this development, the idea of particle-soliton correspondence was advocated by Skyrme who, however, used the purely classical nonquantum approach. Skyrme used the sine-Gordon model [32] and also devised an example of a particle-like solution for the equation in four-dimensional space-time [33].

In the meantime, the inverse scattering method was developing and attracting more and more followers. The number of examples multiplied and technical were tools perfected. One more American group (AKNS) – Ablowitz, Kaup, Newell, Segur [34] – joined the club as well as Flashka [35]; Zakharov got an enthusiastic student Manakov [36]; in my vicinity, Kulish and Korepin joined Takhtajan in our work on the method; S. Novikov entered with powerful ideas applicable to the case of finite space $-L \leqslant x \leqslant L$ accompanied by the periodic boundary conditions [37]. This list of the end of 1973 began to extend very rapidly in the following years.

One of the technical achievements of this time (due to AKNS, Novikov and Zakharov and Shabat [25]) was the substitution of the Lax scheme by the more powerful zero-curvature condition: the nonlinear equation was obtained as a commutativity condition of covariant derivatives $\nabla_1 = \partial_x + U(x,t;\lambda)$ and $\nabla_0 = \partial_t + V(x,t;\lambda)$, where U and V are matrices, parametrized by fields $v_i(x,t)$ and complex parameter λ. The spectral problem of the Lax approach was substituted by an auxiliary linear equation

$$L(\lambda)\Psi \equiv (\partial_x + U(x,\lambda))\Psi = 0,$$

into which the 'spectral parameter' λ could enter nonlinearly. In this formulation, the scattering data were defined as the asymptotic behavior of the holonomy

$$T_L(\lambda) = \overleftrightarrow{\exp}\left\{\int_{-L}^{L} U(x,\lambda)\,\mathrm{d}x\right\},$$

where we use notation $\overleftrightarrow{\exp}$ for the ordered exponential. The introduction of the angle-action variables was based on the possibility of explicit calculation of the Poisson brackets of matrix elements of holonomy (or rather its asymptotic with $L \to \infty$). The sine-Gordon model was treated by Takhtajan and me in [27] exactly along these lines.

An instructive example of L-operator with nonadditive entering of λ was proposed by Takhtajan [39]:

$$L(x) = \frac{\partial}{\partial x} + \frac{S(x)}{\lambda},$$

where

$$S(x) = \begin{pmatrix} S_3 & S_1 - iS_2 \\ S_1 + iS_2 & -S_3 \end{pmatrix}$$

is parametrized by real fields $S_a(x)$, $a = 1, 2, 3$. This L-operator is appropriate for the Heisenberg ferromagnet equation

$$\frac{\partial S^a}{\partial t} = \varepsilon_{abc} S^b_{xx} S^c.$$

Thus, during the middle of the 70s, a well defined procedure to get the angle-action variables for the integrable models (a new term, which gradually sustituted the soliton equations) was devised. The scheme:

nonlinear equation $\longrightarrow$ operator $L(\lambda)$ $\longrightarrow$ holonomy for finite interval $\longrightarrow$ $L < x < L$, large L asymptotics $\longrightarrow$ angle-action variables

was elaborated for all known examples, but was to be worked out separately in every concrete case. The quasiclassical quantization, based on the existence of angle-action variables, led to a particle interpretation of solitons and wave excitations, producing interesting mass spectra. The large-time asymptotic led to rather explicit formulas for the corresponding scattering phase shifts. The results and technical means of this development were reviewed by Korepin and me in [40], where one can find an extensive list of references.

After the completion of [40], it became clear to me that the possibilities of the quasiclassical quantization were exhausted and time came to repeat all the formalisms of the inverse scattering method in the quantum domain. At first sight, the idea looked too formidable. However, two circumstances showed that it could be possible to do the exact quantization of integrable models:

1. The existence of the model, for which the exact quantization was already known, namely the NSE model. The quantum version of its Hamiltonian

$$H = \int_{-\infty}^{\infty} \partial_x \Psi^* \partial_x \Psi \, dx + g \int_{-\infty}^{\infty} (\Psi^*(x))^2 (\Psi(x))^2 \, dx,$$

where the operators $\Psi^*(x)$, $\Psi(x)$ with the commutation relations

$$\Psi^*(x)\Psi(y) - \Psi(y)\Psi^*(x) = \delta(x-y),$$

could be reduced to a family of N-body Schrödinger operators

$$H_N = -\sum \frac{d^2}{dx_i^2} + g \sum_{i<j} \delta(x_i - x_j), \quad i = 1, \dots, N,$$

for nonrelativistic particles with the δ-function pairwise interaction. The Schrödinger equation

$$H_N \Psi = E\Psi$$

was known to be solvable from the papers by McGuire, Berezin, Yang, Brezin and Zinn-Justin mentioned above.

2. The insight of Adler [41] and Kostant [42] that the integrable models allow an interpretation in terms of coadjoint orbits of Lie algebras. Adler treated the KdV equation and the algebra was that of pseudodifferential operators. Kostant worked on the Toda lattice model and the algebra was $\mathrm{SL}(N, \mathbf{C})$. In the orbit method [43], the quantization means representation of the corresponding Lie algebra, which is usually rather explicit; thus this connection showed the exact quantization of a integrable model, i.e. representation of corresponding Lie algebras was quite feasible.

Motivated by these circumstances, I recruited a new student, E. Sklyanin, to work on the exact quantization. As a first step, we conjectured the exact commutation relations for the quantum analogues of the matrix elements of the asymptotical holonomy (monodromy) [44]. The exact eigenvalues of the quantum Hamiltonian and the mass spectrum followed from the relations. This was one more encouragement for more systematic work.

One more indication in favor of the possibility to quantize the inverse scattering method came from a completely different direction – occasionally, I came across the paper by Baxter [12] on two-dimensional lattice models of classical statistical mechanics; several formulae of the so-called transfer matrix method (going back at least to the famous L. Onsager paper on the Ising model) and their use were strikingly similar to those characteristic for the monodromy of the L-operator.

Sklyanin rapidly adapted some ideas of the Baxter approach to the quantum nonlinear Schrödinger equation [45]. The main idea was to write a commutation relation for the matrix elements of a (local) L-operator in such a form that one can use them to derive the commutation relations for the matrix elements of (global) monodromy. Independent work on the quantum interpretation of the scattering data for the Lax operator of NLS equation was done in [46, 47].

The success was quite encouraging. However, Sklyanin heavily used the possibility of introducing the normal ordering of operators $\Psi^*(x)$, $\Psi(x)$. No such ordering was available for the sine-Gordon case. Moreover, the quantum version of the SG model has infinities which are to be eliminated by mass renormalization.

So I decided to try another trick to regularize the quantum field-theoretical problem, namely discretizing the space variables. The auxiliary linear problem then acquires the form

$$\Psi_{n+1} = L_n(\lambda)\Psi_n,$$

where L_n can be interpreted as an 'infinitesimal' holonomy along the segment of length Δ (lattice spacing), so that

$$L_n(\lambda) \cong I + \Delta U(x, \lambda), \quad x = n\Delta.$$

After some experimenting with the appropriate choice of $L_n(\lambda)$ for the sine-Gordon model, which is a 2×2 matrix with elements expressed in terms of the dynamical variables and complex parameter λ, Takhtajan, Sklyanin and I [48] found the local commutation relations in the form

$$R(\lambda - \mu)L_n^1(\lambda)L_n^2(\mu) = L_n^2(\mu)L_n^1(\lambda)R(\lambda - \mu);$$

$$L_n^1(\lambda)L_m^2(\mu) = L_m^2(\mu)L_n^1(\lambda), \quad m \neq n.$$

Here the following notations are used: L^1 and L^2 are 4×4 matrices constructed in terms of the 2×2 matrix $L_n(\lambda)$ according to the rule

$$L^1 = L \otimes I; \qquad L^2 = I \otimes L,$$

where the tensor product of the matrices is defined with respect to the structure $\mathbf{C}^4 = \mathbf{C}^2 \otimes \mathbf{C}^2$. The 4×4 matrix $R(\lambda)$ contains the functions of λ only as the matrix elements; no dynamical variables enter into R. In the following, we shall refer to the relations above as Fundamental Commutation Relations (FCR). After imposing the periodicity condition $L_n \equiv L_{n+N}$ (one can consider it as one more regularization – that of finite volume in space), we consider the monodromy

$$M(\lambda) = \overleftarrow{\prod} L_n(\lambda) = L_N(\lambda) \ldots L_1(\lambda) = \begin{pmatrix} A(\lambda) & B(\lambda) \\ C(\lambda) & D(\lambda) \end{pmatrix}.$$

It follows from the FCR for $L_n(\lambda)$ that $M(\lambda)$ (global object) satisfies the same FCR:

$$R(\lambda - \mu)M^1(\lambda)M^2(\mu) = M^2(\mu)M^1(\lambda)R(\lambda - \mu).$$

Thus it became clear that FCR decode rather cumbersome calculations of Poisson relations between the elements of the scattering matrix performed by Takhtajan and me for the SG case in 1973 with great effort. FCR for $M(\lambda)$ shows that its trace $\mathrm{tr}M(\lambda) = A(\lambda) + D(\lambda)$ commutes for different values of the spectral parameter

$$[\mathrm{tr}\, M(\lambda),\ \mathrm{tr}\, M(\mu)] = 0.$$

So $\mathrm{tr}\, M(\lambda)$ can be considered as a generating function for a family of commuting operators; the physical Hamiltonian belongs to this family (which was to be shown) and so it is accompanied by a host of commuting conservation laws (quantum action variables).

These formulas are very similar to those of classical statistical mechanics on the two-dimensional lattice [12]. The lattice Lax operator is an analogue of the local Boltzmann weight. The fundamental commutation relations are similar to the star-triangle equations. The trace of monodromy $M(\lambda)$ is an analogue of the transfer matrix. The offdiagonal elements of $M(\lambda)$ are not used in statistical mechanics, but are of importance in our story.

Indeed, the offdiagonal matrix elements were shown to be quantum analogues of angle variables. This was realized as follows: we constructed one particular eigenvector Ω of $\mathrm{tr}\, M(\lambda)$ such that $C(\lambda)\Omega = 0$ and showed that

$$\Omega(\{\lambda\}) = B(\lambda_1)\ldots B(\lambda_l)\Omega$$

is also an eigenvector provided that the parameters λ_i, $i = 1, \ldots, l$, satisfy some system of algebraic equations (artifact of finite volume).

With these results obtained, we could not help feeling that the new form of the inverse scattering method being essentially algebrized is quite universal and most natural. Active work in which Korepin and Kulish joined us and new names – Izergin, Reshetikhin, Smirnov, Tarasov – gradually entered, revealing that the formalism which got the name Quantum Inverse Scattering Method (QISM) was powerful indeed. All classical models: NLS, Toda lattice, Heisenberg magnet, etc., were imbedded into the realm of the method. In this last example, the dynamical variables being spin variables, we found that for a spin $1/2$ representation for them the method leads in an algebraic fashion to the results obtained by direct calculation by H. Bethe as early as in 1931. Bethe was solving a system of Schrödinger equations for a particle wave function parametrized by momentum variables $p_1, \ldots, p_l$ and showed that the periodicity condition in coordinates leads to the equation for p_i, $i = 1, \ldots, l$,

$$\mathrm{e}^{ip_i N} = \prod_{i \neq j} S(p_i - p_j),$$

where the $S(p)$ was the scattering amplitude for the two-body case.

The equation for the SG case was essentially of the same form. Moreover, the application of QISM to a spin $1/2$ spin chain leads exactly to the Bethe equation.

Thus one more root to the general formalism was found and we began to call our method the Algebraic Bethe Ansatz (ABA). In this way, our development fused with quite a different one which already had a prominent place in the quantum mechanical many-body problem and led to its new interpretation and generalization. My goal here does not include the description of the applications of QISM or the ABA to concrete physical problems. (For some representative results, see [49, 50].) In this, our Leningrad group worked in parallel with several other groups, notably with the Landau Institute and LPTHE at University Paris VI. We give here some references [51–54] and continue the discussion of the algebraic aspects of the method.

The algebraic treatment of the Bethe example was based on the L_n-operator

$$L_n(\lambda) = \begin{pmatrix} \lambda + iS_n^3 & S_n^1 - iS_n^2 \\ S_n^1 + iS_n^2 & \lambda - iS_n^3 \end{pmatrix},$$

where S_n^a, $a = 1, 2, 3$, $n = 1, \ldots, N$, are spin operators acting in $\Pi \otimes \mathbf{C}^2$

$$S_n^a = I \times I \times \tfrac{1}{2}\sigma^a \times \cdots I,$$

where $\mathbf{C}^2$ is considered as a two-dimensional representation of SU(2) realized by Pauli matrices σ^a:

$$\sigma^1 = \begin{pmatrix} 0 & 1 \\ 1 & 0 \end{pmatrix}, \qquad \sigma^2 = \begin{pmatrix} 0 & -i \\ i & 0 \end{pmatrix}, \qquad \sigma^3 = \begin{pmatrix} 1 & 0 \\ 0 & -1 \end{pmatrix}.$$

It is instructive to compare this L-operator with the classical continuous case of Takhtajan to see that the former is the most naive discretization of the latter. It became clear that the same L-operator can be used also for the higher spin representations for the dynamical variables S_n^a; however a new problem appeared, namely that of defining a suitable system of commuting operators. The particular solution for spin systems by Kulish, Reshetikhin and Sklyanin [55], as well as the inclusion in this family of the natural discrete-space regularization of the NLS system by Korepin and Izergin [56] was elucidated by Tarasov, Takhtajan and me [57]. Our results could be interpreted in the following way: the FCR and concrete choice of L-operator correspond to the representation of the following universal algebraic scheme: an algebra $\mathcal{A}$ is given having representations $\rho^a(\lambda)$, parametrized by 'spin' a and complex parameter λ (a typical example is that of affine algebras). A universal object R is defined as an element in $\mathcal{A} \otimes \mathcal{A}$ subject to the relation

$$R^{12}R^{13}R^{23} = R^{23}R^{13}R^{12}$$

holding in $\mathcal{A} \otimes \mathcal{A} \otimes \mathcal{A}$ with evident notation $R^{12} = R \otimes I$, $R^{12} = i \otimes R$, etc. The L operator appears as

$$L(\lambda - \mu) = (\rho^a(\lambda) \otimes \rho^b(\mu)R,$$

where the representation $\rho^a(\lambda)$ corresponds to the choice of dynamical variables in a given lattice point of a chain and $\rho^b(\mu)$ defines the corresponding matrix (auxiliary) space. In particular case $\rho^a(\lambda) = \rho^b(\lambda) = \rho(\lambda)$, the local dynamical and auxiliary spaces coincide, as happens, for example, in the case of the Bethe spin $1/2$ magnetic chain and the general relation for R leads to

$$R^{12}(\lambda-\mu)R^{13}(\lambda-\sigma)R^{23}(\mu-\delta) = R^{23}(\mu-\sigma)R^{13}(\lambda-\sigma)R^{12}(\lambda-\mu)$$

for

$$R(\lambda-\mu) = (\rho(\lambda)\otimes\rho(\mu))R.$$

It is the last relation which was basic for C. N. Yang (as a relation for the two-body scattering matrix) and R. Baxter (as a relation for the Boltzmann weights of a solvable lattice model of statistical mechanics on the plane). This is why Takhtajan and I proposed to call it the Yang–Baxter relation. With this name it now enters both physical and mathematical literature, being used also for the FCR above. However, it was only the quantum inverse scattering method which made it natural to distinguish the R-matrix and L-operator in FCR leading in this way to the more general scheme of the ABA.

It is instructive to mention here the connection of the Yang–Baxter relation with the braid group. Let σ be a permutation operator in $\mathcal{A}\otimes\mathcal{A}$, $\sigma(a\otimes b) = b\otimes a$. Then $\widehat{R} = R\sigma$ satisfies the relation

$$\widehat{R}_{12}\widehat{R}_{23}\widehat{R}_{12} = \widehat{R}_{23}\widehat{R}_{12}\widehat{R}_{23},$$

which is nothing but the relation for the generators of the braid group. Thus, it is only natural that the ABA soon became useful in the theory of knots.

However, let us return to one more development of the beginning of 80s. Having imbedded into the ABA such models as NLS, SG and HM we were looking for a possible extension of the validity of the method. It was realized by Kulish and Reshetikhin [58] that the following L-operator

$$L_n(\lambda) = \frac{1}{\sin\gamma}\begin{pmatrix} \sinh(\lambda + i\gamma S_n^3) & i\sin\gamma S_n^- \\ i\sin\gamma S_n^+ & \sinh(\lambda - i\gamma S_n^3) \end{pmatrix}$$

is suitable for the anisotropic magnetic chain (the so-called XXZ model), and satisfies FCR, if the dynamical variables S_n obey the relations

$$[S_n^3, S_n^\pm] = \pm S_n^\pm,$$

$$[S_n^+, S_n^-] = \frac{\sin\gamma S_3}{\sin\gamma},$$

which evidently define a deformation of Lie algebra SL(2) with the deformation parameter γ. Then Sklyanin [59] proposed a further deformation of this algebra to the two parameter case, associated with the fully anisotropic XYZ model and developed a representation theory for his algebra. In the meantime, Izergin and Korepin [56, 57] have shown that continuous models such as NLS or SG when properly discretized, could be considered as corresponding to representations of XXX and XXZ models, respectively. Thus, with the development of the quantum inverse scattering method, its algebraic base became more clear and concise. It evidently needed a more abstract algebraic point of view.

This was produced by Drinfeld [61] who generalized the Kulish–Reshetikhin deformation of SL(2) to any semisimple Lie algebra (independently performed by Jimbo [62]) and interpreted it in terms of a special class of Hopf algebras. Due to several associations, Drinfeld called these algebras 'quantum groups'. In his famous address to the ICM–86 in Berkeley, he introduced general definitions, a universal object R mentioned above, and examples of quantum groups as deformations of classical Lie algebras. It was the generalization of the Kulish–Reshetikhin relations rather than FCR which entered into Drinfeld's approach. Thus, we in Leningrad decided to develop an alternative way of presenting Quantum Groups based solely on FCR. This was udertaken by Reshetikhin, Takhtajan and me [64], and our method became quite workable in numerous applications which appeared in physics and mathematics.

Most evident applications: representation theory, differential geometry and the corresponding theory of special functions, have now a lot of people working on them and give a foundation of the various q-analogues with $q = e^{i\gamma}$ of the type developed in combinatorics beginning with Gauss. Quantum Groups, in this domain acquired the same role as classical Lie algebras in the theory of special functions. However, a new peculiarity of the representation theory for special values of q, being the root of unity, leads to the appearance of some interesting arithmetics. Different aspects of this development and references to the literature could be found in monographs [60, 66, 67], collection of papers [68] and conference proceedings [69, 70].

The theory of knots also benefited greatly from this development. An independent approach of V. Jones [71] was shown by Turaev and Reshetikhin [72] to have a natural interpretation in terms of quantum groups and their representations.

But less evident, though important, was application which returned Quantum Groups into the realms of Physics. It was a gradual realization that Quantum Groups are inherently connected with the Conformal Field Theory. This theory, also having its source and important applications in statistical mechanics [73], was shown by Polyakov [74] (who was one of its most active developers) to be an important ingredient of string theory, thus the results of the development of the quantum inverse scattering method entered into the mainstream of Quantum Field Theory.

The impact of all this on many branches in pure mathematics: topology, noncommutative geometry, combinatorics, theory of infinite-dimensional algebraic structures and their representations, complex analysis, is difficult to overestimate. The mathematical community certainly appreciated the development: three laureates of the Fields Medal in ICM–90 in Kyoto are clearly connected with it.

Mathematical Physics gained enormously, it obtained a new method and family of solvable particular problems.

The value of this progress for Physics seems now to depend on the success of string theory. However, different aspects of the general philosophy of Quantum Groups, i.e. the possibility of quantizing the space-time itself deserves more attention (see, e.g., [75])

Thus, the formal trick found in applied plasma physics on the example of the Korteweg–de Vries equation, being gradually developed in a mathematical manner with quantum mechanical applications in view, threw new light on the methods of many-body problems in solid state physics, crystalized into several fundamental formulae and led to a new mathematical structure with a host of unexpected applications including that in the modern physics of elementary particles.

I believe that this process which has been evolving in front of our eyes during last 25 years, is unprecedented so far in the history of relations of Physics and Mathematics. As it is not finished yet, one can expect even more to happen in a near future.

References

1. Korteweg, D. and de Vries, G.: *Phil. Mag.* **39** (1895), 442.
2. Gardner, C. S., Green, J. M., Kruskal, M. D., and Miura, R. M.: *Phys. Rev. Lett.* **19** (1967), 1095.
3. Lieb, E. H. and Matis, D. C. (eds): *Mathematical Physics in One-Dimension*, Academic Press, New York and London, 1966.
4. Bethe, H.: *Zeit. Phys.* **71** (1931), 105.
5. Hulten, L.: *Arkiv. Mat. Astron. Fysik* **26A** (1938), 1.
6. Gaudin, M.: *Phys. Rev. Lett.* **26** (1971), 1301.
7. Des Cloizeaux, J. and Pearson, J. J.: *Phys. Rev.* **150** (1966), 327.
8. Yang, C. N. and Yang, C. P.: *Phys. Rev.* **150** (1966), 131.
9. Lieb, E. H. and Liniger, W.: *Phys. Rev.* **130** (1963), 1605.
10. Onsager, L.: *Phys. Rev.* **65** (1944), 177.
11. Lieb, E. H.: *Phys. Rev.* **162** (1967), 162; *Phys. Rev. Lett.* **19** (1967), 108.
12. Baxter, R. J.: *Ann. Phys. (N. Y.)* **70** (1972), 323; *Ann.Phys. (N. Y.)* **76** (1973), 1; *Ann. Phys. (N. Y.)* **76** (1973), 25; *Ann. Phys. (N. Y.)* **76** (1973), 48.
13. McGuire, I. B.: *J. Math. Phys.* **6** (1965), 432.
14. Berezin, F. A., Pokhil, G. P., and Finkelberg, V. M.: *Vestnik Mosk. Gos. Univ.* **1** (1964), 21.
15. Yang, C. N.: *Phys. Rev. Lett.* **19** (1967), 1312.
16. Brezin, E. and Zinn-Justin, J.: *CR Acad. Sci., Paris*, **B 263** (1966), 670.
17. Gelfand, I. M. and Levitan, B. M.: *DAN SSSR* **88** (1953), 593.
18. Krein, M. G.: *Doklady AN SSSR* **105**(3) (1955), 433.
19. Marchenko, V. A.: *Doklady AN SSSR* **104**(5) (1955), 695.
20. Lax, D.: *Comm. Pure Appl. Math.* **21** (1968), 467.
21. Zakharov, V. E. and Shabat, A. B.: *JETP* **61** (1971), 118; English translation in *Sov. Phys. JETP* **34** (1972), 62.
22. Faddeev, L. D.: *Doklady AN SSSR* **121** (1958), 63.

23. Zakharov, V. E. and Faddeev, L. D.: *Funk. Anal. Pril.* **5** (1972), 18.
24. Zakharov, V. E. and Manakov, S. V.: *Doklady AN SSSR* **19** (1973), 332.
25. Zakharov, V. E., Faddeev, L. D., and Takhtajan, L. A.: *Doklady AN SSSR* **219** (1974), 1334.
26. Takhtajan, L. A. and Faddeev, L. D.: *Uspekhi Math. Nauk* **28** (1974), 249.
27. Takhtajan, L. A. and Faddeev, L. D.: *Theor. Math. Phys.* **21** (1974), 160.
28. Dashen, R. F., Hasslasher, B., and Neveu, A.: *Phys. Rev.* **D11** (1975), 3424.
29. Coleman, S.: *Phys. Rev.* **D11** (1975), 2088.
30. Goldstone, J. and Jackiw, R.: *Phys. Rev.* **D11** (1975), 1486.
31. Frölich, J.: *Comm. Math. Phys.* **47** (1976), 269.
32. Skyrme, T. H. R.: *Proc. Roy. Soc.* **A247** (1958), 260.
33. Skyrme, T. H. R.: *Proc. Roy. Soc.* **A260** (1961), 127.
34. Ablowitz, M. J., Kaup, D. J., Newell, A.C., and Segur H.: *Stud. Appl. Math.* **53**(4) (1973), 249.
35. Flashka, H.: *Phys. Rev.* **B9**(4) (1974), 215.
36. Manakov, S. V.: *Zh. Eksper. Teor. Fiz.* **65** (1973), 505.
37. Novikov, S. P.: *Funk. Anal. Pril.* **8**(3) (1974), 54.
38. Zakharov, V. E. and Shabat, A. B.: *Funk. Anal. Pril.* **13**(3) (1979), 13.
39. Takhtajan, L. A.: *Zapiski LOMI* **37** (1973), 66.
40. Faddeev, L. D. and Korepin, V. E.: *Phys. Rep.* **C42** (1978), 3.
41. Adler, M.: *Invent. Math.* **50**(2) (1979), 219.
42. Kostant, B.: *Adv. Math.* **34** (1979), 195.
43. Kirillov, A. A.: *Elements of Representation Theory*, Nauka, Moscow, 1972 (in Russian).
44. Faddeev, L. D. and Sklyanin, E. K.: *Sov. Akad. Dokl.* **244** (1978), 1337.
45. Sklyanin, E. K.: *Sov. Akad. Dokl.* **244** (1978), 1337.
46. Thacker, H. B. and Wilkinson, D.: *Phys. Rev.* **D19** (1979), 3660.
47. Honerkamp, J., Weber, P., and Weisler, A.: *Nucl. Phys.* **B152** (1979), 266.
48. Faddeev, L. D., Sklyanin, E. K., and Takhtajan, L. A.: *Teor. Mat. Fiz.* **40** (1979), 194.
49. Korepin, V. E., Bogolubov, N. M., and Izergin, A. G.: *Quantum Inverse Scattering Method and Correlation Function*, Cambridge Monograph. Mathematical Physics, Cambridge University Press, 1993.
50. Faddeev, L. D. and Takhtajan, L. A.: *Phys. Lett.* **85A** (1981), 375.
51. Zamolodchikov, A. B. and Zamolodchikov, Al. B.: *Ann. Phys. (N. Y.)* **120** (1979), 253.
52. Belavin, A. A.: *Funk. Anal.* **14** (1980), 18.
53. Babelon, O., De Vega, H. J., and Viallet, C. M.: *Nucl. Phys.* **B220** (1983), 283.
54. De Vega, H. J.: *Integrable Quantum Field Theories and Statistical Models*, World Scientific, Singapore, 1986.
55. Kulish, P. P., Reshetikhin, N. Yu., and Sklyanin, E. K.: *Lett. Math. Phys.* **5** (1981), 393.
56. Izergin, A. G. and Korepin, V. E.: *Doklady AN SSSR* **259** (1981), 76.
57. Tarasov, V. O., Takhtajan, L. A., and Faddeev, L. D.: *Teor. Mat. Fiz.* **57** (1983), 1059.
58. Kulish, P. P. and Reshetikhin, N. Yu.: *Zapiski Nauchn. Semin. LOMI* **101** (1981), 101.
59. Sklyanin, E. K.: *Funk. Anal. Pril.* **16**(4) (1982), 27.
60. Izergin, A. G. and Korepin, V. E.: *Lett. Math. Phys.* **5** (1981), 199.
61. Drinfeld, V. G.: *Doklady AN SSSR* **282** (1985), 1060.
62. Jimbo, M.: *Lett. Math. Phys.* **11** (1986), 247.
63. Drinfeld, V. G.: in *Proc. Int. Congr. of Math.* Berkeley, Vol. 1, p. 798, 1986.
64. Faddeev, L. D., Reshetikhin, N. Yu., and Takhtajan, L. A.: *Leningrad Math. J.* **1** (1990).
65. Shnider, S. and Sternberg, S.: *Quantum Groups*, Int. Press, Boston, 1993.
66. Lusztig, G.: *Introduction to Quantum Groups*, Birkhäuser, Basel, 1993.
67. Chari, V. and Pressley, A. N.: *A Guide to Quantum Groups*, Cambridge University Press, Cambridge, 1994.
68. Yang, C. N. and Ge, M. L. (eds): *Braid Group, Knot Theory and Statistical Mechanics*, Advanced Series Math. Phys., Vols 9 and 17, World Scientific, 1989, 1994.
69. Doebner, H. J. and Dobrev, V. (eds): *Quantum Groups*, Lect. Notes Phys., Vol. 370, 1990.
70. Kulish, P. P. (ed.): *Quantum Groups*, Lect. Notes Math., Vol. 1510, 1992.
71. Jones, V. F. R.: *Bull. Amer. Math. Soc.* **12** (1985), 103.

72. Reshetikhin, N. and Turaev, V.: *Invent. Math.* **103** (1991), 547.
73. Belavin, A. A., Polyakov, A. M., and Zamolodchikov, A. B.: *Nucl. Phys.* **B241** (1984), 333.
74. Polyakov, A. M.: *Phys. Lett.* **103B** (1981), 207.
75. Wess, J. and Zumino, B.: *Nucl. Phys. Suppl.* **19B** (1990), 330.

Acta Applicandae Mathematicae **39:** 85–90, 1995.

Optical Solitons in Communications: From Integrability To Controllability

AKIRA HASEGAWA
Department of Communication Engineering, Faculty of Engineering, Osaka University, Suita 565, Japan

(Received: 1 September 1994)

Abstract. A brief historical review and recent developments of theories and experiments of optical solitons in fibers are presented with particular emphasis on their integrability and controllability in real systems.

Mathematics Subject Classification (1991): 58F07.

Key words: optical solitons, integrability, controllability.

1. Introduction

The optical soliton in fibers represents a rare example of success of achieving a realistic high technology application of the mathematical object called a soliton. Signal transmission by means of optical solitons now surpasses any other conventional means both in the distance and in the bitrate. Needless to say, the developments of low loss fiber, semiconductor lasers, and the discovery of the optical amplifier (the erbium doped fiber amplifier, EDFA) have contributed to its success. However, the new development of theories of optical soliton, its integrability and controllability, have played an essential role helped by the social demand of faster and longer communications with less cost. Optical solitons have made a large impact on high-speed communications because of two important properties, the integrability of the model equation and the controllability of the solitons. The proof of integrability (even in the presence of loss and repeated amplification) is essential in their use for communications in providing a guarantee on the input and output causal relation for light waves in fibers. In addition, it assures that solitons (rather than Fourier modes) are the best means of transmitting signals through optical fibers because any input light waves having a power beyond a critical level develop into solitons. The controllability of solitons warrants error-free signal transmissions even in the presence of perturbations such as noise from amplifiers, mutual interaction among neighboring solitons and other undesirable effects. This property, which is unique to the optical soliton, and is unavailable in linear waves, provides an additional basis of the use of optical solitons for communications.

2. Integrability

In this section, we describe the physical background of the derivation of the nonlinear Schrödinger equation and its soliton-bearing property as well as how the equation is related to optical communications. The electrical properties of an optical fiber can be described by its index of refraction, n. This index has a linear portion, n_0 and a nonlinear portion which, in a symmetrical material such as ordinary glass, is given by $n_2|E|^2$. Here, E is the Fourier amplitude of the light wave electric field, n_0 is generally a function of the light wave frequency, which leads to the wave dispersion, n_2 is called the Kerr coefficient and has a value of approximately $10^{-22}(\mathrm{m/V})^2$. The wave number k of the light wave in the fiber can be expressed in terms of the index of refraction,

$$k = \frac{\omega}{c}\left[n_0(\omega) + n_2|E|^2\right]. \tag{1}$$

In this expression, ω is the angular frequency of the light wave, c is the speed of light in vacuum. The equation which describes the evolution of the complex light wave envelope q in the fiber can be derived by expanding the wave number k around the carrier wave number [1], and is given, in the lowest order of group dispersion and nonlinearity, by

$$\mathrm{i}\,\frac{\partial q}{\partial Z} + \frac{1}{2}\frac{\partial^2 q}{\partial T^2} + |q|^2 q = 0. \tag{2}$$

Here, T is the time on the coordinate moving at the group velocity $\partial\omega/\partial k$ $(= (k')^{-1})$ which is normalized by the pulse width t_0, Z is the distance normalized by the dispersion distance z_0 $(= t_0/k'')$ and q is the normalized electric field given by $(\omega n_2 z_0/2c)^{1/2}E$. In Equation (2), the second term represents the effect of the group dispersion and is proportional to k'' (which is assumed to be negative) and the third term Kerr effect.

Here let us see how a realistic light wave envelope may be properly described by the nonlinear Schrödinger equation. The light wave electric field in a fiber has a value of typically 10^5 V/m for a few milliwatts of power. Thus, the nonlinear change of the index of refraction $n_2|E|^2$ has a value only of the order of 10^{-12}. Then why is such a small change in the index of refraction which contributes significantly to the evolution of the light wavelength so huge. As seen in Equation (1), the nonlinear change in the wave number k is multiplied by ω/c. This means that a small change in the index of refraction of 10^{-12} occurs at every distance of propagation of the light wave wavelength which is around 1.5×10^{-6} m. Thus, over a distance of 1 000 km, the nonlinear effect becomes quite appreciable. The same is true for the dispersion effect and it can be shown that a light pulse of around a few tens of picoseconds is distorted significantly by the group dispersion over the similar distance. In other words, the dispersion distance z_0 becomes around several hundred kilometers for a light pulse of a few tens of picoseconds.

Zakharov and Shabat [2] discovered the Lax pair for the nonlinear Schrödinger equation for the inverse scattering transform. The spectral equation is given by

$$\mathrm{i}\,\frac{\partial\psi_1}{\partial T}+q\psi_2=\zeta\psi_1,$$

$$-\mathrm{i}\,\frac{\partial\psi_2}{\partial T}-q^*\psi_1=\zeta\psi_2. \tag{3}$$

From the property of the Lax pair, the eigenvalue ζ is invariant in Z if q in (3) evolves in Z according to the nonlinear Schrödinger equation (2). As a result, if the initial envelope $q\ (Z=0,T)$ is given, the properties of solitons which emerge from the initial envelope are determined from the eigenvalue ζ. In particular, the one soliton solution is given by

$$q=\eta\sec h[\eta(T+\varkappa Z-T_0)]\exp\left[-\mathrm{i}\varkappa T+\frac{\mathrm{i}}{2}(\eta^2-\varkappa^2)Z+\mathrm{i}\sigma_0\right], \tag{4}$$

where the parameters η and $\varkappa$ are given by the eigenvalue

$$\zeta=\frac{\varkappa+\mathrm{i}\eta}{2}. \tag{5}$$

The integrability provided by the inverse scattering transform warrants the precise input–output relationship of light waves in a fiber in terms of solitons if the fiber is lossless. However, a real fiber has a loss which should be compensated for by amplifiers. In the presence of loss and amplifications, the envelope equation is modified to

$$\mathrm{i}\,\frac{\partial q}{\partial Z}+\frac{1}{2}\frac{\partial^2 q}{\partial T^2}+|q|^2q=-\mathrm{i}[\Gamma-g(Z)]q. \tag{6}$$

Here Γ is the product of the fiber loss per unit distance and the dispersion distance and g represents the gain of the amplifier(s). Since Γ is proportional to the dispersion distance which is proportional to the pulse width squared, its magnitude may exceed unity for a broad soliton. For a nominal fiber, Γ in fact becomes larger than one for a soliton with a pulse width of more than 10 picoseconds. It seems, then, obvious that the integrability of (6) and the soliton solution are lost.

However, interestingly enough, the integrable property is recovered when the light wave is periodically amplified in such a way that on the average the fiber loss is completely compensated for. The proof is made by transforming first q to u through

$$q=ua_0\exp\left[\int\{-\Gamma+g(Z)\}\,\mathrm{d}Z\right] \tag{7}$$

and u to ν through the Lee transformation

$$u=\exp\{\tilde{\phi}\cdot\nabla\}\nu, \tag{8}$$

where ϕ is the generating function which is successively determined at each period of amplification. It is then possible to show [3] that ν satisfies the ideal nonlinear Schrödinger equation (2) to the order $(z_a/z_0)^2$, where z_a is the amplifier spacing. It means that if the amplifier spacing is made much shorter than the dispersion distance, the transformed amplitude ν has soliton solutions even if q varies periodically at every amplifier distance. We call this soliton solution 'the guiding center soliton' in analogy to the guiding center motion of a charged particles in a magnetic field. Consequently, the integrability is warranted even in the presence of fiber loss (with a relatively large Γ) provided that the loss is periodically compensated for by amplifiers.

3. Controllability

There are other nonideal disturbances in a real fiber which produce fluctuation on soliton parameters, such as the amplitude η, the frequency (velocity) $\varkappa$, the position T_0 and/or the phase σ_0. Amplifier noise, Brillouin excited phonons, collision-induced jitters among solitons and higher-order dispersion and nonlinear effects may all contribute to the disturbances and may induce transmission error. However, several methods have been discovered of controlling soliton parameters from these disturbances and reducing their effects. These methods fully utilize properties that are unique to solitons, thus are available only to soliton-based transmission systems. The controllability of the soliton parameters presents an additional unique benefit for soliton-based communications. As an example, let us illustrate a simple method of the use of frequency filters [4, 5]. The effect of the filter may be incorporated by introducing a term given by $\mathrm{i}\beta\partial^2 q/\partial T^2$ to the right-hand side of Equation (6). It also requires an excess frequency independent gain to compensate for the loss created by the filter. The total frequency dependent gain is shown in Figure 1. The variation of the soliton parameters in the presence of these controls may be studied by constructing dynamical equations for the soliton parameters from the perturbed nonlinear Schrödinger equation. The dynamical equations have a structure given by [4],

$$\begin{aligned}\frac{\mathrm{d}\eta}{\mathrm{d}Z} &= G(\eta, \varkappa, Z), \\ \frac{\mathrm{d}\varkappa}{\mathrm{d}Z} &= H(\eta, \varkappa, Z).\end{aligned} \tag{9}$$

Here, G and H are functions which represent effects of filters, excess gain and disturbances. Naturally, if they are zero, the soliton parameters do not change. Figure 2 shows the trajectories of η and $\varkappa$ in the presence of the filter and excess gain (but without any other disturbances). It is clear that the soliton parameters converge to a fixed and stable point in their phase space. It should be pointed our that the control is effective only to solitons because a linear wave

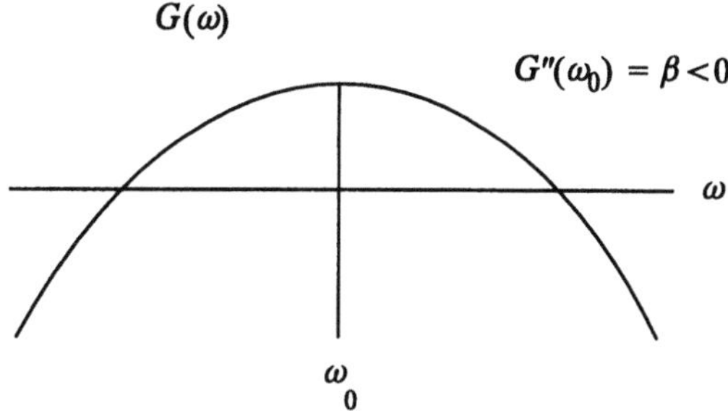

Fig. 1. Frequency characteristic of a filter and excess gain which can control the soliton parameters.

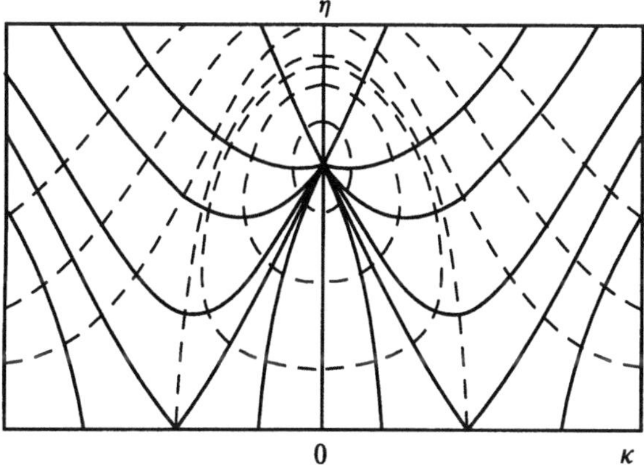

Fig. 2. Phase space trajectory of soliton parameters, the amplitude η and the speed k in the presence of the control filters.

with a certain frequency spectrum becomes unstable due to the excess gain at the center frequency of the filter. The stable soliton behavior results from its property that the product of the amplitude and pulse width is constant. Thus, when the amplitude increases, the width decreases leading to a spread of the frequency spectrum whose edge will be cut by the finite band width of the filters. The controllability of solitons has its basis on the fact that behavior of solitons may be described by only a finite set of parameters, that are the discrete eigenvalues of the spectral equation (3), in spite of the fact that the original wave equation has an infinite degree of freedom.

Mollenaner *et al.* [6] succeeded in transmitting 20 Gigabit/s soliton signals error-free over a distance of 14 000 kilometers in a loop experiment. Here, they used the guiding center solitons and controlling frequency filters. To avoid the accumulation of unstable noise near the central frequency region, they used filters whose central frequency is continuously shifted along the soliton path.

4. Conclusion

The two unique properties of optical solitons, the integrability of the model equation and the controllability of the soliton parameters present a natural reason as to why solitons are most suitable as carriers of high-speed information over extended distances in a fiber having group dispersion and nonlinearity. A hundred years after the discovery of a solitary wave in the KdV equation, a soliton is now known as a real object in the high technology world of high-speed communication.

References

1. Hasegawa, A. and Tappert, F. D.: *Appl. Phys. Lett.* **23** (1973), 142.
2. Zakharov, V. E. and Shabat, A. B.: *Sov. Phys. JETP* **34** (1972), 62.
3. Hasegawa, A. and Kodama, Y.: *Phys. Rev. Lett.* **66** (1991), 161.
4. Kodama, Y. and Hasegawa, A.: *Opt Lett.* **17** (1992), 31.
5. Meccozi, A., Moores, J. D., Haus, H. A., and Lai, Y.: *Opt. Lett.* **16** (1991), 1841.
6. Mollenaner, L. F., Lichtman, E., Neubelt, M. J., and Harvey, G. T.: *Electron. Lett.* **29** (1993), 910.

Acta Applicandae Mathematicae **39:** 91–92, 1995.

Korteweg, de Vries, and Dutch Science at the Turn of the Century

A. J. KOX
Institute of Theoretical Physics, University of Amsterdam, The Netherlands

(Received: 2 January 1995)

The equation that is now known as the Korteweg–de Vries equation first appeared in the doctoral dissertation of D. J. Korteweg's student Gustav de Vries, which was defended on 1 December 1894 at the Amsterdam University [1]. That the centenary of the equation is nevertheless celebrated in 1995, instead of in 1994, has a good reason: the work of Korteweg and de Vries became internationally known through a joint paper they published in May 1895 [2].

Little is known of the further life and career of de Vries (he became a secondary-school teacher); Korteweg, on the other hand, was an important and influential figure in the Dutch scientific world. After his death in 1941 at the age of 93, he was characterized as "an almost legendary figure, being the last survivor of the heroic age of the Amsterdam natural sciences faculty".* The reference is to the period that started around 1875 during which Dutch science acquired sudden international fame after a long period of decline.** This period, also known as the 'Second Golden Age' of Dutch science (following the first one in the seventeenth century), is not only interesting from the point of view of history of science, but also in a more general way, because some very interesting connections can be made between the rise of Dutch science on the one hand and economic and educational factors, such as secondary-school and university reforms, on the other (see, e.g., [3]).

Although formally a mathematician, Korteweg was a typical representative of the new generation of Dutch scientists that made Dutch science internationally famous. He wrote his doctoral dissertation under the guidance of the physicist Van der Waals‡ and continued collaborating with him in later years. Korteweg's work is characterized by a strong emphasis on mathematical physics, and from

* See the obituary in *Nederlandsche Akademie van Wetenschappen. Verslagen van de gewone vergaderingen der Afdeeling Natuurkunde* **50** (1941), 35–37.

** Characteristic is that in the first thirteen years of the existence of the Nobel Prizes five prizes were awarded to Dutch scientists.

‡ The dissertation dealt with the problem of wave propagation in elastic tubes, a problem of physiological interest. See [4].

the very beginning of his career, he emphasized the importance of the application of mathematics to practical problems.* In this respect, he personified an important trait of Dutch science of his days, namely the fruitful interaction between theory and experiment.** His teaching reflected his interests as well: he regularly taught courses on mechanics and astronomy. All this does not mean, however, that Korteweg was not interested in pure mathematics: he supervised the work on many doctoral dissertations in this field. His most famous student was perhaps the pure mathematician L. E. J. Brouwer, whose talent he recognized and stimulated at a very early stage.

What makes Korteweg a unique figure is that in addition to his work in mathematical physics, he was active in other fields as well. For many years, for example, he was the main editor of the *Oeuvres complètes de Christiaan Huygens*, the 22-volume edition of Huygens's works that is still one of the most impressive editions of its kind. From his correspondence with other editors such as Lorentz and Zeeman, it becomes clear how much time and energy he devoted to this work. A totally unrelated project that took much of Korteweg's time and energy was the *International Catalogue of Scientific Literature* (ICSL), an ambitious and eventually failed project to make all international scientific literature accessible through a central catalogue in London. Korteweg headed the Dutch section of the ICSL.

To conclude, Korteweg was a many-sided and important figure in the Dutch scientific community of his days, whose lasting contributions are not limited to the equation that bears his name.

References

1. de Vries, G.: *Bijdrage tot de kennis der lange golven*, Loosjes, Haarlem, 1894.
2. Korteweg, D. J. and de Vries, G.: On the change of form of long waves advancing in a rectangular canal, and on a new type of long stationary waves, *Philosophical Magazine* **39** (1895), 422–443.
3. Willink, B.: Origins of the Second Golden Age of Dutch science after 1860: Intended and unintended consequences of educational reform, *Social Studies of Science* **21** (1991), 503–526.
4. Korteweg, D. J.: *Over voortplantings-snelheid van golven in elastische buizen*, Van Doesburgh, Leiden, 1878.
5. Korteweg, D.J.: *De wiskunde als hulpwetenschap*, Van Heteren, Amsterdam, 1881.

* Already in his inaugural lecture of 1881 he strongly emphasized this point. See [5].

** Characteristic are the collaborations between Van der Waals and Kamerlingh Onnes and between Lorentz and Zeeman.

Acta Applicandae Mathematicae **39:** 93–125, 1995.

Algebraic–Geometrical Methods in the Theory of Integrable Equations and Their Perturbations

I. KRICHEVER
Landau Institute for Theoretical Physics, Kosygina str. 2, 117940 Moscow, Russia

(Received: 10 February 1995)

Mathematics Subject Classification (1991): 58F07.

Key words: integration, finite-gap theory.

1. Introduction

The year of 1995 is not merely the centenary of the Korteweg–de Vries equation which we celebrate at this conference. It is also the year of the 'majority' of the finite-gap or algebraic–geometrical theory of integration of nonlinear equations – one of the most important components of the branch of modern mathematical physics, which is called the theory of integrable systems or the soliton theory. The main goal of this paper is to present the key points of the finite-gap theory and some of its applications. Part of its applications is directly related to the KdV equation, while a part of them lies beyond the framework of not only this particular equation but the theory of soliton equations in general. Corresponding examples refer to the string theory and topological field theory models. Not aspiring to be exhaustive, they manifest versatility of the methods, the origin of which would be forever related to the magic words: *Korteweg–de Vries Equation.*

Same as in the rapidly decreasing case, primerely the program of constructing periodic solutions of KdV equation was completely based on the spectral theory of the Sturm–Liouville operator – one of the so-called Lax operators for the KdV equation.

But unlike the rapidly decreasing case in which the efficiency of direct and inverse spectral transform was sufficient both for constructing multi-soliton solutions and for solving the Cauchy problem in general, in the periodic case the level of efficiency of the corresponding spectral problems was far from sufficient.

The creation of the effective spectral theory of finite-gap Sturm–Liouville operators proposed in the cycle of papers by S. P. Novikov, B. A. Dubrovin, V. B. Matveev and A. R. Its (see their reviews in [1, 2]; part of the corresponding results was obtained a bit later in [3, 4]), made it possible to construct not only a broad class of periodic and quasi-periodic solutions of the KdV equation. It also brought about the reinterpretation of the approach to the spectral theory of ordinary periodic linear differential operators, as whole.

The statement that the Bloch solutions of such operators considered for the arbitrary complex values of spectral parameter E, form a single-valued function on a Riemann surface, quite obvious nowadays, had remained outside the framework of the classical Flouque theory. It turned out that analytical properties of the Bloch functions on this surface are crucial for the solution of the inverse problem of reconstructing the coefficients of operators. In the case when this surface has a finite genus, the solution of the inverse problem is based on the methods of classical algebraic geometry and the theory of theta functions.

The significance of the algebraic–geometrical approach was completely revealed in [5, 6], where the general algebraic–geometrical construction of periodic solutions of two-dimensional soliton equations of the Kadomtsev–Petviashvili type was proposed. This construction is based on the concept of the Baker–Akhiezer function $\psi(x,y,t,Q)$, which is uniquely determined by their analytical properties on the auxiliary Riemann surface Γ, $Q\in\Gamma$. The corresponding analytical properties naturally generalize the analytical properties of the Bloch functions of ordinary linear periodic differential operators. Their peculiarity is that for any function possessing these properties there always exist differential operators L and A of the form

$$L=\sum_{i=0}^{n} u_i(x,y,t)\partial_x^i, \qquad A=\sum_{j=0}^{n} v_j(x,y,t)\partial_x^j, \quad \partial_x=\frac{\partial}{\partial x}. \tag{1.1}$$

such that the Baker–Akhiezer function is the common solution of the linear equations

$$(\partial_y-L)\psi(x,y,t,Q)=0, \qquad (\partial_t-A)\psi(x,y,t,Q)=0. \tag{1.2}$$

The compatibility conditions of the overdetermined system of the linear problems (1.2) imply the operator equation

$$[\partial_y-L,\partial_t-A]=0 \longleftrightarrow \partial_t L-\partial_y A+[L,A]=0 \tag{1.3}$$

that is equivalent to the system of nonlinear partial differential equations on the coefficients of the operators L and A.

For example, if L and A have the form

$$L=\sigma^{-1}(\partial_x^2-u(x,y,t)), \qquad A=\partial_x^3-\tfrac{3}{2}\,u\partial_x+w(x,y,t), \tag{1.4}$$

then (1.3) is equivalent to the KP equation

$$\tfrac{3}{4}\sigma^2 u_{yy}=\left(u_t-\tfrac{3}{2}\,uu_x+\tfrac{1}{4}\,u_{xxx}\right)_x. \tag{1.5}$$

If $u=u(x,t)$ is independent of the second variable y and $w=\frac{3}{4}u_x(x,t)$, then (1.3) is equivalent to the KdV equation

$$u_t-\tfrac{3}{2}\,uu_x+\tfrac{1}{4}\,u_{xxx}=0. \tag{1.6}$$

We present the general algebraic–geometrical construction of exact solutions of the soliton equations in Section 2. It should be specially emphasized that this construction is purely local and is a sort of inverse transform: from a set of algebraic–geometrical data to solutions of the integrable nonlinear partial differential equations

$$\{\text{algebraic–geometrical data}\} \longmapsto \{\text{solutions of NLPDE}\} \tag{1.7}$$

In a generic case, the space of algebraic–geometrical data is a union for all g of the spaces

$$\widetilde{M}_{g,N} = \{\Gamma_g, P_\alpha, k_\alpha^{-1}(Q), \gamma_1, \dots, \gamma_g\}, \quad \alpha = 1, \dots, N, \tag{1.8}$$

where Γ_g is an algebraic curve of genus g with fixed local coordinates $k_\alpha^{-1}(Q)$, $k_\alpha^{-1}(P_\alpha) = 0$, in neighborhoods of N punctures P_α, and $\gamma_1, \dots, \gamma_g$ are points of Γ_g in a general position. (It is to be mentioned that $\widetilde{M}_{g,N}$ are 'universal' data. For the given nonlinear integrable equation, the corresponding subset of data has to be specified.)

A posteriory, it can be shown that these solutions can be expressed in terms of the corresponding Riemann theta functions and are quasi-periodic functions of all variables. Within this approach, it is absolutely impossible to give an answer to the basic questions: "How many algebraic–geometrical solutions are there? And what is their role in the solution of the periodic Cauchy problem for two-dimensional equations of the KP type?"

The answer to the corresponding question in lower dimensions is as follows. For finite-dimensional $(0+1)$-systems, a typical Lax representation has the form

$$\partial_t U(t, \lambda) = \left[U(t, \lambda),\ V(t, \lambda)\right], \tag{1.9}$$

where $U(t, \lambda)$ and $V(t, \lambda)$ are matrix functions that are rational (or sometimes elliptic) functions of the spectral parameter λ. In that case, *all* the general solutions are algebraic–geometrical and can be represented in terms of Riemann theta functions.

For special one-dimensional evolution equations of the KdV type ($(1+1)$-systems), the existence of direct and inverse spectral transform allow one to prove (though it is not always the rigorous mathematical statement) that algebraic–geometrical solutions are dense in the space of all periodic (in x) solutions.

It turns out that the situation for two-dimensional integrable equations is much more complicated. For one of the real forms of the KP equation that is called the KP-2 equation and which corresponds to $\sigma = 1$ in (1.5), the algebraic–geometrical solutions are dense in the space of all periodic (in x and y) solutions [7]. It seems, that the same statement for the KP-1 equation ($\sigma = i$) is wrong. One of the most important problems in the theory of two-dimensional integrable systems which is still unsolved is 'in what sense' the KP-1 equation that has the operator representation (1.3) and for which a wide class of periodic solution has been constructed, is an '*nonintegrable*' system.

The proof of the integrability of the periodic problem for the KP-2 equation is based on the spectral Floque theory of the parabolic operator

$$M = \partial_y - \partial_x^2 + u(x,y), \tag{1.10}$$

with periodic potential $u(x+l_1,y) = u(x,y+l_2) = u(x,y)$. This theory is presented in Section 3. It is a natural generalization of the spectral theory of the periodic Sturm–Liouville operator. We would like to mention that despite its application to the theory of nonlinear equations and related topics, the structure of the Riemann surface of Bloch solutions of the corresponding linear equation that was found in [7] has been used as a starting point for the abstract definition of the Riemann surfaces of the infinite genus [9].

In the last section, we present the algebraic–geometrical perturbation theory of soliton equations and its application to the topological models of quantum field theory.

2. The Baker–Akhiezer Functions. General Scheme

Let Γ be a nonsingular algebraic curve of genus g with N punctures P_α and fixed local parameters $k_\alpha^{-1}(Q)$ in neighborhoods of the punctures. For any set of the points $\gamma_1,\ldots,\gamma_g$ in a general position, there exists a unique (up to constant factor $c(t_{\alpha,i})$) function $\psi(t,Q)$, $t = (t_{\alpha,i})$, $\alpha = 1,\ldots,N$; $i = 1,\ldots,$ such that:

(i) the function ψ (as a function of the variable Q which is a point of Γ) is meromorphic everywhere except for the points P_α and has at most simple poles at the points $\gamma_1,\ldots,\gamma_g$ (if all of them are distinct);

(ii) at the neighborhood of the point P_α the function ψ has the form

$$\psi(t,Q) = \exp\left(\sum_{i=1}^{\infty} t_{\alpha,i}k_\alpha^i\right)\left(\sum_{s=0}^{\infty}\xi_{s,\alpha}(t)k_\alpha^{-s}\right), \quad k_\alpha = k_\alpha(Q). \tag{2.1}$$

This is the most general definition of a scalar *multi-puncture* and *multi-variable* Baker–Akhiezer function. It depends on the variables $t = \{t_{1,i},\ldots,t_{N,i}\}$ as on external parameters.

From the uniqueness of the Baker–Akhiezer function, it follows that for each pair (α,n) there exists a unique operator $L_{\alpha,n}$ of the form

$$L_{\alpha,n} = \partial_{\alpha,1}^n + \sum_{j=1}^{n-1} u_j^{(\alpha,n)}(t)\partial_{\alpha,1}^j \tag{2.2}$$

(where $\partial_{\alpha,i} = \partial/\partial t_{\alpha,i}$) such that

$$(\partial_{\alpha,i} - L_{\alpha,n})\psi(t,Q) = 0. \tag{2.3}$$

The idea of the proof of theorems of this type which was proposed in [5, 6] is universal.

For any formal series of the form (2.1), their exists a unique operator $L_{\alpha,n}$ of the form (2.2) such that

$$(\partial_{\alpha,i} - L_{\alpha,n})\psi(t,Q) = O(k^{-1})\exp\left(\sum_{i=1}^{\infty} t_{\alpha,i}k_{\alpha}^{i}\right). \tag{2.4}$$

The coefficients of $L_{\alpha,n}$ are differential polynomials with respect to $\xi_{s,\alpha}$. They can be found after substitution of the series (2.1) into (2.4).

It turns out that if the series (2.1) is not formal but is an expansion of the Baker–Akhiezer function in the neighborhood of P_α, the congruence (2.4) becomes an equality. Indeed, let us consider the function

$$\psi_1 = (\partial_{\alpha,n} - L_{\alpha,n})\psi(t,Q). \tag{2.5}$$

It has the same analytical properties as ψ except for one. The expansion of this function in the neighborhood of P_α starts from $O(k^{-1})$. From the uniqueness of the Baker–Akhiezer function, it follows that $\psi_1 = 0$ and the equality (2.3) is proved.

COROLLARY 2.1. *The operators $L_{\alpha,n}$ satisfy the compatibility conditions*

$$[\partial_{\alpha,n} - L_{\alpha,n}, \partial_{\alpha,m} - L_{\alpha,m}] = 0. \tag{2.6}$$

Remark. Equations (2.6) are gauge invariant. For any function $g(t)$ operators

$$\widetilde{L}_{\alpha,n} = gL_{\alpha,n}g^{-1} + (\partial_{\alpha,n}g)g^{-1} \tag{2.7}$$

have the same form (2.2) and satisfy the same operator equations (2.6). The gauge transformation (2.7) corresponds to the gauge transformation of the Baker–Akhiezer function

$$\psi_1(t,Q) = g(t)\psi(t,Q). \tag{2.8}$$

EXAMPLE (one-puncture Baker–Akhiezer function). In the one-puncture case, the Baker–Akhiezer function has an exponential singularity at a single point P_1 and depends on a single set of variables. Let us choose the normalization of the Baker–Akhiezer function with the help of the condition $\xi_{1,0} = 1$, i.e. an expansion of ψ in the neighborhood of P_1 equals

$$\psi(t_1, t_2, \ldots, Q) = \exp\left(\sum_{i=1}^{\infty} t_i k^i\right)\left(1 + \sum_{s=1}^{\infty} \xi_s(t)k^{-s}\right). \tag{2.9}$$

In this case, operator L_n has the form

$$L_n = \partial_1^n + \sum_{i=o}^{n-2} u_i^{(n)} \partial_1^i. \tag{2.10}$$

For example, for $n = 2, 3$ after redefinition $x = t_1$ we have $L_2 = \sigma L,\ L_3 = A$, where L and A are differential operators (1.4) and

$$u(x, t_2, \ldots) = 2\partial_x \xi_1(x, t_2, \ldots), \tag{2.11}$$

Therefore, if we define $y = \sigma^{-1} t_2$, $t = t_3$, then $u(x, y, t, t_4, \ldots)$ satisfies the KP Equation (1.5).

It should be emphasized that a algebraic–geometrical construction is not a sort of abstract 'existence' and 'uniqueness' theorems. It provides the exact formulae for solutions in terms of the Riemann theta functions. They are the corollary of the exact formula for the Baker–Akhiezer function

$$\psi(t, Q) = \exp\left(\sum_{i,\alpha} t_{i,\alpha} \Omega_{i,\alpha}(P)\right) \frac{\theta(A(P) + \sum_{i,\alpha} U_{i,\alpha} t_{i,\alpha} + Z)}{\theta(A(P) + Z)}, \tag{2.12}$$

Here:

$\theta(z) = \theta(z \mid B)$ is the Riemann theta function – the entire function of g complex variables $z = (z_1, \ldots, z_g)$ that is defined by the matrix B of b-periods of normalized holomorphic differentials $\mathrm{d}\omega_i$, $i = 1, \ldots, g$, on Γ:

$$\theta(z_1, \ldots, z_g) = \sum_{m \in Z^g} \mathrm{e}^{2\pi i (m,z) + \pi i (Bm,m)}; \tag{2.13}$$

$\Omega_{i,\alpha}(P)$ is an Abelian integral

$$\Omega_{i,\alpha}(P) = \int^P \mathrm{d}\Omega_{i,\alpha}, \tag{2.14}$$

corresponding to the unique normalized

$$\oint_{a_k} \mathrm{d}\Omega_{i,\alpha} = 0, \tag{2.15}$$

meromorphic differential on Γ with the only pole of the form

$$\mathrm{d}\Omega_{i,\alpha} = \mathrm{d}k_\alpha^i (1 + \mathrm{O}(k_\alpha^{-i-1})) \tag{2.16}$$

at the puncture P_α;

$2\pi i U_{j,\alpha}$ is the vector of b-periods of the differential $\mathrm{d}\Omega_{j,\alpha}$

$$U_{j,\alpha}^k = \frac{1}{2\pi i} \oint_{b_k} \mathrm{d}\Omega_{j,\alpha}; \tag{2.17}$$

$A(P)$ is a vector with coordinates

$$A(P) = \int^P d\omega_k \tag{2.18}$$

(formula (2.18) is called the Abel transformation);

Z is an arbitrary vector (it corresponds to the divisor of poles of the Baker–Akhiezer function).

It follows from (2.11) that in order to get the solution of the KP equation, it is enough to take the derivative of the first coefficient of the expansion at the puncture of the ratio of theta functions in the formula (2.12). The final formula for the algebraic–geometrical solutions of the KP hierarchy has the form

$$u(t_1, t_2, \dots) = 2\partial_1^2 \ln \theta \left(\sum_{i=1}^{\infty} U_i t_i + Z \right) + \text{const}, \tag{2.19}$$

(see details in [6]).

Formula (2.19) that was derived in [6], has lead to one of the most important pure mathematical applications of the theory of nonlinear integrable systems. This is the solution of the famous Riemann–Shottky problem.

According to the Torrelli theorem, the matrix of b-periods of normalized holomorphic differentials uniquely defines the corresponding algebraic curve. The Riemann–Shottky problem is to describe symmetrical matrices with the positive imaginary part that are the matrices of b-periods of normalized holomorphic differentials on algebraic curves. Novikov conjected that the function

$$u(x, y, t) = 2\partial_1^2 \ln \theta(Ux + Vy + Wt + Z \mid B) \tag{2.20}$$

is a solution of the KP equation iff the matrix B that defines the theta function is the matrix of b-periods of normalized holomorphic differentials on an algebraic curve and U, V, W are vectors of the b-periods of corresponding normalized meromorphic differentials with the only pole at a point of this curve. This conjecture was proved in [10].

Equations (2.6) for $n = 2$, $m > 3$ describe evolutions of $u(x, y, t, t_4, \dots)$ with respect to 'higher times' or, equivalently, the whole KP hierarchy. Here it is necessary to make a few comments. In the original form, Equations (2.6) are a set of nonlinear equations on the coefficients $u_i^{(n)}$ and do not have the form of evolution equations. It can be shown (see [11]) that they are equivalent to the evolution system in the form which was proposed by the Kyoto group [12]. We show this equivalence for algebraic–geometrical solutions.

For any formal series of the form (2.9), there exists a unique pseudo-differential operator $\mathcal{L}$

$$\mathcal{L} = \partial_1 + \sum_{i=1}^{\infty} u_i(t_1, \dots) \partial_1^{-i}, \tag{2.21}$$

such that

$$\mathcal{L}\psi(t,k) = k\psi(t,k). \tag{2.22}$$

Then the operators L_n which are uniquely defined from the congruence (2.4) are equal to

$$L_n = [\mathcal{L}^n]_+, \tag{2.23}$$

where $[\ldots]_+$ stands for the differential part of the pseudo-differential operator. From (2.3), it follows that if $\psi(t,k)$ is an expansion of the Baker–Akhiezer function, then

$$(\partial_n - [\mathcal{L}^n]_+)\psi(t,Q) = 0. \tag{2.24}$$

The compatibility conditions of (2.22) and (2.24) imply the evolution equations

$$\partial_n \mathcal{L} = [[\mathcal{L}^n]_+, \mathcal{L}] \tag{2.25}$$

on the coefficients $u_i(t_1,\ldots)$ of $\mathcal{L}$. The Equations (2.25) are the Sato form for the KP hierarchy.

At the end of this section, we shall make a few comments about the multi-puncture case. For each α Equations (2.6), up to the gauge transformation, are equivalent to the KP hierarchy corresponding to each set of variables $\{t_{\alpha,i}\}$. What is the interaction between two different KP hierarchies?

As was found in [13], for the two-puncture case, a full set of equations can be represented in the following form

$$[\partial_{\alpha,n} - L_{\alpha,n}, \partial_{\beta,n} - L_{\beta,n}] = \mathrm{D}_{N,m}^{\alpha,\beta} H^{\alpha,\beta}, \tag{2.26}$$

where $H^{\alpha,\beta}$ is the two-dimensional Schrödinger operator in a magnetic field

$$H^{\alpha,\beta} = \frac{\partial^2}{\partial_{\alpha,1}\partial_{\beta,1}} + v_1^{\alpha,\beta}\partial_{\alpha,1} + v_2^{\alpha,\beta}\partial_{\alpha,2} + u^{\alpha,\beta} \tag{2.27}$$

and operators $\mathrm{D}_{N,m}^{\alpha,\beta}$ are differential operators in the variables $t_{\alpha,1}, t_{\beta,1}$.

The sense of (2.26) is as follows. For the given operator $H^{\alpha,\beta}$, any differential operator D in the variables $t_{\alpha,1}$, $t_{\beta,1}$ can be uniquely represented in the form

$$\mathrm{D} = \mathrm{D}_1 H^{\alpha,\beta} + \mathrm{D}_2 + \mathrm{D}_3, \tag{2.28}$$

where D_2 is a differential operator with respect to the variable $t_{\alpha,1}$ only and D_3 is a differential operator with respect to the variable $t_{\beta,1}$ only. Equations (2.26) imply that the second and third terms in the corresponding representation for the left-hand side of (2.26) are equal to zero. This implies $n+m-1$ equations on $n+m$ unknown functions (the coefficients of operators $L_{\alpha,n}$ and $L_{\beta,m}$). Equations

(2.26) are gauge invariant. That's why the number of equations equals a number of unknown functions. Therefore, the operator equations (2.26) are equivalent to the well-defined system of nonlinear partial differential equations.

3. Spectral Theory of Two-Dimensional Periodic Operators

As was mentioned in the introduction, nowadays the origin of the Riemann surfaces in the spectral theory of periodic ordinary linear differential operators looks more or less obvious.

Indeed, let L be an ordinary differential operator

$$L = \sum_{i=0} u_i(x)\partial_x^i \tag{3.1}$$

with periodic coefficients $u_i(x+l) = u_i(x)$ (that are scalar or matrix functions). Then the monodromy operator

$$\widehat{T}\colon\ y(x) \longmapsto y(x+l) \tag{3.2}$$

induces the finite-dimensional linear operator

$$\widehat{T}(E)\colon\ \mathcal{L}(E) \longmapsto \mathcal{L}(E) \tag{3.3}$$

on the space $\mathcal{L}(E)$ of solutions of the ordinary differential equation

$$Ly(x) = Ey(x), \tag{3.4}$$

where E is a complex spectral parameter.

The characteristic equation

$$R(w, E) = \det(w - \widehat{T}(E)) \tag{3.5}$$

defines the Riemann surface Γ of the Bloch solutions of Equation (3.4), i.e. the solutions $\psi(x, Q)$ of (3.4) that are eigenfunctions of the monodromy operator

$$\psi(x+l, Q) = w\psi(x, Q), \tag{3.6}$$

Here Q is a pair $Q = (w, E)$ such that (3.5) is fulfilled, i.e. a point of the corresponding Riemann surface $Q \in \Gamma$.

Equation (3.5) represents the Riemann surface of the Bloch solutions as an N-sheet covering of the complex plane of the spectral parameter E. In [7] another representation of the Riemann surface of the Bloch solutions was proposed.

Let us consider as an example a nonstationary Schrödinger operator (1.10). The solutions $\psi(x, y, w_1, w_2)$ of the nonstationary Schrödinger equation

$$(\sigma\partial_y - \partial_x^2 + u(x, y))\psi(x, y, w_1, w_2) = 0 \tag{3.7}$$

with a periodic potential $u(x, y) = u(x + a_1, y) = u(x, y + a_2)$ are called the Bloch solutions, if they are eigenfunctions of the monodromy operators, i.e.

$$\psi(x + a_1, y, w_1, w_2) = w_1\psi(x, y, w_1, w_2), \tag{3.8}$$

$$\psi(x, y + a_2, w_1, w_2) = w_2 \psi(x, y, w_1, w_2). \tag{3.9}$$

The Bloch functions will always be assumed to be normalized so that $\psi(0, 0, w_1, w_2) = 1$. The set of pairs $Q = (w_1, w_2)$, for which there exists such a solution is called the Floque set and will be denoted by Γ. The multivalued functions $p(Q)$ and $E(Q)$ such that

$$w_1 = e^{ipa_1}, \qquad w_2 = e^{iEa_2},$$

are called quasi-momentum and quasi-energy, respectively.

The gauge transformation $\psi \to e^{h(y)}\psi$, where $\partial_y h(y)$ is a periodic function, transfers the solutions of (3.7) into solutions of the same equation but with another potential $\tilde{u} = u - \sigma \partial_y h$. Consequently, the spectral sets corresponding to the potentials u and $\tilde{u}$ are isomorphic. Therefore, in what follows we restrict ourselves to the case of periodic potentials such that

$$\int_0^{a_1} u(x, y)\, dx = 0. \tag{3.10}$$

To begin with, let us consider as a basic example the 'free' operator

$$M_0 = \sigma \partial_y - \partial_x^2 \tag{3.11}$$

with zero potential $u(x, y) = 0$. The Floque set of this operator is parametrized by the points of the complex plane of the variable k

$$w_1^0 = e^{ika_1}, \qquad w_2^0 = e^{-\sigma^{-1}k^2 a_2} \tag{3.12}$$

and the Bloch solutions have the form

$$\psi(x, y, k) = e^{ikx - \sigma^{-1}k^2 y}. \tag{3.13}$$

The functions

$$\psi^+(x, y, k) = e^{-ikx + \sigma^{-1}k^2 y} \tag{3.14}$$

are Bloch solutions of the formal ajoint operator

$$(\sigma \partial_y + \partial_x^2)\psi^+ = 0. \tag{3.15}$$

Formulae (3.12) define the map

$$k \in C \longmapsto (w_1^0, w_2^0) \in C^2. \tag{3.16}$$

Its image is the Floque set for the free operator M_0. It is the Riemann surface with self-intersections that correspond to the pairs $k \neq k'$ such that

$$w_i^0(k) = w_i^0(k'), \quad i = 1, 2. \tag{3.17}$$

From (3.12) it follows that

$$k - k' = \frac{2\pi N}{a_1}, \tag{3.18}$$

$$k^2 - (k')^2 = \frac{\sigma 2\pi i M}{a_2}, \tag{3.19}$$

where N and M are integers. Hence, all the resonant points have the form

$$k = k_{N,M} = \frac{\pi N}{a_1} - \frac{\sigma i M a_1}{N a_2}, \quad N \neq 0, \ k' = k_{-N,-M}. \tag{3.20}$$

The basic idea of the construction of the Riemann surface of Bloch solutions of Equation (3.7) that was proposed in [7] is to consider (3.7) as a perturbation of the free operator (3.7), assuming that the potential $u(x,y)$ is formally small.

For any $k_0 \neq k_{N,M}$ it is easy to construct a formal Bloch solution $\widetilde{\psi}$ of the Equation (3.7) as a formal series

$$\widetilde{\psi} = \sum_{s=0}^{\infty} \widetilde{\phi}_s(x,y,k_0), \quad \widetilde{\phi}_0(x,y,k_0) = \psi(x,y,k_0) = \psi_0. \tag{3.21}$$

This series describes a ‘perturbation’ of the Bloch solution ψ_0 of the nonperturbed equation.

LEMMA 3.1. *If $k_0 \neq k_{N,M}$, then there exists a unique formal series*

$$F(y,k_0) = \sum_{s=1}^{\infty} F_s(y,k_0) \tag{3.22}$$

such that the equation

$$(\sigma\partial_y - \partial_x^2 + u(x,y))\Psi(x,y,k_0) = F(y,k_0)\Psi(x,y,k_0) \tag{3.23}$$

has a formal solution of the form

$$\Psi(x,y,k_0) = \sum_{s=0}^{\infty} \phi_s(x,y,,k_0), \quad \phi_0 = \psi_0, \tag{3.24}$$

satisfying the conditions

$$\langle \psi_0^+ \Psi \rangle_x = \langle \psi_0^+ \psi_0 \rangle_x, \qquad \psi_0^+ = \psi^+(x,y,k_0), \tag{3.25}$$

(here and below $\langle f(x) \rangle_x$ stands for the mean value in x of the corresponding periodic function f)

$$\Psi(x+a_1,y,k_0) = w_{10}\Psi(x,y,k_0), \quad w_{10} = w_1^0(k_0), \tag{3.26}$$

$$\Psi(x,y+a_2,k_0)=w_{20}\Psi(x,y,k_0),\quad w_{20}=w_2^0(k_0). \tag{3.27}$$

The corresponding solution is unique and is given by the recursion formulae (3.28)–(3.32).

$$\phi_s=\sum_{n\neq 0}c_n^s(y,k_0)\psi_n(x,y),\quad s>1, \tag{3.28}$$

$$\psi_n=\psi_n(x,y)=\psi(x,y,k_n),$$

$$\psi_n^+=\psi^+(x,y,k_n),\qquad k_n=k_0+\frac{2\pi n}{a_1}, \tag{3.29}$$

$$\begin{aligned} c_n^s(y,k_0) &= \sigma^{-1}\frac{w_{2n}}{w_{20}-w_{2n}}\times \\ &\quad\times\int_y^{y+a_2}\left(\sum_{i=1}^{s-1}F_ic_n^{s-i}-\frac{\langle\psi_n^+u\phi_{s-1}\rangle_x}{\langle\psi_n^+\psi_n\rangle_x}\right)\mathrm{d}y', \end{aligned} \tag{3.30}$$

$$w_{2n}=w_2^0(k_n), \tag{3.31}$$

$$F_s(y,k_0)=\frac{\langle\psi_0^+u\phi_{s-1}\rangle_x}{\langle\psi_0^+\psi_0\rangle_x}. \tag{3.32}$$

From (3.23), (3.26), (3.27), it follows that the formula

$$\widetilde{\psi}(x,y,k_0)=\frac{\Psi(x,y,k_0)}{\Psi(0,0,k_0)}\exp\left\{-\sigma^{-1}\int_0^y F(y',k_0)\,\mathrm{d}y'\right\} \tag{3.33}$$

defines the formal Bloch solution of Equation (3.7):

$$\widetilde{\psi}(x+a_1,y,k_0)=w_{10}\psi(x,y,k_0), \tag{3.34}$$

$$\widetilde{\psi}(x,y+a_2,k_0)=\widetilde{w}_{20}\psi(x,y,k_0), \tag{3.35}$$

where the corresponding Bloch multiplier is equal to

$$\widetilde{w}_{20}=w_{20}\exp\left\{-\sigma^{-1}\int_0^{a_2}F(y',k_0)\,\mathrm{d}y'\right\}. \tag{3.36}$$

For sufficiently small $u(x,y)$, it is not too hard to show that the above constructed series of the perturbation theory converge outside some neighborhoods of the resonant points (3.20) and determine there a function $\widetilde{\psi}(x,y,k_0)$ which is analytical in k_0. This is true for any σ. The principle distinction between the cases $\operatorname{Re}\sigma=0$ and $\operatorname{Re}\sigma\neq 0$ is revealed under an attempt to extend $\widetilde{\psi}$ to a 'resonant' domain. In the case $\operatorname{Re}\sigma = 0$, the resonant points are dense on the real axis. In the case $\operatorname{Re}\sigma \neq 0$, there is only a finite number of the resonant points (3.20) in any finite domain of the complex plane. The discreteness of the resonant points in the last case is crucial for the extension of $\widetilde{\psi}$ to a 'resonant' domain (and for the proof of the approximation theorem).

In the stationary case, when u does not depend on y, the preceding formulae turn into the usual formulae of the perturbation theory of eigenfunctions corresponding to simple eigenvalues. The condition

$$w_{2n} \neq w_{20} \longleftrightarrow k_0 \neq k_{NM} \tag{3.37}$$

is a simple analog of an eigenvalue of an operator. In cases when it is violated, it is necessary to proceed along the same lines as in the perturbation theory of multiple eigenvalues.

As the set of indices corresponding to the resonances, we can take an arbitrary set of integers $I \in Z$ such that

$$w_{2\alpha} \neq w_{2n}, \quad \alpha \in I, \ n \notin I. \tag{3.38}$$

LEMMA 3.2. *There are unique formal series*

$$F^{\alpha}_{\beta}(y, w_1) = \sum_{s=1}^{\infty} F^{\alpha}_{\beta,s}(y, w_1) \tag{3.39}$$

such that the equations

$$(\sigma\partial_y - \partial_x^2 + u)\Psi^{\alpha}(x, y, w_1) = \sum_{\beta} F^{\alpha}_{\beta}(y, w_1)\Psi^{\beta}(x, y, w_1) \tag{3.40}$$

have unique formal Bloch solutions of the form

$$\Psi^{\alpha}(x, y, w_1) = \sum_{s=0}^{\infty} \phi^{\alpha}_s(x, y, w_1), \quad \phi^{\alpha}_0 = \psi_{\alpha} = \psi(x, y, k_{\alpha}). \tag{3.41}$$

$$\Psi^{\alpha}(x + a_1, y, w_1) = w_1 \Psi^{\alpha}(x, y, w_1), \tag{3.42}$$

$$\Psi^{\alpha}(x, y + a_2, w_1) = w_{2\alpha}\Psi^{\alpha}(x, y, w_1). \tag{3.43}$$

such that

$$\langle \psi^{+}_{\beta} \Psi^{\alpha} \rangle_x = \delta_{\alpha,\beta} \langle \psi^{+}_{\alpha} \psi_{\alpha} \rangle_x. \tag{3.44}$$

The corresponding formulae for F^{α}_{β} and Ψ^{α} are the matrix generalization of formulae (3.28), (3.29) (see details in [7]).

Let us define the matrix $T = T^{\alpha}_{\beta}(y, w_1)$ by the equation

$$\sigma\partial_t T + TF = 0, \quad T(0, w_1) = 1. \tag{3.45}$$

The functions

$$\widehat{\Psi}^{\alpha}(x,y,w_1)=\sum_{\beta}T^{\alpha}_{\beta}(y,w_1)\Psi^{\beta}(x,y,w_1) \tag{3.46}$$

are solutions of (3.7). Under the translation by the period in x, they are multiplied by w_1, while under the translation by the period in y, they are transformed as follows

$$\widehat{\Psi}^{\alpha}(x,y+a_2,w_1)=\sum_{\beta}\widehat{T}^{\alpha}_{\beta}(w_1)w_{2\beta}\widehat{\Psi}^{\beta}(x,y,w_1), \tag{3.47}$$

where

$$\widehat{T}^{\alpha}_{\beta}(w_1)=T^{\alpha}_{\beta}(a_2,w_1). \tag{3.48}$$

It is natural to call a finite set of the formal solutions $\widehat{\Psi}^{\alpha}$ *quasi-Bloch*, since it remains invariant under the translation by the periods in x and y.

The characteristic equation

$$R(w_1,\widetilde{w}_2)=\det\left(\widetilde{w}_2\delta_{\alpha,\beta}-\widehat{T}^{\alpha}_{\beta}(w_1)w_{2,\beta}\right)=0 \tag{3.49}$$

is an analog of the 'secular equation' in the ordinary perturbation theory of multiple eigenvalues.

Let $h_{\alpha}(w_1,\widetilde{w}_2)$ be an eigenvector of the matrix $\widehat{T}^{\alpha}_{\beta}(w_1)w_{2,\beta}$ normalized so that

$$\sum_{\alpha}h_{\alpha}(\widetilde{Q})\widehat{\Psi}^{\alpha}(0,0,w_1)=1,\quad \widetilde{Q}=(w_1,\widetilde{w}_2), \tag{3.50}$$

then

$$\widetilde{\psi}(x,y,\widetilde{Q})=\sum_{\alpha}h_{\alpha}(\widetilde{Q})\widehat{\Psi}^{\alpha}(x,y,w_1) \tag{3.51}$$

is the formal Bloch solution of (3.7) with multipliers w_1 and $\widetilde{w}_2$, normalized in the standard way. The last statement means that the Bloch solutions are defined (locally) on the Riemann surface (3.49).

3.1. Structure of the 'global' Riemann surface of Bloch solutions

To begin with, we shall give here an explanation of the structure of the 'global' Riemann surface of Bloch solutions in the case of small u. Let us consider some neighborhoods $R_{N,M}$ and $R_{-N,-M}$ of the resonant pair of the points $k_{N,M}$ and $k_{-N,-M}$, respectively. The function $w_1(k)$ (3.12) identifies them with some neighborhood $\widehat{R}_{N,M}$ of the point $w_1(k_{N,M})=w_1(k_{-N,-M})$ on the complex plane of the variable w_1. The series (3.28)–(3.32) of the nonresonant perturbation theory diverge in $R_{N,M}$ and $R_{-N,-M}$, but it turns out that the series of the Lemma 3.2 converge in $\widehat{R}_{N,M}$ and in this domain define quasi-Bloch solutions of (3.7) which are analytical in w_1. The characteristic Equation (3.49) in this case has the form

$$\widetilde{w}_2^2 - f_1(w_1)\widetilde{w}_2 + f_2(w_1) = 0 \tag{3.52}$$

and defines two-sheet covering $\widetilde{R}_{N,M}$ over $\widehat{R}_{N,M}$ on which the Bloch solutions of (3.7) are defined. The boundary of $\widetilde{R}_{N,M}$ can be naturally identified with the boundaries of $R_{N,M}$ and $R_{-N,-M}$. Hence, the structure (local) of the Riemann surface Γ of the Bloch functions looks as follows. Let us cut out $R_{N,M}$ and $R_{-N,-M}$ from the complex plane and, instead of them, glue a corresponding piece of the Riemann surface $\widetilde{R}_{N,M}$. From the topological point of view, this surgery is a glueing of a 'handle' between two resonant points.

The remarkable thing is that the perturbation approach works even when $u(x,y)$ is not small. Of course, in that case, the estimations of the perturbation theory series are much more complicated. In [7], it was proved that if the potential $u(x,y)$ can be analytically extended into a domain

$$|\mathrm{Im}\ x| < \tau_1, \qquad |\mathrm{Im}\ y| < \tau_2, \tag{3.53}$$

for some τ_1, τ_2, then the perturbation series for the nonresonant case converge outside some central finite domain R_0 and outside $R_{N,M}$ for $k_{N,M} \notin R_0$. Outside R_0, we again have to perform a surgery of the previous type ('glue' handles between $k_{N,M}$ and $k_{-N,-M}$ for $k_{N,M} \notin R_0$). In the central domain R_0, we have to glue some finite genus piece of the corresponding Riemann surface $\widetilde{R}_0$ instead of disc R_0. As a result, we obtain the global Riemann surface Γ of the Bloch solutions of Equation (3.7) with Re $\sigma \neq 0$.

THEOREM 3.3. *If the potential* $u(x,y)$ *of Equation* (3.20) *can be analytically extended into the domain* (3.53), *then the Riemann surface* Γ *of the Bloch solutions of this equation is a result of the above-defined glueing of the three types of 'pieces':*

1°. *A complex plane of the variable* k *without small neighborhoods of the finite or infinite set of points* $k_{N,M}, k_{-N,-M}$ *and without some central domain* $|k| > K_0$*;*

2°. *A set of 'handles'* $\widehat{R}_{|N,M|}$ *that are defined by equations of the form* (3.52) *as the two-sheets covering of the small neighborhoods of the pairs* $k_{N,M}$, $k_{-N,-M}$*;*

3°. *A Riemann surface* $\widetilde{R}_0$ *(with the boundary) of a finite genus* g_0.

The Bloch solutions of (3.7) $\psi(x,y,Q)$, $Q \in \Gamma$, *that are normalized by the condition* $\psi(0,0,Q) = 1$ *are meromorphic on* Γ. *Their poles do not depend on* x,y. *It has one simple pole in each of the domains* $\widehat{R}_{|N,M|}$. *In the domain* $\widehat{R}_0$, *it has* g_0 *poles, where* g_0 *in general positions (when* $\widehat{R}_0$ *is smooth) equals the* *genus of $\widehat{R}_0$). Outside these domains, the function ψ is holomorphic and has no zeros.*

If there is a finite number of handles that are glued then the corresponding curve is compactified by one point and the corresponding Bloch function is the Baker–Akhiezer function on the compactified Riemann surface.

In the case of real and smooth $u(x,y)$ for $\sigma = 1$ the final form of the Floque set can be represented in the following form [7]. Let us fix some finite or infinite subset S of integer pairs $(N > 0,\ M)$. The set of pairs of complex numbers $\pi = \{p_{s,1}, p_{s,2}\}$, where $s \in S$ would be called 'admissible' if

$$\text{Re } p_{s,i} = \frac{\pi N}{a_1}, \qquad |p_{s,i} - k_s| = \mathrm{o}(|k_s|^{-1}), \quad i = 1,2, \tag{3.54}$$

and the intervals $[p_{s,1}, p_{s,2}]$ do not intersect. (Here k_s are resonant points (3.20), $s = (N, M)$.)

Let us define the Riemann surface $\Gamma(\pi)$ for any admissible set π. It is obtained from the complex plane of the variable k by cutting it along the intervals $[p_{s,1}, p_{s,2}]$ and $[-\bar{p}_{s,1}, -\bar{p}_{s,2}]$ and by sewing after that the left side of the first cut with the right side of the second cut and vice versa. After this surgery, for any cut $[p_{s,1}, p_{s,2}]$ there corresponds a nontrivial cycle a_s on $\Gamma(\pi)$.

THEOREM 3.4. *For any real periodic potential $u(x,y)$ which can be analytically extended into some neighborhood of the real values x, y, the Bloch solutions of Equation* (3.7) *with $\sigma = 1$ are parametrized by points Q of the Riemann surface $\Gamma(\pi)$ corresponding to some admissible set π. The function $\psi(x,y,Q)$ which is normalized by the condition $\psi(0,0,Q) = 1$ is meromorphic on Γ and has a simple pole γ_s on each cycle a_s. If the admissible set π contains only a finite number of pairs, then $\Gamma(\pi)$ has finite genus and is compactified by only one point P_1 $(k = \infty)$, in the neighborhood of which the Bloch function ψ has the form* (2.9).

The potentials u for which $\Gamma(\pi)$ has finite genus are called finite-gap potentials and as it follows from the last statement of the theorem, they coincide with the algebraic–geometrical potentials. The following theorem states that the finite-gap potentials are dense in the space of all periodic smooth functions in two variables [7].

THEOREM 3.5. *Each smooth periodic potential $u(x,y)$ of Equation* (3.7) *with* $\mathrm{Re}\,\sigma \neq 0$ *analytically extendable to a neighbourhood of real x, y can be approximated by finite-gap potentials uniformly with any number of derivatives.*

4. Algebraic–Geometrical Perturbation Theory of Integrable Systems

The nonlinear WKB (or Whitham) method for the construction of the asymptotic solutions may be applied for any nonlinear system that has exact quasi-periodic solutions of the form

$$u(x,y,t) = u_0(Ux + Vy + Wt \mid I), \tag{4.1}$$

where $u_0(z_1, \dots, z_g \mid I)$ is a periodic function in variables z_i depending on a set of parameters $I = (I_1, \dots, I_N)$; $U = U(I)$, $V = V(I)$, $W = W(I)$ are vectors depending on the same set of data.

Let us consider the asymptotical solutions of the same equation of the form

$$\begin{aligned} u(x,y,t) &= u_0(\varepsilon^{-1}S(X,Y,T) \mid I(X,Y,T)) + \\ &\quad + \varepsilon u_1(x,y,t) + \varepsilon^2 u_2(x,y,t), \end{aligned} \tag{4.2}$$

where $X = \varepsilon x,\ Y = \varepsilon y,\ T = \varepsilon t$ are 'slow variables'.

Remark. Parameters I are integrals of the corresponding nonlinear partial differential equation. As usual, in the perturbation theory the 'integrals' of the initial equation become functions of the slow variables. We would like to emphasize that for partial differential equations one of the possible types of perturbation (that does not exist in the finite-dimensional case of the classical Hamiltonian integrable systems) is not the perturbation of the equation by itself, but the perturbation of a class of initial or boundary conditions. The construction of the asymptotic solutions of the form (4.2) corresponds to the perturbation when instead of a class of periodic or quasi-periodic functions one considers functions with slowly modulated periods.

If the vector $S(X,Y,T)$ satisfies the relations

$$\partial_X S = U(I(X,Y,T)) = U(X,Y,T),$$

$$\partial_Y S = V(X,Y,T), \qquad \partial_T S = W(X,Y,T), \tag{4.3}$$

then the main term u_0 in the expansion (4.2) satisfies the initial equation up to the first order in ε. After that, all the other terms of (4.2) are defined from the nonhomogeneous linear equations. They can be easily solved if a full set of solutions for a homogeneous linear equation are known.

The asymptotic solutions of the form (4.2) can be constructed with an arbitrary dependence of the parameters I_k on the slow variables. In this case, the expansion (4.2) will be valid on a scale of order 1. The right-hand side of the nonhomogeneous linear equation for u_1 contains the first derivatives of the parameters I_k. Therefore, the choice of the dependence of I_k on slow variables can be used for the cancellation of the 'secular' term in u_1.

Let us show briefly the way to derive necessary conditions for the existence of the asymptotic solutions of the form (4.2) with uniformly bounded first-order term (u_1) for an equation that has a zero-curvature representation of the form (1.3).

Let L_0, A_0 be the linear operators corresponding to an exact algebraic–geometrical solution of the corresponding equation. We consider the asymptotic solutions of (1.3)

$$L = L_0 + \varepsilon L_1 + \varepsilon^2 L_2 + \cdots, \qquad A = A_0 + \varepsilon A_1 + \varepsilon^2 A_2 + \cdots, \tag{4.4}$$

assuming that the algebraic–geometrical data that define L_0, A_0 are functions of the slow variables. The corresponding dependence of L_0, A_0 defines the right-hand side $F(L_0, A_0)$ of the nonhomogeneous equation

$$\partial_t L_1 - \partial_y A_1 + [L_1, A_0] + [L_0, A_1] = F(L_0, A_0). \tag{4.5}$$

for the first-order terms of the asymptotic solution (4.4). (The left-hand side of (4.5) is a linearization of (1.3) on the background of the exact solution L_0, A_0.)

Let ψ be a solution of the auxiliary linear problems

$$(\partial_y - L_0)\psi = 0, \qquad (\partial_t - A_0)\psi = 0, \tag{4.6}$$

and ψ^+ be a solution of the adjoint equations

$$\psi^+(\partial_y - L_0) = 0, \qquad \psi^+(\partial_t - A_0) = 0, \tag{4.7}$$

then (4.5), (4.6) imply that

$$\partial_t(\psi^+ L_1 \psi) - \partial_y(\psi^+ A_1 \psi) - (\psi^+ F \psi) = \partial_x(\psi^+ \mathrm{D} \psi), \tag{4.8}$$

where D is a differential operator with coefficients that are differential polynomials in the coefficients of the operators L_0, L_1, A_0, A_1.

The equality (4.8) implies that if the product $\psi^+\psi$ is quasi-periodic in the variables x, y, t, then

$$\langle(\psi^+ F \psi)\rangle_{x,y,t} = 0, \tag{4.9}$$

where $\langle\cdot\rangle_{x,y,t}$ stands for the mean value in the variables x, y, t of the corresponding function.

In [8] it was found that (4.9) and the compatibility conditions of (4.3) are equivalent to a well-defined system of partial differential equations on the moduli space of curves with punctures and fixed local coordinates in the neighborhoods of these punctures.

4.1. Whitham Equations

To begin with, we shall give an algebraic form of the 'universal' Whitham hierarchy (see [14]). Let $\Omega_A(k, T)$ be a set of holomorphic functions of the variable k (which is defined in some complex domain $\mathcal{D}$), depending on a finite or infinite number of variables t_A, $T = \{t_A\}$. (We keep the same notation t_A

for slow variables εt_A because we are not going to consider 'fast' variables in this section any more.) Let us introduce a one-form

$$\omega = \sum_A \Omega_A(k,T)\,\mathrm{d}t_A, \tag{4.10}$$

onto the space with coordinates (k, t_A). Its full external derivative equals

$$\delta\omega = \sum_A \delta\Omega_A(k,T) \wedge \mathrm{d}t_A, \tag{4.11}$$

where

$$\delta\Omega_A = \partial_k\Omega_A\,\mathrm{d}k + \sum_B \partial_B\Omega_A\,\mathrm{d}t_B, \quad \partial_k = \partial/\partial k,\ \partial_A = \partial/\partial t_A. \tag{4.12}$$

The equation

$$\delta\omega \wedge \delta\omega = 0 \tag{4.13}$$

we shall call, by definition, *the Whitham hierarchy*.

The 'algebraic' form (4.13) of the Whitham equations is equivalent to a set of partial differential equations that have to be fulfilled for any triple A, B, C

$$\sum_{\{A,B,C\}} \varepsilon^{\{A,B,C\}}\partial_A\Omega_B\partial_k\Omega_C = 0 \tag{4.14}$$

(summation in (4.14) is taken over all permutations of indices A, B, C and $\varepsilon^{\{A,B,C\}}$ is a sign of permutation).

Equations (4.13) are invariant with respect to an invertable change of variable

$$k = k(p,T), \qquad \partial_p k \neq 0. \tag{4.15}$$

Let us fix an index A_0 and denote the corresponding function by

$$p(k,T) = \Omega_{A_0}(k,T). \tag{4.16}$$

At the same time, we introduce a special notation for the corresponding 'time'

$$t_{A_0} = x. \tag{4.17}$$

After that, all Ω_A can be considered as functions of the new variable p, $\Omega_A = \Omega_A(p,T)$. Equations (4.14) for A, B, $C = A_0$ after this change of variable k have the form

$$\partial_A\Omega_B - \partial_B\Omega_A + \{\Omega_A, \Omega_B\} = 0, \tag{4.18}$$

where $\{f, g\}$ stands for the usual Poisson bracket on the space of functions of the two variables x, p

$$\{f, g\} = \partial_x f \partial_p g - \partial_x g \partial_p f. \tag{4.19}$$

The Whitham equations were obtained in [8] in the form (4.14). In [7], it was found that they can be represented in the algebraic form (4.13). (We would like to mention here the papers [15, 16] where it was shown that the algebraic form of the Whitham equations leads directly to the semiclassical limit of 'string' equations.)

The Whitham equations in the form (4.14) are equations on the set of functions $\Omega_A(p,T)$ and give but a certain 'shape' that has to be filled with a real content. It is necessary to show that they do define correct systems of equations on the moduli spaces $\widehat{M}_{g,N}$ of smooth algebraic curves Γ_g of genus g with local coordinates $k_\alpha^{-1}(P)$ in neighborhoods of N punctures P_α ($k_\alpha^{-1}(P_\alpha)=0$)

$$\widehat{M}_{g,N}=\{\Gamma_g,\ P_\alpha,\ k_\alpha^{-1}(P),\ \alpha=1,\ldots,N\}. \tag{4.20}$$

Let us consider as the first example the zero-genus case ($g=0$). In this case, a point of the 'phase space' $\widehat{M}_{g=0,N}$ is a set of points p_α, $\alpha=1,\ldots,N$, and a set of formal local coordinates $k_\alpha^{-1}(p)$:

$$k_\alpha(p)=\sum_{s=-1}^{\infty} v_{\alpha,s}(p-p_\alpha)^s \tag{4.21}$$

('formal local coordinate' means that the r.h.s of (4.21) is considered as a formal series without any assumption of its convergency). Hence, $\widehat{M}_{0,N}$ is a set of sequences

$$\widehat{M}_{0,N}=\{p_\alpha,\ v_{\alpha,s},\ \alpha=1,\ldots,N,\ s=-1,0,1,2,\ldots\}. \tag{4.22}$$

The Whitham equations define a dependence of points of $\widehat{M}_{0,N}$ with respect to the variables t_A, where the set of indices $\mathcal{A}$ is as follows

$$\begin{aligned}\mathcal{A} = \{&A=(\alpha,i),\ \alpha=1,\ldots,N,\ i=1,2,\ldots\\ &\text{and for } i=0,\ \alpha\neq 1\}.\end{aligned} \tag{4.23}$$

As it was explained above, we can fix one of the points p_α with the help of an appropriate change of the variable p. Let us choose $p_1=\infty$.

Let us introduce meromorphic functions $\Omega_{\alpha,i}(p)$ for $i>0$ with the help of the following conditions:

$\Omega_{\alpha,i>0}(p)$ has a pole only at p_α and coincides with the singular part of an expansion of $k_\alpha^i(p)$ near this point, i.e.

$$\Omega_{\alpha,i}(p)=\sum_{s=1}^{i} w_{\alpha,i,s}(p-p_\alpha)^{-s}=k_\alpha^i(p)+O(1),$$

$$\Omega_{\alpha,i}(\infty)=0,\quad \alpha\neq 1. \tag{4.24}$$

$$\Omega_{1,i}(p) = \sum_{s=1}^{i} w_{1,i,s} p^s = k_1^i(p) + \mathrm{O}(k_1^{-1}). \tag{4.25}$$

These polynomials can be written in the form of the Cauchy integrals

$$\Omega_{\alpha,i}(p,T) = \frac{1}{2\pi i} \oint_{C_\alpha} \frac{k_\alpha^i(z_\alpha, T)\,\mathrm{d}z_\alpha}{p - z_\alpha}. \tag{4.26}$$

Here C_α is a small cycle around the point p_α.

The functions $\Omega_{\alpha,i=0}(p)$, $\alpha \neq 1$, are equal to

$$\Omega_{\alpha,0}(p) = -\ln(p - p_\alpha). \tag{4.27}$$

Remark. The asymmetry of the definitions of $\Omega_{\alpha,i}$ reflects our intention to choose the index $A_0 = (1,1)$ as a 'marked' index.

The coefficients of $\Omega_{\alpha,i>0}(p)$ are polynomial functions of $v_{\alpha,s}$. Therefore, the Whitham equations (4.12) (or (4.18)) can be rewritten as equations on $\widehat{M}_{0,N}$. But still it has to be shown that they can be considered as a correctly defined system.

THEOREM 4.1. *The zero-curvature form* (4.18) *of the Whitham hierarchy in the zero-genus case is equivalent to the Sato-form that is a compatible system of evolution equations*

$$\partial_A k_\alpha = \{k_\alpha, \Omega_A\}. \tag{4.28}$$

Let us demonstrate a few examples.

EXAMPLE 1 (Khokhlov–Zabolotskaya hierarchy). The Khokhlov–Zabolotskaya hierarchy is the particular $N = 1$ case of our considerations. Any local coordinate $K^{-1}(p)$ near the infinity ($p_1 = \infty$)

$$K(p) = p + \sum_{s=1}^{\infty} v_s p^{-s} \tag{4.29}$$

defines a set of polynomials:

$$\Omega_i(p) = [K^i(p)]_+, \tag{4.30}$$

here $[\ldots]_+$ denotes a nonnegative part of Laurent series. For example,

$$\Omega_2 = k^2 + u, \quad \Omega_3 = k^3 + \tfrac{3}{2} uk + w, \tag{4.31}$$

where $u = 2v_1$, $w = 3v_2$. If we denote $t_2 = y$, $t_3 = t$, then Equation (4.18) for $A = 2,\ B = 3$ gives

$$w_x = \tfrac{3}{4} u_y, \qquad w_y = u_t - \tfrac{3}{2} uu_x, \tag{4.32}$$

from which the dispersionless KP (dKP) equation (Khokhlov–Zabolotskaya equation) is derived:

$$\tfrac{3}{4}\, u_{yy} + \left(u_t - \tfrac{3}{2}\, u u_x\right)_x = 0. \tag{4.33}$$

The Khokhlov–Zabolotskaya equation is a partial differential equation and though it has no pure evolution form, one can expect that its solutions are to be uniquely defined by their Cauchy data $u(x, y, t = 0)$, that is a function of the two variables x, y. Up to now, it is not clear if this two-dimensional equation can be considered as the third equivalent form of the Whitham hierarchy (we recall that solutions of the hierarchy (4.28) formally depend on an infinite number of functions of one variable).

EXAMPLE 2 (longwave limit of 2D Toda lattice). The hierarchy of the longwave limit of a two-dimensional Toda equation is the particular $N = 2$ case of our considerations. There are two local parameters. One of them is near the infinity $p_1 = \infty$ and one is near a point $p_2 = a$. They depend on two sets of variables $t_{\alpha,s}$, $\alpha = 1, 2$, $s = 1, 2, \ldots$, and also on the variable t_0. We shall present here only the basic two-dimensional equation of this hierarchy (an analogue of the Khokhlov–Zabolotskaya equation).

Consider three variables $t = t_0$, $x = t_{1,1}$, $y = t_{2,1}$. The corresponding functions are

$$\Omega_0 = \ln(p - a), \qquad \Omega_{1,1} = p, \qquad \Omega_{2,1} = \frac{v}{p - a}. \tag{4.34}$$

Their substitution into the zero-curvature Equation (4.18) gives

$$v_x = a_t v, \qquad v_t + a_y = 0, \qquad w_t = 0. \tag{4.35}$$

From (4.35), it follows that

$$\partial^2_{xy}\phi + \partial^2_t e^{\phi} = 0, \tag{4.36}$$

where $\phi = \ln v$. This is the longwave limit of the 2D Toda lattice equation

$$\partial^2_{xy}\varphi_n = e^{\varphi_{n-1} - \varphi_n} - e^{\varphi_n - \varphi_{n+1}} \tag{4.37}$$

corresponding to the solutions that are slow functions of the discrete variable n, which is replaced by the continuous variable t. Equation (4.36) has arisen independently in general relativity, the theory of wave phenomena in shallow water, long radio-relay lines, and so on. A bibliography can be found in [17] where a representation of solutions of (4.36) in terms of convergent series was proposed.

EXAMPLE 3 (N-layer solutions of the Benny equation). This example corresponds to a general $N+1$ points case, but we consider only one zero-curvature equation. Let us choose three functions

$$\Omega_1 = p, \qquad \Omega_2 = p + \sum_{i=1}^{N} \frac{v_i}{p - p_i}, \qquad \Omega_3 = p^2 + u, \tag{4.38}$$

which are coupled with the variables x, y, t, respectively. In our standard notations, they are

$$x = t_{1,1}, \qquad y = \sum_{\alpha=1}^{N+1} t_{\alpha,1}, \qquad t = t_{1,2}. \tag{4.39}$$

The zero-curvature Equation (4.18) gives the system

$$p_{it} - (p_i^2)_x + u_x = 0, \qquad v_{it} = 2(v_i p_i)_x,$$

$$u_y - u_x + 2\sum_i v_{ix} = 0. \tag{4.40}$$

Solutions of this system that do not depend on y are N-layer solutions of the Benny equation. As it was noticed in [18], the corresponding system

$$p_{it} - (p_i^2)_x + u_x = 0, \qquad v_{it} = 2(v_i p_i)_x, \qquad u = 2\sum_i v_i \tag{4.41}$$

is a classical limit of the vector nonlinear Schrödinger equation

$$i\psi_{it} = \psi_{i,xx} + u\psi_i, \qquad u = \sum_i |\psi_i|^2. \tag{4.42}$$

(Using this observation in [18], the integrals of (4.41) were found.)

In the case of the arbitrary genus curves, the *basic* Whitham hierarchy is generated by Equations (4.13) or (4.18), where the set of indices $\mathcal{A}$ is the same as in genus zero case. The corresponding functions for $A = (\alpha, i > 0)$ are defined by (2.14)–(2.16). The functions $\Omega_{\alpha,0}$ are integrals of the normalized Abelian differentials $\mathrm{d}\Omega_{\alpha,0}$, $\alpha \neq 1$, of the third kind with simple poles at the points P_1 and P_α with residues 1 and -1, respectively,

$$\mathrm{d}\Omega_{\alpha,0} = \mathrm{d}k_\alpha (k_\alpha^{-1} + \mathrm{O}(k_\alpha^{-1})),$$

$$\mathrm{d}\Omega_{\alpha,0} = -\mathrm{d}k_1 (k_1^{-1} + \mathrm{O}(k_1^{-1})); \tag{4.43}$$

Remark. It should be mentioned here that the corresponding system of equations is defined on the covering $\widehat{M}^*_{g,N}$ of the moduli space $\widehat{M}_{g,N}$ that corresponds to the choice of canonical basis of cycles on the curve. In order to have the system of equations that are defined on $\widehat{M}_{g,N}$ but not on its covering, it is necessary to change the normalization conditions (2.15) by the conditions that do not depend on the choice of basic cycles. The corresponding conditions are

$$\operatorname{Im} \oint_c \mathrm{d}\Omega_A = 0, \quad c \in H_1(\Gamma_g, Z). \tag{4.44}$$

This normalization conditions were obtained for the Whitham hierarchy in [8]. Below, for the simplicity, we consider the Whitham hierarchy on $\widehat{M}^*_{g,N}$, only.

4.2. Algebraic orbits and exact solutions of the Whitham equations

In [7] a construction of exact solutions of the Whitham equations corresponding to 'algebraic orbits' was proposed. (This construction is a generalization and effectivization of the scheme [19] where 'generalized godograph' transformation was proposed for hydrodynamical-type diagonalizable Hamiltonian systems; see [20–22].)

Let $\Omega_A(k, T)$ be a solution of the general zero-curvature Equation (4.18)

$$\partial_A \Omega_B - \partial_B \Omega_A + \{\Omega_A, \Omega_B\} = 0. \tag{4.45}$$

They are the compatibility conditions for the equations

$$\partial_A E = \{E, \Omega_A\}, \tag{4.46}$$

Therefore, an arbitrary function $E(p, x)$ defines (at least locally) the corresponding solution $E(p, T)$ of (4.46), $E(p, x) = E(p,\ t_{A_0} = x,\ t_A = 0, A \neq A_0)$. In the domain where $\partial_p E(p, T) \neq 0$, we can use a variable E as a new coordinate, $p = p(E, t)$. The derivatives for fixed E and p are interrelated with the help of the following formula

$$\partial_A F(p, t) = \partial_A f(E, t) + \frac{\mathrm{d}f}{\mathrm{d}E} \partial_A E(p, t). \tag{4.47}$$

Therefore, in the new coordinate, Equations (4.45) are equivalent to the equations

$$\partial_A \Omega_B(E, T) = \partial_B \Omega_A(E, T). \tag{4.48}$$

Hence, there exists a potential $S(E, T)$ such that

$$\Omega_A(E, T) = \partial_A S(E, T). \tag{4.49}$$

Due to this potential, the one-form ω (4.10) can be represented as

$$\omega = \delta S(E, T) - Q(E, T)\, \mathrm{d}E, \tag{4.50}$$

where

$$Q(E,T) = \frac{\partial S(E,T)}{\partial E}. \tag{4.51}$$

Hence,

$$\delta\omega = \delta E \wedge \delta Q. \tag{4.52}$$

Formula (4.52) implies that the functions E and Q as functions of two variables p, x satisfy the classical string equation

$$\{Q, E\} = 1. \tag{4.53}$$

They show that

$$\partial_A Q = \{Q, \Omega_A\}, \tag{4.54}$$

also.

A set of the pairs of functions $Q(p,x)$, $E(p,x)$ satisfying the string equation is a group with respect to the composition, i.e. if $Q(p,x)$, $E(p,x)$ and $Q_1(p,x)$, $E_1(p,x)$ are solutions of (4.53) then the functions

$$\widetilde{Q}(p,x) = Q_1(Q(p,x),\ E(p,x)); \quad \widetilde{E}(p,x) = E_1(Q(p,x), E(p,x)) \tag{4.55}$$

are a solution of (4.53), as well. The Lie algebra of this group is the algebra $\mathrm{SDiff}(T^2)$ of two-dimensional vector-fields preserving an area. The action of this algebra on the potential, τ-function (and so on) within the framework of the longwave limit of a 2D Toda lattice was considered in [15, 16].

By definition, the 'algebraic orbits' of the Whitham hierarchy are solutions such that there exists a 'global' solution of Equation (4.46). In the case of genus-zero Whitham equations, 'global' means that $E(p,T)$ *is a meromorphic solution of Equations* (4.46) *such that*

$$\{E(p,T), k_\alpha(p,T)\} = 0. \tag{4.56}$$

The last equality implies that there exist functions $f_\alpha(E)$ of one variable such that

$$k_\alpha(p,T) = f_\alpha(E(p,T)). \tag{4.57}$$

In order not to be lost in a too general setting right at the beginning, let us start with an example.

EXAMPLE (Lax reductions, $N = 1$). Consider solutions of the dKP hierarchy such that some power of local parameter (4.29) is a polynomial in p, i.e.

$$E(p,T) = p^n + u_{n-2}p^{n-2} + \cdots + u_0 = K^n(p,T). \tag{4.58}$$

The relation (4.58) implies that only a few first coefficients of the local parameter are independent. All of them are polynomials with respect to the coefficients u_i of the polynomial $E(p,T)$. The corresponding solutions of dKP hierarchy can be described in terms of dispersionless Lax equations

$$\partial_i E(p,T) = \{E(p,T), \Omega_i(p,T)\}, \tag{4.59}$$

where

$$\Omega_i(p,T) = \left[E^{i/n}(p,T)\right]_+ \tag{4.60}$$

(as before, $[\dots]_+$ denotes a nonnegative part of corresponding Laurent series). These solutions of KP hierarchy can be also characterized by the property that they do not depend on the variables $t_n, t_{2n}, t_{3n}, \dots$.

The construction of the exact solutions of the dispersionless Lax equations can be presented in the following form. For each polynomial $E(p)$ of the form (4.58) and each formal series

$$Q(p) = \sum_{j=1}^{\infty} b_j p^j, \tag{4.61}$$

the formula

$$t_i = \frac{1}{i} \operatorname{res}_\infty \left(K^{-i}(p) Q(p)\, \mathrm{d}E(p)\right) \tag{4.62}$$

defines the variables

$$t_k = t_k(u_i, b_j), \quad i = 0, \dots, n-2,\ j = 0, \dots . \tag{4.63}$$

as functions of the coefficients of E, Q. Consider the inverse functions

$$u_i = u_i(t_1, \dots), \qquad b_j = b_j(t_1, \dots). \tag{4.64}$$

Remark. In order to be more precise, let us consider a case when Q is a polynomial, i.e. $b_j = 0, j > m$. From (4.62), it follows that $t_k = 0, k > n+m-1$. Therefore, we have $n+m-1$ 'times' t_k, $k = 1, \dots, n+m-1$, that are linear functions of b_j, $j = 1, \dots, m$, and polynomials in u_i, $i = 0, \dots, n-2$. So, locally the inverse functions (4.64) are well defined.

THEOREM 4.2. *The functions $u_i(T)$ are solutions of dispersionless Lax Equation* (4.59). *Any other solutions of* (4.59) *are obtained from this particular one with the help of translations, i.e.* $\tilde{u}(t_i) = u(t_i - t_i^0)$.

The construction of solutions of the dispersionless Lax equations that was presented above manifest the general form of the construction of exact solutions of universal Whitham hierarchy on all spaces $\widehat{M}^*_{g,N}$.

The key elements of this construction are as follows:

1. First of all there is the 'big phase space', that is the moduli space

$$\widetilde{\mathcal{N}}_g(n_\alpha) = \{\Gamma_g,\ \mathrm{d}Q,\ \mathrm{d}E,\ a_i, b_i \in H_1(\Gamma_g, Z)\} \tag{4.65}$$

of curves with a fixed canonical basis of cycles, with a fixed normalized meromorphic differential $\mathrm{d}E$ having poles of orders $n_\alpha + 1$ at points P_α and with a fixed normalized differential $\mathrm{d}Q(P)$ that is holomorphic outside the punctures.

2. The second step is the definition of a set of functions on this moduli space. They are given by the formulae

$$\begin{aligned} t_{\alpha,i} &= \frac{1}{i}\operatorname{res}_{P_\alpha}\left(k_\alpha^{-i}(p)Q(p)\,\mathrm{d}E(p)\right), \quad i > 0,\ k_\alpha^{n_\alpha}(p) = E(p); \\ t_{\alpha,0} &= \operatorname{res}_{P_\alpha}\left(Q(p)\,\mathrm{d}E(p)\right) \end{aligned} \tag{4.66}$$

and the formulae

$$t_{h,i} = \oint_{a_i} \mathrm{d}S, \quad i = 1, \ldots, g, \qquad \mathrm{d}S = Q\,\mathrm{d}E, \tag{4.67}$$

$$t_{Q,i} = -\oint_{b_i} \mathrm{d}E, \qquad t_{E,i} = \oint_{b_i} \mathrm{d}Q, \quad i = 1, \ldots, g. \tag{4.68}$$

Let us introduce the differentials that are 'coupled' to the new type of the variables t_A. Namely, to the variables $t_{h,k}$, $t_{Q,k}$, $t_{E,k}$.

1. $\mathrm{d}\Omega_{h,k}$ is normalized

$$\oint_{a_i} \mathrm{d}\Omega_{h,k} = \delta_{ik} \tag{4.69}$$

holomorphic differentials.

2. The differentials $\mathrm{d}\Omega_{E,i}$, $\mathrm{d}\Omega_{Q,i}$ are holomorphic on the curve Γ_g everywhere except for the a-cycles where they have 'jumps'. Their boundary values on a_j cycle satisfy the relations

$$\mathrm{d}\Omega_{E,i}^+ - \mathrm{d}\Omega_{E,i}^- = \delta_{ij}\,\mathrm{d}E, \tag{4.70}$$

$$\mathrm{d}\Omega_{Q,i}^+ - \mathrm{d}\Omega_{Q,i}^- = \delta_{ij}\,\mathrm{d}Q, \tag{4.71}$$

These differentials are uniquely normalized by the conditions

$$\oint_{a_i} \mathrm{d}\Omega_{E,j} = \oint_{a_i} \mathrm{d}\Omega_{Q,j} = 0. \tag{4.72}$$

Let

$$\mathcal{D}_g(n_\alpha) \subset \widetilde{\mathcal{N}}_g(n_\alpha) \tag{4.73}$$

be an open subset of the big phase space such that the corresponding differentials $\mathrm{d}E$, $\mathrm{d}Q$ have no common zeros on Γ_g.

THEOREM 4.3. *The map*

$$\mathcal{D}_g(n_\alpha) \longmapsto (T = \{t_A\}) \tag{4.74}$$

that is defined by formulae (4.66)–(4.68) *is nondegenerate, i.e. the set of functions* $\{t_A\}$ *defines a system of coordinates on* $\mathcal{D}_g(n_\alpha)$. *The corresponding dependence of the curve* $\Gamma_g(T)$ *and the differential* $\mathrm{d}E(T)$ *with respect to these coordinates define a solution of the universal Whitham hierarchy on the moduli space* $\widehat{M}^*_{g,N}$.

In other words, the Whitham hierarchy can be considered as a way to define the special system of coordinates on the moduli space of curves with punctures and with jets of local coordinates in neighborhoods of the punctures. The n_α-jets of the local coordinates are the equivalence classes of the coordinates. Two local coordinates $k_{1,\alpha}(P)^{-1}$ and $k_\alpha(P)$ are equivalent iff

$$k_{1,\alpha}(P) = k_\alpha(P) + \mathrm{O}(k_\alpha(P)^{-n_\alpha - 1}). \tag{4.75}$$

The formula

$$k_\alpha^{n_\alpha}(P) = E(P), \tag{4.76}$$

where $E(P)$ is the integral of the differential $\mathrm{d}E$ defines locally a map from the moduli space of $\{\Gamma_g, \mathrm{d}E\}$ of curves with fixed normalized meromorphic differentials of the second kind with poles of orders $n_\alpha + 1$ at points P_α to the moduli space $\widehat{M}^*_{g,N}$. This maps is a local isomorphism with the moduli space $M_{g,N}(n_\alpha)$ of curves with punctures and with fixed n_α jets of local coordinates.

4.3. Topological Minimal Models

In [23], it was notice that the calculations of [25] of the perturbed primary rings of the so-called topological minimal A_n-models can be identified with the construction of the solutions of the first n equations of the dispersionless Lax hierarchy (4.59). This fact made it possible to include the corresponding deformations of the primary rings into a hierarchy of an infinite number of commuting flows. The calculations in [25] of partition function for perturbed A_n models gave an impulse for the introduction a τ-function for dispersionless Lax equation. The corresponding results were generalized in [14, 24, 26, 27].

Two- and three-points correlation functions

$$\langle \phi_\alpha \phi_\beta \rangle = \eta_{\alpha\beta}, \qquad c_{\alpha\beta\gamma} = \langle \phi_\alpha \phi_\beta \phi_\gamma \rangle \tag{4.77}$$

of any topological field theory with N primary fields $\phi_1, \ldots, \phi_N$ define associative algebra

$$\phi_\alpha \phi_\beta = c^\gamma_{\alpha\beta} \phi_\gamma, \qquad c^\gamma_{\alpha\beta} = c_{\alpha\beta\mu} \eta^{\gamma\mu}, \qquad \eta_{\alpha\mu} \eta^{\mu\beta} = \delta^\beta_\alpha, \tag{4.78}$$

with a unit ϕ_1

$$\eta_{\alpha\beta} = c_{1\alpha\beta}. \tag{4.79}$$

It turns out that there exists N parametric deformations of the theory such that 'metric' $\eta_{\alpha\beta}$ is a constant and three-point correlators are given by the derivatives of free energy $F(t)$ of the deformed theory

$$c_{\alpha\beta\gamma}(t) = \partial_{\alpha\beta\gamma}F(t), \qquad \eta_{\alpha\beta} = \partial_{1\alpha\beta}F(t) = \text{const}. \tag{4.80}$$

Such deformations are called *potential Fröbenius deformations*. (We recall that the commutative associative algebra with the bilinear form $\langle\cdot,\cdot\rangle$ that is compatible with multiplication, i.e. $\langle ab, c\rangle = \langle a, bc\rangle$ is called the Fröbenius algebra.)

The associativity conditions of algebra (4.78) with structural constants (4.80) are equivalent to a system of partial differential equations on F (WDVV equations). In [28], it was noticed that these equations are equivalent to the equations

$$[\partial_\alpha - \lambda C_\alpha(t), \partial_\beta - \lambda C_\beta(t)] = 0, \tag{4.81}$$

where $C_\alpha(t)$ is a matrix with entries $c^\gamma_{\alpha\beta}$. In other words, the structure constants of the algebra of deformed topological theories define a flat connection on the space of deformed theories with a spectral parameter λ. The coupling constant are flat coordinates for the space of deformed theory.

The topological minimal A_n models are particular examples of such models. Topological states for twisted $N = 2$ superconformal Landau–Ginsburg models are defined by the superpotential E. In case of the A_{n-1} model, the deformed potential has the form

$$E(p) = p^n + u_{n-2}p^{n-2} + \cdots + u_0. \tag{4.82}$$

The coefficients u_i can be considered as coordinates on the space of the deformed topological minimal models. Each polynomial E defines a factor ring

$$\mathcal{R}_E = \frac{C[p]}{E' = 0}, \qquad E'(p) = \frac{\partial E}{\partial p}. \tag{4.83}$$

The formula

$$\langle f, g\rangle = \sum_{q_s} \text{res}_{q_s}\left(\frac{f(p)g(p)}{E_p}\,dp\right), \quad f(p),\ g(p) \subset \mathcal{R}_E, \tag{4.84}$$

defines a nondegenerate bilinear form on $\mathcal{R}_E$, i.e. supplies $\mathcal{R}_E$ with the structure of the Fröbenius algebra.

In [25], it was shown that the functions $t_i(u_0, \ldots, u_{n-2})$, $i = 1, \ldots, n-1$, that are defined by formula (4.62) for $Q(p) = p$ are flat coordinates for the bilinear form (4.84) and define the potential Fröbenius deformation of the algebra (4.83).

THEOREM 4.4 ([23]). *Let $E(p, t_1, t_2, \ldots, t_{n-1}, t_{n+1}, \ldots)$ be the solution of the dispersionless Lax hierarchy which was constructed above. Then the superpotential of the perturbed A_{n-1} topological minimal model is equal to*

$$W(p, t_1, \ldots, t_{n-1}) = E(p, t_1, \ldots, t_{n-1}, 1, 0, 0, \ldots). \tag{4.85}$$

The free energy of this model is equal to

$$F = F(t_1, t_2, \ldots, t_{n-1}, 1, 0, 0, \ldots), \tag{4.86}$$

where $F(t_1, t_2, \ldots.)$ is given by the formula

$$F(T) = \operatorname{res}_\infty\left(\sum_{i=1} t_i K^i \, \mathrm{d}S\right), \quad \mathrm{d}S(p) = Q(p)\,\mathrm{d}E(p). \tag{4.87}$$

The function $\tau(T) = \exp(F(T))$ is the τ-function of the dispersionless Lax hierarchy. The general definition of such a function for algebraic orbits of the universal Whitham hierarchy for all genera is given by the formula

$$\ln \tau = \int_{\Gamma_g} \bar{\mathrm{d}}S \wedge \mathrm{d}S, \quad \mathrm{d}S = Q\,\mathrm{d}E. \tag{4.88}$$

Important remark. The integral (4.88) is not equal to zero, because $S(p, T)$ is holomorphic on Γ except for the punctures P_α and some contours, where it has 'jumps'. Therefore, the integral over Γ_g equals to a sum of the residues at P_α and the contour integrals of the corresponding one-forms.

The τ-function is a function of the variables t_A, $\tau = \tau(T)$. As it was shown in [14] it contains full information about the corresponding solutions $\Omega_A(p, T)$ of the Whitham equations. For $g > 0$ in τ a geometry of the moduli spaces is encoded.

At the end of this section we present the results of [14] where it was shown that the open domain $\mathcal{N}_g(n_\alpha)$ of the 'small' phase space

$$\mathcal{N}_g(n_\alpha) = \{\Gamma_g, \ \mathrm{d}E, \ a_i, b_i \in H_1(\Gamma_g, Z)\} \tag{4.89}$$

that is the moduli space of marked smooth genus g algebraic curves with fixed normalized Abelian differential $\mathrm{d}E$ which has poles of orders $n_\alpha + 1$ at some points P_α, is a Fröbenius foliation, i.e. a smooth foliation with leaves such that the fibers of the tangent bundle has a structure of the Fröbenius algebra and they form the potential deformation of these algebras.

The leaves $\mathcal{N}_g^V(n_\alpha)$ of the foliation are defined by the periods of $\mathrm{d}E$

$$\oint_{b_k} \mathrm{d}E = V_k, \tag{4.90}$$

and are parametrized by constants $V = (V_k)$. Let us us denote the normalized differential with the only pole at the point P_1 of the form (4.91) by $\mathrm{d}p$;

$$\mathrm{d}p = \mathrm{d}(E^{1/n_1} + \mathrm{O}(1)). \tag{4.91}$$

From Theorem 4.3, it follows that the functions

$$t_{\alpha,i} = \frac{1}{i}\operatorname{res}_{P_\alpha}\left(k_\alpha^{-i}(p)p\,\mathrm{d}E(p)\right), \qquad \alpha=1,\ i=1,\dots,n_1-1,\quad \alpha>1,\ i=1,\dots,n_\alpha, \tag{4.92}$$

$$t_{\alpha,0} = \operatorname{res}_\alpha(p\,\mathrm{d}E(p)); \tag{4.93}$$

$$t_{h,k} = \oint_{a_k} p\,\mathrm{d}E, \quad k=1,\dots,g; \tag{4.94}$$

$$t_{E,k} = \oint_{b_k} \mathrm{d}p, \quad k=1,\dots,g; \tag{4.95}$$

define a system of the coordinates on the leaves $\mathcal{N}_g^V(n_\alpha)$ of the foliation $\mathcal{N}_g(n_\alpha)$. Therefore, the vectors $\partial_a = \partial/\partial t_a$ form the basis in the tangent space of $\mathcal{N}_g^V(n_\alpha)$.

THEOREM 4.5. *The formulae*

$$\langle\partial_a,\partial_b\rangle = \sum_{q_s}\operatorname{res}_{q_s}\left(\frac{\mathrm{d}\Omega_a\,\mathrm{d}\Omega_b}{\mathrm{d}E}\right), \quad \mathrm{d}E(q_s)=0, \tag{4.96}$$

$$C_{abc} = \sum_{q_s}\operatorname{res}_{q_s}\left(\frac{\mathrm{d}\Omega_a\,\mathrm{d}\Omega_b\,\mathrm{d}\Omega_c}{\mathrm{d}p\,\mathrm{d}E}\right), \tag{4.97}$$

(*where the differentials* $\mathrm{d}\Omega_a$ *are defined by* (2.15, 2.16, 4.43, 4.69–4.72)) *define on* $\mathcal{N}_g^V(n_\alpha)$ *the structure of the Fröbenius manifold. The vector* $\Psi(t,\lambda)$ *with coordinates*

$$\Psi_a(t) = \int \mathrm{d}\Omega_a\,\mathrm{e}^{\lambda E} = \partial_a\int \mathrm{d}p\,\mathrm{e}^{\lambda E}, \tag{4.98}$$

are horizontal sections of the flat connection (4.81), *i.e.*

$$(\partial_a + \lambda C_a(t))\Psi(t,\lambda) = 0. \tag{4.99}$$

Within the framework of the theory of topological models (see [28]), this result may be presented in the following way. The linear space spanned by the differentials $\mathrm{d}\Omega_a$ is isomorphic to H^1 cohomology of the operator $\mathrm{D}_1 = \bar\partial + \mathrm{d}E\wedge$ acting in the space of smooth differential forms on the curve Γ_g. The bilinear form (4.96) on the space of such differentials coincides with the natural pairing in the middle dimensional cohomologies of this operator. The periods (4.98) define the class of equivariant cohomologies of the operator

$$\bar\partial + \mathrm{d}E + \lambda\partial \tag{4.100}$$

that are horizontal sections of the flat connection defined with the help of (4.97).

References

1. Dubrovin, B., Matveev, V., and Novikov, S.: Non-linear equations of Korteweg–de Vries type, finite zone linear operators and Abelian varieties, *Uspekhi Mat. Nauk* **31**(1) (1976), 55–136.
2. Zakharov, V., Manakov, S., Novikov, S., and Pitaevski, L.: *Soliton Theory*, Nauka, Moscow, 1980.
3. Lax, P.: Periodic solutions of Korteweg–de Vries equation, *Comm. Pure Appl. Math.* **28** (1975), 141–188.
4. McKean, H. and van Moerbeke, P.: The spectrum of Hill's equation, *Invent. Math.* **30** (1975), 217–274.
5. Krichever, I. M.: The algebraic–geometrical construction of Zakharov–Shabat equations and their periodic solutions, *Doklady Akad. Nauk USSR* **227**(2) (1976), 291–294.
6. Krichever, I.: The integration of non-linear equations with the help of algebraic–geometrical methods, *Funk. Anal. Pril.* **11**(1) (1977), 15–31.
7. Krichever, I.: Spectral theory of two-dimensional periodic operators and its applications, *Uspekhi Mat. Nauk* **44**(2) (1989), 121–184.
8. Krichever, I.: Averaging method for two-dimensional integrable equations, *Funk. Anal. Pril.* **22**(3) (1988), 37–52.
9. Feldmann, J., Knorrer, H., and Trubowitz, E.: Riemann surfaces of the infinite genus, Preprint ETH, Zurich.
10. Shiota, T.: Characterization of Jacobian varieties in terms of soliton equations, *Invent. Math.* **83** (1986), 333–382.
11. Takebe, T.: Representation theoretical meaning for the initial value problem for the Toda lattice hierarchy: I, *Lett. Math. Phys.* **21** (1991), 77–84.
12. Sato, M.: Soliton equations and universal Grassmann manifold, *Math. Lect. Notes Ser. 18*, Sophia University, Tokyo, 1984.
13. Dubrovin, B., Krichever, I., and Novikov, S.: The Schrödinger equation in a periodic field and Riemann surfaces, *Soviet Dokl.* **229**(1) (1976), 15–18.
14. Krichever, I.: The τ-function of the universal Whitham hierarchy, matrix models and topological field theories, *Comm. Pure Appl. Math.* **47** (1994), 437–475.
15. Takasaki, K. and Takebe, K.: SDiff(2) Toda equation hierarchy, tau-function and symmetries, Preprint RIMS-790, Kyoto.
16. Takebe, K.: Area-preserving diffeomorphisms and nonlinear integrable systems, in: *Proceedings of Topological and Geometrical Methods in Field Theory*, May 1991, Turku, Finland.
17. Saveliev, M.: On the integrability problem of the continuous long wave approximation of the Toda lattice, Preprint ENSL, Lyon, 1992.
18. Zakharov, V.: Benney equations and quasi-classical approximation in the inverse problem method, *Funk. Anal. Pril.* **14**(2) (1980), 89–98.
19. Tsarev, S.: The geometry of Hamiltonian systems of hydrodynamic type, *Izvest. USSR, Ser. Matem.* **54**(5) (1990), 96–154.
20. Dubrovin, B. and Novikov, S.: The Hamiltonian formalism of one-dimensional systems of hydrodynamic type and Bogolyubov–Whitham averaging method, *Sov. Math. Dokl.* **27** (1983) 665–669.
21. Dubrovin, B. and Novikov, S.: Hydrodynamics of weakly deformed soliton lattices. Differential geometry and Hamiltonian theory, *Russ. Math. Surveys* **44**(6) (1989), 35–124.
22. Novikov, S.: The geometry of conservative systems of hydrodynamics type. The method of averaging for field-theoretical systems, *Russ. Math Surveys* **40**(4) (1985), 85–98.
23. Krichever, I.: The dispersionless Lax equations and topological minimal models, *Comm. Math. Phys.* **143**(2) (1992), 415–429.
24. Krichever, I.: Whitham theory for integrable systems and topological field theories, to appear in: *Proceedings of Summer Cargese School*, July, 1991.

25. Verlinder, E. and Verlinder, H.: A solution of two-dimensional topological quantum gravity, Preprint IASSNS-HEP-90/40, PUPT-1176, 1990.
26. Dubrovin, B.: Hamiltonian formalism of Whitham-type hierarchies and topological Landau–Ginsburg models, *Comm. Math. Phys.* **145**(1) (1992), 195–207.
27. Dubrovin, B.: Differential geometry of moduli spaces and its application to soliton equations and to topological conformal field theory, Preprint No. 117, Scuola Normale Superiore, Pisa, November 1991.
28. Blok, B. and Varchenko, A.: Topological conformal field theories and the flat coordinates, *Int. J. Modern Phys. A* **7**, 1467–1490.

Acta Applicandae Mathematicae **39:** 127–132, 1995.

An ODE to a PDE: Glories of the KdV Equation. An Appreciation of the Equation on Its 100th Birthday!

MARTIN D. KRUSKAL
Rutgers University, Piscataway, New Jersey 08855-0909, USA

(Received: 2 March, 1995)

Though you may not believe me, I am normally of a shy and retiring disposition, modest to a fault, even timid at times. Therefore it is difficult for me to bring myself to make the following bold claim, but I must do so in the interest of the Higher Truth.

Together with Norman J. Zabusky, I am the person who, more than anyone else, resuscitated (perhaps revitalized is the right word) the Korteweg–deVries equation after its long period of, if not oblivion, at least neglect.

What is the KdV equation, which we are celebrating at this conference? It appears deceptively simple. Stripped of all unnecessary coefficients, which can be removed, or restored, by simple scaling of the variables, it is the partial differential equation (PDE)

$$u_t + uu_x = u_{xxx},$$

where the names of the independent variables t and x reflect their physical origins as time and space coordinates, and u is the dependent variable; as is customary the subscripts denote partial differentiation, so for example

$$u_{xxx} \equiv \frac{\partial^3 u}{\partial x^3}.$$

Now one might say, so what? This is just another equation, among innumerable other equations. What is special about it, that we're making such a fuss over it? Has there been a conference to celebrate the Laplace equation? The heat equation?

Well, many of these other equations, including the two just mentioned, are *linear*, and also in fact homogeneous. They enjoy the special *superposition property*, according to which one can obtain new solutions by forming linear combinations of known solutions (multiplying by constants, adding or integrating). While not all questions about linear equations are easy because of this, there is no doubt but that the mathematician or scientist studying linear equations starts out with an enormous advantage.

On the other hand, nonlinear equations are traditionally recalcitrant to exact analysis. Yes, they can be solved approximately by various numerical methods – but as everyone who works with serious scientific problems knows, numerical solutions are not a substitute for analytic understanding, but a complement.

In fact, the theme that numerics and analysis cooperate in mathematics much as experiment and theory do in science is by now almost a truism, in large part because of the early and repeated insistence on it by my colleague Norman Zabusky, who was perhaps the first to use the term *synergism* in this context.

Since I have had occasion to mention him already more than once, let me say that though we were colleagues and collaborators in the 1960's, when together we did the earliest modern work on the KdV equation, and are happily also colleagues now, 30 years later, our paths diverged in between, and our collegiality has been at two different institutions and came about through different circumstances. Thirty years ago we were together at Princeton University, in particular at the Princeton Plasma Physics Laboratory, which was at that time called Project Matterhorn. Today we are colleagues at Rutgers University.

But the question I wanted to discuss is, why did the KdV equation languish so long, in low appreciation? This can be thought of as one aspect of a broader question, why were (and still are) the *third order* PDEs relatively neglected? Why do I say that they are?

Nearly 50 years ago, my mentor and nominal thesis advisor was Richard Courant at New York University, which he had almost singlehandedly built up from an institution of no mathematical distinction to one of the world's leading centers, which it still remains today so many years later. And the most influential early book on PDEs as well as other topics (for long virtually the only one of significance) was the famous 'Courant–Hilbert.' It was a two-volume work called *Methods of Mathematical Physics,* which Courant wrote largely based on notes of courses he took from the great mathematician David Hilbert.

In that work, after disposing of ODEs (ordinary differential equations, differential equations with only one independent variable), the authors turn their attention to PDEs (two or more independent variables). They start off by analyzing *first order* PDEs, those involving only first derivatives of the unknown (dependent) variable. For these there is a quite complete and satisfying theory, as long as everything is regular, called the method of characteristics, which reduces the PDE to a family of ODEs, which are at least theoretically much more tractable. This holds even when the PDE is nonlinear, though in a more complicated way, of course. I might mention in passing, however, that the occurrence of singularities raises very interesting questions that are under active investigation today: the so-called caustics that arise in ray optics and their relationship to an underlying wave theory is one example.

After the first order PDEs, naturally one studies the second order PDEs. Here a slew of new phenomena emerge, and even the simplest such equations, linear, in two independent variables, and with only two terms, must be separated into different types that have quite different properties. Now one finds the first distinction between *elliptic, hyperbolic,* and *parabolic* equations, as exemplified in their respective prototypes, the *potential* equation,

$$u_{xx} + u_{yy} = 0,$$

the *wave* equation,

$$u_{xx} - u_{tt} = 0,$$

and the *heat* equation:

$$u_t - u_{xx} = 0.$$

These differ characteristically in the boundary conditions which make them *well-posed*, which in part means having a unique solution.

There are of course many much more complicated second order PDEs, with more independent variables, with more terms, lower order ones or of the same order, nonlinear, variety enough to keep a mathematician busy for a lifetime (and many do devote their lives to the study of such equations).

But now comes a strange phenomenon. Surely, one would think, after the first order and the second order equations, the next in line would be the third order? But no, that did not happen. The third order is simply skipped over, and next come the *fourth order* PDEs! Among many others, for example, the biharmonic equation

$$u_{xxxx} + 2u_{xxyy} + u_{yyyy} = 0$$

and the beam equation

$$u_{tt} + u_{xxxx} = 0,$$

as well as all sorts of embellishments of them as before.

Why was the third order ignored? Well, it will not have escaped your notice that all these equations have names, and notations, showing that their usefulness in describing particular physical (or biological or chemical or economic) situations was what led to their importance. In other words, they were studied for their physical significance, not their intrinsic mathematical interest.

The real mystery here is why physics and other sciences so rarely lead to third order PDEs (or fifth and higher order ones, either). One evident reason is that, because the independent variable so often represents physical length, what we have called x throughout here, and space shows itself to us as symmetric (why that should be the case is a much deeper philosophical, perhaps metaphysical question that I have no comment on), only *even* derivatives with respect to x will occur in our more basic equations. Otherwise phenomena would look different from left to right than from right to left, contradicting our common experience. (Another deep mystery is why so many fundamental phenomena are describable by differential equations at all.)

What is more surprising is that large parts of fundamental physics are also *time reversible*, so that only even order time derivatives occur in the governing differential equations. (The Schrödinger equation of quantum mechanics, certainly one of the most fundamental equations we know, is only an apparent exception: it is first order in t, true, but it is a complex equation, and the dependent variable ψ is complex; if one separates the real and imaginary parts and eliminates one of them to get a single real equation, it turns out to be of even order and time reversible after all.) The curiosity is that some very basic physics is *not* time reversible: the heat equation, and the increase of entropy which its solutions exhibit, the directedness of 'time's arrow', is the most obvious and notable example. Some physicists consider the law of increase of entropy to be *the* most fundamental law of physics.

Well, now we have the converse problem. If odd order PDEs are, for the most part, so intrinsically unphysical, how in the world did Korteweg and de Vries come up with their now famous equation? The answer is, by a process of *reduction*.

Asymptotics is one of the fundamental mathematical concepts and procedures in physics and the other sciences, as well as in mathematics. In the present context it means looking at a limiting case, finding the equations that govern a situation in a well defined extreme situation.

All the famous reductions of the more exact and hard sciences proceed this way. Fluid dynamics and thermodynamics are 'reduced' to statistical physics by kinetic molecular theory. Kirchoff's laws for electrical circuits are 'reduced' to Maxwell's equations of electromagnetism taken in the asymptotic limit of thin wires. And so on and on.

The physical situation faced by Korteweg and deVries was much simpler than these grand theme problems that I have just mentioned; more like the kind of problem many of us come upon frequently. But it had a special feature making it very tricky, for its time. Nowadays what they did is quite a standard technique, which we teach to our students regularly. but theirs was one of the first, maybe *the* first, nontrivial real life employment of it. It is now called the *multiscale* or *secularity-free* expansion method, and it is a staple of the branch of asymptotics called singular perturbation theory.

Korteweg and deVries studied the motion of long, small amplitude water waves in a shallow channel. The terms 'long' and 'shallow' are relative to each other: the large parameter driving the asymptotics is their ratio, the ratio of the longitudinal wavelength (or typical scale of variation) of the waves to the depth of water in the channel. Another asymptotic parameter, small in this case (but you can take its reciprocal if you want a large parameter), is the amplitude of the surface disturbance, taken as a ratio to the depth.

In order to study the phenomenon of a solitary wave (which in some cases is what we now call a soliton), they performed the appropriate asymptotic analysis for a traveling wave form that changes shape only gradually – that's where the multiple scales come in – due to dispersive and nonlinear effects, both small under the assumed conditions.

Starting from an established PDE called the Boussinesq equation, which has a second order time derivative and a fourth order space derivative and is, therefore, a usual kind of PDE in the sense discussed above, the reduction process led them systematically to an equation with one less order of differentiation in each variable, their eponymous equation.

I came to the same equation in a very similar way, starting from a similar fourth order PDE arising in an analysis of the socalled Fermi–Pasta–Ulam recurrence phenomenon; the history of this has often been described. Norman Zabusky played a large rôle in the early description and analysis of this phenomenon too.

Now, it turns out that the KdV equation has a number of very special and remarkable properties, which the other speakers at this conference will certainly be talking about. Here is where it parts company with any run-of-the-mill third order equation.

For one thing it has a lot of symmetries, some obvious and some hidden. The obvious ones include space translation, time translation, scaling symmetry, and Galilean invariance. It is arguably the simplest and most symmetric nonlinear third order PDE.

The big and fruitful surprise was the discovery, at first numerical but later shown to be mathematically exact, of solitons as distinct and highly stable ingredients or entities in its solutions. A *soliton* is a solitary traveling wave, but only in a solution of an equation special in the right way to support waves with the strong stability, or persistence of form through nonlinear interactions, that led Norman and me to give them such a particle-like name.

The KdV is the first equation discovered to have such properties, and is still the prototypical soliton equation in many ways. Depending on how you count, there are hundreds known now ... though many have been found to be obtainable as reductions of others, and it may turn out in the end that there is only one gigantic soliton equation which contains all the others within it in some sense.

Actually, another very basic soliton equation, called the sine–Gordon equation, a particular kind of nonlinear wave equation, namely

$$u_{tt} - u_{xx} + \sin u = 0,$$

was studied by differential geometers even earlier, but although they discovered some of its special properties, they didn't discern the solitons.

In any case, one can not say that K and dV studied their equation deeply in the sense we think of today (nor would it have been possible for them to do so). They were largely content to show how water waves could travel without change in shape, and that was quite enough of a challenge.

To bring out the point how little attention was paid to their equation, let me mention that after I derived it on my own and only then discovered it to be a 'known' and named equation, I asked one of the greatest analysts of the time, Solomon Bochner, a Princeton mathematics professor, what he could tell me about it. The answer was, nothing. I do not recall now whether he had even ever heard of it, but it was clear that he had never thought about it. I do not mean this to sound at all critical; there were undoubtedly good reasons for the lack of attention. But what is understandable one mathematician at a time is stranger when considered collectively.

I would like to end this discussion by summing up what I have said, but my presentation has been so rambling that it is difficult to do that. I would, however, like to emphasize the serendipity of scientific and mathematical research, how one never knows what will lead where, and how one never knows what surprises may lie just below the surface. The story of the KdV equation and the many directions in which it sprouted and played a significant role shows once again that science and mathematics are great adventures in which nothing can be taken for granted, and the most remarkable conceptual reversals can be forced upon you. My colleague at Rutgers, the great mathematician Israel Gel'fand, tells how he thought he knew all there was to know about the Schrödinger equation and the associated scattering problem, a subject in which he is one of the world's leading experts; but then he learned from one of the 'sprouts' that the spectrum of the Schrödinger equation is invariant if the potential evolves according to the KdV equation, which he never would have suspected or even dreamed of looking into, but which was of great importance for both KdV theory and Schrödinger theory.

Acta Applicandae Mathematicae **39:** 133–158, 1995.

The Discrete Korteweg–de Vries Equation

FRANK NIJHOFF
Department of Applied Mathematical Sciences, University of Leeds, Leeds LS2 9JT, U.K.

and

HANS CAPEL
Institute of Theoretical Physics, University of Amsterdam, Valckenierstraat 65, 1018 XE Amsterdam, The Netherlands

(Received: 9 January 1995)

Abstract. We review the different aspects of integrable discretizations in space and time of the Korteweg–de Vries equation, including Miura transformations to related integrable difference equations, connections to integrable mappings, similarity reductions and discrete versions of Painlevé equations as well as connections to Volterra systems.

Mathematics Subject Classification (1991): 58F07.

Key words: lattice equations, linearization, discretization, Miura transformations, similarity reductions.

1. Introduction

When Korteweg and de Vries published their famous equation in the *Philosophical Magazine* [1], exactly one hundred years ago on the occasion of this conference, they were probably far from being aware of the enormous ramifications this nonlinear partial differential equation would have, and whose revived interest was marked at the end of the sixties by the celebrated paper by Gardner, Green, Kruskal and Miura [2]. What is now called the KdV equation was the drosophila of a class of nonlinear evolution equations, which triggered the development of new methods and structures (inverse scattering transform, Bäcklund transformations, bi-Hamiltonian structures, master-symmetries, etc.), applicable not only to the KdV equation itself, but also to a host of other systems as well (cf., e.g., [3, 4]). Independently, and curiously enough at roughly the same time as Korteweg and de Vries were studying models for shallow water waves, there was an important development in geometry, marked by the classic works of Bäcklund [5], Darboux [6], and Bianchi [7]. It was only in this century that people realized how relevant the works of these classical geometers were to the study of the much more applied problem of the behaviour of water in canals. It is not surprising from a modern perspective why it took almost eighty years before this connection was fully understood: the road from geometry to the analysis of water waves goes via algebra, and a kind of algebra that was invented only sixty years later, namely the type of infinite-dimensional Lie algebras that are nowadays known as Kac–Moody algebras. In this talk we do not want to go into even more recent and more exotic connection, such as with knot theory, quantum groups or moduli spaces of curves. Who could guess that (via string theory and conformal field theory) the small and beautiful equation that Korteweg and de Vries wrote down in their paper [1] would stand at the very foundations of present-day descriptions of the universe?

Let us now come to the other theme of our talk: difference or discrete systems. Korteweg and de Vries lived in an era that had no knowledge of computers. In their days, nonetheless, from the point of view of classical education in mathematics, differential analysis started with difference equations (cf., e.g., [8]). And it was thought to be very natural always to link differential and difference equations together. It was the time when another famous Dutch mathematician, Thomas Stieltjes, developed theories of continued fractions, recursion relations and orthogonal polynomials, thus standing in a long and rich tradition of 18th- and 19th-century classical analysis. Little did Korteweg and de Vries realize that also these subjects were intimately linked to their water-wave equation. The study of linear difference equations was being developed in the beginning of this century, mainly by the propagents of the Birkhoff school (cf., e.g., [9]). Unfortunately, the good tradition of treating differential equations always alongside their difference counterparts, and vice versa, seemed to have got lost during the second half of this century, especially after the second world war.

Today, the advent of computers provides new motivations to look at difference equations. Neural networks, cellular automata, self-organizing phenomena and complexity are some of the key words of modern developments inspired by the birth and growth of newer and newer generations of computers. On the other hand, numerical analysis, and its applications to all branches of sciences where computer calculations play a role, relies heavily on discreteness in all its aspects: interpolations, convergence algorithms, and their analytical bases: orthogonal polynomials, recurrence relations. It is unlikely that Korteweg and de Vries could envisage how important their equation was to these aspects of modern-day research. Little also do the people who use these tools and methods, know that the small and elegant equation that Korteweg and de Vries wrote down in order to capture phenomena of wave motion is of relevance to them.

In this paper we want to tie some of these different threads together. We will look into exact discretizations of the KdV equation and highlight a few important aspects of them. A starting point is an integrable lattice version of the KdV equation, which is a nonlinear partial *difference* equation, i.e. a system in which both the spatial- as well as the time-variable is discrete. Such systems were systematically studied in a number of papers, e.g., in [10–18] (cf., also [19–23]). The lattice model is, in a sense, more fundamental than the original continuous equation, as one can always retrieve the latter by applying an appropriate continuum limit on the lattice system. However, the reverse is not true: to obtain a discretization of the KdV equation which retains its essential integrability characteristics, is a highly nontrivial undertaking. Once having it at its disposal, one can use it as a universal model to study a number of features that, as we will see, become more transparent on the level of the difference system.

The outline of this paper is as follows. In Section 2 we will exhibit the lattice version of the KdV equation, or rather of its potential (i.e. integrated) version. We will show what continuum limits to apply in order to get back to the continuous KdV equation and what are its relations to other integrable lattice systems. In Section 3, we will list some other integrable lattice equations that are less directly related to the KdV equation, but still are of importance. Next, we will look into special solutions of the KdV equation. First, in Section 4, solutions arising from periodic initial value problems in the lattice will be considered, leading to a reduction to integrable finite-dimensional mappings. Second, in Section 5, we will exhibit similarity type of solutions which lead to lattice and difference versions of the second Painlevé transcendent. Finally, we give some insight into the algebraic background of these lattice equations, deriving them from the so-called direct linearization approach method. In this approach, briefly reviewed in Section 6, the relevant equations are embedded in an infinite matrix structure, which gives a powerful handle on establishing the interconnection of the different (Miura-) related systems. For example, a discrete-time version of the Volterra model arises naturally in this way, thus leading possibly to connections with matrix models in two-dimensional quantum gravity and string theory, but also with numerical schemes, Padé approximants and convergence algorithms.

2. The KdV Lattice and Related Lattice Equations

The lattice version of the KdV equation that we want to investigate in this paper is the following nonlinear partial difference equation

$$(p - q + u_{n,m+1} - u_{n+1,m})\,(p + q - u_{n+1,m+1} + u_{n,m}) \;=\; p^2 - q^2. \quad (2.1)$$

To explain the notations in Equation (2.1), we mention that $u = u_{n,m}$ is the dynamical (field) variable at site (n, m), $n, m \in \mathbf{Z}$, $p, q \in \mathbf{C}$ are lattice parameters. Equation (2.1) was derived in [10] from the direct linearization approach (see Section 5 below), and further studied in a number of papers [11–14].

We mention first a few important aspects of Equation (2.1). First, we note that, in contrast to the continuum KdV equation, the lattice equation is covariant with respect to the interchange of the two discrete variables, n and m, interchanging also the lattice parameters p and q. This might come as a surprise, but as we shall see below, we must conclude that the noncovariance of the continuous KdV might be considered to be an artefact of the continuum limit. Actually, the discrete variables being on the same footing, is a very nice feature of the equation which allows us to play some games on the lattice (like making particular choices of initial value- and boundary-value problems) which would be quite awkward in the continuum.

Another feature of the equation is that it arises from a discrete action principle. The action for the KdV lattice (2.1) reads

$$\mathcal{S} = \sum_{n,m\in\mathbf{Z}} \left[u_{n,m+1}\left(u_{n+1,m+1} - u_{n,m}\right) + \right.$$
$$\left. +\varepsilon\delta\ln\left(\varepsilon + u_{n,m} - u_{n+1,m+1}\right)\right], \tag{2.2}$$

where $\delta = p - q$, $\varepsilon = p + q$. The Euler–Lagrange equations for (2.2), which are obtained by variation of $\mathcal{S}$ with respect to the variables $u_{n,m}$, i.e.

$$\frac{\delta\mathcal{S}}{\delta u_{n,m}} = 0,$$

leads to the equation

$$u_{n+1,m} - u_{n,m+1} + u_{n-1,m} - u_{n,m-1} +$$
$$+\frac{\varepsilon\delta}{\varepsilon + u_{n,m} - u_{n+1,m+1}} - \frac{\varepsilon\delta}{\varepsilon + u_{n-1,m-1} - u_{n,m}} = 0, \tag{2.3}$$

which is a consequence of Equation (2.1). In fact, as we shali demonstrate below, Equation (2.1) goes to the potential (i.e. integrated) KdV equation, whereas Equation (2.3) goes to the KdV equation itself. However, by abuse of terminology, we will refer to (2.1) as the lattice KdV equation, because we will be mainly concerned with that equation.

A third important aspect of the lattice Equation (2.1), directly related to the integrability of ths system, is that it arises as the compatibility condition of an overdetermined linear system (Lax or Zakharov–Shabat pair). In fact, the Lax pair takes the form of a set of two equations for the translations of a 'wavevector' ϕ_k in the n- and m-directions,

$$(p-k)\phi_k(n+1,m) = L_k \cdot \phi_k(n,m), \tag{2.4a}$$

$$(q-k)\phi_k(n,m+1) = M_k \cdot \phi_k(n,m), \tag{2.4b}$$

where L_k is given by

$$L_k = \begin{pmatrix} p - u_{n+1,m} & 1 \\ k^2 - p^2 + (p+u_{n,m})(p - u_{n+1,m}) & p + u_{n,m} \end{pmatrix}, \tag{2.5}$$

and where M_k is given by a similar matrix obtained from (2.5) by making the replacements $p \to q$ and $(n+1,m) \to (n,m+1)$. The subscript k in (2.4) and (2.5) denotes the dependence on the spectral parameter. Establishing a Lax pair and an action for the KdV system ensures us that, in principle, we can expect many of the integrability characteristics to hold for this system, such as the existence of an infinite number of integrals of the discrete motion, of an inverse scattering scheme and of a canonical formalism. However, to make these things precise and to be able to solve suitable initial-value problems on the lattice, we should supply the KdV system with appropriate boundary conditions. Below, we shall investigate solution and initial value problems in two specific situation: (i) periodic initial data on well-chosen configurations of lattice points ('staircases' in the lattice). This will lead to exactly integrable *mappings*, i.e. discrete-time systems with a finite number of degrees of freedom; (ii) localized initial data under the condition of scaling invariance leading to lattice and discrete versions of the second Painlevé transcendent.

Now why is it justified to refer to (2.1) as the lattice KdV equation? Well, let us investigate what happens under a continuum limit, bringing us back to the continuum situation. Since we have two discrete variables in the lattice equation, namely n and m, we have to perform the continuum limit in two steps: one letting the variable m become continuous, reducing our equation to a *differential-difference equation*, i.e. an equation with one discrete and one continuous variable, and a second step in which the remaining discrete variable will become continuous. Both steps are achieved by shrinking the corresponding lattice step (encoded in the parameters p and q) to zero. The most convenient way of doing this is first, on the lattice, to do a change of discrete variables, namely $u_{n,m} =: u_{n'}(m)$, and then doing the limit by taking

$$\delta \equiv p - q \longmapsto 0, \quad m \mapsto \infty, \quad \delta m \mapsto \tau, \tag{2.6}$$

where $n' = n + m$ is to remain fixed. This limit is motivated from the behaviour of discrete plane-wave factors

$$\rho_k(n, m) = \left(\frac{p+k}{p-k}\right)^n \left(\frac{q+k}{q-k}\right)^m, \tag{2.7}$$

which, in fact, are related to solutions of (2.4) for the linearized problem, namely

$$\left(\frac{p+k}{p-k}\right)^n \left(\frac{q+k}{q-k}\right)^m \longmapsto \left(\frac{p+k}{p-k}\right)^{n'} \exp\left\{\frac{2k\tau}{p^2-k^2}\right\} \tag{2.8}$$

(cf. [10, 11, 14]). By this limit, the lattice KdV equation (2.1) goes over into the following differential-difference equation (we omit the prime of the n' variable)

$$1 + \partial_\tau u_n = \frac{2p}{2p - u_{n+1} + u_{n-1}}. \tag{2.9}$$

This differential-difference equation is related to the Kac–van Moerbeke–Volterra equation [24]. Next, the second continuum limit is performed by taking

$$p \longrightarrow \infty, \quad n \longrightarrow \infty, \quad \tau \longrightarrow \infty, \tag{2.10a}$$

such that

$$2\frac{n}{p}+2\frac{\tau}{p^2}\longmapsto x,\qquad \frac{2}{3}\frac{n}{p^3}+2\frac{\tau}{p^4}\longmapsto t, \tag{2.10b}$$

in which case (2.9) goes over into the potential KdV equation

$$u_t=u_{xxx}+3u_x^2, \tag{2.11}$$

which is the integrated version of the KdV equation.

Before looking into special reductions of the lattice KdV equation, let us focus for a moment on a very important aspect of such integrable systems, namely that they are not isolated objects: they have a number of companion systems which are related to the original equation by some kind of implicit transformation, usually referred to as Miura transformation. This holds true for the original KdV equation, but also for its lattice version (2.1). First, there is the discrete analogue of the *modified* Korteweg–de Vries (MKdV) equation, which reads

$$\begin{aligned}(p-r)\frac{v_{n,m+1}}{v_{n+1,m+1}}-(q-r)\frac{v_{n+1,m}}{v_{n+1,m+1}}\\ =(p+r)\frac{v_{n+1,m}}{v_{n,m}}-(q+r)\frac{v_{n,m+1}}{v_{n,m}},\end{aligned} \tag{2.12}$$

which we will refer to as the lattice modified KdV equation.

The Miura transformation between the two Equations (2.12) and (2.1) can be obtained from the relations

$$(p-r)v_{n,m}-(p+u_{n,m})v_{n+1,m}=t_{n+1,m}, \tag{2.13a}$$

$$(p-u_{n+1,m})v_{n,m}-(p+r)v_{n+1,m}=t_{n,m}, \tag{2.13b}$$

in which $t_{n,m}$ and $v_{n,m}$ are some new functions of the lattice sites. By the general covariance of the lattice systems, Equations (2.13) hold also for the other lattice direction, i.e. with the replacements: $(n+1,m)\mapsto(n,m+1)$ and $p\mapsto q$. Thus, by combining these relations we can eliminate, for instance, $t_{n,m}$ entirely, to obtain the Miura transformation between (2.1) and (2.12) consisiting of the relations

$$\begin{aligned}p-q+u_{n,m+1}-u_{n+1,m}\\ =(p-r)\frac{v_{n,m+1}}{v_{n+1,m+1}}-(q-r)\frac{v_{n+1,m}}{v_{n+1,m+1}},\end{aligned} \tag{2.14a}$$

$$\begin{aligned}p+q+u_{n,m}-u_{n+1,m+1}\\ =(p-r)\frac{v_{n,m}}{v_{n+1,m}}+(q+r)\frac{v_{n+1,m+1}}{v_{n+1,m}}.\end{aligned} \tag{2.14b}$$

The lattice MKdV Equation (2.12) also carries a Lax pair which is again of the form (2.4), together with

$$L_k = \begin{pmatrix} p-r & v_{n+1,m} \\ (k^2-r^2)\dfrac{1}{v_{n,m}} & (p+r)\dfrac{v_{n+1,m}}{v_{n,m}} \end{pmatrix}, \tag{2.15}$$

instead of (2.5), and where again M_k is given by a similar matrix obtained from (2.15) by making the same replacements $p \to q$ and $(n+1,m) \to (n,m+1)$. Again, there is an action principle, now not directly in terms of the variable $v_{n,m}$, but in terms of its logarithm. Thus writing $v_{n,m} = \mathrm{e}^{y_{n,m}}$, we have the action

$$\begin{aligned} \mathcal{S} \;=\; \sum_{n,m\in\mathbf{Z}} \Big[& y_{n,m+1}\left(y_{n+1,m+1} - y_{n,m}\right) + F\left(y_{n,m} - y_{n+1,m+1} + \sigma\right) - \\ & - F\left(y_{n,m} - y_{n+1,m+1} + \rho\right)\Big], \end{aligned} \tag{2.16}$$

in which the function F is given by

$$F(x) \equiv \int_{-\infty}^{x} \mathrm{d}\xi \; \log(1+\mathrm{e}^{\xi}), \tag{2.17}$$

which, in fact, is directly related to the Euler dilogarithm function, and where σ and ρ are parameters related to the lattice parameters p and q, namely by

$$\mathrm{e}^{\sigma} = \frac{q-r}{p+r}, \qquad \mathrm{e}^{\rho} = \frac{p-r}{q+r}.$$

So, everything that holds true for the lattice KdV equation holds true also for the lattice MKdV equation. Performing the continuum limit (2.6) also to this case, we obtain in the first step

$$p\,\partial_\tau y_n = \tanh\left[\tfrac{1}{2}(y_{n+1} - y_{n-1})\right], \quad y_n \equiv \ln v_n, \tag{2.18}$$

which is a differential-difference version of the MKdV equation, and in the next step we get

$$y_t = y_{xxx} + 2y_x^3, \tag{2.19}$$

which is the potential MKdV equation.

There is, however, an even more fundamental lattice equation, related to both the lattice KdV and MKdV equations, which we write in terms of a new object $s_{n,m}$, related to the MKdV field by

$$1 - (p+r')s_{n+1,m} + (p-r)s_{n,m} = v_{n+1,m}w_{n,m}, \tag{2.20}$$

in which $w_{n,m}$ is the MKdV field for the choice r' as the fixed parameter. For $s_{n,m}$ we have the lattice equation

$$\frac{1-(p+r')s_{n+1,m}+(p-r)s_{n,m}}{1-(q+r')s_{n,m+1}+(q-r)s_{n,m}} = \frac{1-(q+r)s_{n+1,m+1}+(q-r')s_{n+1,m}}{1-(p+r)s_{n+1,m+1}+(p-r')s_{n,m+1}}. \tag{2.21}$$

We will refer to Equation (2.21) as the lattice Krichever–Novikov equation. It was first derived in [10], where it was shown that, in fact, the other lattice equations are included in this more general equation for special choices of the parameters r and r'. Furthermore, continuum limits yield the various differential-difference counterparts of the KdV equation, as well as of the MKdV equation, as was shown in [11]. Finally, one can derive a Lax pair for Equation (2.21) which is of the form

$$L_k = \begin{pmatrix} p-r & P'_{n+1,m} \\ (k^2-r^2)\dfrac{1}{P_{n,m}} & (p+r)\dfrac{P'_{n,m}}{P_{n,m}} \end{pmatrix}, \tag{2.22}$$

with

$$\begin{aligned} P_{n,m} &= 1-(p+r)s_{n+1,m}+(p-r')s_{n,m}, \\ P'_{n,m} &= 1-(p+r')s_{n+1,m}+(p-r)s_{n,m}. \end{aligned}$$

The equation for s is a very rich equation, and contains many parameter-subcases. In the continuum limit we obtain

$$\partial_\tau s_n = \frac{s_{n-1}-s_{n+1}-2p(s_n^2+s_{n+1}s_{n-1})+(r+r')s_n(s_{n+1}-s_{n-1})}{2p+(p-r')(p-r)s_{n-1}-(p+r')(p+r)s_{n+1}}, \tag{2.23}$$

respectively,

$$s_t = s_{xxx} + 3\frac{(s_{xx}+r's_x)(s_{xx}+rs_x)}{1-(r+r')s-2s_x}. \tag{2.24}$$

in the two subsequent steps. In the special case that $r = r' = 0$, Equation (2.24) reduces to another very important equation in the KdV family: the Schwarzian KdV equation, or Krichever–Novikov equation (cf. [25]),

$$\psi_t = \psi_x S(\psi), \quad S(\psi) \equiv \frac{\psi_{xxx}}{\psi_x} - \frac{3}{2}\frac{\psi_{xx}^2}{\psi_x^2}, \tag{2.25}$$

which is an equation invariant under Möbius transformations

$$\psi \longmapsto \frac{a\psi+b}{c\psi+d}, \tag{2.26}$$

i.e. the group $SL(2,\mathbf{R})$. In that special case the lattice Equation (2.21) can be easily seen to reduce to an even more simple equation, namely

$$\frac{(z_{n,m}-z_{n+1,m})(z_{n,m+1}-z_{n+1,m+1})}{(z_{n,m}-z_{n,m+1})(z_{n+1,m}-z_{n+1,m+1})}=\frac{q^2}{p^2}, \tag{2.27}$$

in which

$$s_{n,m}=z_{n,m}+\frac{n}{p}+\frac{m}{q}.$$

The left-hand side of Equation (2.27) is the conformally invariant cross-ratio of four point in the complex plane, which – as was noted before in, e.g., [26] (cf. also [27]) – can be considered to be the discrete analogue of the Schwarzian derivative S. It is remarkable that the equation: cross-ratio equal to constant can be considered to be an integrable lattice equation carrying a Lax pair.

3. Other Lattice Equations

Of course, the lattice KdV is not the only equation for which all of this holds: there are many other integrable lattice equations known, and they were constructed in a number of papers [10–17], as well as in the literature [19–23]. We cannot go here into the details of all these equations, as the focuspoint in this celebration paper is the Korteweg–de Vries example. However, to demonstrate the wide variety of equations, let us just list a few of them.

Probably the best known example of an integrable lattice equation is the lattice analogue of the sine-Gordon equation,

$$\begin{aligned}&\sin\left(\theta_{n,m}+\theta_{n+1,m}+\theta_{n,m+1}+\theta_{n+1,m+1}\right)\\ &\quad=\frac{p}{q}\sin\left(\theta_{n,m}-\theta_{n+1,m}-\theta_{n,m+1}+\theta_{n+1,m+1}\right),\end{aligned} \tag{3.1}$$

is directly related to the lattice MKdV for fixed parameter $r=0$ by a simple transformation of the form

$$v_{n,m}\longmapsto\exp\left\{2i(-1)^m\theta_{n,m}\right\}.$$

The lattice sine-Gordon equation was first presented by Hirota [20], and therefore sometimes referred to as the Hirota equation. Actually, it was already present in the work of Bianchi [7], appearing as permutability condition of Bäcklund transformations for the usual sine-Gordon equation, which has a great significance in the classical differential geometry of surfaces in $\mathbf{R}^3$.

Another important class of lattice equations is obtained by extending the lattice KdV equation to higher-order partial difference systems. This is the so-called lattice Gel'fand–Dikii hierarchy, presented first in [17], which is the discrete analogue of the integrable family of partial differential equations associated with the higher-order differential spectral problem (cf., e.g., [28]). The lattice KdV equation is naturally embedded as the lowest member in this class of equations, which is labelled by the roots of unity, $\omega \equiv \exp(2\pi i/N)$, $N \in \mathbf{Z}$. Let us mention only the first member after the KdV lattice (which is the case $N = 2$), leading to a lattice analogue of the Boussinesq (BSQ) equation ($N = 3$). The linear system in this case is given by

$$(p + \omega k)\phi_k(n+1, m) = L_k \cdot \phi_k(n, m), \tag{3.2a}$$

$$(q + \omega k)\phi_k(n, m+1) = M_k \cdot \phi_k(n, m), \tag{3.2b}$$

in which

$$L_k = \begin{pmatrix} p - u_{n+1,m} & 1 & 0 \\ -v_{n+1,m} & p & 1 \\ k^3 + p^3 + * & w_{n,m} & p + u_{n,m} \end{pmatrix}, \tag{3.3}$$

and again M_k is obtained by replacing $p \to q$ and $(n+1, m) \to (n, m+1)$. In (3.3) the v and w are auxiliary fields (not to be confused with the variables appearing in the previous section), and the term $*$ in the left-lower corner of the matrix L_k is determined by the condition that the determinant $\det(L_k) = p^3 + k^3$.

The compatibility relations of (3.3) lead to the relations

$$\begin{aligned} v_{n,m+1} - v_{n+1,m} &= u_{n+1,m+1}\left(p - q + u_{n,m+1} - u_{n+1,m}\right) + \\ &\quad + q\, u_{n+1,m} - p\, u_{n,m+1}, \\ w_{n,m+1} - w_{n+1,m} &= -u_{n,m}\left(p - q + u_{n,m+1} - u_{n+1,m}\right) + \\ &\quad + p\, u_{n+1,m} - q\, u_{n,m+1}, \\ v_{n+1,m+1} - w_{n,m} &= pq - \left(p + q + u_{n,m}\right)\left(p + q - u_{n+1,m+1}\right) + \\ &\quad + \frac{p^3 - q^3}{p - q + u_{n,m+1} - u_{n+1,m}}, \end{aligned} \tag{3.4}$$

from which one obtains the lattice BSQ equation

$$\begin{aligned} &\frac{p^3 - q^3}{p - q + u_{n+1,m+1} - u_{n+2,m}} - \frac{p^3 - q^3}{p - q + u_{n,m+2} - u_{n+1,m+1}} - \\ &\quad - u_{n,m+1}u_{n+1,m+2} + u_{n+1,m}u_{n+2,m+1} + \\ &\quad + u_{n+2,m+2}\left(p - q + u_{n+1,m+2} - u_{n+2,m+1}\right) + \\ &\quad + u_{n,m}\left(p - q + u_{n,m+1} - u_{n+1,m}\right) \\ &= (2p + q)\left(u_{n+1,m} + u_{n+1,m+2}\right) - \\ &\quad - (p + 2q)\left(u_{n,m+1} + u_{n+2,m+1}\right). \end{aligned} \tag{3.5}$$

From (3.5) by appropriate continuum limits, we recover the continuum BSQ equation. An intermediate continuum limit of the lattice BSQ yields the following differential-difference equation

$$3p^2\partial_\tau\left(\frac{1}{1+\dot{u}_n}\right) = (1+\dot{u}_{n+1})(u_{n+2}-u_{n-1}-3p) - -(1+\dot{u}_{n-1})(u_{n+1}-u_{n-2}-3p), \tag{3.6}$$

whereas the second limit yields the (potential) BSQ equation. There is also an action for the BSQ lattice, given by (here $\varepsilon \equiv p^2+pq+q^2$, δ as before)

$$\mathcal{S} = \sum_{n,m\in\mathbf{Z}} \Big[\varepsilon\delta \ln\left(\delta + u_{n,m+1} - u_{n+1,m}\right) + q\, u_{n,m}\, u_{n,m+1} - p\, u_{n,m}\, u_{n+1,m} - -(p+q+u_{n,m})\left(p+q-u_{n+1,m+1}\right)\left(\delta+u_{n,m+1}-u_{n+1,m}\right)\Big]. \tag{3.7}$$

So, in principle, the situation for the lattice BSQ equation is similar as for the KdV lattice, except that here we are dealing with a system in which there are not only nearest-neighbour terms in the equation, but also next-nearest-neighbour terms. Similar to the KdV situation, there is also here a Miura transformation to a modified equation, which is the discrete analogue of the modified BSQ equation (the continuum modified BSQ was first derived in [29]), and it reads

$$\frac{(p^2+pr+r^2)\, v_{n+1,m+1} - (q^2+qr+r^2)\, v_{n,m+2}}{(p-r)\, v_{n,m+2} - (q-r)\, v_{n+1,m+1}} \frac{v_{n+1,m+2}}{v_{n,m+1}} - - \frac{(p^2+pr+r^2)\, v_{n+2,m} - (q^2+qr+r^2)\, v_{n+1,m+1}}{(p-r)\, v_{n+1,m+1} - (q-r)\, v_{n+2,m}} \frac{v_{n+2,m+1}}{v_{n+1,m}} = (p-r)\left(\frac{v_{n,m}}{v_{n+1,m}} - \frac{v_{n+1,m+2}}{v_{n+2,m+2}}\right) - -(q-r)\left(\frac{v_{n,m}}{v_{n,m+1}} - \frac{v_{n+2,m+1}}{v_{n+2,m+2}}\right). \tag{3.8}$$

Apart from the above examples of lattice equations related in some way to our original lattice Equation (2.1), there are also lattice equations belonging to other types of spectral problems. Lattice systems related to the AKNS scheme [4], for PDE's include a complicated partial difference version of the nonlinear Schrödinger equation (NLS) given in [19]. However, a more elegant version was derived in [11] which reads

$$|\alpha|^2+|\beta|^2+2\mathrm{Re}\left(\theta\alpha\beta\,\phi_{n,m}\phi^*_{n,m+1}\right) = |\alpha|^2(1+|\phi_{n,m+1}|^2)\frac{\beta^*\phi_{n+1,m+1}-\theta\beta\phi_{n,m}}{\theta\alpha\phi_{n+1,m}-\alpha^*\phi_{n,m+1}} + +|\beta|^2(1+|\phi_{n,m}|^2)\frac{\theta\alpha\phi_{n+1,m}-\alpha^*\phi_{n,m+1}}{\beta^*\phi_{n+1,m+1}-\theta\beta\phi_{n,m}}, \tag{3.9}$$

in which α and β are parameters with $\mathrm{Re}(\alpha)=\mathrm{Re}(\beta)$, depending on the lattice parameters, θ is a parameter with modulus 1, $|\theta|=1$, and the $*$ denotes complex conjugation. This equation is Miura-related to a lattice version of the isotropic Heisenberg spin chain (IHSC) equation, also derived in [11], reads

$$\begin{aligned}
&\mathbf{S}_{n+1,m+1} - \mathbf{S}_{n+1,m} + \mathbf{S}_{n,m+1} - \mathbf{S}_{n,m} \\
&\quad = \Delta\left(\frac{\gamma(\mathbf{S}_{n,m})(\mathbf{S}_{n,m} + \mathbf{S}_{n,m+1}) + \alpha\mathbf{S}_{n,m+1}\mathbf{S}_{n,m}}{1 + \mathbf{S}_{n,m+1}\cdot\mathbf{S}_{n,m}}\right), \\
&\gamma(\mathbf{S}_{n,m})^2 = (\mathbf{S}_{n,m+1}\cdot\mathbf{S}_{n,m})^2 + \beta\mathbf{S}_{n,m+1}\cdot\mathbf{S}_{n,m} + \beta - \alpha^2 - 1, \\
&|\mathbf{S}_{n,m}|^2 = 1,
\end{aligned} \tag{3.10}$$

where $(\Delta f)_{n,m} \equiv f_{n+1,m} - f_{n,m}$, and α, β are arbitrary real constants. Another lattice version of the IHSC was presented in [23], but that model seem to carry complex spins. An anisotropic version of (3.10), i.e. the lattice Landau–Lifschitz equation, was presented in [15].

Furthermore, there are lattice equations in $(2+1)$ dimensions, which involve three lattice directions associated with three lattice parameters p, q, and r. Specific examples are the lattice KP equation [12, 13],

$$\begin{aligned}
&\frac{p - r + u_{n,m+1,k+1} - u_{n+1,m+1,k}}{p - r + u_{n,m,k+1} - u_{n+1,m,k}} \\
&\quad = \frac{q - r + u_{n+1,m,k+1} - u_{n+1,m+1,k}}{q - r + u_{n,m,k+1} - u_{n,m+1,k}},
\end{aligned} \tag{3.11}$$

and the lattice sine-Gordon equation in $(2+1)$ dimensions [20], which is a coupled system, namely

$$\begin{aligned}
&r\left(v_{n,m+1,k+1}w_{n,m+1,k} - w_{n+1,m,k}v_{n+1,m,k+1}\right) + \\
&\quad + p\left(\frac{v_{n+1,m,k+1}}{v_{n,m,k+1}} - \frac{w_{n,m+1,k}}{w_{n+1,m+1,k}}\right) + q\left(\frac{w_{n+1,m,k}}{w_{n+1,m+1,k}} - \frac{v_{n,m+1,k+1}}{v_{n,m,k+1}}\right) = 0, \\
&p\left(\frac{v_{n+1,m,k}}{v_{n,m,k}} - \frac{w_{n,m+1,k}}{w_{n+1,m+1,k}}\right) + q\left(\frac{w_{n+1,m,k}}{w_{n+1,m+1,k}} - \frac{v_{n,m+1,k}}{v_{n,m,k}}\right) = 0,
\end{aligned} \tag{3.12}$$

and the lattice modified KP (MKP) equation [13],

$$\begin{aligned}
&(p+s)\left(\frac{v_{n+1,m,k+1}}{v_{n,m,k+1}} - \frac{v_{n+1,m+1,k}}{v_{n,m+1,k}}\right) + \\
&\quad + (q+s)\left(\frac{v_{n+1,m+1,k}}{v_{n+1,m,k}} - \frac{v_{n,m+1,k+1}}{v_{n,m,k+1}}\right) + \\
&\quad + (r+s)\left(\frac{v_{n,m+1,k+1}}{v_{n,m+1,k}} - \frac{v_{n+1,m,k+1}}{v_{n+1,m,k}}\right) = 0,
\end{aligned} \tag{3.13}$$

in which s is an additional fixed parameter. For the special choice $s = -r$ this equation reduces to what is referred to in [23] as the discrete KP equation, and which is related to the DAGTE (discrete analogue of generalized Toda equation) of [20]. We can of course add other equations to this small list, such as the IHSC in $(2+1)$ dimensions, the lattice Davey–Stewartson equation and generalizations to matrix systems. A systematic construction of lattice equations in $(2+1)$ dimensions was presented in [16]. Although these lattice equations share with their continuum counterparts many of the characteristics of integrable systems, such as the existence of Lax pairs and discrete inverse scattering scheme, many of the features of these systems, still have to be investigated. There is to date little systematic study about solutions of such equations. Since a few years a number of people are engaged in developing the theory of integrable lattice equations further, i.e. the search for analytic solutions of initial and boundary problems, similarity reductions and the problem of quantization. In this contribution, we will highlight for the lattice counterparts of the KdV equation what has been done so far. In particular, we will exhibit two special classes of solutions (reductions): solutions of periodic initial-value problems leading to integrable mappings, the lattice analogue of similarity solutions leading to lattice versions of the second Painlevé equation. A third class of solutions of the lattice equations, in particular of the lattice KP equation, consists of pole-solutions (cf. [30]), of rational, trigonometric (hyperbolic) or elliptic type. As was shown in a number of recent papers, [31–33], these lead to integrable discrete-time many-body problems of (relativistic or nonrelativistic) Calogero–Moser type. We will not have the opportunity of displaying these solutions here.

4. Mappings of KdV Type

A first type of solutions of the lattice KdV equation (2.1) that we will discuss here, are solutions arising from a periodic initial-value problem on the lattice. We will show here, that these solutions give a rise of a reduction to integrable dynamical *mappings* with $2P$ degrees of freedom ($P = 1, 2, \ldots$). An important feature of the lattice KdV equation (2.1) is that the variables in the equation are arranged along an elementary square or 'plaquette' in the lattice, i.e. involving the lattice points (n, m), $(n+1, m)$, $(n, m+1)$, and $(n+1, m+1)$ only. Thus, a 'local' initial-value problem for the lattice KdV can be given by specifying values $a_0, a_1, a_2, a_3, \ldots$ of the field u on a staircase consisting of alternating horizontal and vertical steps (cf. [34]), see Figure 1.

Choosing such initial data for the variables along such a staircase on the lattice, namely

$$u_{j,j} =: a_{2j}, \qquad u_{j+1,j} =: a_{2j+1}, \quad j \in \mathbf{Z},$$

we perform iterations by updating the lattice variables u along a vertical shift in the m-direction, i.e.

$$u_{j,j+1} = a'_{2j}, \qquad u_{j+1,j+1} = a'_{2j+1},$$

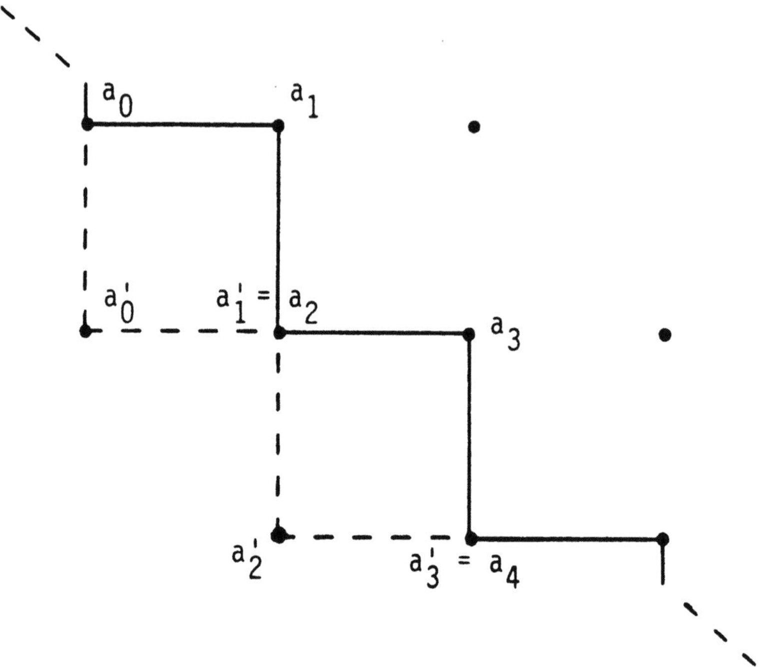

Fig. 1. Configuration of periodic initial data on the lattice.

using the lattice KdV equation (2.1). In this way we obtained the following mapping

$$a'_{2j} = a_{2j+1} - \delta + \frac{\varepsilon\delta}{\varepsilon - a_{2j+2} + a_{2j}}, \qquad a'_{2j+1} = a_{2j+2}, \tag{4.1}$$

where $\delta = p - q$, $\varepsilon = p + q$, which by imposing periodic initial conditions on the staircase, i.e.

$$a_{2(j+P)} = a_{2j}, \qquad a_{2(j+P)+1} = a_{2j+1},$$

reduces to a finite-dimensional mapping of dimension $2P$. Introducing the variables for the differences on odd and even sites of the staircase, i.e.

$$x_j \equiv \varepsilon + a_{2j-1} - a_{2j+1}, \qquad y_j \equiv \varepsilon + a_{2j} - a_{2j+2}, \tag{4.2}$$

the mapping can be even further reduced to a $(2P-2)$-dimensional one which reads

$$\begin{cases} x'_j = y_j, \\ y'_j = x_{j+1} - \frac{\varepsilon\delta}{y_{j+1}} + \frac{\varepsilon\delta}{y_j}, \end{cases} \qquad j = 1, \ldots, P, \tag{4.3}$$

with the Casimirs

$$\sum_{j=1}^{P} x_j = C_1, \qquad \sum_{j=0}^{P-1} y_j = C_2,$$

C_1 and C_2 being constant equal to $P\varepsilon$.

The mapping (4.3) is a multidimensional generalization of the McMillan mapping [35], and it arises as the compatibility condition of a linear (Zakharov–Shabat type of) problem, which is obtained using a special property of the matrices L_k and M_k of (2.4), (2.5) and performing at each site of the staircase a gauge transformation (cf. [34, 36]). The compatibility equation for the mapping (4.3) in terms of the reduced variables x_j and y_j is

$$L_j'(k) \cdot M_j(k) = M_{j+1}(k) \cdot L_j(k), \tag{4.4}$$

where

$$\begin{aligned} L_j(k) &= \begin{pmatrix} 0 & 1 \\ ld + \varepsilon\delta & x_{j+1} \end{pmatrix} \cdot \begin{pmatrix} 0 & 1 \\ ld & y_j \end{pmatrix}, \\ M_j(k) &= \begin{pmatrix} -\varepsilon\delta/y_j & 1 \\ ld & y_j \end{pmatrix}, \end{aligned} \tag{4.5}$$

in which $\lambda = k^2 - p^2$. From this reduced linear system one can construct a monodromy matrix $T(k)$ by gluing the elementary translation matrices L_j along the staircase over one period P, leading to

$$T(k) = \prod_{j=0}^{P-1} L_j(k). \tag{4.6}$$

The trace of the monodromy matrix (4.6) is invariant under the mapping as a consequence of the periodicity. Thus, by expanding the trace in powers of k^2 we obtain $P - 1$ nontrivial independent invariants.

The generalized McMillan mappings, viewed as periodic reductions of the lattice KdV equation (2.1), can be integrated using the finite-gap integation technique (cf. [37]). For this purpose, it is useful to reformulate the linear system in a different way, namely in terms of $2P \times 2P$ matrices as follows

$$\mathbf{L} \cdot \varphi = \lambda\varphi, \quad \mathbf{L} = \begin{pmatrix} 0 & -y_0 & 1 & & & 0 \\ 0 & \varepsilon\delta & -x_1 & 1 & & 0 \\ \vdots & & 0 & -y_1 & \ddots & \vdots \\ \vdots & & & \ddots & \ddots & 1 \\ \eta & 0 & & \cdots & 0 & -y_{P-1} \\ -\eta x_0 & \eta & \cdots & \cdots & 0 & \varepsilon\delta \end{pmatrix}, \tag{4.7a}$$

and

$$\varphi' = \mathbf{M} \cdot \varphi, \quad \mathbf{M} = \begin{pmatrix} -\varepsilon\delta/y_0 & 1 & & & & \\ 0 & 0 & 1 & & & \\ \vdots & 0 & -\varepsilon\delta/y_1 & 1 & & \\ \vdots & & \ddots & \ddots & \ddots & \\ 0 & & \cdots & 0 & -\varepsilon\delta/y_{P-1} & 1 \\ \eta & 0 & \cdots & & 0 & 0 \end{pmatrix}. \tag{4.7b}$$

It is may be one of the advantages of the lattice systems that one can take advantage of these kind of reformulations. The compatibility of the system (4.7)

$$\mathbf{L}' = \mathbf{M} \cdot \mathbf{L} \cdot \mathbf{M}^{-1}, \tag{4.8}$$

again yields the mapping (4.3). The Floquet parameter η, arising from the quasi-periodicity of the eigenfunctions ϕ_k of (4.7), in conjunction with the spectral parameter λ, form the local coordinates on the following invariant spectral curve

$$\det(\mathbf{L} - \lambda \mathbf{1}) = \lambda^P (\lambda + \varepsilon\delta)^P + \eta^2 - \eta \operatorname{tr}(T(k)) . \tag{4.9}$$

The explicit solutions of the mappings (4.3) can be expressed in terms of the theta-functions associated with the hyperelliptic curve (4.9). Furthermore, the invariants, arising from the coefficients of (4.9) are in involution with respect to the Poisson brackets

$$\{x_j\,,\, y_{j'}\} = \delta_{j,j'} - \delta_{j,j'+1}, \qquad \{x_j\,,\, x_{j'}\} = \{y_j\,,\, y_{j'}\} = 0. \tag{4.10}$$

The involution property of the invariants which was proven in [36], follows also from the Poisson bracket

$$\{\operatorname{tr} T(k),\ \operatorname{tr} T(k')\} = 0, \tag{4.11}$$

which also follows as a consequence of a nonultralocal Yang–Baxter equation for the mappings established in [38]. The mapping (4.3) is a canonical transformation and can be inferred also from a generating function which we obtained in [36] from the action of the lattice KdV equation. This is another important advantage of the discrete situation: starting from the generalized McMillan mappings, one has an unambiguous quantization procedure, leading to what we have called a theory of integrable quantum mappings (cf. [39–44]). In fact, as was exhibited in [41, 42], there is a very interesting new type of quantum Yang–Baxter structure underlying these discrete-time systems. As we can associate with any periodic solution of the lattice KdV equation a McMillan mapping of arbitrary dimension, and as the lattice KdV equation can be considered to be a lattice regularization of the continuum KdV equation, we can adopt the point of view that this quanization procedure of the corresponding McMillan mappings amounts to a quantization of the KdV system. We will not go into any more details on this issue here, but refer to the paper by Prof. Faddeev in this symposium for a treatment of similar issues (cf., also [45]).

As we have demonstrated above, the lattice KdV equation yields integrable mappings of arbitrary dimension as special reductions. This is very useful for nonlinear dynamics, as such maps can be used as a starting point for the study of near-integrable phenomena (e.g., by studying small perturbations around the integrable situation). It is only relatively recently that integrable maps have been systematically studied (cf. [46–51]), for other treatments in the literature. Possibly, this offers a hand to people working in the arena of chaotic phenomena, who now have at their disposal a large number of exactly solvable, yet highly nontrivial, discrete dynamical systems (other than the simple linear ones that were are traditionally used), to study the more analytical aspects of transitions from integrability to chaos.

5. Similarity Reductions of Integrable Lattices

Similarity solutions of the lattice equations provide another class of solutions that we want to consider in this talk. As is well-known the similarity solutions of integrable nonlinear partial differential equations (PDEs) give rise to the Painlevé transcendents. For instance, the Painlevé II equation

$$y'' = 2y^3 + \xi y + \mu, \tag{5.1}$$

in which ξ is the independent variable, is related to the KdV equation as well as to the MKdV equation by similarity reduction (for the connection between Painlevé equations and soliton equations, see [52–54] and references therein). The Painlevé equations were discovered in the beginning of this century in connection with the problem of classifying all second-order ordinary differential equations without movable singularities other than poles (cf. [55–57]). They were encountered in physics from the seventies onwards in many places, e.g., in the correlation functions of quantum exactly solvable (spin) models and, more recently, in connection with 2D quantum gravity and random matrix models. The issue of constructing difference analogues of the Painlevé equations has been outstanding for a long time, but in the last few years major progress in this direction has been achieved (see [58] for a review). One of the methods that has been applied to obtain such discrete Painlevé equations was the method of similarity reduction, extended to the lattice situation in [59].

The similarity reduction in the lattice case exhibits some new features with respect to the continuum case. Let us illustrate our idea by the analogy to that case. For example, the reduction to similarity solutions for the potential KdV can be formulated in terms of a system consisting of the equation itself and an integrable constraint, namely

$$0 = u + xu_x + 3tu_t, \tag{5.2}$$

from which one derives the similarity variable $\xi \equiv x/3(t^{1/3})$, i.e. $u(x,t) = t^{-1/3} = u(\xi)$. As was shown in [59], in the lattice case we supplement the original lattice equation, which is the lattice (potential) KdV Equation (2.1), with an integrable constraint

$$0 = u + pn\,\frac{u_{n+1,m} - u_{n-1,m}}{2p - u_{n+1,m} + u_{n-1,m}} + qm\,\frac{u_{n,m+1} - u_{n,m-1}}{2q - u_{n,m+1} + u_{n,m-1}}, \tag{5.3}$$

which, in contrast to the continuous case, is nonlinear. Although it might not be possible to solve explicitly for a similarity variable, nevertheless it provides us with a *system* of difference equations, which carries an associated isomonodromic deformation problem. In fact, the similarity reduction can be obtained from the Lax pair of the lattice KdV equation (2.4), together with

$$\begin{aligned} k\frac{\mathrm{d}}{\mathrm{d}k}\phi_k &= S_-\cdot\phi_k + pn\left[-F+\frac{1}{R_p}\left(\frac{V_p}{p-k}+\frac{V_p'}{p+k}\right)\right]\cdot\phi_k + \\ &\quad + qm\left[-F+\frac{1}{R_q}\left(\frac{V_q}{q-k}+\frac{V_q'}{q+k}\right)\right]\cdot\phi_k. \end{aligned} \tag{5.4}$$

In Equation (5.4) we used the following notations:

$$\begin{aligned} R_p &= 2p - u_{n+1,m} + u_{n-1,m}, \qquad R_q = 2q - u_{n,m+1} + u_{n,m-1}, \\ V_p &= \begin{pmatrix} p - u_{n+1,m} & 1 \\ * & p + u_{n-1,m} \end{pmatrix}, \\ V_p' &= \begin{pmatrix} -(p + u_{n-1,m}) & 1 \\ * & -(p - u_{n+1,m}) \end{pmatrix}, \end{aligned} \tag{5.5}$$

and V_q, resp., V_q' denote the matrices obtained from (5.5) by interchanging $p \leftrightarrow q$ and $(n \pm 1, m) \leftrightarrow (n, m \pm 1)$. $*$ denotes in (5.5) the product of the diagonal entries, and finally the matrices F and S_- are given by

$$F = \begin{pmatrix} 0 & 0 \\ 1 & 0 \end{pmatrix}, \qquad S_- = \begin{pmatrix} 0 & 0 \\ 0 & 1 \end{pmatrix}.$$

Clearly, by imposing the constraint (5.3) arising from the compatibility of the linear system (2.4) of the lattice KdV equation and the isomonodromy problem (5.4), we have an integrable reduction corresponding to the lattice analogue of a similarity reduction. In order to give an interpretation of this reduction from the point of view of the initial-value problem, let us note that global solutions can be obtained by imposing initial data on localized configurations of sites ('embryons') on the lattice (cf. [59]). This contrasts the situation of the mappings treated in the previous sections, where we have a periodic initial-value problem on an infinite path in the two-dimensional lattice. The similarity reduction is completely analogous to the continuous case, in which the PDE allows for similarity solutions that are obtained from an ordinary differential equation that requires initial data at a single point in spacetime.

So far, we have discussed the similarity reduction of the lattice KdV equation. However, to get a more direct connection to the Painlevé II equation, it is more convenient to work with the lattice MKdV equation, this providing an alternative way of obtaining a lattice discretization thereof. In that case, the compatibility condition for (2.4) and (5.4) will again lead to an integrable constraint which is

$$n\,\frac{v_{n+1,m}-v_{n-1,m}}{v_{n+1,m}+v_{n_1,m}}+m\,\frac{v_{n,m+1}-v_{n,m-1}}{v_{n,m}+v_{n,m-1}}=0. \tag{5.6}$$

Again, the system consisting of the lattice MKdV Equation (2.12) together with the similarity constraint (5.6) can be viewed as a lattice analogue of the Painlevé II equation. It is this system that we shall refer to as the lattice P-II system.

We shall now discuss the continuum limits that enables one to recover Painlevé II from the lattice P-II system. In Section 2, we already discussed the continuum limit that leads from the lattice equation to the (potential) KdV equation. As the lattice P-II is written as a system containing the original equation plus a similarity constraint, it suffices to look at what happens under this limit with the similarity constraint. Again, we have to perform the limit in two steps: the first leading to the differential-difference equation, and then the next step to the partial differential equation. Performing the first step, (2.6), we obtain from (5.3) the following nonautonomous differential-difference equation

$$\begin{aligned} 0 \;=\; u_n + &\frac{(pn-\tau)\,(u_{n+1}-u_{n-1})\,+\,p\tau\,\partial_\tau(u_{n+1}+u_{n-1})}{2p-u_{n+1}\,+\,u_{n-1}} + \\ &+\frac{2p\tau\,(u_{n+1}-u_{n-1})}{(2p-u_{n+1}\,+\,u_{n-1})^2}. \end{aligned} \tag{5.7}$$

The subsequent next step, performing the limit (2.10) leads to the similarity constraint (5.2). Thus, by two subsequent limits, the lattice system goes over to the system containing the potential KdV equation and the *linear* similarity constraint, which is then related to the Painlevé II equation by a Miura transformation.

To get the more direct connection to Painlevé II, we can also use the lattice MKdV rather than the lattice KdV. In that case, the similarity constraint (5.6) has a limit

$$0=2\tau\,\partial_\tau(y_{n+1}+y_{n-1})+n\left(\mathrm{e}^{y_{n+1}-y_{n-1}}-\mathrm{e}^{y_{n-1}-y_{n+1}}\right), \tag{5.8}$$

and the second continuum limit leads to the MKdV and similarity constraint

$$xy_x+3ty_t=0, \tag{5.9}$$

which immediately leads to the Painlevé II equation (for special value of the parameter $\mu = 0$) after solving the linear constraint in terms of a similarity variable.

A more direct way of obtaining Painlevé type equations on the differential-difference level, is by eliminating from (5.7) and (5.8) the derivatives with respect to τ by using Equations (2.9), resp., (2.18), which in the case of the MKdV equation, leads to an ordinary difference equation, namely

$$z_{n+1}+z_{n-1}+\frac{2p}{\tau}\,\frac{n\,z_n}{1-z_n^2}=0, \qquad z_n\equiv\tanh\left[\tfrac{1}{2}\,(y_{n+1}-y_{n-1})\right], \tag{5.10}$$

which under the above limit also leads to Painlevé II. It is this equation that we would like to call nowadays the *discrete Painlevé II equation*. Other discrete Painlevé equations, i.e. nonautonomous ordinary difference equations that tend to the Painlevé transcendents in a proper continuum limit and that share many of the integrability characteristics with the original Painlevé equations, have been derived and studied in the recent literature (cf. [60–63]). This is an area of intensive research and many new and exciting results are to be expected here.

6. Direct Linearization

Let us now explain how the lattice equations arise from the direct linearization approach. This generalization of the inverse scattering approach, first introduced by Fokas and Ablowitz in [71] and generalized in [72–74] (see also [18] for a review), employs singular linear integral equations with general integration measure and contour. The integral equation introduced in [71] is of the form

$$u_k + \rho_k \int_C \mathrm{d}\lambda(\ell)\, \frac{u_\ell}{k+\ell} = \rho_k. \tag{6.1}$$

Here u_k is a wave function to be solved from the integral equation depending on a complex spectral parameter k and on the coordinates of the system. As we shall note below, these coordinates can be chosen to be discrete as well as continuous, and it is the freedom in this choice that makes integral equations of the type (6.1) a convenient tool to develop discrete integrable systems. Furthermore, in Equation (6.1) C is a contour in the complex k-plane and $\mathrm{d}\lambda(k)$ is a suitably chosen integration measure, whereas ρ_k is a free-wave function depending in a given way on k and on the coordinates of the system. The contour C and measure $\mathrm{d}\lambda(k)$ need to be chosen to be such that the solution u_k of the integral equation for given ρ_k is unique.

Consider now, for example, the free-wave function

$$\rho_k = \mathrm{e}^{kx+k^3 t}, \tag{6.2}$$

and let u_k be the solution of (6.1) for some choice of C and $\mathrm{d}\lambda(k)$, then the potential

$$u(x,t) = \int_C \mathrm{d}\lambda(k)\, u_k(x,t) \tag{6.3}$$

with the same C and $\mathrm{d}\lambda(k)$ satisfies

$$(\partial_t - \partial_x^3)u - 3(\partial_x u)^2 = 0, \tag{6.4}$$

i.e. $v = \partial_x u$ satisfies the Korteweg–de Vries equation.

The integral Equation (6.1) is particularly useful to obtain the lattice discretization of the KdV equation. For this we need to see how Bäcklund transformations

(BTs) arise in the direct linearization approach. The BTs are generated by singular transformations of the measure, or equivalently by a transformation of the free-wave function

$$\rho_k \longrightarrow \tilde{\rho}_k = \frac{p+k}{p-k}\,\rho_k. \tag{6.5}$$

In fact, it can be shown that $\tilde{u} = \int_C \mathrm{d}\lambda(k)\,\tilde{u}_k$ with $\tilde{u}_k$ being the solution of (6.1) with ρ_k replaced by $\tilde{\rho}_k$, is again a solution of (6.4). Considering now two different BTs, one given by $\tilde{}$ and p as paremeter as in Equation (6.5), and the other one being given by replacing the $\tilde{}$ and p replaced by $\hat{}$ and q, we obtain by combining the two BTs

$$(p+q-\hat{\tilde{u}}+u)(p-q+\hat{u}-\tilde{u}) = p^2 - q^2, \tag{6.6}$$

which is the so-called Bianchi identity expressing the commutativity of the two BTs. The connection with the lattice KdV is now clear: the two BTs can be independently iterated retaining the commutativity, leading to a *lattice* of Bäcklund transformed fields u, and the Bianchi identity (6.6) can be interpreted as a consistency condition on a lattice in the form of a partial *difference* equation. Thus, associating the two BTs $\tilde{}$ and $\hat{}$ with two basic translations on a two-dimensional lattice, one finds the lattice KdV equation (2.1). Thus, this partial difference equation is integrable in the sense that solutions can be obtained solving a linear integral equation.

Now, to see the interconnection between the various different lattice equations, we introduce a slight generalization of the integral Equation (6.1), namely

$$\mathbf{u}_k + \rho_k \int_C \mathrm{d}\,\lambda(\ell)\,\frac{\mathbf{u}_\ell}{k+\ell} = \rho_k \mathbf{c}_k, \tag{6.7}$$

where $\mathbf{u}_k$ is now a vector solution, each entry being associated with the corresponding entry of the vector $\mathbf{c}_k$ in the inhomogeneous term, which is given by $(\mathbf{c}_k)_i = k^i$, $i \in \mathbf{Z}$. Thus, we get an infinite vector of solutions, and for this 'wave'-vector we can derive an infinite-matrix system of equations. In fact, introducing the infinite-component 'potential' matrix $\mathbf{U}$, with entries $u_{i,j}$, $i, j \in \mathbf{Z}$, defined as

$$\mathbf{U} = \int_C \mathrm{d}\,\lambda(\ell)\,\mathbf{u}_\ell\,{}^t\mathbf{c}_\ell, \tag{6.8}$$

in which ${}^t\mathbf{c}_\ell$ denotes the adjoint vector to $\mathbf{c}_\ell$ and considering again a BT generated by (2.7), we can derive the following relations for the transformed wave-vectors $\tilde{\mathbf{u}}_k$

$$(p-k)\tilde{\mathbf{u}}_k = (p + \mathbf{\Lambda} - \tilde{\mathbf{U}}\cdot\mathbf{O})\cdot\mathbf{u}_k,$$

$$(p+k)\mathbf{u}_k = (p - \mathbf{\Lambda} + \mathbf{U}\cdot\mathbf{O})\cdot\tilde{\mathbf{u}}_k. \tag{6.9a}$$

The matrix $\mathbf{\Lambda}$ and its transposed ${}^t\mathbf{\Lambda}$ that we have introduced in Equation (6.9) are index-raising operators acting from the left, respectively, from the right, i.e. it acts on, e.g., $\mathbf{u}_k$ by $(\mathbf{\Lambda}\cdot\mathbf{u}_k)^{(i)} = u_k^{(i+1)}$ and $\mathbf{O}$ is a projection matrix singling out the central element of the infinite-component matrix, i.e. for example $(\mathbf{O}\cdot\mathbf{u}_k)^{(i)} = u_k^{(0)}\delta_{i,0}$. By integrating Equations (6.9) over the region C with the same measure $\mathrm{d}\lambda(\ell)$, we arrive at a system of equations for $\mathbf{U}$

$$\widetilde{\mathbf{U}}\cdot(p-{}^t\mathbf{\Lambda}) = (p+\mathbf{\Lambda}-\widetilde{\mathbf{U}}\cdot\mathbf{O})\cdot\mathbf{U}. \tag{6.9b}$$

The system of Equations (6.9b) is an infinitely coupled system, and from it one can derive an inifitely coupled system of *nonlinear* equations for the matrix $\mathbf{U}$. However, it can be shown that by properly combining (6.9b) for different lattice directions and different lattice parameters $p, q, \ldots$, we can get rid of the index-raising operators $\mathbf{\Lambda}$ and find closed partial difference equations, together with their Lax pairs, by using Equations (6.9) to obtain finite-dimensional matrix systems for suitably chosen subsets of components of the vector $\mathbf{u}_k$. Thus, we can obtain from (6.9) the basic set of Equations (2.13), as well as (2.14) and (2.20) of Section 2, by making the following identifications

$$u \equiv U_{0,0}, \qquad v \equiv 1-\left(\frac{1}{r+\mathbf{\Lambda}}\cdot\mathbf{U}\right)_{0,0},$$

$$w \equiv 1-\left(\mathbf{U}\cdot\frac{1}{r'+{}^t\mathbf{\Lambda}}\right)_{0,0}, \tag{6.10a}$$

$$s \equiv \left(\frac{1}{r+\mathbf{\Lambda}}\cdot\mathbf{U}\cdot\frac{1}{r'+{}^t\mathbf{\Lambda}}\right)_{0,0}, \qquad t \equiv \left(\frac{1}{r+\mathbf{\Lambda}}\cdot\mathbf{U}\cdot{}^t\mathbf{\Lambda}\right)_{0,0}. \tag{6.10b}$$

Considering these objects as functions of the lattice sites (n,m), we obtain the corresponding lattice variables, respectively $u_{n,m}$, $v_{n,m}$, $w_{n,m}$, $s_{n,m}$, and $t_{n,m}$ of Section 2. Introducing, furthermore, the variables depending on the spectral parameter k

$$v_k \equiv \left(\frac{1}{r+\mathbf{\Lambda}}\cdot\mathbf{u}_k\right)^{(0)}, \qquad u_k \equiv (\mathbf{u}_k)^{(0)}, \tag{6.11}$$

we can easily derive from having the general system (6.9) the set of equations

$$(p-k)\tilde{v}_k = (p-r)v_k + \tilde{v}u_k, \tag{6.12a}$$

$$(p+k)v_k = (p+r)\tilde{v}_k - v\tilde{u}_k, \tag{6.12b}$$

from which one can also derive the Lax representations presented in Section 2. As a particular interesting special example included in this entire parameter family of related objects, we look into the variables for $r=-p$, $r'=-p$. In that case, we have a very simple situation and, in fact, the system (6.12) leads to the following linear system

$$\kappa P_{n,m}(\kappa) = P_{n+1,m}(\kappa) + R_{n,m}P_{n-1,m}(\kappa), \tag{6.13a}$$

$$P_{n,m+1}(\kappa) = P_{n,m}(\kappa) + \frac{1}{\kappa}Q_{n,m}P_{n-1,m}(\kappa), \tag{6.13b}$$

in terms of the objects

$$\begin{aligned} P_{n,m}(\kappa) &= \left(\frac{p+k}{p-k}\right)^{-n/2}\left(\frac{q+p}{q-k}\right)^{-m/2} v_k(n,m), \\ R_{n,m} &\equiv \frac{h_{n,m}}{h_{n-1,m}}, \qquad Q_{n,m} \equiv -\frac{2p}{p+q}\frac{h_{n,m+1}}{h_{n-1,m}}, \end{aligned} \tag{6.14}$$

where $h_{n,m} = 1 + 2ps_{n,m}$, and where $\kappa = 2p\left(p^2 - k^2\right)^{-1/2}$ is the 'effective' spectral parameter. The compatibility conditions of the system (6.13) leads to the following discrete equation for the variable $Q_{n,m}$

$$\frac{Q_{n,m+1}}{Q_{n+1,m}} = \frac{\alpha Q_{n,m} - 1}{\alpha Q_{n+1,m+1} - 1}, \tag{6.15}$$

in which $\alpha = (q^2 - p^2)/(4p^2)$ arises as an integration constant. Equation (6.15) can be considered to be a discrete version of the Volterra–Kac–van Moerbeke equation (cf. [24]), which is recovered after a suitable continuum limit. In [63] another lattice discretization of the Volterra equation was discussed in connection with a q-deformation of the discrete Painlevé I equation, arising in connection with orthogonal polynomials. We mention that the continuous-time Volterra system plays a role in matrix models for two-dimensional quantum gravity (cf., e.g., [60, 64, 65]). We expect, that the discrete Volterra Equation (6.15) would play a similar role in a discrete variant of that theory. Furthermore, such discrete lattice equations are also occurring in connection with convergence acceleration algorithms in numerical analysis (cf., e.g., [66]) and also [67], as well as in the QD-algorithm and the theory of Padé approximants (cf. [68–70]). This shows the ubiquitousness of the KdV equation in its discrete form: one hundred years after its appearance we have not yet ceased to marvel at its great variety of disguises.

References

1. Korteweg, D. J. and de Vries, G.: On the change of form of long waves advancing in a rectangular canal, and on a new type of long stationary waves, *Philosophical Magazine* **39** (1895), 422–443.
2. Gardner, C. S., Greene, J. M., Kruskal, M. D., and Miura, R. M.: Method for solving the Korteweg–de Vries equation, *Phys. Rev. Lett.* **19** (1967), 1095–1097.
3. Zakharov, V. E. and Shabat, A. B.: Exact theory of two-dimensional self-focussing and one-dimensional self-modulation of waves in nonlinear media, *Sov. Phys. JETP* **34** (1972), 62–69.

4. Ablowitz, M. J., Kaup, D., Newell, A. C., and Segur, H.: The inverse scattering transform – Fourier analysis for nonlinear problems, *Stud. Appl. Math.* **53** (1974), 249–315.
5. Bäcklund, A. V.: Über Flachentransformationen, *Math. Ann.* **9** (1876), 207–320; Zur Theorie der Flachentransformationen, *Math. Ann.* **19** (1882), 387–422.
6. Darboux, G.: *Leçons sur la théorie générale des surfaces et les applications géométriques du calcul infinitesimal*, Vols 1–4, Gauthier-Villars, Paris, 1887–1896.
7. Bianchi, L.: *Lezioni di geometria differenziale*, Enrico Spoerri, Pisa, 1894.
8. Nörlund, N. E.: *Vorlesungen über Differenzrechnung*, Kopenhagen, 1923.
9. Birkhoff, G. D.: General theory of linear difference equations, *Trans. Amer. Math. Soc.* **12** (1911), 243–284; The generalized Riemann problem for linear differential equations and the allied problems for linear difference and q-difference equations, *Proc. Am. Acad. Arts Sci.* **49** (1913), 521–568.
10. Nijhoff, F. W., Quispel, G. R. W., and Capel, H. W.: Direct linearization of Nonlinear difference-difference equations, *Phys. Lett.* **97A** (1983), 125–128.
11. Quispel, G. R. W., Nijhoff, F. W., Capel, H. W., and van der Linden, J.: Linear integral equations and nonlinear difference-difference equations, *Physica* **125A** (1984), 344–380.
12. Nijhoff, F. W., Capel, H. W., Wiersma, G. L., and Quispel, G. R. W.: Bäcklund transformations and three-dimensional lattice equations, *Phys. Lett.* **105A** (1984), 267–272.
13. Nijhoff, F. W., Capel, H., and Wiersma, G. L.: Integrable lattice systems in two- and three-dimensions, in R. Martini (ed.), *Geometric Aspects of the Einstein Equations and Integrable Systems*, Springer Lect. Notes Phys. Vol. 239 (1985), pp. 263–302.
14. Wiersma, G. and Capel, H. W.: Lattice equations, hierarchies and Hamiltonian structures, *Physica* **142A** (1987), 199–244; *ibid.* **149A** (1988), 49–74.
15. Nijhoff, F. W. and Papageorgiou, V. G.: Lattice equations associated with the Landau–Lifschitz equations, *Phys. Lett.* **141A** (1989), 269–274.
16. Nijhoff, F. W. and Capel, H. W.: The direct linearisation approach to hierarchies of integrable PDE's in $2+1$ dimensions: I. Lattice equations and the differential-difference hierarchies, *Inv. Problems* **6** (1990), 567–590.
17. Nijhoff, F. W., Papageorgiou, V. G., Capel, H. W., and Quispel, G. R. W.: The lattice Gel'fand–Dikii hierarchy, *Inv. Problems* **8** (1992), 597–621.
18. Capel, H. W. and Nijhoff, F. W.: Integrable lattice equations, in A. S. Fokas and V. E. Zakharov (eds), *Important Developments in Soliton Theory*, Lect. Notes in Nonlinear Dynamics, Springer, 1993, pp. 38–57.
19. Ablowitz, M. J. and Ladik, F. J.: A nonlinear difference scheme and inverse scattering, *Stud. Appl. Math.* **55** (1976), 213–229; On the solution of a class of nonlinear partial difference equations, *ibid.* **57** (1977), 1–12.
20. Hirota, R.: Nonlinear partial difference equations, I–III, *J. Phys. Soc. Japan* **43** (1977), 1424–1433, 2074–2089.
21. Hirota, R.: Discrete analogue of generalized Toda equation, *J. Phys. Soc. Japan* **50** (1981), 3785–3791.
22. Levi, D., Pilloni, L., and Santini, P. M.: Integrable three-dimensional lattices, *J. Phys. A: Math. Gen.* **14** (1981), 1567–1575.
23. Date, E., Jimbo, M., and Miwa, T.: Method for generating discrete soliton equations, I–V, *J. Phys. Soc. Japan* **51** (1982), 4116–4131, **52** (1983), 388–393, 761–771.
24. Kac, M. and van Moerbeke, P.: On an explicitly soluble system on nonlinear differential equations related to certain Toda lattices, *Adv. Math.* **16** (1975), 160–169.
25. Krichever, I. M. and Novikov, S. P.: Holomorphic bundles over algebraic curves and nonlinear equations, *Russ. Math. Surv.* **35** (1980), 53–79.
26. Faddeev, L. D. and Tahktajan, L. A.: *Liouville Model on the Lattice*, Lect. Notes Phys., Vol 246, Springer, 1986, pp. 166–179.
27. Volkov, A. Yu.: Miura transformation on the lattice, *Theor. Math. Phys.* **74** (1988), 96–99.
28. Drinfel'd, V. G. and Sokolov, V. V.: Equations of KdV type and simple Lie algebras, *Sov. Math. Dokl.* **23** (1981), 457–462; *J. Sov. Math.* **30** (1985), 1975.
29. Quispel, G. R. W., Nijhoff, F. W., and Capel, H. W.: Linearization of the Boussinesq equation and of the modified Boussinesq equation, *Phys. Lett.* **91A** (1982), 143–145.

30. Airault, H., McKean, H., and Moser, J.: Rational and elliptic solutions of the Korteweg–de Vries equation and a related many-body problem, *Comm. Pure Appl. Math.* **30** (1977), 95–148.
31. Nijhoff, F. W. and Pang, G. D.: A time-discretized version of the Calogero–Moser model, *Phys. Lett.* **191A** (1994), 101–107.
32. Nijhoff, F. W. and Pang, G. D.: Discrete-time Calogero–Moser model and lattice KP equations, in D. Levi, L. Vinet and P. Winternitz (eds), *Proc. of the Int. Workshop on Symmetries and Integrability of Difference Equations*, CRM Lecture Notes and Proceedings, to appear, hep-th/9409071.
33. Nijhoff, F. W., Ragnisco, O., and Kuznetsov, V.: Integrable time-discretisation of the Ruijsenaars–Schneider model, Preprint No. 94-27, University of Amsterdam, hep-th/9412170.
34. Papageorgiou, V. G., Nijhoff, F. W., and Capel, H. W.: Integrable mappings and nonlinear integrable lattice equations, *Phys. Lett.* **147A** (1990), 106–114.
35. McMillan, E. M.: in W. E. Brittin and H. Odabasi (eds), *Topics in Physics*, Colorado Associated Univ. Press, Boulder, 1971, p. 219.
36. Capel, H. W., Nijhoff, F. W., and Papageorgiou, V. G.: Complete integrability of Lagrangian mappings and lattices of KdV type, *Phys. Lett.* **155A** (1991), 377–387.
37. Belokolos, E. D., Bobenko, A. I., Enolskii, V. Z., Its, A. R., and Matveev, V. B.: *Algebro–Geometric Approach to Nonlinear Evolution Equations*, Springer-Verlag, 1994.
38. Nijhoff, F. W., Papageorgiou, V. G., and Capel, H. W.: Integrable time-discrete systems: lattices and mappings, in P. P. Kulish (ed.), *Quantum Groups*, LNM, Vol. 1510, Springer, 1992, pp. 312–325.
39. Nijhoff, F. W., Capel, H. W., and Papageorgiou, V. G.: Integrable quantum mappings, *Phys. Rev.* **A46** (1992), 2155–2158.
40. Quispel, G. R. W. and Nijhoff, F. W.: Integrable two-dimensional quantum mappings, *Phys. Lett.* **161A** (1991), 419–422.
41. Nijhoff, F. W. and Capel, H. W.: Integrable quantum mappings and non-ultralocal Yang–Baxter structures, *Phys. Lett.* **A163** (1992), 49–56.
42. Nijhoff, F. W. and Capel, H. W.: Integrability and fusion algebra for quantum mappings, *J. Phys.* **A26** (1993), 6385–6407.
43. Nijhoff, F. W. and Capel, H. W.: Integrable quantum mappings and quantization aspects of discrete-time integrable systems, in P. A. Clarkson (ed.), *Applications of Analytic and Geometric Methods to Nonlinear Differential Equations*, NATO ASI Series C, Vol. 413, Kluwer Acad. Publ., Dordrecht, 1993, pp. 163–182.
44. Nijhoff, F. W. and Capel, H. W.: Quantization of integrable mappings, *Springer Lect. Notes Phys.* **424** (1993), 187–211.
45. Faddeev, L. D. and Volkov, A. Yu.: Quantum inverse scattering method on a space-time lattice, *Theor. Math. Phys.* **92** (1992), 837–842.
46. Quispel, G. R. W., Roberts, J. A. G., and Thompson, C. J.: Integrable mappings and soliton equations, *Phys. Lett.* **A126** (1988), 419–421; *ibid. II, Physica* **D34** (1989), 183–192.
47. Veselov, A. P.: Integrable maps, *Russ. Math. Surv.* **46**(5) (1991), 1–51.
48. Bruschi, M., Ragnisco, O., Santini, P. M., and Tu, G.-Z.: Integrable symplectic maps, *Physica* **49D** (1991), 273–294.
49. Veselov, A. P.: Growth and integrability in the dynamics of maps, *Comm. Math. Phys.* **145** (1992), 181–193.
50. Veselov, A. P.: Integrable Lagrangian correspondences and the factorization of matrix polynomials, *Funct. Anal. Appl.* **25** (1991), 112–122;
Deift, P. A., Li, L. C., and Tomei, C.: Matrix factorizations and integrable systems, *Comm. Pure Appl. Math.* **42** (1989), 443–521;
Moser, J. and Veselov, A. P.: Discrete versions of some classical integrable systems and factorization of matrix polynomials, *Comm. Math. Phys.* **139** (1991), 217–243.
51. Suris, Yu. B.: Integrable mappings of the standard type, *Funct. Anal. Appl.* **23** (1989), 74–79; Generalized Toda lattices in discrete time, *Leningrad Math. J.* **2** (1991), 339–352.
52. Ablowitz, M. J. and Segur, H.: *Solitons and the Inverse Scattering Transform*, SIAM, Philadelphia, 1981.

53. Its, A. R. and Novokshenov, V. Y.: *The Isomonodromic Deformation Theory in the Theory of Painlevé Equations*, Lect. Notes Math., Vol. 1191, Springer-Verlag, Berlin, 1986.
54. Ablowitz, M. J. and Clarkson, P. A.: *Solitons, Nonlinear Evolution Equations and Inverse Scattering*, LMS Lect. Notes, Vol. 149, Cambridge University Press, 1991.
55. Painlevé, P.: Memoire sur les équations différentielles dont l'intégrale générale est uniforme, *Bull. Soc. Math. France* **28** (1900), 201–261; Sur les équations différentielles du second ordre et d'ordre supérieur dont l'intégrale générale est uniforme, *Acta Math.* **25** (1902), 1–85.
56. Gambier, B.: Sur les équations différentielles du second ordre et du premier degré 'dont l'intégrale générale est à points critiques fixés, *Acta Math.* **33** (1909), 1–55.
57. Ince, E. L.: *Ordinary Differential Equations*, Dover Publ., New York, 1956.
58. Grammaticos, B. and Ramani, A.: Discrete Painlevé equations: Derivation and properties, in P. A. Clarkson (ed.), *Applications of Analytic and Geometric Methods to Nonlinear Differential Equations*, NATO ASI Series C, Vol. 413, Kluwer Acad. Publ., Dordrecht, 1993, pp. 299–314.
59. Nijhoff, F. W. and Papageorgiou, V. G.: Similarity reductions of integrable lattices and discrete analogues of the Painlevé II equation, *Phys. Lett.* **153A** (1991), 337–344.
60. Fokas, A. S., Its, A. R., and Kitaev, A. V.: Discrete Painlevé equations and their appearance in quantum gravity, *Comm. Math. Phys.* **142** (1991), 313–344.
61. Ramani, A., Grammaticos, B., and Hietarinta, J.: Discrete versions of the Painlevé equations, *Phys. Rev. Lett.* **67** (1991), 1829–1832.
62. Papageorgiou, V. G., Nijhoff, F. W., Grammaticos, B., and Ramani, A.: Isomonodromic deformation problems for discrete analogues of Painlevé equations, *Phys. Lett.* **A164** (1992), 57–64.
63. Nijhoff, F. W.: On a q-deformation of the discrete Painlevé I equation and q-orthogonal polynomials, *Lett. Math. Phys.* **30** (1994), 327–336.
64. Gross, D. J. and Migdal, A. A.: Non-perturbative two-dimensional quantum gravity, *Phys. Rev. Lett.* **64** (1990), 127; *Nucl. Phys.* **B340** (1990), 333.
65. Alvarez, O. and Windey, P.: Universality in two-dimensional quantum gravity, *Nucl. Phys.* **B348** (1991), 490.
66. Papageorgiou, V. G, Grammaticos, B., and Ramani, A.: Integrable lattices and convergence acceleration algorithms, *Phys. Lett.* **179A** (1993), 111–115.
67. Brezinski, C.: *Accélération de la Convergence en Analyse Numérique*, Lecture Notes in Mathematics, Vol. 584, Springer, 1970.
68. Papageorgiou, V. G, Grammaticos, B., and Ramani, A.: Orthogonal polynomial approach to discrete Lax pairs for initial-value-boundary problems of the QD-algorithm, to appear in *Lett. Math. Phys.*
69. Brezinski, C.: *Padé-Type Approximation and General Orthogonal Polynomials*, ISNM, Vol. 50, Birkhäuser Verlag, 1980.
70. Gragg, W. B.: The Padé table and its relation to certain algorithms of numerical analysis, *SIAM Rev.* **14** (1972), 1–62.
71. Fokas, A. S. and Ablowitz, M. J.: Linearization of the Korteweg–de Vries and painlevé II equations, *Phys. Rev. Lett.* **47** (1981), 1096–1100.
72. Nijhoff,F. W., Quispel, G. R. W., van der Linden, J., and Capel, H. W.: On some linear integral equations generating solutions of nonlinear partial differential equations, *Physica* **119A** (1983), 101–142.
73. Nijhoff, F. W.: Linear integral transformations and hierarchies of integrable nonlinear evolution equations, *Physica* **D31** (1988), 339–388.
74. Santini, P. M., Fokas, A. S., and Ablowitz, M. J.: The direct linearization of a class of nonlinear evolution equations, *J. Math. Phys.* **25** (1984), 2614–2619.

Acta Applicandae Mathematicae **39:** 159–172, 1995.

Coherent Structure Visiometrics: From the Soliton to HEC*

NORMAN J. ZABUSKY
Laboratory for Visiometrics and Modeling, Department of Mechanical and Aerospace Engineering, Rutgers University, Piscataway, New Jersey 08855-0909, USA

(Received: 30 January 1995)

Abstract. In this paper I give a brief personal review of the history of the early soliton days and how they led to the *visiometrics and reduced modeling* paradigm that has been a part of my approach to nonlinear science in the last three decades. I illustrate it with HEC (Hybrid Elliptic-Contour): a fast, minimal, asymptotically motivated model for unforced, 2-dimensional incompressible weakly dissipative turbulence (U2DIT).

Mathematics Subject Classification (1991): 58F07.

Key words: visiometrics, reduced modeling, asymptotics, fluid dynamics.

1. Introduction and Overview

Thirty years ago, to the season, Martin Kruskal and I introduced the soliton paradigm for the Korteweg–de Vries (KdV) nonlinear partial differential equation of 1895 [12], while studying phenomena observed in the Fermi–Pasta–Ulam nonlinear lattice simulations. We focussed on the space-time interactions of coherent structures and were able to 'understand' the near-recurrence of low harmonic initial states in terms of their disassembly into solitons and their long-time rearrangement and reassembly (or focussing). Our numerical simulation results were visualized by Gary Deem and myself in a cinema, the first for a nonlinear partial differential equation [4]. It was made available by the Bell Telephone Laboratories film library in the 1960's through the 1980's and seen by many.

This work came about through the effective collaboration of Kruskal (*asymptotologist* at the Princeton Plasma Physics Laboratory and Deem (*computationalist*) and myself (*'visiometricist'*) at Bell Telephone Laboratories in Whippany, New Jersey. These *titles* are to highlight what I believe to be a modus for productive collaborations. I will illustrate this approach by presenting recent accomplishments on unforced, 2-dimensional incompressible weakly dissipative turbulence (U2DIT) done in collaboration with D. G. Dritschel.

* The work on reduced modeling of two-dimensional turbulence is on-going with David G. Dritschel and Hongbing Yao (incompressible); and Jaideep Ray, Thomas Scheidegger and Ravi Samtaney (compressible).

Fifteen years later, when I wrote my first two review papers on the subject [22, 23], I realized that we had provided an answer to the oft-quoted von Neumann [19] 1946 questions, namely:

1) "To what extent can human reasoning in the sciences be more efficiently replaced by mechanisms?

2) What phases of pure and applied mathematics can be furthered by the use of large-scale, automatic computing instruments?"

After some dramatically bold remarks,* he promulgated his peak prescience:

> "We conclude by remarking that really efficient high-speed computing devices may, in the field of nonlinear partial differential equations as well as in many other fields which are now difficult or entirely denied of access, provide us with those heuristic hints which are needed in all parts of mathematics for genuine progress...."

The visiometric approach, about which I have written previously, and which I illustrate below is a beginning answer to the first question. The introduction of the soliton paradigm in 1965 and its overflowing fundamental and applied consequences in mathematics, physics and engineering** is an answer to the second question. Together, they affirm von Neumann's conjectures.

2. Reduced Modeling

Nonlinear processes are usually unstable, chaotic and turbulent. Often they are unpredictable beyond 'early' times, and features evolve into complex topologies with near-singular space-time events. To acquire deeper intuition and understanding of these phenomena, we often find it useful to formulate low degree-of-freedom models from insights gained by juxtaposing a large number of many degree-of-freedom computer simulation solutions with laboratory and field experiments. If the latter experiments are not available, then great care must be taken to elucidate the content of the massive data sets obtained from numerical simulations, as described below. A practical goal is to obtain 'efficient' computer

* "Our present analytical methods seem unsuitable for the solution of the important problems arising in connection with nonlinear partial differential equations and, in fact, with virtually all types of nonlinear problems in pure mathematics. The truth of this statement is particularly striking in the field of fluid dynamics. Only the most elementary problems have been solved analytically in this field.... The advance of analysis is, at the moment, stagnant along the entire front of nonlinear problems... This phenomenon is not of a transient nature, but... we are up against an important conceptual difficulty... yet no decisive progress has been made... which could be rated as important by the criteria that are applied in other, more successful (linear!) parts of mathematical physics...."

** The application of the soliton concept to a successful commercial venture – the optical fiber transmission line, is taking place as a result of the persistent and combined efforts of ingenious theorists and laser experimentalists, materials scientists and simulation specialists. They have built on the early ideas for optical solitons of Frederick Tappert and Akira Hasegawa, who both were at Bell Labs in the early soliton days.

models that aid the prediction and design of natural and man-made processes.

By juxtaposition, I mean visualized and quantified comparisons among side-by-side or overlaid data sets. These comparisons are of similar-or-different functions at the same-or-different times from simulations of the same or linked mathematical models in a relevant domain of parameters (initialization). This is a step in the *visiometric* process: i.e. *the visualization and quantification of evolving amorphous structures within space-time data sets.*

For nonlinear phenomena, one must be aware, particularly at long times, solution (or system response) to forcing and to dissipative and dispersive regularizations at small length scales, including artifactual numerical structures. This information will allow a validation of candidate mathematical models [2, 18, 25]. The goal is to interactively and automatically juxtapose adequately visualized and quantified evolutionary data sets from various experiments and simulations and become aware of their heuristic-innovative content. In essence, visiometrics is the way to mathematical metaphors, the mother lode referred to in the opening quotation [21].

I propose a tentative definition:

A reduced model is one in an asymptotically linked hierarchy, which is used to obtain the solution to a cogent problem and minimizes an error-cost function. Cogent means within a narrow-but-useful physical parameter range. The error-cost could be the product (solution error)$\times$(computational cost), where error is a combination of truncation, filtering, round-off and physical (omission of processes) and computational cost relates to algorithm and parallelization, etc.

For example, an explicit formula is usually good for steady processes, e.g., the Kolmogoroff spectral indices for homogeneous, isotropic forced incompressible turbulence. A set of linear simultaneous equations without-and-with constraints may be the next level. This is followed in complexity by: sets of nonlinear equations; single or coupled sets of linear ordinary, partial or integro-differential equations (e.g., linear stability theory); single or coupled sets of nonlinear ordinary, partial or integro-differential equations, and particle-continuum hybrid systems.

The derivation by Korteweg and de Vries of the (1+1) nonlinear partial differential equation from the (2+1) Euler equations in an ideal shallow water environment is an example of low-degree-of-freedom modeling. Kruskal's formulation of the KdV equation from the one-dimensional cubic nonlinear lattice, follows in this tradition.

3. Computational Fluid Dynamics

The field of fluid dynamics has dominated the contemporary application of computers, from the practical to the fundamental, e.g., from the design of aircraft and Tokamak fusion devices to the understanding and prediction of large-scale geophysical and astrophysical phenomena. This arena of endeavor has seen the interaction of basic and applied engineers, physicists and mathematicians.

In this brief survey of reduced modeling, I wish to focus on a problem which has been revived recently by several groups: modeling vortex interactions in nearly-inviscid U2DIT. Unfortunately, because the simulations have not been adequately diagnosed, unrealistic impressions of the nature of elementary vortex interactions have been used for model building. To correct this situation, one must develop tools for perusing and quantifying massive data sets, as described below.

The simulation data sets must be computed beyond some 'early' time. For example, a time interval associated with several linear instability growth times or *five* eddy turn around times of some intermediate-sized vortex in a turbulent flow. As the simulation time interval increases one must be wary of phenomena which are manifestations of the initial and boundary condition or artifacts of the numerical codes. For example: the dissipation operator (real and regularized at large and small scale); order of discretization in space and time; and filtering (e.g., dealiasing in spectral codes).

I define more precisely some terms in frequent use when elucidating the properties of data sets.

- *Global quantification*: Measuring total quantities, some of which are invariant in some region of parameter space. This includes: energy, enstrophy, helicity; drag force vector on an object; correlation functions; slopes and variances of power density spectra; fractal dimension; moments of statistical quantities (e.g., skewness, kurtosis, etc); Lyapunov exponents; probability distribution functions, etc.
- *Detailed quantification*: Characterizing space-time topologies, including the evolution of critical and 'extremal' points and surfaces of scalar, vector and tensor fields in appropriate moving frames; (e.g., the rate of strain tensor and its diagonal representation at at every point in the field, etc); space-time or parameter-time projections of localized functions and their gradients; slopes, normals, curvatures and torsions of lines and surfaces; low-order moment and skeletal description of coherent structures; distances between extrema (or centroids) and critical points; temporal variation of modal amplitudes and phases; etc.
- *Coherent vortex structure (CVS) or 'vortex'*: A localized region ('core' or 'layer') of fluid which persists for many periods of some relevant time interval (see above). We believe that an optimal definition of a coherent object will allow one to to track the largest fraction of the content of the object (for vortices, the circulation) for the longest time. That is, when a CVS is distorting due to the presence of a strong strain field, one should be able to track it until a complex interaction configuration is formed and then afterward when another set of 'outgoing' coherent vortex states appear. The time element should be considered in defining a coherent structure. In

1987, Melander, McWilliams and I [16] made quantifications of the time-dependent 'axisymmetrization' of a single, elliptically shaped CVS in 2D in a time varying corotating reference frame. We introduced *diagnostic ellipses* to quantify domains that are near extrema of the function.

4. An Asymptotically Motivated Model for U2DIT: The Hybrid Elliptic-Contour Model (HEC)

4.1. Introduction to Contour Dynamics/Surgery and the Elliptical/Moment Models

Despite massive computer power, the nearly-inviscid regime of U2DIT at intermediate and late times (after hundreds to thousands of vortex rotation periods) has remained largely inaccessible to standard numerical methods [9]. Some believe that the use of hyperviscosity (so common today in numerical models) has given access to this regime. However, recent works [14, 20] have uncovered serious numerical artifacts associated with its use and have put into question the results of many simulations of U2DIT beyond early times. Hence, we propose a dual representation: the Contour Dynamics/Surgery model [24, 5] with one contour per vortex (the curve is discretized by a prescribed finite number of nodes which govern a truncation error) and the Elliptical/Moment model [15, 7, 13] with one ellipse per vortex (represented by: center; area; aspect ratio and phase).

The contour dynamics method [24] treats the evolution of the 2-dimensional incompressible Euler equations, initialized with *piecewise-constant* distributions of vorticity (single or nested contours) and the 'surgery' is a dissipative regularization which alters the topology of the contours at length scales up to two orders of magnitude smaller than treated by standard (e.g., pseudospectral) numerical methods. Essentially, identical magnitude vorticity regions are joined or split if a small scale criterion is satisfied and small-scale filaments are cut-away or joined to larger vortex domains. At present, the small cut-away filaments are removed from the flow domain (i.e. moved to an uncoupled 'sea' of well-mixed small-scale vorticity). Most important, vorticity gradients do not dissipate.

A substantial number of works have been devoted to the study of vortex interactions in U2DIT [8]. However, remarkably little quantitative evidence has been given for what constitutes the generic vortex interaction in weakly dissipative U2DIT. That is, there has been little work on the study of vortex behaviour in a many-vortex environment. *Ad-hoc* models for the evolution of vortex properties have been built on the idea that the generic interaction at any Reynolds number is the merging of two like-signed vortices [1, 3]. These models have been criticized [8, 10, 14, 20] for their: over-simplicity; lack of asymptotic validity; implicitly large-dissipative behavior; and inability to explain vortex statistical behaviour at very high Reynolds numbers.

It is indeed tempting to have a vortex-based theory for U2DIT, and two crucial questions are:

(1) Does their exist a small set of generic vortex collisions (or ingoing–outgoing states)?

(2) Can one prescribe an asymptotically motivated set of transformation rules among models in a hierarchy to relate the ingoing and outgoing collision states?

We discuss below why the answers to (1) and (2) are no and yes, respectively.

4.2. Vortex Tracking, Panning and Zooming in an Interactive Environment

Recently, an automatic-and-interactive vortex-tracking algorithm has been developed for weakly disipative 2D flows [11, 6]. This begins to make possible a detailed analysis of the complexity of vortex interactions in U2DIT and, generally, in layerwise-2D geophysical flows. This new tool uses and extends the CS algorithm itself and enables the topological reconnection, fission or fusion, of contiguous regions of vorticity, rather than just contours (vorticity field isolevels). This extension is non-trivial due to the frequent appearance, at high Reynolds number, of entrained 'pockets' of weak vorticity within contiguous regions of higher vorticity.

The algorithm, presently designed for simulations on a *spherical* surface, gives a list of all inelastic interactions between vortices that result in a specified change in circulation for one of the vortices involved. This information is then used to select 'interesting' vortices (e.g., those undergoing strong inelastic interactions) and time periods for visualizing and quantifying. The algorithm pans the selected vortex by placing it at the center of the users field of view, and renders the top side of the sphere with vortices in color or both sides with vortices as contours (front side as solid curves and the back or hidden side as dashed curves), as shown in Figures 1 and 2.

Figure 1 shows an example of the colored vortices at a particular time for $N_0 = 100$, $f_0 = 0.1$, where N_0 is the number of positive and negative vortices and f_0, the vortex fractional area or 'diluteness', is defined as the fractional area of the surface covered initially by vorticity. We have enhanced this figure by zooming in (lower right) on the tracked vortex (No. 15) and also a complex interaction (lower left). Also indicated is $N = 124$, the total number of vorticies (contours) at that time. In Figure 2, in addition to the contours, we also show at the centroid of the tracked contour: a diagnostic ellipse (obtained by using second-order moments of the tracked contour); an orthogonal coordinate system rendered by small solid and dashed line segments (obtained by diagonalizing the local strain tensor), and the relative velocity vector of the contour (not shown here).

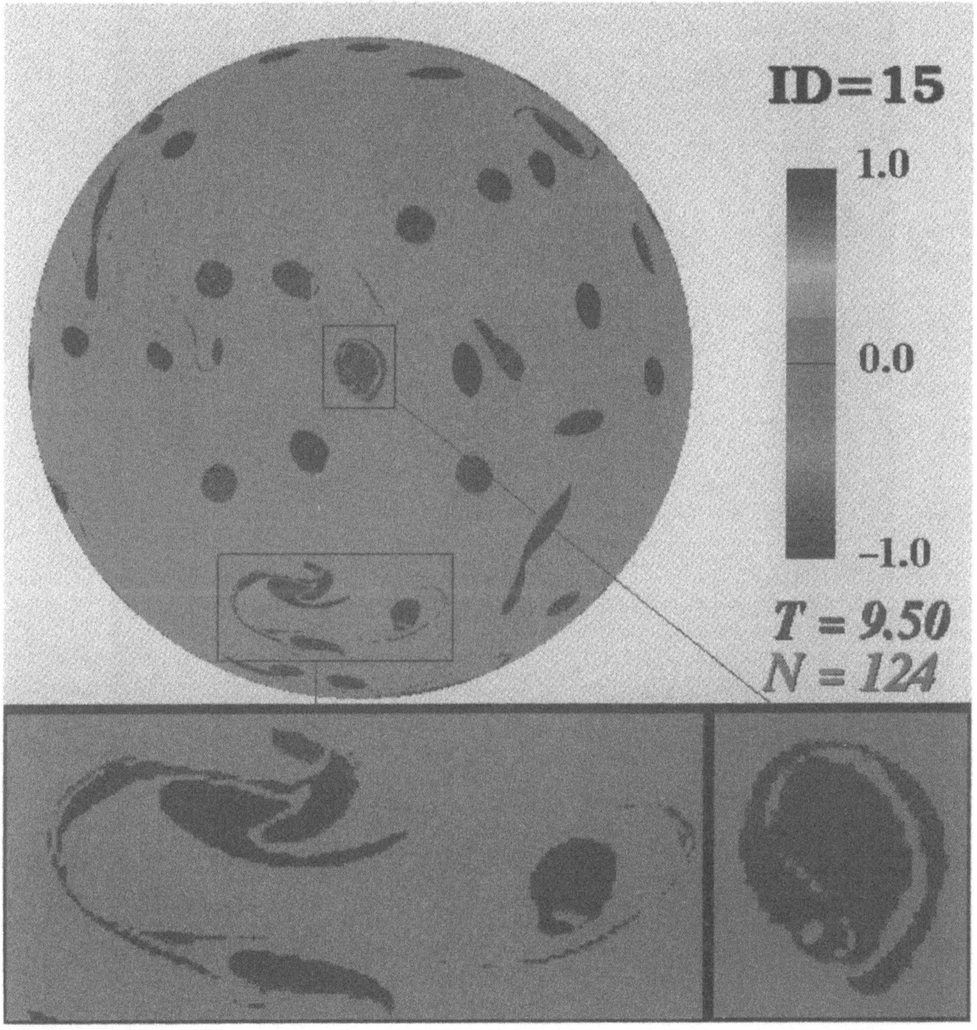

Fig. 1. Vortices in color at a particular time for $N_0 = 100$, $f_0 = 0.1$, where N_0 is the initial number of positive and negative vortices and f_0, the initial vortex density or diluteness, is defined as the fractional area of the surface initially covered by vorticity. We have enhanced the figure by zooming in on the tracked vortex, lower right (No. 15) and a complex interaction, lower left. Also indicated is $N = 124$, the total number of vorticies (contours) at that time.

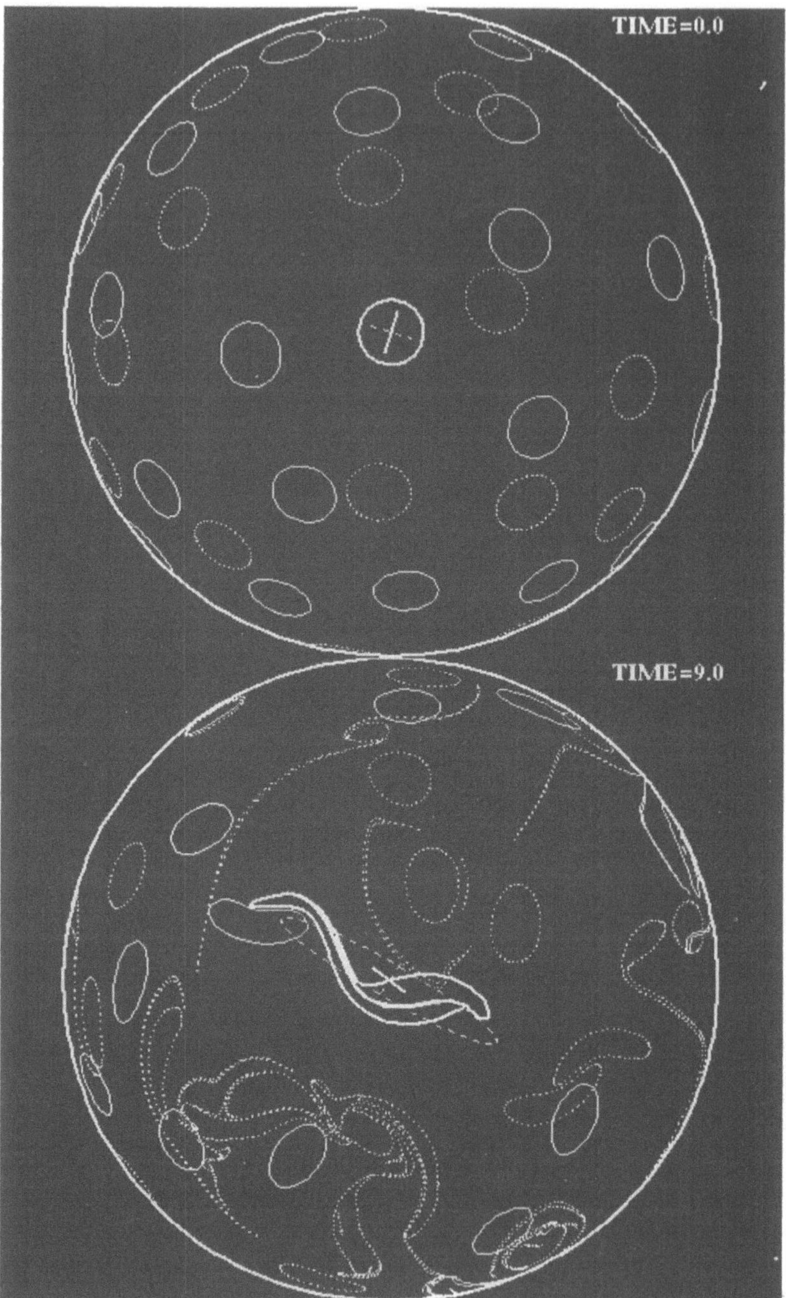

Fig. 2. Vortices as contours on both sides of the sphere at two different times for $N_0 = 40$, $f_0 = 0.1$. (The front side vortices are solid curves and the back or hidden side are dashed curves.) Also shown at the centroid of the tracked contour: an ellipse (obtained by using second-order moments of the contour and rendered as a dashed line); the centroid of the tracked contour: an ellipse (obtained by using second-order moments of the contour and rendered as a dashed line); the orthogonal diagonal elements of the local strain tensor (rendered as as solid and dashed line segments).

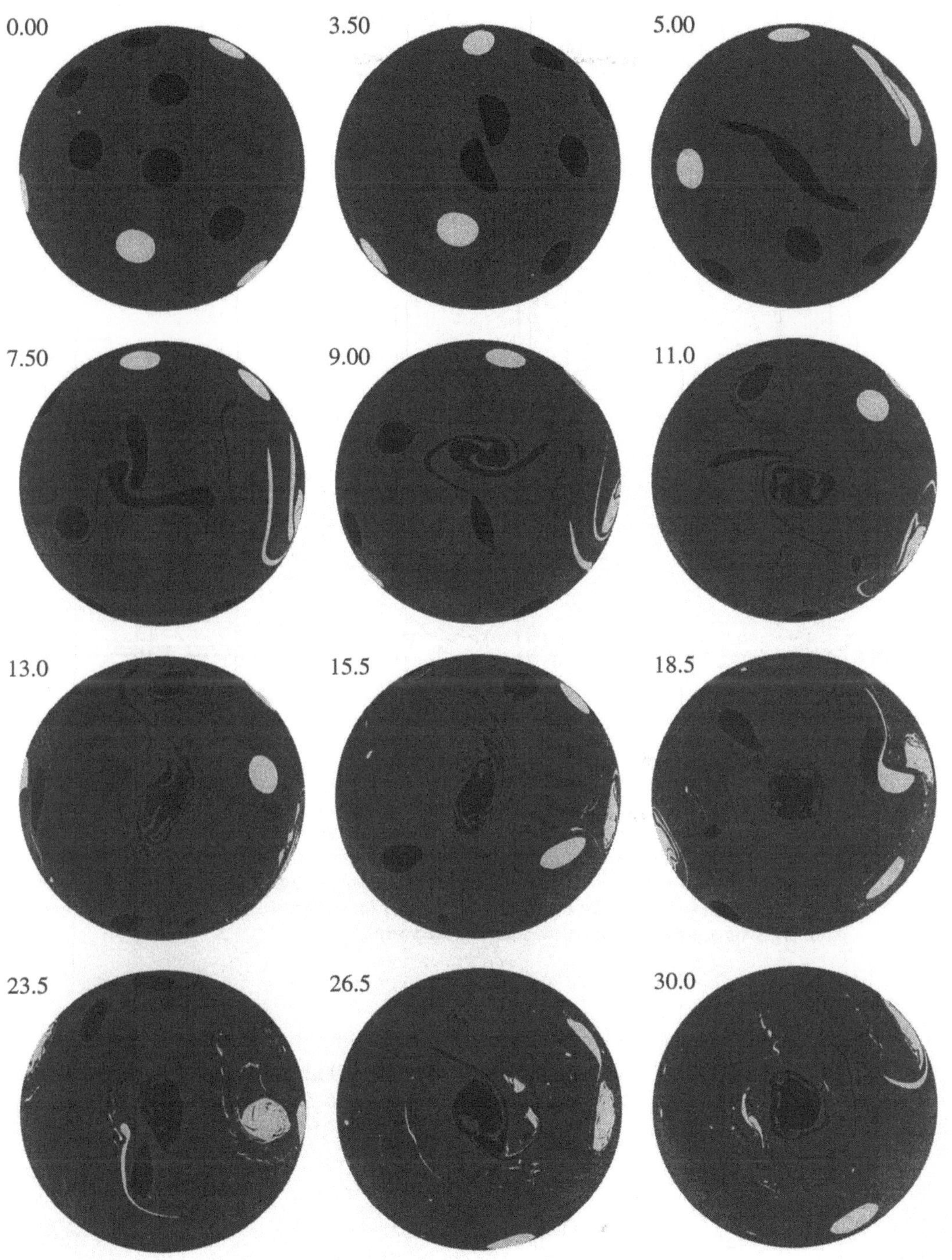

Fig. 3. A gray scale version of 12 frames of the top side for $N_0 = 20$, $f_0 = 0.1$.

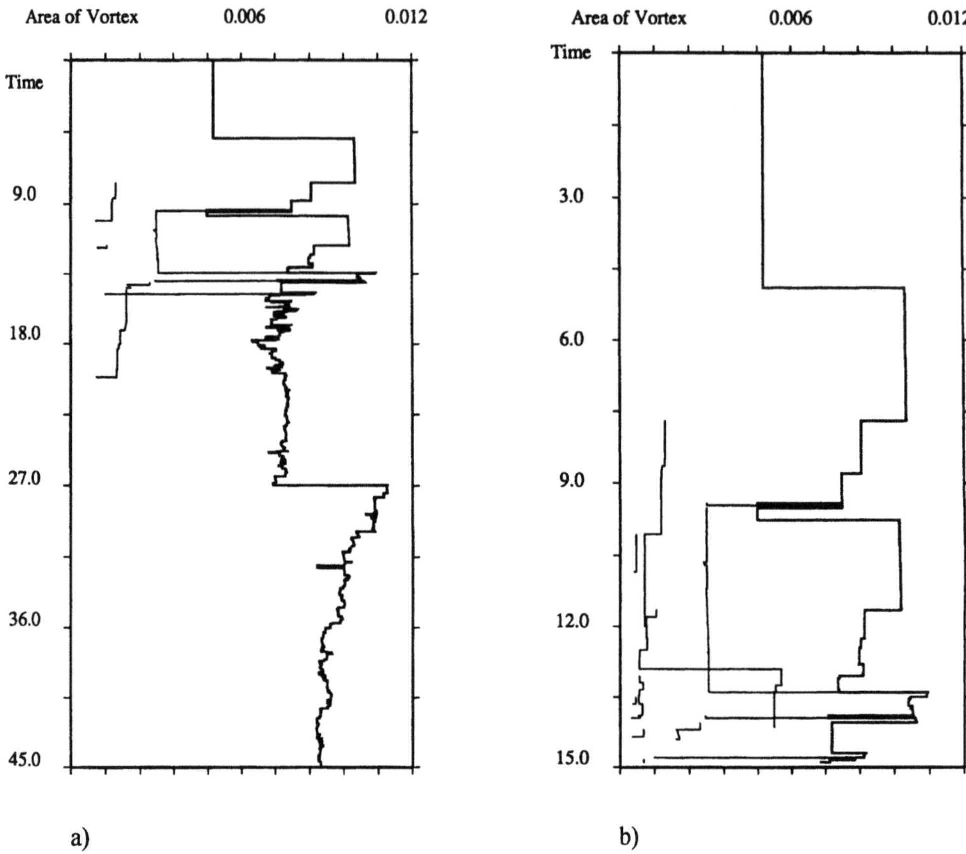

Fig. 4. Cascade (vortex-area/time) diagram for the simulation exhibited in Figure 3 (for $N_0 = 20$, $f_0 = 0.1$) with two different time scales. In the diagrams the horizontal axis is the area of a preselected or tracked vortex (renderered as a bold line) and its progeny (renderered as light lines). Movement to the right (left) indicates merger with (splitting-off of) progeny vortex regions. In the latter case the progeny region is (or regions are) tracked until the remnants fall below a threshold or until they merge. In this figure we also see a remerger of a progeny vortex with the tracked vortex.

The algorithm also produces a *cascade* or *a–t* diagram, where the preselected vortex *area* is plotted vs *time* (downward). This diagram allows us to obtain a quantitative assessment of the complexity of an interaction. Figures 4a and 4b are a–t diagrams, which correspond to the evolution of Figure 3, a gray scale version containing 12 frames of the front side of the sphere for $N_0 = 20$, $f_0 = 0.1$. (Note, Figure 4b expands the time axis of Figure 4a by a factor of 3.)

A horizontal movement to the right (left) of the bold line indicates merger with (splitting-off of) another or progeny vortex region. In the latter case, the progeny region is (or regions are) rendered to the left (with a lighter line) if the magnitude of the split-off area is greater than a preassigned threshold or until they merge with another vortex (a light line to the right and they lose their identity).

The tracking continues until the remnants fall below a preassigned threshold. In these figures, we see progeny mergers with the tracked vortex.

4.3. DISCUSSION OF RESULTS

We now focus on some inelastic vortex interactions in three recent long-time CS simulations. In each case at the initial time, an equal number of positive and negative equal-vorticity magnitude and nonoverlapping circular patches (single-contour vortices containing uniform vorticity) were placed randomly on the surface of a sphere of unit radius. The patches were constrained to be separated initially by a minimum distance which is large compared to an average vortex radius. (The ratio: [minimum distance (along the chord) between the centers of any pair of vortices / average of their radii] is large).

The vortices were distributed in sizes according to the number density distribution $n(A) \propto A^{-1.8}$. Note, such a number density distribution (with a slope -1.8 ± 0.1) has been observed to arise during long-time evolutions [8] among all but the largest-scale vortices.

In our lowest density simulation ($f_0 = 0.02$), there were 3 inelastic interactions resulting in a fractional area change $\Delta f / f_0 > 10^{-2}$ during 250 vortex rotation periods of evolution (10 000 time steps). This very low collision frequency shows that we are truly in the dilute U2DIT limit. If generic vortex interactions with 'simple' characteristics are to be found, then one may expect to observe them in this context for the effects of distant vortices are weak.

At higher vortex densities (e.g., $f_0 = 0.05$ and $f_0 = 0.1$), one often sees relatively simple, approximately isolated *three-vortex* interactions. That is like-signed vortices may be driven together by the strain field exerted by the opposite-signed vortex, a phenomenon analyzed by Gröbli, Novikov and Aref for the collapse of three point vortices.

The simulations often exhibit the 'milling' or rapid squeezing of small vortices and filamentary structures as they pass between the dominantly bound or 'condensing' dipolar vortex structures. The milled structures are thinned by two to three orders of magnitude in a single vortex rotation period, a mechanism for intermittent cascade to small scales.

In conclusion, and our answer to the first question above: there is no simple set of 'generic' vortex interactions in U2DIT and the *time-dependent strain field* of the surrounding vortices cannot be neglected in the interaction between two vortices.

4.4. THE MINIMAL MODEL FOR U2DIT

All recent 'physical' models for U2DIT [1, 3, 17] are ad-hoc, since they are not asymptotically verifiable and therefore they lack a predictive capability. Their parametric switching ('punctuation') choices are inadequately quantified and have not been justified beyond early times, especially with respect to the effects of dissipation. David Dritschel and I now propose [11] a numerically-efficient *minimal model*, called the Hybrid Elliptical-Contour (HEC) model. It is asymptotic to the full CD/CS algorithm during collisions or periods of strong strain and otherwise reduces to elliptical patches.

Fortunately, aspects of the reduced model are already embodied in the *accelerated* CS code (MACS) [8] which has been used to study U2DIT. Here, the vortices are simply patches of uniform vorticity and a multipole expansion is used to accelerate the calculation. This expansion means that the velocity field at a point due to a distant vortex is decomposed into a leading-order 'point-vortex' contribution, followed by higher-order moment contributions (e.g., the next one is due to the vortex's elliptical shape.) Locally, the given vortex sees a superposition of simple velocity fields, the dominant being a mean, uniform flow, followed by a simple straining field, etc. In this algorithm, the full shape of each vortex is retained at all time.

The switch from a contour representation to an elliptical representation is made whenever the mean strain rate experienced by a vortex falls below a prescribed threshold value, γ_c, and when the *signed-area difference* between a vortex and the best-fit ellipse is sufficiently small (how small is controlled by γ_c). A switch back to a contour representation occurs when the mean strain rate climbs above γ_c. The large computational savings arises because many of the vortices are representable by ellipses. The resulting numerical algorithm is expected to be 10–100 times faster than MACS, or 10^3–10^4 times faster than the original CD/CS algorithm. It is asymptotic to this algorithm in the limit $\gamma_c \rightarrow 0$.

The proposed model, our answer to the second question above, is asymptotic, that is predictive in the sense that it can be used to explore other types of 2DIT, e.g., vortices of varying vorticity magnitudes, distributed (multi-contour) vortices, conservative forcing, and rotating 2DIT, all at a level of dissipation much below that which can be achieved by any other algorithm. Furthermore, it may be linked to a model of the 'sea' of small-scale, well-mixed vorticity generated by surgery; this sea is modeled by weak, fluctuating velocity components obtained from a statistical construct.

5. Future Prospects for Visiometrics on Massive Data Sets

The march of technology has enhanced the physics, dimensions and scale sizes that we have been able to study computationally. It is more difficult to fathom how this technology when harnessed by visiometric processes of juxtaposition, feature extraction, tracking, and quantification will enhance our ability to produce mathematically cogent models. The ability to sit in an environment (e.g., at a network-linked terminal) that will be sensitive to keyboard, body movement, voice and tactile inputs and steer the computer through the parameter space of a model should synergize an enhanced productivity for the visiometrically trained scientist.

Acknowledgement

The author acknowledges the collaboration of David Dritschel in the generation and visiometrics of U2DIT data and the formulation of the HEC model. The work was accomplished during the authors visit to the Department of Applied Mathematics and Theoretical Physics and St Catherines College at Cambridge University in the fall of 1994. This work has been made possible by the continuing support of grants from ONR (Grant No. N00014–93–1–0132 monitored by Dr. M. Fiadeiro) and by ARPA (Grant DAST 63–93–C–0064 monitored by Lt. Col. John Toole). Computations were done on the Cray Y-MP at the Pittsburgh Supercomputer Center and and the CM-5 at The National Center for Supercomputer Applications, both supported by the National Science Foundation. We also acknowledge the support of Center for Computer Aids for Industrial Productivity at Rutgers University.

References

1. Benzi, R., Briscolini, M., Colella, M., and Santangelo, P.: A simple point vortex model for two-dimensional decaying turbulence, *Phys. Fluids A* **4** (1992), 1036–1039.
2. Bitz, F. J. and Zabusky, N. J.: David and 'visiometrics': Visualizing and quantifying evolving amorphous objects, *Computers in Physics*, Nov./Dec. (1990), 603–614.
3. Carnevale, G. F., McWilliam, J. C., Pomeau, Y., Weiss, J. B., and Young, W. R.: Evolution of vortex statistics in two-dimensional turbulence, *Phys. Rev. Lett.* **66** (1992), 2735–2737.
4. Deem, G. S., Zabusky, N. J., and Kruskal, M. D.: Formation, propagation and interaction of 'solitons' in nonlinear dispersive media (numerical solutions of differential equations describing wave motion in nonlinear dispersive media), Bell Telephone Laboratories Film Library, 1965–1975.
5. Dritschel, D. G.: Contour dynamics and contour surgery: numerical algorithms for extended, high-resolution modelling of vortex dynamics in two-dimensional, inviscid, incompressible flows, *Comp. Phys. Rep.* **10** (1989), 77–146.
6. Dritschel, D. G.: Vortex feature tracking in high reynolds number two-dimensional flows, in preparation, 1995.
7. Dritschel, D. G. and Legras, B.: The elliptical model of two-dimensional vortex dynamics, II. Disturbance equations, *Phys. Fluids A* **3** (1991) 855–869.
8. Dritschel, D. G.: Vortex properties of two-dimensional turbulence, *Phys. Fluids A* **5** (1993), 984–997.
9. Dritschel, D. G.: A general theory for two-dimensional vortex interaction, submitted to *J. Fluid Mech.*, 1995.
10. Dritschel, D. G. and Waugh, D. W.: Quantification of the inelastic interaction of two asymmetric vortices in two-dimensional vortex dynamics, *Phys. Fluids A* **4** (1992), 1737–1744.
11. Dritschel, D. G and Zabusky, N. J.: Toward a reduced model for nearly-inviscid unforced incompressible two-dimensional turbulence, letter submitted to *Phys. Fluids*, 1995.
12. Korteweg, D. J. and de Vries, G.: On the change of form of long waves advancing in a rectangular canal, and on a new type of long stationary waves, *Phil. Mag.* **39** (1895), 422–443.
13. Legras, B. and Dritschel, D. G.: The elliptical model of two-dimensional vortex dynamics, I. The basic state, *Phys. Fluids A* **3** (1991), 845–854.

14. Mariotti, A., Legras, B., and Dritschel, D. G.: Vortex stripping and the erosion of coherent structures in two-dimensional flows, *Phys. Fluids* **6** (1994), 3954–3962.
15. Melander, M., Zabusky, N. J., and Styczek, A. S.: A moment model for vortex interactions of the 2d Euler equations and computational validation of an elliptical representation, *J. Fluid Mech.* **167** (1986), 95–115.
16. Melander, M. V., McWilliams, J. C., and Zabusky, N. J.: Axisymmetrization and gradient-intensification of an isolated 2d vortex through filamentation, *J. Fluid Mech.* **178** (1987), 137–159.
17. Piva, R., Riccardi, G., and Benzi, R.: A physical model for merging in 2d decaying turbulence, Preprint, 1995.
18. Silver, D. and Zabusky, N. J.: Quantifying visualizations for reduced modeling in nonlinear science: Extracting structures from data sets, *J. Visual Comm. Image Representation* **4** (1993), 46–61.
19. Taub, A. (ed.): *The Collected Works of John von Neumann*, Macmillan, New York, 1963.
20. Yao, H. B., Dritschel, D. G., and Zabusky, N. J.: High-gradient phenomena in 2d vortex interaction, to appear in *Phys. Fluids*, 1995.
21. Zabusky, N. J.: Coherent structures in fluid dynamics, in *The Significance of Nonlinearity in the Natural Sciences*, Plenum Press, New York, 1977, pp. 145–205.
22. Zabusky, N. J.: Computational synergetics and mathematical innovation, *J. Comput. Phys.* **43** (1981), 195–249.
23. Zabusky, N. J.: Computational synergetics, *Phys. Today* **37** (1984), 36–46.
24. Zabusky, N. J., Hughes, M., and Roberts, K. V.: Contour dynamics for the Euler equations in two dimensions, *J. Comp. Phys.* **30** (1979), 96–106.
25. Zabusky, N. J., Silver, D., Pelz, R., and Vizgroup '93: Visometrics, juxtaposition, and modeling, *Phys. Today* (1993), 24–31.

Part II: Invited Contributions

Acta Applicandae Mathematicae **39:** 175–192, 1995.

The KPI Equation with Unconstrained Initial Data*

M. BOITI, F. PEMPINELLI, and A. POGREBKOV**
Dipartimento di Fisica dell'Università and Sezione INFN, 73100 Lecce, Italia

(Received: 12 September 1994)

Abstract. The solution $u(t,x,y)$ of the Kadomtsev–Petviashvili I (KPI) equation with given initial data $u(0,x,y)$ belonging to the Schwartz space is considered. No additional special constraints, usually considered in the literature as $\int \mathrm{d}x u(0,x,y) = 0$ are required to be satisfied by the initial data. The spectral theory associated with KPI is studied in the space of the Fourier transform of the solutions. The variables $p = \{p_1, p_2\}$ of the Fourier space are shown to be the most convenient spectral variables to use for spectral data. Spectral data are shown to decay rapidly at large p but to be discontinuous at $p = 0$. Direct and inverse problems are solved with special attention to the behavior of all the quantities involved in the neighborhood of $t = 0$ and $p = 0$. It is shown in particular that the solution $u(t,x,y)$ has a time derivative discontinuous at $t = 0$ and that at any $t \neq 0$ it does not belong to the Schwartz space no matter how small in norm and rapidly decaying at large distances the initial data are chosen.

Mathematics Subject Classifications (1991). 35Q53, 58F07.

Key words: Kadomtsev–Petviashvili equation, spectral transform.

1. Introduction

We consider the initial value problem for the Kadomtsev–Petviashvili equation [1–3] in its version called KPI

$$(u_t - 6uu_x + u_{xxx})_x = 3u_{yy}, \qquad u = u(t,x,y),$$
$$u(0,x,y) = u(x,y), \tag{1.1}$$

for $u(t,x,y)$ real. As early as 1974 [4] it was acknowledged to be integrable since it can be associated to a linear spectral problem and, precisely, to the nonstationary Schrödinger equation

$$(-i\partial_y + \partial_x^2 - u(x,y))\Phi = 0. \tag{1.2}$$

However, to building a complete and coherent theory for the spectral transform of the potential $u(x,y)$ in (1.2) that could be used to linearize the initial value

* Work supported in part by Ministero delle Università e della Ricerca Scientifica e Tecnologica, Italia.

** Permanent address: Steklov Mathematical Institute, Vavilov Str. 42, Moscow 117966, GSP-1, Russia; e-mail POGREB@QFT.MIAN.SU.

problem of (1.1) turned out to be unexpectedly difficult. The real breakthrough has been the discovery that the problem is solvable via a nonlocal Riemann–Hilbert formulation [5, 6].

Subsequently further progress has been made. The characterization problem for the spectral data was solved in [7]. The extension of the spectral transform to the case of a potential $u(x,y)$ approaching zero in every direction except a finite number has been faced in [8]. The questions of the associated conditions (often called 'constraints') and how to choose property ∂_x^{-1} in the evolution form of KPI

$$u_t(t,x,y) - 6u(t,x,y)u_x(t,x,y) + u_{xxx}(t,x,y) = 3\partial_x^{-1}u_{yy}(t,x,y) \quad (1.3)$$

have been studied in [9]. Some relevant points on constraints were known earlier, but have not been published [10]. See also [11]. In the article [12] some rigorous results on the existence and uniqueness of solutions of the direct and inverse problem have been proved.

In this article, as well as in the previous ones [13, 14] by the same authors, the description of the KPI equation by means of the inverse spectral transform is revisited. We think that this is necessary since some specific properties of the solutions of the initial value problem (1.1) and, more generally, of the $2+1$-integrable equations are not properly (or in any case not exhaustively) described in the literature. Moreover, these properties are specially interesting since they seem to be peculiar of the integrable equations in $2+1$ dimensions. Our study is essentially based on some new theoretical tools in the theory of the spectral transform developed in [8, 15, 16].

In this article we consider the equation (1.1) for initial data $u(x,y)$ belonging to the Schwartz space $\mathcal{S}$. The main results we get regarding the properties of the solutions and the theory of the spectral transform can be summarized in the following points (the integrations when it is not differently indicated are performed all along the real axis from $-\infty$ to $+\infty$).

1. The solution $u(t,x,y)$ immediately leaves the Schwartz space since at any time $t \neq 0$ develops a tail slowly decreasing for $x \to t\infty$ according to the following formula

$$u(t,x,y) = \frac{-1}{4\pi\sqrt{3tx}\,|x|} \iint \mathrm{d}x'\,\mathrm{d}y' u(x',y') + \mathrm{o}(|x|^{-3/2}). \quad (1.4)$$

   Only in the opposite direction, i.e. for $x \to -t\infty$, it decreases rapidly.
2. The problem of the proper choice of ∂_x^{-1} in (1.3) requires a special investigation in the neighborhood of the initial time $t = 0$. The final answer is that, for initial data $u(x,y) \in \mathcal{S}$ which are arbitrary chosen and, therefore, not necessarily subjected to the constraint

$$\int \mathrm{d}x\, u(x,y) = 0, \quad (1.5)$$

the function $u(t,x,y)$ reconstructed by solving the inverse spectral problem for (1.2) evolves in time according to the equation

$$\begin{aligned} &u_t(t,x,y) - 6u(t,x,y)u_x(t,x,y) + u_{xxx}(t,x,y) \\ &\quad = 3\int_{-t\infty}^{x} dx'\, u_{yy}(t,x',y'). \end{aligned} \tag{1.6}$$

Notice that the possibility that the inversion of ∂_x can depend on t has already been mentioned in the literature for the Davey–Stewartson equation [17].

3. The time derivative $u_t(t,x,y)$ has different left and right limits at $t=0$ and the condition

$$\int dx\, u(t,x,y) = 0 \tag{1.7}$$

is dynamically generated by the evolution equation at times $t \neq 0$, i.e. for these times we recover the result obtained in [9].

4. The primitives $\int_{\pm\infty}^{x}$ or their antisymmetric combination $\frac{1}{2}(\int_{-\infty}^{x} - \int_{x}^{+\infty})$ cannot be exchanged with the limit $t \to 0$. In fact we have

$$\begin{aligned} \lim_{t\to+0} \int_{-\infty}^{x} dx'\, u(t,x',y) &= -\lim_{t\to+0} \int_{x}^{+\infty} dx'\, u(t,x',y) \\ &= \lim_{t\to+0} \frac{1}{2}\left(\int_{-\infty}^{x} - \int_{x}^{+\infty}\right) dx'\, u(t,x',y) \\ &= \int_{-\infty}^{x} dx'\, u(x',y) \end{aligned} \tag{1.8}$$

and

$$\begin{aligned} \lim_{t\to-0} \int_{-\infty}^{x} dx'\, u(t,x',y) &= -\lim_{t\to-0} \int_{x}^{+\infty} dx'\, u(t,x',y) \\ &= \lim_{t\to-0} \frac{1}{2}\left(\int_{-\infty}^{x} - \int_{x}^{+\infty}\right) dx'\, u(t,x',y) \\ &= -\int_{x}^{+\infty} dx'\, u(x',y). \end{aligned} \tag{1.9}$$

Thus the two forms of writing the evolution form of the Kadomtsev–Petviashvili equation suggested in [9] and in (1.6) are equivalent, but only in the last formulation can the integration over x and the limit $t \to 0$ be exchanged, showing explicitly the discontinuity at $t=0$.

5. Conditions ('constraints') higher than (1.7) can be obtained. But some special care is needed due to the asymptotic behavior of u as given in (1.4). For example, in the condition

$$\int dx\, x u_y(t,x,y) = 0, \quad t \neq 0, \tag{1.10}$$

the y-derivative cannot be extracted from the integral. Higher conditions are nonlinear and can be given by means of recursion relations.

6. In spite of the dynamic condition (1.7) we have that at all times

$$\int \mathrm{d}x \int \mathrm{d}y\, u(t,x,y) = \int\int \mathrm{d}x\, \mathrm{d}y\, u(0,x,y). \tag{1.11}$$

The integrations in the l.h.s. are performed in the order explicit indicated. Since no any constraint is imposed to the initial data this quantity is not necessarily equal to zero and gives a nontrivial integral of motion.

7. The properties of the Jost solutions of the equation (1.2) essentially differs at $t = 0$ and $t \neq 0$. In particular, at $t = 0$ the coefficients of the asymptotic $1/\mathbf{k}$-expansion, where $\mathbf{k}$ denotes the spectral parameter, depend on the sign of the imaginary part $\mathbf{k}_\Im$ (cf. [10] and [16]), while at $t \neq 0$ these coefficients in the limit $t \to 0$ depend on the sign of t.
8. Spectral data do not belong to the Schwartz space $\mathcal{S}$ and no small norm or smoothness condition can improve this situation.

Throughout our paper we find it convenient to consider the problem in terms of the Fourier transform of u

$$v(t,p) \equiv v(t,p_1,p_2) = \frac{1}{(2\pi)^2}\int\int \mathrm{d}x\, \mathrm{d}y\, \mathrm{e}^{i(p_1x-p_2y)}u(t,x,y), \tag{1.12}$$

where p denotes a 2-component vector

$$p \equiv \{p_1, p_2\}. \tag{1.13}$$

Then, the initial value problem (1.1) for the KPI equation can be rewritten as $(\mathrm{d}^2q \equiv \mathrm{d}q_1\, \mathrm{d}q_2)$

$$\begin{aligned} p_1\frac{\partial v(t,p)}{\partial t} &= -3ip_1^2\int \mathrm{d}^2q\, v(t,q)v(t,p-q) - i(p_1^4+3p_2^2)v(t,p), \\ v(0,p) &= v(p) \in \mathcal{S}, \end{aligned} \tag{1.14}$$

where $v(p)$ is the Fourier transform (1.12) of the initial data $u(x,y)$. The problem of property defining ∂_x^{-1} in (1.3) is reformulated as the problem of properly regularizing the distribution p_1^{-1}. In terms of the Fourier transform $v(t,p)$ we have instead of (1.6)

$$\frac{\partial v(t,p)}{\partial t} = -3ip_1\int \mathrm{d}^2q\, v(t,q)v(t,p-q) - i\frac{p_1^4+3p_2^2}{p_1+i0t}v(t,p) \tag{1.15}$$

and condition (1.7) takes the form

$$\lim_{p_1\to 0} v(t,p_1,p_2) = 0 \quad \text{for } t \neq 0, \tag{1.16}$$

where the limit is understood in the sense of distributions in the p_2 variable.

According to the usual scheme of the spectral transform theory we expect that the spectral data evolve in time as the Fourier transform of the linear part of (1.1). This makes clear that the special behavior in time of the solutions of KPI is just complicated by the nonlinearity of the KPI but is, in fact, inherent to the singular character of its linear part.

In Section 2 the inverse spectral transform method for the KPI equation is revisited. We use a specific form of the equations of the inverse problem suggested in [8]. The advantages of this formulations as well as the connections with the results obtained in [13] are given in Section 3. Some propositions on distributions that are of common use in the article are reported in the appendix. For a proof see [14].

2. The Spectral Transform of KPI

2.1. Initial Remarks and Notions

The analysis of the linearized version of the KPI equation (see [14]) indicates that it is convenient to study the KPI equation in the space $p = \{p_1, p_2\}$ of the Fourier transform of the solutions

$$v(t,p) \equiv v(t,p_1,p_2) = \frac{1}{(2\pi)^2} \iint \mathrm{d}x\,\mathrm{d}y\,\mathrm{e}^{i(p_1x-p_2y)} u(t,x,y). \tag{2.1}$$

Correspondingly, the nonstationary Schrödinger equation, too,

$$\big(-i\partial_y + \partial_x^2 - u(x,y)\big)\Phi = 0, \tag{2.2}$$

which is the spectral equation associated to KPI, has to be stated in the Fourier space of the variables $p = \{p_1, p_2\}$.

If $\Phi(x, y, \mathbf{k})$ is the Jost solution of (2.2) the corresponding Jost solution in the p-space can be defined by first shifting $\Phi(x, y, \mathbf{k})$ and then by taking the Fourier transform according to the following formula

$$\nu(p \mid \mathbf{k}) = \frac{1}{(2\pi)^2} \int \mathrm{d}x\,\mathrm{d}y\,\mathrm{e}^{ip_1x-ip_2y}\mathrm{e}^{i\mathbf{k}x-i\mathbf{k}^2y}\Phi(x,y,\mathbf{k}). \tag{2.3}$$

The specific role played by the spectral parameter $\mathbf{k}$ is emphasized by separating it from the p variable by a $|$ and a **bold** font is used to indicate that $\mathbf{k}$ belongs to the complex plane.

Then the spectral equation (2.2) takes the form

$$[\mathcal{L}(p) - 2p_1\mathbf{k}]\nu(p \mid \mathbf{k}) = \int \mathrm{d}^2p'\, v(p-p')\nu(p' \mid \mathbf{k}), \tag{2.4}$$

where

$$\mathcal{L}(p) = p_2 - p_1^2. \tag{2.5}$$

The integral equation determining the Jost solution [5] has, in the Fourier space, the form

$$\nu(p \mid \mathbf{k}) = \delta^2(p) + \frac{\rho(p \mid \mathbf{k})}{\mathcal{L}(p) - 2p_1\mathbf{k}}, \tag{2.6}$$

where

$$\rho(p \mid \mathbf{k}) = \int \mathrm{d}^2p' \, v(p - p')\nu(p' \mid \mathbf{k}) \tag{2.7}$$

and $\delta^2(p) \equiv \delta(p_1)\delta(p_2)$. The new function

$$\rho(p \mid \mathbf{k}) = [\mathcal{L}(p) - 2p_1\mathbf{k}]\nu(p \mid \mathbf{k}) \tag{2.8}$$

will play a crucial role in the following. Let us introduce the boundary values of ν and ρ on the real axis of the $\mathbf{k}$-plane

$$\begin{aligned} \nu^{\pm}(p \mid k) &= \lim_{\mathbf{k}_{\Im} \to \pm 0} \nu(p \mid \mathbf{k}), \\ \rho^{\pm}(p \mid k) &= \lim_{\mathbf{k}_{\Im} \to \pm 0} \rho(p \mid \mathbf{k}), \quad k = \mathbf{k}_{\Re} \end{aligned} \tag{2.9}$$

and write

$$\nu^{\pm}(p \mid k) = \delta^2(p) + \frac{\rho^{\pm}(p \mid k)}{\mathcal{L}(p) - 2p_1k \mp i0p_1}. \tag{2.10}$$

Then the spectral data are most naturally obtained (see [16]) by computing $\rho(p \mid k)$ at the special value of k that vanishes the denominator

$$\rho^{\pm}\left(p \,\middle|\, \frac{\mathcal{L}(p)}{2p_1}\right) = \int \mathrm{d}^2p' \, v(p - p')\nu^{\pm}\left(p' \,\middle|\, \frac{\mathcal{L}(p)}{2p_1}\right). \tag{2.11}$$

We conclude that the variables $p = \{p_1, p_2\}$ can be considered as the spectral variables of the spectral data. For their introduction in the general scheme of the resolvent approach [8, 15, 16] and for analogous variables used in the KPII case see [18].

The spectral data $\rho^{\pm}$ evolve in time as follows

$$\rho^{\pm}\left(t, p \,\middle|\, \frac{\mathcal{L}(p)}{2p_1}\right) = \exp\left(-it\frac{p_1^4 + 3p_2^2}{p_1}\right)\rho^{\pm}\left(p \,\middle|\, \frac{\mathcal{L}(p)}{2p_1}\right), \tag{2.12}$$

i.e. in the same way as the linearized KPI equation in the Fourier transformed space. In fact this is the deep reason of the privileged role played by the variables $\{p_1, p_2\}$ in all the theory.

For the characterization conditions satisfied by the spectral data see [7] and [14].

2.2. Properties of Jost Solutions and Spectral Data

2.2.1. $1/\mathbf{k}$-*expansion of the Jost Solution at* $t = 0$

To get the asymptotic $1/\mathbf{k}$-expansion of the Jost solution we consider the expansion of the kernel $(\mathcal{L}(p) - 2p_1\mathbf{k})^{-1}$ of the integral equation (2.6). By considering it as a distribution in the variables $\mathcal{L}(p)$ and p_1 we have that

$$\frac{1}{\mathcal{L}(p) - 2p_1\mathbf{k}} = -\sum_{n=0}^{\infty} \frac{\mathcal{L}(p)^n}{(2\mathbf{k})^{n+1}(p_1 \pm i0\mathcal{L}(p))^{n+1}}, \quad \mathbf{k} \to \infty, \ \pm\mathbf{k}_{\Im} > 0. \tag{2.13}$$

Note that the coefficients of the expansion depend on the half-plane in which $\mathbf{k}$ tends to infinity. Consequently, also the expansion of the Jost solution at large $\mathbf{k}$ depends on the half plane in which the limit is performed. If we note the coefficients of the asymptotic expansion as follows

$$\begin{aligned} \nu(p \mid \mathbf{k}) &= \sum_{n=0}^{\infty} \frac{\nu_n^{\pm}(p)}{(2\mathbf{k})^n}, \\ \rho(p \mid \mathbf{k}) &= \sum_{n=0}^{\infty} \frac{\rho_n^{\pm}(p)}{(2\mathbf{k})^n}, \quad \mathbf{k} \to \infty, \ \pm\mathbf{k}_{\Im} > 0 \end{aligned} \tag{2.14}$$

we have from (2.6) that

$$\nu_0^{\pm}(p) = \delta^2(p), \qquad \rho_0^{\pm}(p) = v(p), \tag{2.15}$$

$$\nu_1^{\pm}(p) = -\frac{v(p)}{p_1 \pm i0\mathcal{L}(p)}, \qquad \rho_1^{\pm}(p) = -\int \mathrm{d}^2q \, \frac{v(p-q)v(q)}{q_1 \pm i0\mathcal{L}(q)}. \tag{2.16}$$

2.2.2. *Properties of Spectral Data*

In order to study the properties of the spectral data we, first, substitute (2.6) into (2.7) getting an integral equation for $\rho(p \mid \mathbf{k})$

$$\rho(p \mid \mathbf{k}) = v(p) + \int \frac{\mathrm{d}^2q}{\mathcal{L}(q) - 2q_1\mathbf{k}} v(p-q)\rho(q \mid \mathbf{k}). \tag{2.17}$$

Performing the limiting procedure (2.9) we have

$$\rho^{\pm}(p \mid \mathbf{k}) = v(p) + \int \frac{\mathrm{d}^2q}{\mathcal{L}(q) - 2q_1k \mp i0q_1} v(p-q)\rho^{\pm}(q \mid k) \tag{2.18}$$

and then, recalling definition (2.11), the following representation for the spectral data

$$\rho^{\pm}\left(p \,\middle|\, \frac{\mathcal{L}(p)}{2p_1}\right) = v(p) + p_1 \int \frac{\mathrm{d}^2 q}{p_1\mathcal{L}(q) - q_1\mathcal{L}(p) \mp i0p_1q_1} v(p-q)\rho^{\pm}\left(q \,\middle|\, \frac{\mathcal{L}(p)}{2p_1}\right). \tag{2.19}$$

We deduce that for $v(p)$ vanishing rapidly at large p also the spectral data $\rho^{\pm}(p \mid \frac{\mathcal{L}(p)}{2p_1})$ vanish rapidly at large p. However, the spectral data are discontinuous at $p = 0$. In fact if the limit $p \to 0$ is taken along the line $p_2 = 2\beta p_1$, with β an arbitrary constant, we get from (2.19)

$$\lim_{\substack{p\to 0\\ p_2=2\beta p_1}} \rho^{\pm}\left(p \,\middle|\, \frac{\mathcal{L}(p)}{2p_1}\right) = \rho^{\pm}(0 \mid \beta) = v(0) + \int \frac{\mathrm{d}^2 q\, \overline{v(q)}\rho^{\pm}(q \mid \beta)}{\mathcal{L}(q) - 2\beta q_1 \mp i0q_1}, \tag{2.20}$$

i.e. the limit depends on β.

2.2.3. $1/\mathbf{k}$-*expansion of the Jost Solution for* $t \neq 0$

Since the properties of the Jost solution depend on the specific form of the spectral equation (2.4) we would expect that the introduction of a parametric dependence on t in $v(p)$ will not change the asymptotic $1/\mathbf{k}$-expansion of $\nu(p \mid \mathbf{k})$ or, more precisely, we would expect that the only difference would be a parametric dependence on t of the coefficients of the expansion. However, we will see that the time evolution of a solution $v(t,p)$ of the KPI equation has special properties in the neighborhoods of $t = 0$ and we have to take into account that not $v(t,p)$ but $\tilde{v}(t,p)$ as defined in the following equation

$$v(t,p) = \mathrm{e}^{-it\frac{p_1^4+3p_2^2}{p_1}}\tilde{v}(t,p) \tag{2.21}$$

is continuous at $p = 0$ for $t \neq 0$.

Therefore in writing the integral equation (2.6) for the Jost solution it is convenient to explicitly factorize the singular behavior in time as follows

$$\nu(t,p \mid \mathbf{k}) = \delta^2(p) + \frac{1}{\mathcal{L}(p) - 2p_1\mathbf{k}} \int \mathrm{d}^2 q\, \mathrm{e}^{-it\frac{q_1^4+3q_2^2}{q_1}} \tilde{v}(t,q)\nu(t,p-q \mid \mathbf{k}). \tag{2.22}$$

We get

$$\lim_{\mathbf{k}\to\infty} \mathbf{k}[\nu(t,p \mid \mathbf{k}) - \delta^2(p)] = \tilde{v}(t,p) \lim_{\mathbf{k}\to\infty} \frac{\mathbf{k}}{\mathcal{L}(p) - 2p_1\mathbf{k}} \mathrm{e}^{-it\frac{p_1^4+3p_2^2}{p_1}}, \tag{2.23}$$

and using proposition A.1 in the Appendix

$$\begin{aligned}\lim_{\mathbf{k}\to\infty} \mathbf{k}[\nu(t,p \mid \mathbf{k}) - \delta^2(p)] &= \frac{-1}{2(p_1 + i0t)} e^{-it\frac{p_1^4+3p_2^2}{p_1}} \tilde{v}(t,p) \\ &= \frac{-1}{2} \frac{v(t,p)}{p_1 + i0t}. \end{aligned} \tag{2.24}$$

Comparing this result with (2.16) we see that the coefficients of $1/\mathbf{k}$ in the asymptotic expansions coincide only if

$$\operatorname{sgn} \mathbf{k}_{\Im} = \operatorname{sgn}\left(t\mathcal{L}(p)\right), \tag{2.25}$$

or in other words only under this condition the limits $\mathbf{k} \to \infty$ and $t \to 0$ commute. Notice that this condition is formulated in terms of the variables p corresponding to the fact that the special behavior of solutions and Jost solutions in the neighborhoods of $t = 0$ are manageable only in the transformed Fourier space.

2.3. Inverse problem and time evolution

2.3.1. *Formulation of the Inverse Problem*

By working in the general framework of the resolvent approach [8, 15] we obtained in [16] the following coupled system of two linear integral equations for the boundary values of the Jost solutions on the real k-axis

$$\begin{aligned}\nu^{\pm}(p \mid k) &= \delta^2(p) + \frac{1}{2} \int \frac{\mathrm{d}^2 q}{\mathcal{L}(-q) + 2kq_1 \pm iq_1 0} \times \\ &\quad \times \sum_{\sigma=+,-} \overline{\rho^{-\sigma}}\left(q \,\middle|\, \frac{\mathcal{L}(q)}{2q_1}\right) \nu^{\sigma}\left(p+q \,\middle|\, \frac{\mathcal{L}(q)}{2q_1}\right). \end{aligned} \tag{2.26}$$

This formula is the main tool to be used for solving the inverse problem, i.e. for reconstructing the potential $v(p)$ starting from the spectral data and for proving that spectral data evolving in time as indicated in (2.12) generate potentials $v(t,p)$ that are solutions of the KPI equation.

In this respect two properties of this system are crucial. First, the system is closed in the sense that the spectral data $\rho^{\pm}$ are reductions of the Jost solutions themselves via equations (2.8), (2.9) and (2.11). Second, it depends linearly on the spectral data, in contrast with the usual equations written in the literature that are quadratic in the spectral data. To our knowledge the only alternative integral equations for solving the inverse problem, which are linear in the spectral data, are given in [12], but they are not closed in the sense that they are written in terms of the so-called advanced/retarded solutions.

We suppose that the system (2.26) has unique solution in the considered class of spectral data. Then the r.h.s. of (2.26) can be continued analytically in the complex $\mathbf{k}$-plane as follows

$$\nu(p\mid\mathbf{k}) = \delta^2(p)+\frac{1}{2}\int\frac{\mathrm{d}^2q}{\mathcal{L}(-q)+2\mathbf{k}q_1}\times \\ \times\sum_{\sigma=\pm}\overline{\rho^{-\sigma}}\left(q\,\Big|\,\frac{\mathcal{L}(q)}{2q_1}\right)\nu^{\sigma}\left(p+q\,\Big|\,\frac{\mathcal{L}(q)}{2q_1}\right). \tag{2.27}$$

For getting from (2.27) an integral equation for $\rho(p,\mathbf{k})$ it is convenient to use equation (2.6) computed on the real axis and to rewrite the Jost solution in the r.h.s. of (2.27) as

$$\nu^{\sigma}\left(p+q\,\Big|\,\frac{\mathcal{L}(q)}{2q_1}\right) \\ =\delta^2(q+p)+\frac{q_1\rho^{\sigma}(p+q\mid\frac{\mathcal{L}(q)}{2q_1})}{q_1\mathcal{L}(p)+p_1\mathcal{L}(-q)-i\sigma 0q_1(q_1+p_1)}. \tag{2.28}$$

Then it is easy to check that

$$\frac{\nu^{\sigma}(p+q\mid\frac{\mathcal{L}(q)}{2q_1})}{\mathcal{L}(-q)+2\mathbf{k}q_1} \\ =\frac{1}{\mathcal{L}(p)-2\mathbf{k}p_1}\left[-\frac{p_1}{q_1}\nu^{\sigma}\left(p+q\,\Big|\,\frac{\mathcal{L}(q)}{2q_1}\right)+\frac{\rho^{\sigma}(p+q\mid\frac{\mathcal{L}(q)}{2q_1})}{\mathcal{L}(-q)+2\mathbf{k}q_1}\right], \tag{2.29}$$

where the ratio p_1/q_1 multiplying ν in the r.h.s. is well defined due to (2.28). By inserting this expression in (2.27) and by recalling definition (2.8) we get the following integral equation for $\rho(p\mid\mathbf{k})$

$$\rho(p\mid\mathbf{k}) \\ =-\int\mathrm{d}^2q\frac{p_1}{2q_1}\sum_{\sigma=\pm}\overline{\rho^{-\sigma}}\left(q\,\Big|\,\frac{\mathcal{L}(q)}{2q_1}\right)\nu^{\sigma}\left(p+q\,\Big|\,\frac{\mathcal{L}(q)}{2q_1}\right)+ \\ +\frac{1}{2}\int\frac{\mathrm{d}^2q}{\mathcal{L}(-q)+2\mathbf{k}q_1}\sum_{\sigma=\pm}\overline{\rho^{-\sigma}}\left(q\,\Big|\,\frac{\mathcal{L}(q)}{2q_1}\right)\rho^{\sigma}\left(p+q\,\Big|\,\frac{\mathcal{L}(q)}{2q_1}\right). \tag{2.30}$$

It has a kernel similar to the kernel of the integral equation (2.6), (2.7) obtained for $\nu(p\mid\mathbf{k})$ in the direct problem. Therefore its asymptotic expansion at large $\mathbf{k}$ can be computed by using the same methods used in the previous section. In particular we get that the first term in the r.h.s. due to (2.14) and (2.16) is just the potential

$$v(p)=-\int\mathrm{d}^2q\frac{p_1}{2q_1}\sum_{\sigma=\pm}\overline{\rho^{-\sigma}}\left(q\,\Big|\,\frac{\mathcal{L}(q)}{2q_1}\right)\nu^{\sigma}\left(p+q\,\Big|\,\frac{\mathcal{L}(q)}{2q_1}\right), \tag{2.31}$$

that is therefore explicitly reconstructed in terms of the spectral data and Jost solutions.

From (2.27) by using the expansion (2.13) we can get the coefficients of the asymptotic expansion at large $\mathbf{k}$ of $\nu(p \mid \mathbf{k})$. Of special interest is the coefficients of $1/\mathbf{k}$. Recalling (2.16) we have

$$\frac{v(p)}{p_1 \pm i0\mathcal{L}(p)} = -\int \frac{d^2 q}{2(q_1 \mp i0\mathcal{L}(-q))} \sum_{\sigma=\pm} \overline{\rho^{-\sigma}\left(q \,\middle|\, \frac{\mathcal{L}(q)}{2q_1}\right)} \nu^{\sigma}\left(p+q \,\middle|\, \frac{\mathcal{L}(q)}{2q_1}\right). \quad (2.32)$$

2.3.2. *Jost Solutions for $t \neq 0$*

We choose spectral data to evolve as in (2.12)

$$\rho^{\pm}\left(t,p \,\middle|\, \frac{\mathcal{L}(p)}{2p_1}\right) = \exp\left(-it\frac{p_1^4 + 3p_2^2}{p_1}\right)\rho^{\pm}\left(p \,\middle|\, \frac{\mathcal{L}(p)}{2p_1}\right), \quad (2.33)$$

where $\rho^{\pm}(p \mid \frac{\mathcal{L}(p)}{2p_1})$ denotes the initial value of spectral data determined in terms of $v(p)$ and Jost solutions $\nu^{\pm}(p \mid k)$ according to formula (2.11).

Then the Jost solution at any time is reconstructed just by inserting these spectral data into (2.27)

$$\begin{aligned}\nu(t,p \mid \mathbf{k}) &= \delta^2(p) + \frac{1}{2}\int \frac{d^2 q e^{it\frac{q_1^4+3q_2^2}{q_1}}}{\mathcal{L}(-q) + 2\mathbf{k}q_1} \times \\ &\quad \times \sum_{\sigma=\pm} \overline{\rho^{-\sigma}\left(q \,\middle|\, \frac{\mathcal{L}(q)}{2q_1}\right)} \nu^{\sigma}\left(t,p+q \,\middle|\, \frac{\mathcal{L}(q)}{2q_1}\right). \end{aligned} \quad (2.34)$$

In the limit $\mathbf{k} \to k \pm i0$ (cf. (2.26)) we obtain the integral equations for the boundary values on the real axis, i.e. for the functions $\nu^{\pm}(t,p \mid k)$. In what follows we always suppose that these equations are uniquely solvable.

We have, first, to prove that this function $\nu(t,p \mid \mathbf{k})$ is indeed the Fourier transformed of the Jost solution of (1.2), i.e. that it satisfies (2.4), and, second, we have to find its time evolution equation, which is needed to know in order to show that the reconstructed $v(t,p)$ satisfies KPI.

2.3.3. *The Reconstructed Spectral Equation*

First we prove that ν defined by (2.34) obeys the spectral equation (2.4) at any time. Let us define $\rho(t,p \mid \mathbf{k})$ by (2.8) rewritten for any time

$$\rho(t,p \mid \mathbf{k}) = [\mathcal{L}(p) - 2p_1\mathbf{k}]\nu(t,p \mid \mathbf{k}). \quad (2.35)$$

Then, by the same procedure used to get (2.30) from (2.27) we derive from (2.34) that

$$\rho(t,p\mid\mathbf{k}) = v(t,p)+\frac{1}{2}\int\frac{\mathrm{d}^2q\mathrm{e}^{it\frac{q_1^4+3q_2^2}{q_1}}}{\mathcal{L}(-q)+2\mathbf{k}q_1}\times \\ \times\sum_{\sigma=\pm}\overline{\rho^{-\sigma}}\left(q\,\Big|\,\frac{\mathcal{L}(q)}{2q_1}\right)\rho^{\sigma}\left(t,p+q\,\Big|\,\frac{\mathcal{L}(q)}{2q_1}\right), \tag{2.36}$$

where we have introduced

$$v(t,p) \\ = -\int\mathrm{d}^2q\frac{p_1}{2q_1}\mathrm{e}^{it\frac{q_1^4+3q_2^2}{q_1}}\sum_{\sigma=\pm}\overline{\rho^{-\sigma}}\left(q\,\Big|\,\frac{\mathcal{L}(q)}{2q_1}\right)\nu^{\sigma}\left(t,p+q\,\Big|\,\frac{\mathcal{L}(q)}{2q_1}\right). \tag{2.37}$$

Now due to the assumption on unique solvability of (2.34) we have from (2.36) that

$$\rho(t,p\mid\mathbf{k}) = \int\mathrm{d}^2q\,v(t,p-q)\nu(t,q\mid\mathbf{k}),$$

i.e. (2.7) is satisfied for any t. Thanks to (2.8) this proves that (2.4) is satisfied at any time.

To prove that ν defined by (2.34) obeys the integral equation (2.6) at any time we use (2.34), (2.36) and (2.37) to get

$$\nu(t,p\mid\mathbf{k})-\delta^2(p)-\frac{\rho(t,p\mid\mathbf{k})}{\mathcal{L}(p)-2p_1\mathbf{k}} \\ = \frac{1}{2}\int\frac{\mathrm{d}^2\,q\mathrm{e}^{it\frac{q_1^4+3q_2^2}{q_1}}}{\mathcal{L}(-q)+2\mathbf{k}q_1}\sum_{\sigma=\pm}\overline{\rho^{-\sigma}}\left(q\,\Big|\,\frac{\mathcal{L}(q)}{2q_1}\right)\times \tag{2.38} \\ \times\left[\nu^{\sigma}\left(t,p+q\,\Big|\,\frac{\mathcal{L}(q)}{2q_1}\right)\frac{\mathcal{L}(p+q)-2(p_1+q_1)\frac{\mathcal{L}(q)}{2q_1}}{\mathcal{L}(p)-2\mathbf{k}p_1}-\frac{\rho^{\sigma}(t,p+q\mid\frac{\mathcal{L}(q)}{2q_1})}{\mathcal{L}(p)-2\mathbf{k}p_1}\right].$$

Then (2.6) at any t follows by using definition (2.35) of ρ^{σ} in the r.h.s. This proves that $\nu(t,p\mid\mathbf{k})$ in (2.34) is the Fourier transform of the Jost solution of the equation (1.2) with potential the Fourier transform of $v(t,p)$ defined in (2.37).

To prove that $v(t,p)$ satisfies equation (2.21) with $\tilde{v}(t,p)$ continuous at $p=0$ it is convenient to insert (2.28) (that we proved to be satisfied at any time) into (2.37) getting

$$v(t,p) = \mathrm{e}^{it\frac{p_1^4+3p_2^2}{p_1}}\frac{1}{2}\sum_{\sigma=\pm}\overline{\rho^{-\sigma}}\left(-p\,\Big|\,\frac{\mathcal{L}(-p)}{-2p_1}\right)- \tag{2.39} \\ -\frac{1}{2}\int\mathrm{d}^2q\,p_1\mathrm{e}^{it\frac{q_1^4+3q_2^2}{q_1}}\sum_{\sigma=\pm}\frac{\overline{\rho^{-\sigma}}(q\mid\frac{\mathcal{L}(q)}{2q_1})\rho^{\sigma}(t,p+q\mid\frac{\mathcal{L}(q)}{2q_1})}{q_1\mathcal{L}(p)+p_1\mathcal{L}(-q)-i\sigma 0q_1(q_1+p_1)}.$$

The time evolution of $\rho^\sigma(t,p+q \mid \frac{\mathcal{L}(q)}{2q_1})$ in the r.h.s. is not explicitly known. However, it is useful to factorize the time evolution it would have in the limit $v(p) \to 0$ by writing (see also (2.56) below)

$$\rho(t,p \mid \mathbf{k}) = \mathrm{e}^{-2it[3(\mathbf{k}+p_1)\mathcal{L}(p)+2p_1^3]}\tilde{\rho}(t,p \mid \mathbf{k}). \tag{2.40}$$

Note that for the special reduction furnishing the spectral data we have that

$$\tilde{\rho}^\sigma\left(t,q \,\middle|\, \frac{\mathcal{L}(q)}{2q_1}\right) = \rho^\sigma\left(q \,\middle|\, \frac{\mathcal{L}(q)}{2q_1}\right) \tag{2.41}$$

is time independent.

Finally we can write by using the identity

$$\begin{aligned}\frac{q_1^4+3q_2^2}{q_1} &= 6\mathcal{L}(p)\frac{\mathcal{L}(-q)}{2q_1} - 2[3p_1\mathcal{L}(p)+2p_1^3] + \\ &\quad +2\left[3\left(\frac{\mathcal{L}(q)}{2q_1}+q_1+p_1\right)\mathcal{L}(p+q)+2(p_1+q_1)^3\right]\end{aligned} \tag{2.42}$$

that

$$\begin{aligned}&\tilde{v}(t,p) - v(p) \\ &\quad = -\frac{1}{2}\int \mathrm{d}^2q\, p_1 \sum_{\sigma=\pm} \frac{\overline{\rho^{-\sigma}}(q \mid \frac{\mathcal{L}(q)}{2q_1})}{q_1\mathcal{L}(p)+p_1\mathcal{L}(-q)-i\sigma 0 q_1(q_1+p_1)} \times \\ &\quad \times \left[\exp\left(6it\mathcal{L}(p)\left(\frac{\mathcal{L}(p)}{2p_1}+\frac{\mathcal{L}(-q)}{2q_1}\right)\right)\times\right. \\ &\quad \left.\times\, \tilde{\rho}^\sigma\left(t,p+q \,\middle|\, \frac{\mathcal{L}(q)}{2q_1}\right) - \rho^\sigma\left(p+q \,\middle|\, \frac{\mathcal{L}(q)}{2q_1}\right)\right].\end{aligned} \tag{2.43}$$

At the special values of p and q for which the denominator in the r.h.s. vanishes

$$p+q|_{q_1\mathcal{L}(p)+p_1\mathcal{L}(-q)=0} = \{\beta-\alpha, \beta^2-\alpha^2\}, \tag{2.44}$$

with

$$\alpha = \frac{\mathcal{L}(q)}{2q_1}, \qquad \beta = \frac{\mathcal{L}(-p)}{-2p_1} \tag{2.45}$$

and we obtain that, due to (2.41), $\tilde{\rho}$ and ρ reduce to the spectral data

$$\rho^\sigma(\beta-\alpha, \beta^2-\alpha^2 \mid \alpha). \tag{2.46}$$

Therefore, at $q_1\mathcal{L}(p)+p_1\mathcal{L}(-q)=0$ the function in the square brackets in the r.h.s. vanishes and p_1 can be extracted from the integral. We conclude that

$$\lim_{p\to 0}\tilde{v}(t,p) = v(0). \tag{2.47}$$

The continuity of $\tilde{v}(t,p)$ at $p=0$ ensures us that the exponent in (2.21) is the dominant term in the time evolution of $v(t,p)$ and therefore we get for $u(t,x,y)$ at large x the behavior in (1.4).

2.3.4. $1/\mathbf{k}$-*expansion in the Inverse Problem*

One can compute the coefficient of $1/\mathbf{k}$ in the asymptotic expansion at large $\mathbf{k}$ of $\nu(t,p\mid\mathbf{k})$ also by using the Jost solution as reconstructed from spectral data via the integral equation (2.34).

Thanks to proposition A.1 in the Appendix we have that

$$\lim_{\mathbf{k}\to\infty}\mathbf{k}\left[\nu(t,p\mid\mathbf{k})-\delta^2(p)\right]$$
$$=\frac{1}{4}\int\frac{\mathrm{d}^2q\,e^{it\frac{q_1^4+3q_2^2}{q_1}}}{q_1-i0t}\sum_{\sigma=\pm}\overline{\rho^{-\sigma}}\left(q\,\Big|\,\frac{\mathcal{L}(q)}{2q_1}\right)\nu^{\sigma}\left(t,p+q\,\Big|\,\frac{\mathcal{L}(q)}{2q_1}\right).\tag{2.48}$$

As mentioned in the Appendix, the $i0t$ term is not relevant for $t\neq 0$ and, therefore, we get from (2.37)

$$\lim_{\mathbf{k}\to\infty}\mathbf{k}\left[\nu(t,p\mid\mathbf{k})-\delta^2(p)\right]=-\frac{v(t,p)}{2p_1}\quad\text{for } t\neq 0.\tag{2.49}$$

The singular function $1/p_1$ does not need any regularization at $t\neq 0$ since it is smoothed by the oscillating behavior in time of $v(t,p)$ indicated in (2.21). On the other side at $t=0$ (2.48) is discontinuous

$$\lim_{t\to\varepsilon 0}\lim_{\mathbf{k}\to\infty}\mathbf{k}\left[\nu(t,p\mid\mathbf{k})-\delta^2(p)\right]\tag{2.50}$$
$$=\frac{1}{4}\sum_{\sigma=\pm}\int\frac{\mathrm{d}^2q}{q_1-i\varepsilon 0}\overline{\rho^{-\sigma}}\left(q\,\Big|\,\frac{\mathcal{L}(q)}{2q_1}\right)\nu^{\sigma}\left(p+q\,\Big|\,\frac{\mathcal{L}(q)}{2q_1}\right),\quad \varepsilon=+,-.$$

Substituting $(q_1-i\varepsilon 0)^{-1}=(q_1\mp i\mathcal{L}(-q)0)^{-1}+2i\pi\varepsilon\delta(q_1)\vartheta(\pm\varepsilon q_2)$ we can use (2.32) and obtain

$$\lim_{t\to\varepsilon 0}\lim_{\mathbf{k}\to\infty}\mathbf{k}\left[\nu(t,p\mid\mathbf{k})-\delta^2(p)\right]=-\frac{v(p)}{2(p_1\pm ip_2 0)}+\tag{2.51}$$
$$+\frac{i\pi\varepsilon}{2}\sum_{\sigma=\pm}\int\mathrm{d}q_2\lim_{q_1\to 0}\overline{\rho^{-\sigma}}\left(q\,\Big|\,\frac{\mathcal{L}(q)}{2q_1}\right)\nu^{\sigma}\left(p+q\,\Big|\,\frac{\mathcal{L}(q)}{2q_1}\right)\vartheta(\pm\varepsilon q_2),\quad\varepsilon=+,-.$$

Due to (2.28) the last term is equal to $i\pi\varepsilon\vartheta(\mp\varepsilon p_2)\delta(p_1)v(0,p_2)$ and consequently,

$$\lim_{t\to\varepsilon 0}\lim_{\mathbf{k}\to\infty}\mathbf{k}\left[\nu(t,p\mid\mathbf{k})-\delta^2(p)\right]=-\frac{v(p)}{2(p_1+i\varepsilon 0)}.\tag{2.52}$$

Combining this result with the previous formula (2.49) we recover the same coefficient (2.24) computed by using the integral equation for $\nu(t,p \mid \mathbf{k})$ of the direct problem and, by recalling (2.48) and (2.49), we obtain that

$$\begin{aligned}\frac{v(t,p)}{p_1+i0t} &= -\frac{1}{2}\int \frac{\mathrm{d}^2 q\, \mathrm{e}^{it\frac{q_1^4+3q_2^2}{q_1}}}{q_1-i0t}\times \\ &\times \sum_{\sigma=\pm} \overline{\rho^{-\sigma}}\left(q \,\Big|\, \frac{\mathcal{L}(q)}{2q_1}\right)\nu^{\sigma}\left(t,p+q \,\Big|\, \frac{\mathcal{L}(q)}{2q_1}\right). \qquad (2.53)\end{aligned}$$

2.3.5. *The Reconstructed Time Evolution of the Jost Solution*

To derive the time evolution equation of the Jost solution we differentiate with respect to t equation (2.34). Proposition A.1 in the Appendix furnishes the derivative of the kernel of this equation

$$\begin{aligned}\frac{\partial}{\partial t}\frac{\mathrm{e}^{it\frac{q_1^4+3q_2^2}{q_1}}}{\mathcal{L}(-q)+2\mathbf{k}q_1} &= 3i\mathcal{L}(-q)\frac{\mathrm{e}^{it\frac{q_1^4+3q_2^2}{q_1}}}{q_1-i0t}+ \\ &+2i(2q_1^3+3(q_1-\mathbf{k})\mathcal{L}(-q))\frac{\mathrm{e}^{it\frac{q_1^4+3q_2^2}{q_1}}}{\mathcal{L}(-q)+2\mathbf{k}q_1}. \qquad (2.54)\end{aligned}$$

By using (2.53) and the identity

$$\begin{aligned}\frac{q_1^4+3q_2^2}{q_1} &= 6\mathcal{L}(p)\left(\mathbf{k}+\frac{\mathcal{L}(-q)}{2q_1}\right)-2[3(\mathbf{k}+p_1)\mathcal{L}(p)+2p_1^3]+ \\ &+2\left[3\left(\frac{\mathcal{L}(q)}{2q_1}+q_1+p_1\right)\mathcal{L}(p+q)+2(p_1+q_1)^3\right] \qquad (2.55)\end{aligned}$$

we obtain an integral equation of the type (2.34) for the quantity $\partial\nu(t,p \mid \mathbf{k})/\partial t+2i\nu(t,p \mid \mathbf{k})\ [3(\mathbf{k}+p_1)\mathcal{L}(p)+2p_1^3]$ with the inhomogeneous term equal to $-3iv(t,p)\frac{\mathcal{L}(p)}{p_1+i0t}$. Thanks to the assumption of uniqueness of the solution of this equation we get, finally, the searched time evolution equation for the Jost solution

$$\begin{aligned}\frac{\partial\nu(t,p \mid \mathbf{k})}{\partial t} &= -2i\nu(t,p \mid \mathbf{k})[3(\mathbf{k}+p_1)\mathcal{L}(p)+2p_1^3]- \\ &-3i\int \mathrm{d}^2 q\, v(t,q)\frac{\mathcal{L}(q)}{q_1+i0t}\nu(t,p-q \mid \mathbf{k}). \qquad (2.56)\end{aligned}$$

As expected the time derivative of $\nu(p \mid \mathbf{k})$ is discontinuous at $t=0$

$$\begin{aligned}\lim_{t\to\pm 0}\frac{\partial\nu(t,p \mid \mathbf{k})}{\partial t} &= -2i\nu(p \mid \mathbf{k})[3(\mathbf{k}+p_1)\mathcal{L}(p)+2p_1^3]- \\ &-3i\int \mathrm{d}^2 q\, v(q)\frac{\mathcal{L}(q)}{q_1\pm i0}\nu(p-q \mid \mathbf{k}). \qquad (2.57)\end{aligned}$$

Note that at large $\mathbf{k}$ the leading term in the l.h.s. is of order $1/\mathbf{k}$ while in the r.h.s. it is of order $\mathbf{k}^0$ and, due to (2.16), it is equal to

$$3iv(p)\left(\frac{\mathcal{L}(p)}{p_1 + i\sigma\mathcal{L}(p)0} - \frac{\mathcal{L}(p)}{p_1 \pm i0}\right), \quad \sigma = \operatorname{sgn}\mathbf{k}_\Im. \tag{2.58}$$

This term, if the initial data $v(p)$ do not satisfy the constraint $v(0, p_2) = 0$, is zero only if $\operatorname{sgn}\mathbf{k}_\Im = \operatorname{sgn}(t\mathcal{L}(p))$. Once more we recover the relevant fact that the two limits $t \to 0$ and $\mathbf{k} \to \infty$ can be exchanged only if condition (2.25) is satisfied.

2.3.6. *The Reconstructed KPI Equation*

Recalling (2.8), that we proved to be satisfied at all times, we deduce from (2.56) that

$$\begin{aligned}
&\frac{\partial \rho(t, p \mid \mathbf{k})}{\partial t} \\
&\quad = -6i\mathcal{L}(p)\mathbf{k}[\rho(t, p \mid \mathbf{k}) - v(t, p)] - \\
&\quad\quad -3i\frac{\mathcal{L}(p)^2}{p_1 + i0t}v(t, p) - 2ip_1[3\mathcal{L}(p) + 2p_1^2]\rho(t, p \mid \mathbf{k}) - \\
&\quad\quad -3i[\mathcal{L}(p) - 2p_1\mathbf{k}]\int \mathrm{d}^2q\, v(t, q)\frac{\mathcal{L}(q)}{q_1 + i0t}[\nu(t, p - q \mid \mathbf{k}) - \delta^2(p - q)].
\end{aligned} \tag{2.59}$$

The time evolution equation of $v(t, p)$ can be obtained from this equation by computing the coefficient of the $1/\mathbf{k}$ term in its asymptotic expansion at large $\mathbf{k}$.

We first note that from (2.24), by using (2.7) that we proved to be satisfied at all times, we get

$$\lim_{\mathbf{k}\to\infty} \mathbf{k}[\rho(t, p \mid \mathbf{k}) - v(t, p)] = -\int \mathrm{d}^2q \frac{v(t, p - q)v(t, q)}{2(q_1 + i0t)}. \tag{2.60}$$

Then we multiply (2.59) by $\mathbf{k}$ and compute the limit for $\mathbf{k} \to \infty$ by inserting (2.24) and (2.60). We obtain

$$\begin{aligned}
\frac{\partial v(t, p)}{\partial t} &= 3i\mathcal{L}(p)\int \mathrm{d}^2q \frac{v(t, p - q)v(t, q)}{q_1 + i0t} - \\
&\quad -i\left(\frac{3\mathcal{L}(p)^2}{p_1 + i0t} + 6p_1\mathcal{L}(p) + 4p_1^3\right)v(t, p) - \\
&\quad -3ip_1\int \mathrm{d}^2q\, v(t, q)\frac{\mathcal{L}(q)}{q_1 + i0t}\frac{v(t, p - q)}{p_1 - q_1 + i0t},
\end{aligned} \tag{2.61}$$

which can be easily transformed into the evolution form of the KPI equation (1.15) reported in the introduction.

Appendix

PROPOSITION A.1. *The function* $\mathrm{e}^{-3it\frac{p_2^2}{p_1}}/(p_2-2\mathbf{k}p_1)$ *defines a distribution in the Schwartz space of the variables* p_1, p_2 *depending continuously on the parameter* t.

(i) *This distribution is continuously differentiable in* t *for any* $t \neq 0$ *and has well defined right/left limits at* $t = 0$ *according to the following formula*

$$\frac{\partial}{\partial t}\frac{\mathrm{e}^{-3it\frac{p_2^2}{p_1}}}{p_2-2\mathbf{k}p_1} = -\frac{6i\mathbf{k}p_2}{p_2-2\mathbf{k}p_1}\mathrm{e}^{-3it\frac{p_2^2}{p_1}} - \frac{3ip_2}{p_1+i0t}\mathrm{e}^{-3it\frac{p_2^2}{p_1}}, \tag{A.1}$$

(ii)

$$\lim_{\mathbf{k}\to\infty}\frac{\mathbf{k}}{p_2-2\mathbf{k}p_1}\mathrm{e}^{-3it\frac{p_2^2}{p_1}} = -\frac{1}{2}\frac{\mathrm{e}^{-3it\frac{p_2^2}{p_1}}}{2p_1+i0t}. \tag{A.2}$$

Note that, due to (2.13),

$$\lim_{\mathbf{k}\to\infty}\frac{\mathbf{k}}{p_2-2\mathbf{k}p_1} = -\frac{1}{2(p_1+i0\sigma p_2)}, \qquad \sigma = \operatorname{sgn}\mathbf{k}_{\Im}, \tag{A.3}$$

and therefore we get from the previous proposition:

PROPOSITION A.2. *The limits in the following formula*

$$\lim_{t\to\pm 0}\lim_{\mathbf{k}\to\infty}\frac{\mathbf{k}}{p_2-2\mathbf{k}p_1}\mathrm{e}^{-3it\frac{p_2^2}{p_1}} \tag{A.4}$$

can be exchanged if and only if

$$\operatorname{sgn}\mathbf{k}_{\Im} = \operatorname{sgn} tp_2. \tag{A.5}$$

We also have

PROPOSITION A.3. *For any distribution defined by*

$$U(t,x,y) = \int \mathrm{d}^2p\,\mathrm{e}^{-ip_1x+ip_2y}\exp\left(-it\frac{p_1^4+3p_2^2}{p_1}\right)V(p) \tag{A.6}$$

with $V(p)$ *an arbitrary function belonging to the Schwartz space*

$$\lim_{t\to\pm 0}\int_{\infty}^{x}\mathrm{d}x'\,U(t,x',y) = \int_{\mp\infty}^{x}\mathrm{d}x'\,U(0,x',y), \tag{A.7}$$

$$\lim_{t\to\pm 0}\int_{-\infty}^{x}\mathrm{d}x'\,U(t,x',y) = \int_{\mp\infty}^{x}\mathrm{d}x'\,U(0,x',y). \tag{A.8}$$

Therefore the operator inverse of ∂_x defined by $\int_{-t\infty}^{x} \mathrm{d}x'$ commutes with the limit $t \to 0$

$$\lim_{t\to\pm 0} \int_{-t\infty}^{x} \mathrm{d}x'\, U(t,x',y) = \int_{\mp\infty}^{x} \mathrm{d}x'\, U(0,x',y) \tag{A.9}$$

while the antisymmetrical operator satisfies

$$\lim_{t\to\pm 0} \frac{1}{2}\left(\int_{-\infty}^{x} + \int_{+\infty}^{x}\right)\mathrm{d}x'\, U(t,x',y) = \int_{\mp\infty}^{x} \mathrm{d}x'\, U(0,x',y). \tag{A.10}$$

References

1. Kadomtsev, B. B. and Petviashvili, V. I.: *Sov. Phys. Doklady* **192** (1970), 539.
2. Novikov, S. P., Manakov, S. V., Pitaevskii, L. P., and Zakharov, V. E.: *Theory of Solitons. The Method of Inverse Scattering*, Plenum, New York, 1984.
3. Ablowitz, M. J. and Segur, H.: *J. Fluid Mech.* **92** (1979), 691.
4. Dryuma, V.: *Sov. Phys. J. Exp. Theor. Phys. Lett.* **19** (1974), 381.
5. Zakharov, V. E. and Manakov, S. V.: *Sov. Sci. Rev. – Phys. Rev.* **1** (1979), 133; Manakov, S. V.: *Physica* **D3** (1981), 420.
6. Fokas, A. S. and Ablowitz, M. J.: *Stud. Appl. Math.* **69** (1983), 211.
7. Boiti, M., Léon, J., and Pempinelli, F.: *Phys. Lett.* **A141** (1989), 96.
8. Boiti, M., Pempinelli, F., Pogrebkov, A. K., and Polivanov, M. C.: *Inverse Problems* **8** (1992), 331.
9. Ablowitz, M. J. and Villarroel, J.: *Stud. Appl. Math.* **85** (1991), 195.
10. Manakov, S. V.: personal communication.
11. Bakurov, V. G.: *Phys. Lett.* **A 160** (1991), 367.
12. Xin Zhou: *Comm. Math. Phys.* **128** (1990), 551.
13. Boiti, M., Pempinelli, F., and Pogrebkov, A.: *Inverse Problems* **10** (1994), 505.
14. Boiti, M., Pempinelli, F., and Pogrebkov, A.: *J. Math. Phys.* **35** (1994), 4683.
15. Boiti, M., Pempinelli, F., Pogrebkov, A. K., and Polivanov, M. C.: in Boiti, M., Martina, L., and Pempinelli F. (Eds), *Nonlinear Evolution Equations and Dynamical Systems*, World Scientific Pub. Co., Singapore, 1992, pp. 97.
16. Boiti, M., Pempinelli, F., Pogrebkov, A. K., and Polivanov, M. C.: *Teoret. i Matemat. Fizika* **93** (1992), 181.
17. Ablowitz, M. J., Manakov, S. V., and Shultz, C. L.: *Phys. Lett.* **A148** (1990), 50.
18. Fokas, A. S. and Sung, L.-Y.: *Inverse Problems* **8** (1992), 673.

Acta Applicandae Mathematicae **39:** 193–228, 1995.

Solitons and the Korteweg–de Vries Equation: Integrable Systems in 1834–1995

R. K. BULLOUGH and P. J. CAUDREY
Department of Mathematics, UMIST, PO Box 88, Manchester M60 1QD, U.K.

(Received: 15 February 1995)

Abstract. We trace the connections back from the paper by D. J. Korteweg and G. de Vries of 1895 to the sequence of events which began in August 1834 when J. S. Russell observed the soliton and so explain why KdV's paper was so much concerned with '*change of form* of long waves'. From this we trace a forward path which since 1895 has seen the discovery of the inverse method, associated with this a far-reaching mathematical structure, and at 1995 found a basis for two-dimensional quantum gravity intimately connected with the Korteweg–de Vries equation.

Mathematics Subject Classification (1991): 58F07.

Key words: integrable systems, solitons.

1. Introduction and the Work of Korteweg and de Vries

In this paper for the Centenary meeting of the now famous paper by Dr D. J. Korteweg and Dr G. de Vries [1], we want to trace the history of the soliton from the time at which it first appeared in the form of the 'solitary wave' of J. S. Russell [2, 3] following his observation of it 'in the month of August 1834' [3] on the surface of the water in a canal in Scotland,[1] use this history to illuminate the controversy which surrounded Russell's work until 1895 when Korteweg and de Vries (KdV) [1] settled that controversy entirely in favour of Russell, use the *Korteweg–de Vries equation* published in the paper [1] as the introduction it actually was to 'solitons' and the dynamics of 'integrable systems' [4–6], use this introduction to understand the mathematical structure of integrable systems, quantum and classical, as we now do understand this structure, and then, finally, to use this structure to sketch the relevance of the Korteweg–de Vries equation to current understanding of two-dimensional quantum gravity! A 'chart' or 'map', first published in its present form in [7, 8] summarising some of this understanding is presented as Figure 1 which follows, and although the KdV equation can hardly be seen on this map, experts will at least detect it embedded in the KP-equations (Kadomtsev–Petviashvili equations [9]) in the 'box' marked "$U(x, y, t)$ solves integrable model in $(2+1)$ dimensions" in the third column from the right

[1] This canal is believed to be the Union Canal with passes below the campus of Heriot-Watt University, Scotland (see [4]).

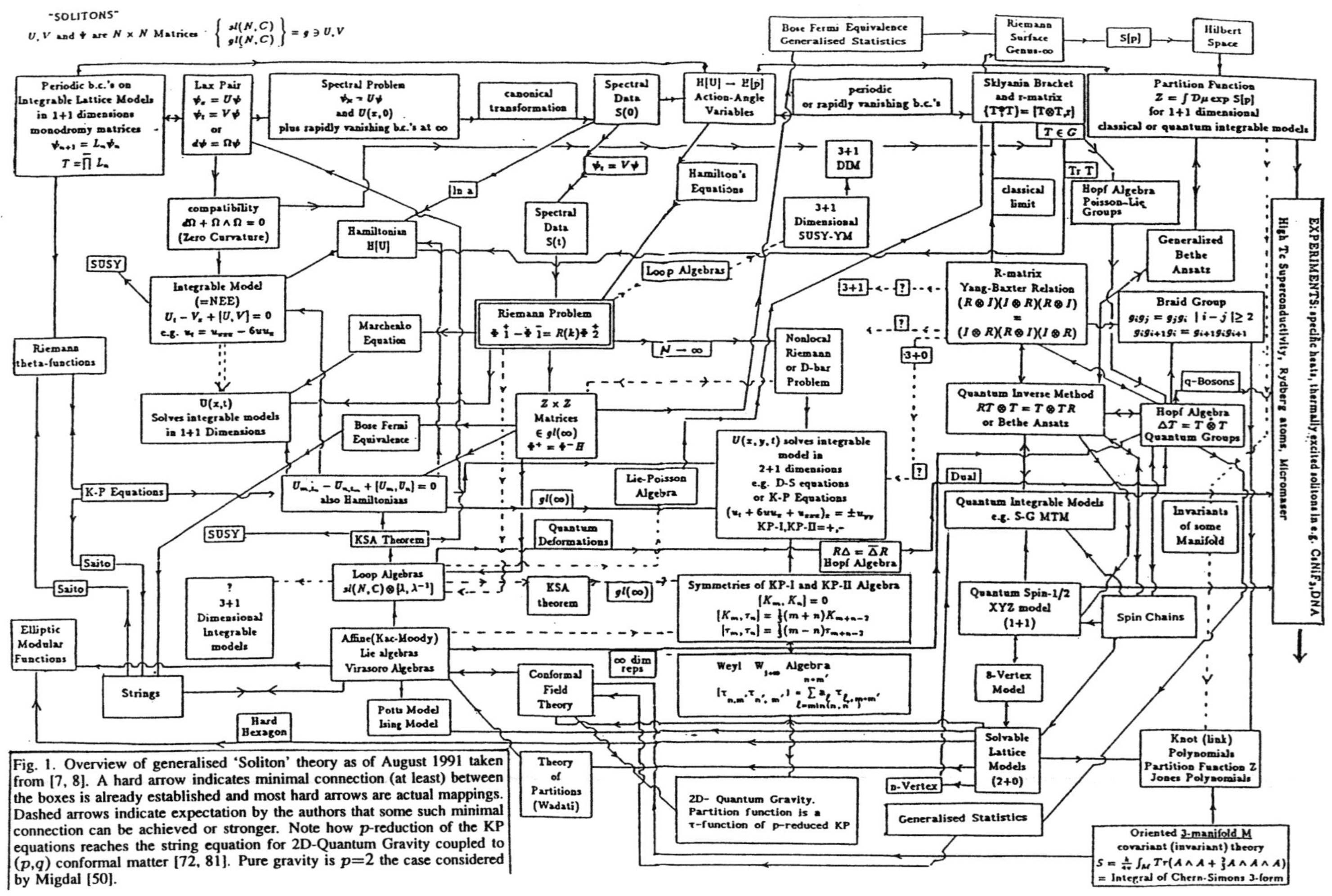

Fig. 1. Overview of generalised 'Soliton' theory as of August 1991 taken from [7, 8]. A hard arrow indicates minimal connection (at least) between the boxes is already established and most hard arrows are actual mappings. Dashed arrows indicate expectation by the authors that some such minimal connection can be achieved or stronger. Note how p-reduction of the KP equations reaches the string equation for 2D-Quantum Gravity coupled to (p,q) conformal matter [72, 81]. Pure gravity is p=2 the case considered by Migdal [50].

of this map, while closer scrutiny will show that it appears explicitly in the 'box' "Integrable Model (= NEE)" second column from the left on this map.

These remarks make plain that our title[2] is indeed very relevant to this centenary of the now famous paper of 1895 by Korteweg and de Vries published in the London, Edinburgh and Dublin *Philosophical Magazine* and *Journal of Science*, Vol. XXXIX, Fifth Series, January–June 1895 in London by Taylor and Francis, Red Lion Court, Fleet Street, and called '*On the Change of Form of Long Waves Advancing in a Rectangular Canal, and on a New Type of Long Stationary Waves*' [1]. Particularly the word 'soliton' itself was coined by Martin Kruskal and Norman Zabusky in a similarly famous paper of 1965 [5] which in essence found such 'solitons' numerically as solutions of a Korteweg–de Vries equation derived in continuum limit from the dynamics of an anharmonic lattice with cubic nonlinearity.

We now know that the equation discovered by Korteweg and de Vries [1] is '*generic*' for a large class of physical systems, namely for the weakly nonlinear-weakly dispersive plane wave systems; and the solitary wave solutions, which depend on a single argument $x-ct$, where x and t are space and time coordinates and c is a velocity, each balance the nonlinearity against the dispersion.[3] A well-known example of the KdV equation occurs in describing ion acoustic waves in a plasma [10] – while a more fanciful and much earlier application to sound waves is in one of Russell's papers [11]!

In their article of 1895, KdV were concerned with the physical problem of *gravity waves* (but not yet quantum gravity – see later in Section 6!) in water of finite depth ℓ; and they gave there their famous equation 'the Korteweg–de Vries equation'; now usually called just KdV as[4]

$$\frac{\mathrm{d}\eta}{\mathrm{d}t} = \frac{3q_0}{2\ell}\frac{\partial}{\partial x}\left(\frac{1}{2}\eta^2 + \frac{2}{3}\alpha\eta + \frac{1}{3}\sigma\frac{\partial^2\eta}{\partial x^2}\right) \tag{1}$$

with $\sigma = \frac{1}{3}\ell^3 - T\ell/\rho g$; ℓ is the depth of the liquid (water), T its surface tension, ρ its density and g the acceleration due to gravity, while the number α proved [1] to be $\alpha = \pm h$, where $+h$ is the height of the disturbance excited above, or $-h$ the depth of the disturbance excited below, the level of the undisturbed water. Elsewhere [12] we took this Equation (1) in the form

$$\frac{\partial u}{\partial t} + \frac{3}{2}\frac{c_s}{h}u\frac{\partial u}{\partial \xi} = \frac{1}{2}c_s h^2\left(\frac{\gamma}{\rho g h^2} - \frac{1}{3}\right)\frac{\partial^3 u}{\partial \xi^3}. \tag{2}$$

---

[2] An original title was '*Aspects of the History of the Soliton*'.

[3] In KdV, Equation (1) below, in particular, with $u_x \equiv \partial u/\partial x$ etc. in modern notation (see below), $u_x + u_t = \mathrm{O}(\varepsilon^3)$ while the nonlinearity and the dispersion are each $\mathrm{O}(\varepsilon^5)$ with ε a multiple scales parameter: $\varepsilon^2 = k\ell^{-1} = h^2\ell^{-2}$, where k is a wave amplitude; ℓ is a wavelength and h the depth of undisturbed water. In connection with their solutions Equations (3) and (4) below, KdV [1] point out that for the analysis of Stokes to be valid (see below, Section 2) $\ell^2 kh^{-3}$ must be small: but $\ell^2 kh^{-3} = k\ell^{-1}\ell^3 h^{-3} = \varepsilon^2\varepsilon^{-3} = \varepsilon^{-1}$ and is *not* small (*N.B.* KdV use h, k, and ℓ differently in Equations (1), (3), (4)).

[4] KdV [1] use both $\partial\eta/\partial t$ and $\mathrm{d}\eta/\mathrm{d}t$ in different places in expressing this equation.

And here the independent variable ξ is $x - c_s t$, so the equation is in a frame moving at the 'sound speed', or linearised velocity, $c_s = \sqrt{gh}$, where h is now (and see later) the depth of undisturbed water (KdV's ℓ) and γ (= T) is the surface tension.

KdV gave the stationary wave solution of their Equation (1) as

$$\eta = h\,\mathrm{sech}^2 x\sqrt{\frac{h}{4\sigma}} \tag{3}$$

and gave the general solution as

$$\eta = h\,\mathrm{cn}^2 x\sqrt{\frac{h+k}{4\sigma}} \quad \left(\mathrm{mod}\ M = \sqrt{\frac{h}{h+k}}\right), \tag{4}$$

in terms of the Jacobian elliptic function cn, to which they attached the name "cnoidal wave" (in analogy with sinusoidal waves). The number k determines wave length and for $k \to 0$, solution (4) is the solution (3); and, as already noted, there is the negative solution from (3) $\eta = -h\,\mathrm{sech}^2 x\sqrt{h/-4\sigma}$; so η in (3) is entirely positive for $\gamma/\rho g h^2 < 1/3$, entirely negative for $\gamma/\rho g h^2 > 1/3$, while for $\gamma/\rho g h^2 \approx 1/3$ a higher-order derivative $\partial^5 u/\partial\xi^5$ would need to be included in Equation (2) and one in $\partial^4\eta/\partial x^4$ inside the bracket in Equation (1).

The solitary wave solution (3) of Equation (1) corresponds to liquid at infinity brought to rest in a frame moving at a speed

$$q = \sqrt{g\ell}\,(1 + h/2\ell). \tag{5}$$

This solitary wave solution, translating with permanent shape $h\,\mathrm{sech}^2$, is actually the 'one-soliton' solution of KdV Equation (1) as we now know this [6]. We believe [13] that the 'soliton' of [5, 6] was originally to be 'solitron' as in 'electron', 'positron' and 'neutron',[5] and because of the particle-like character of the *multi*-soliton solutions first described numerically in [5]; but certain commercial activities at that time (1965) precluded such a usage. We also believe that 'soliton' and 'solitron' both derive naturally from the 'well-known solitary wave solution' of [1], and this usage goes back to J. S. Russell's paper [2] in particular where he used the phrase 'a large solitary progressive wave', and was apparently the first person to do so.

In this paper, as already noted, we want to go back, first of all, to Russell in order to show why KdV were so much concerned with 'change of form' and thereby why the KdV paper [1] was consequently a landmark in the early history of the soliton. That early history goes back to Russell's first observation of the solitary wave in August 1834 and continues controversially up to KdV's paper [1]. Then this paper served to eliminate all controversy in favour of Russell's original [2, 3] understanding of his own observations. After this, we go on to

[5] 'Electron' is due to George Johnstone Stoney, Vice-President, Royal Dublin Society.

the extraordinary achievements since 1895 in the field of nonlinear wave theory stimulated very much by KdV's paper [1].

2. J. S. Russell's Work 1834–1844 and Its Relation to KdV 1895

Russell, in the period 1834–1844 culminating in his paper [3], was almost wholly concerned to establish the validity of his solitary wave, as he described it in the papers [2, 3], as a genuine feature of the science of hydrodynamics. That science, as a practical science, was in its infancy in 1834, although at a theoretical level, Stokes [14] could make an impressive handling of Euler's equations in 1847 – even though, and see below, he misinterpreted his results in just the point which was really at issue, the strictly nonlinear character of Russell's solitary wave.

Russell himself apparently had a useful knowledge of classical texts like Lagrange's [15], Laplace's [16], and even Poisson's [17], while he particularly appreciated [4] a work [18] by the two brothers Weber, Professors at Leipzig and Halle, respectively. Still, and despite many other talents [4, 12, 19], Russell's mathematical skills were rather elementary and he tried to validate his *observations* of the solitary wave [2, 3] (which incidentally include [2–4] both soliton break-up and soliton collision) by justifying by actual experiment his proposed formula for the speed c of his solitary wave

$$c^2 = g(h + k), \tag{6}$$

where it is k which is KdV's h the height of the solitary wave, and h is KdV's ℓ the depth of undisturbed water. Note the explicit dependence on k (KdV's h), a typical nonlinear feature. This formula led Russell into bitter controversy with the then Astronomer Royal G. B. Airy [20] and his interpretation was strikingly at variance, precisely in the explicit k dependence, with the conclusions reached by the great Sir George Stokes as reported in [14]. And although Boussinesq [21] and Rayleigh [22] actually justified all of Russell's conclusions later, including [4] especially the velocity formula (6), argument continued all the way to 1895 and KdV's paper [1] when, as we show they settled it definitively in favour of Russell.

Russell was actually *classifying* waves in his 1844 paper [3] and used the classification 'solitary wave' there: 'a large progressive solitary wave' is already in [2]. He evidently preferred 'Wave of Translation' or 'The Wave' or "The Wave" "*par excellence*", the solitary progressive wave of equilibrium of the fluid rather than 'solitary wave' [4], and in his posthumous book [23], he used the title '*The Wave of Translation in the Oceans of Water, Air and Ether*'. In the last, 'Ether', he perhaps already impinges on quantum gravity (!), although in fact he was concerned to apply his formula (6) to these different phenomena, the waves of the sea, sound waves in the air (cf. [10]) and 'ether waves' in the universe. From the last he found a figure for the depth h of the universe as 5×10^{17} miles – wrong only by about five orders of magnitude [4].

The issue with Airy surrounded the problem of waves of permanent profile (permanent shape). Such waves are necessarily $u(x,t) = u(x - ct)$, and Russell believed his wave was in this category and by implication was the only such wave being *necessarily* everywhere positive and lump shaped as experiments on the actual profile of the wave in the laboratory convincingly showed [4]. In [20], Airy found a velocity formula

$$c^2 = gm^{-1}(\tanh mh) \tag{7}$$

for harmonic waves of wave number m, and this reduces to c_s^2 for $m \ll h^{-1}$. This formula disagrees substantially with Russell's formula (6) being independent of the height k, the height of the disturbance above the undisturbed water. Moreover, Airy [20] states, e.g., that

> "We are not disposed to recognize this wave (Russell's) as deserving the epithets 'great' or 'primary' ... and we conceive that ever since it was known that the theory of shallow waves of great length was contained in the relation $\mathrm{d}^2X/\mathrm{d}t^2 = g\varkappa\,\mathrm{d}^2X/\mathrm{d}x^2 \ldots$ the theory of the solitary wave has been well-known! Here the solutions was Sir John Herschel's comment "It is merely half of a common wave that has been cut off (!)" [24]."

However, real difficulty for Russell's viewpoint emerged when Stokes [14] appeared to show from Euler's equations that the only waves of permanent shape were, as was quoted by KdV [1] in the rest frame, the sequence of harmonic waves

$$\eta = h\cos(2\pi x/\lambda) - (3h^2\lambda^2/64\pi^2\ell^3)\cos(4\pi x/\lambda); \tag{8}$$

h is KdV's h.

Earlier Stokes [25] had said

> "It is the opinion of Mr Russell that the solitary wave is a phenomenon *sui generis* in nowise deriving its character from the circumstances of the generation of the wave. His experiments seem to render this conclusion probable. Should it be correct the analytical character of the solitary wave remains to be discovered."

However, in the paper read on 1 March 1847 and published in the *Trans. Camb. Phil. Soc.* "On the theory of oscillatory waves" [14] he effectively concludes (for axes x horizontal and y vertically *downwards*) that

$$y = a\cos m(x - ct) - Ka^2\cos 2m(x - ct) \tag{9}$$

(with $a = -mAcg^{-1}$ and m arbitrary, but K then fixed and depending on a, m and h and c given by Equation (7))

> "is the only form of wave which possesses the property of being propagated with a constant velocity and without change of form – so that a solitary

wave cannot be propagated in this manner. Thus the degradation observed by Russell is not due to the imperfect fluidity of the fluid and its adhesion to the sides and bottom of the canal but is an essential characteristic of the solitary wave."

The formula (9) is of course the formula (8) quoted by KdV [1] now translating up x at velocity c as given by Equation (7). We summarize Stokes's argument as it is sketched in [4], since it is very relevant to the comment KdV make upon it in their paper [1].

Euler's equation for the conservation of momentum under pressure gradients $-\nabla p$, and body forces $-\nabla\Omega$, in a incompressible fluid of density ρ is (in modern notation)

$$\frac{\mathrm{D}}{\mathrm{D}t}\rho\mathbf{u} = \rho(\mathbf{u}_t + (\mathbf{u}\cdot\nabla)\mathbf{u}) - \nabla p - \nabla\Omega. \tag{10}$$

Conservation of mass is

$$\nabla\cdot\mathbf{u} = 0. \tag{11}$$

In the case of irrotational motion $u = \nabla\phi$, and a first integral of Equation (10) as used by Stokes is

$$p = g\rho y - \rho\frac{\mathrm{d}\phi}{\mathrm{d}t} - \frac{1}{2}\rho\left[\left(\frac{\mathrm{d}\phi}{\mathrm{d}x}\right)^2 + \left(\frac{\mathrm{d}\phi}{\mathrm{d}y}\right)^2\right] \tag{12}$$

with continuity

$$\frac{\mathrm{d}^2\phi}{\mathrm{d}x^2} + \frac{\mathrm{d}^2\phi}{\mathrm{d}y^2} = 0 \tag{13}$$

and with

$$\frac{\mathrm{d}\phi}{\mathrm{d}y} = 0 \quad \text{when} \quad y = h,$$

$$\frac{\mathrm{d}p}{\mathrm{d}t} + \frac{\mathrm{d}\phi}{\mathrm{d}x}\frac{\mathrm{d}p}{\mathrm{d}x} + \frac{\mathrm{d}\phi}{\mathrm{d}y}\frac{\mathrm{d}p}{\mathrm{d}y} = 0 \quad \text{when} \quad p = 0, \tag{14}$$

the condition for the free surface remaining the free surface throughout the motion. The velocities $u = \mathrm{d}\phi/\mathrm{d}x$ and $v = \mathrm{d}\phi/\mathrm{d}y$, and p, depend on $x - ct$, the condition for permanent profile, *together with* y. From Equation (12) we can admit in ϕ a term linear in t to fix p so

$$\phi = f(x - ct) + Ct \tag{15}$$

and $x - ct$ is called x (as quoted by KdV [1]) while C is set to $-gk$. Then Equation (12) is

$$p = g\rho(y + k) + c\rho\frac{\mathrm{d}\phi}{\mathrm{d}x} - \frac{1}{2}\rho\left[\left(\frac{\mathrm{d}\phi}{\mathrm{d}x}\right)^2 + \left(\frac{\mathrm{d}\phi}{\mathrm{d}y}\right)^2\right] \tag{16}$$

so that $u = v = 0$, $p = 0$, when $y = -k$. Also from the dependence on $x - ct$, Equation (14) is

$$\left(\frac{\mathrm{d}\phi}{\mathrm{d}x} - c\right)\frac{\mathrm{d}p}{\mathrm{d}x} + \frac{\mathrm{d}\phi}{\mathrm{d}y}\frac{\mathrm{d}p}{\mathrm{d}y} = 0 \quad \text{when} \quad p = 0, \tag{17}$$

and substituting Equation (16) into Equation (17) gives

$$\begin{aligned} g\frac{\mathrm{d}\phi}{\mathrm{d}y} &- c^2\frac{\mathrm{d}^2\phi}{\mathrm{d}x^2} + 2c\frac{\mathrm{d}\phi}{\mathrm{d}x}\frac{\mathrm{d}^2\phi}{\mathrm{d}x^2} + \frac{\mathrm{d}\phi}{\mathrm{d}y}\frac{\mathrm{d}^2\phi}{\mathrm{d}y^2} - \\ &- \left[\frac{\mathrm{d}\phi}{\mathrm{d}x}\right]^2\frac{\mathrm{d}^2\phi}{\mathrm{d}x^2} - 2\frac{\mathrm{d}\phi}{\mathrm{d}x}\frac{\mathrm{d}\phi}{\mathrm{d}y}\frac{\mathrm{d}^2\phi}{\mathrm{d}x\,\mathrm{d}y} - \left[\frac{\mathrm{d}\phi}{\mathrm{d}y}\right]^2\frac{\mathrm{d}^2\phi}{\mathrm{d}y^2} = 0 \end{aligned} \tag{18}$$

when

$$g(y + k) + c\frac{\mathrm{d}\phi}{\mathrm{d}x} - \frac{1}{2}\left\{\left(\frac{\mathrm{d}\phi}{\mathrm{d}x}\right)^2 + \left(\frac{\mathrm{d}\phi}{\mathrm{d}y}\right)^2\right\} = 0. \tag{19}$$

This result is exact, but at this point Stokes neglects certain nonlinear terms [4] and iterates to get, with $y = 0$ the undisturbed surface,

$$\begin{aligned} &g\phi_y - c^2\phi_{xx} - g(\phi_{yy} - c^2\phi_{xxy})\left[k + cg^{-1}\phi_x\right] + \\ &\quad +2c(\phi_x\phi_{xx} + \phi_y\phi_{xy}) = 0 \end{aligned} \tag{20}$$

in which $\phi_y = [\mathrm{d}\phi/\mathrm{d}y]_{y=0}$, etc. Then ϕ satisfies Equation (13), with Equation (14) at $y = b$, and with Equation (20) at $y = 0$. When ϕ is determined the ordinate y of the surface is found [4] and k is fixed by the condition that the mean value of y shall be zero.

The first approximation to Equation (20) is

$$g\frac{\mathrm{d}\phi}{\mathrm{d}y} - c^2\frac{\mathrm{d}^2\phi}{\mathrm{d}x^2} = 0 \quad \text{when } y = 0. \tag{21}$$

So put [4, 14]

$$\phi = \sum_{m^2+n^2=0} A\mathrm{e}^{(mx+ny)} \tag{22}$$

which satisfies Equation (13): the boundary conditions are chosen to make m imaginary – so the behaviour along x *is matched to a Fourier series and periodic boundary conditions have entered* – though Stokes does not comment on this. Then for this Fourier series

$$\phi = \sum\left[\mathrm{e}^{m(h-y)} + \mathrm{e}^{-m(h-y)}\right](A\sin mx + b\cos mx) \tag{23}$$

satisfies Equation (14). Then *one* nontrivial contribution to Equation (23) satisfies the condition Equation (7) on c, and Stokes proves [4] a unique m satisfies this condition.

For a second approximation he gets

$$g\phi_y - c^2\phi_{xx} - 2cA^2c^2m^2\sin 2mx = 0, \tag{24}$$

and thus reaches (4) the ordinate y of the surface to second approximation

$$y = a\cos mx - Ka^2\cos 2mx. \tag{25}$$

Then to go further, he chooses $h \gg \lambda$ ($m = 2\pi\lambda^{-1}$) and gets

$$y = a\cos mx - \tfrac{1}{2}\, ma^2 2mx + \tfrac{3}{8}\, m^2a^2\cos 3mx, \tag{26}$$

and it is only at this order that the amplitude a of the wave enters [4] the expression for c. The result Equation (25) is the result Equation (9), and this is KdV's result Equation (8).

One of the most important points made by KdV [1], and early in their paper, concerns their cnoidal wave solution Equation (4):

> "For very large values of k these cnoidal waves coincide with the train of oscillatory waves of unchanging shape discovered by Stokes [14], which therefore *in the theory of long waves*[6] constitutes a particular case of the cnoidal form. Indeed the equation obtained by Stokes, when written in our notation becomes ... (Equation (8) above) ... but, as Sir G. Stokes remarks, in order that the method of approximation adopted by him may be legitimate, $\lambda^2 h\ell^3$ must be a small fraction.[7]"

This way they show that when capillarity is neglected the modulus M in Equation (4) is small and the cnoidal wave approximates to sinusoidal. Then in this case

> "the equation of their surface may be developed in a rapidly convergent Fourier series of which Sir G. Stokes has given the first two terms."

In short Stokes's result Equations (8) or (9) is correct, but his *conclusion*, as quoted below Equation (9) is *wrong*! Stokes has actually found the first two terms of the Fourier series for the elliptic function solution Equation (4) of the KdV equation under periodic boundary conditions, and if he had carefully taken that period to *infinity* he would be sending modulus M to unity and would have reached the $h\,\mathrm{sech}^2$ wave solution Equation (3) and its translations!

Stokes acknowledged error later [26], while it is also plain why KdV gave the title of their paper as they did, and why they devoted so much of this paper to the study of the change of shape (change of form) of the solitary and the other waves they considered.

KdV start their paper by referring to "such excellent treatises as those of Lamb and Basset" (cf. [4]) as stating that long waves in a rectangular canal "must necessarily change their form as they advance becoming steeper in front and less steep behind" also quoting Airy [20] here:

[6] The theory of long waves is due to Lagrange [15] as is described in [4].

[7] See Footnote 3 above.

> "Yet since the investigations of de Boussinesq [21], Lord Rayleigh [22], and St. Venant [27] on the solitary wave, there has been some cause to doubt the truth of this assertion. Indeed, if the reasons advanced were really decisive, it is difficult to see why the solitary wave should make an exception."

They then note that even Lord Rayleigh and McCowan [4, 28] are not quite decisive here and

> "It is the desire to settle this question definitively which has led us into the somewhat tedious calculations which are to be found at the end of this paper,"

the paper [1].

Thus in the Section IV of [1] KdV consider 'Deformation of Non-stationary Waves' in some detail. And they find, e.g., for the solitary wave Equation (3), namely $\eta = h\,\mathrm{sech}^2 px$ initially at rest in the moving frame, that for $h > 4\sigma p^2$ the wave becomes steeper in front, while for $h < 4\sigma p^2$ it becomes steeper behind, only the former agreeing with Airy [20]: for $h = 4\sigma p^2$ the wave is stationary in the moving frame. The cnoidal wave is similar, but sinusoidal waves always steepen in front – here but only here agreeing with Airy.

KdV finish with two Sections V and VI where

> "In order to remove every doubt as to the existence of absolutely stationary waves, we will show how by development in a rapidly convergent series the state of motion of the fluid belonging to such wave-motion may be calculated."

3. The Work of Boussinesq (1872) and of Rayleigh (1876)

Let us here quickly see how in "the investigations of de Boussinesq [21], Lord Rayleigh [22] and St. Venant [27]" these authors actually found the solitary waves which are now known to be one-soliton solutions. We focus on de Boussinesq and then briefly on Rayleigh. The paper by de Boussinesq [21] will be of interest later, because his equation is integrable in the way the KdV Equation (1) is integrable; and both KdV and the Boussinesq equation figure in the theory of 'the string equation' for two-dimensional quantum gravity as described in Section 6 below.

From Equations (10) and (11), Boussinesq [21] reached the partial differential equation

$$u_{tt} = c_s^2 u_{xx} + c_s^2 \left[\tfrac{3}{2}\, u^2 h^{-1} + \tfrac{1}{3}\, h^2 u_{xx} \right]_{xx}, \tag{27}$$

where h is the depth of undisturbed water, and $c_s = \sqrt{gh}$. The notation is the modern notation $u_{tt} \equiv \partial^2 u/\partial t^2$, $u_{xx} = \partial^2 u/\partial x^2$, etc. This system is invariant under time reversal, and the square brackets otherwise balance the nonlinearity against the dispersion as in the one-way going KdV equation. Boussinesq gave the solitary wave solution of Equation (27), namely

$$u(x,t) = k\,\mathrm{sech}^2\left[\frac{1}{2}\sqrt{\frac{3k}{h^3}}\,(x-ct)\right]. \tag{28}$$

The height at peak of this solution is k, the notation as used by Russell [3]. The velocity c now proves to be *exactly* that proposed by Russell, namely Equation (6). One easily scales Equation (28) to

$$u_{tt} = u_{xx} + \left[\tfrac{3}{2}\,u^2 + u_{xx}\right]_{xx} \tag{29}$$

called 'the Boussinesq equation' and the solitary wave solution is then

$$u = \lambda^2\,\mathrm{sech}^2\left[\tfrac{1}{2}\,\lambda(x-ct)\right] \tag{30}$$

while $c^2 = 1 + \lambda^2$. As noted, the Boussinesq equation is integrable, and $u(x,t)$ Equation (28) is a soliton, but the *spectral problem* which solves it by the inverse method (see Section 4) is a third-order (or 3×3 matrix) spectral problem [29, 30], and so it is rather more complicated than the Schrödinger Sturm–Liouville spectral problem (Section 4) which solves the KdV equation. It thus becomes fascinating to speculate on the history of the soliton itself if Boussinesq's paper of 1872 had then, and subsequently, received the attention it deserved. Would we even be gathered together at this Centenary for KdV 1895: we both sincerely hope so![8]

Of course [4] the Boussinesq equation, Equation (29), 'contains' the KdV equation. For in reaching Equation (27) terms effectively of order $\mathrm{O}(k^2h^{-2})$ are dropped.[9] And to this order Equation (27) can be 'factorized' so that Equation (29) is written as

$$\left[\frac{\partial}{\partial t} - \frac{\partial}{\partial x}\right]\left(\frac{\partial}{\partial t} + \frac{\partial}{\partial x} + \frac{1}{2}\frac{\partial}{\partial x}\left[\frac{3}{2}\,u + \frac{\partial^2}{\partial x^2}\right]\right)u = 0. \tag{31}$$

A solution of KdV, now scaled to

$$u_t + u_x + \tfrac{1}{2}\left[u^2 + u_{xx}\right]_x = 0 \tag{32}$$

solves Equation (31), and such a solution is

$$u = \lambda^2\,\mathrm{sech}^2 \tfrac{1}{2}\,\lambda(x - c_1 t) \tag{33}$$

with

$$c_1 = 1 + \tfrac{1}{2}\,\lambda^2. \tag{34}$$

---

[8] See the similar comment directed still elsewhere in [4].

[9] See Footnote 3 above: $k = \varepsilon^2\ell$ means $k^2h^{-2} = \ell^2\varepsilon^4h^{-2} = h^2\ell^{-2}$.

Evidently

$$c_1^2 = 1 + \lambda^2 + \tfrac{1}{4}\lambda^4 \tag{35}$$

and c_1^2 differs from $c^2 = 1+\lambda^2$ from below Equation (30) by $O(\lambda^4) = O(k^2h^{-2})$. On the other hand, Equation (33) is the solitary wave solution of KdV and c_1 is KdV's velocity expression Equation (5). Thus Equation (35) makes the point that KdV's soliton actually moves at a speed slightly different from Russell's formula (6)!

Still it was at $O(\lambda^2)$ that the argument actually raged, and $\lambda^2 \approx 1/5$ in Russell's first experiments [2] on his canal. Larger values of λ^2 arise in the more precise laboratory experiments subsequently reported in [3], however: the largest value quoted seems to be $\lambda^2 = 1/3$ ([3, Table VII, p. 336]), but values of $\lambda = 1$ were examined by Russell [3] for he certainly knew that the wave broke in this case as he points out (e.g., [3, p. 340]). Rayleigh [22] quotes Airy [20] as saying the wave always broke in this case and shows why.

Rayleigh's main result was to show that the ordinate y of the (positive) free surface was, in the stationary frame,

$$\begin{aligned} y - h &= k\,\mathrm{sech}^2\left(\tfrac{1}{2}\,xb^{-1}\right), \\ b^2 &= h^2(h+k)/3k, \end{aligned} \tag{36}$$

and that the velocity of translation c satisfies Russell's formula Equation (6) *exactly*. The result (36) differs from Boussinesq's result by a scaling $\sqrt{h/(h+k)}$ on x [4]. But this scaling has the same order of discrepancy as does the KdV solitary wave Equation (33), for

$$\sqrt{1/(1+kh^{-1})} \approx 1 - \tfrac{1}{2}\,(kh^{-1}).$$

But now the exact result Equation (6) is found for c (to within the order worked). Thus, by the very direct methods Rayleigh used [4], he further justifies Boussinesq's analysis and wholly justifies Equation (6) and so, *for Russell*, his whole position. Even so, and despite this striking conclusion it still remained for KdV to close up the whole matter definitively as both they and we have noted.[10]

Rayleigh finishes with remarks on the *breaking* of the solitary wave and goes on to consider Periodic Waves in Deep Water. Further details on Rayleigh's deduction of Equations (36) and (6) are in [4], of course in [22] and in [31].[11]

With the matter of Russell's solitary wave settled by KdV in 1895 we can turn from the *reasons for* the paper [1] of KdV and turn to the *influence of* that paper on what has happened since. First we work from 1895 to 1967 and the discovery of the inverse method for exactly solving the KdV equation.

[10] Basset [1] treats the solitary wave but does not refute the opinion of Airy [20]. Lamb [31] in 6th edition (1952) gives extensive coverage including also Rayleigh [22] but not apparently in the 2nd edition (1895).

[11] See Footnote 10.

4. The KdV Equation and the Discovery of the Inverse Method for Solving It of 1967

The discovery of the inverse method for solving the KdV equation in 1967 [6] became the discovery of 'the integrable systems' soon afterwards: this section concerns these two aspects. Work by Stan Ulam and colleagues (c. 1940) at Los Alamos on the Fermi–Pasta–Ulam (FPU) problem [33] as it was elucidated in [5] by Zabusky and Kruskal was the immediate source of the classical inverse method for solving the KdV equation. But work done by Painlevé and his contemporaries [32] some forty years before FPU [33], has since proved significant to our understanding of these 'integrable systems', while much other work done in the early part of this century is only now being rediscovered and reinterpreted in this context. Darboux, for example, [34–36] might be particularly mentioned here for his book [34] is a source from which much is being recovered only now [35, 37]. Painlevé's papers of 1900, 1902 [37] proves peculiarly relevant (Section 6) to the current understanding of two-dimensional quantum gravity so we start with this: the rest of this section is then concerned with the discovery of the inverse method and the consequent discovery of the integrable systems as these are described in our book [12].

Painlevé, Gambier and Fuchs in particular [32, 38] were concerned to determine all of those ordinary differential equations of the form

$$\mathrm{d}^2 w/\mathrm{d}z^2 = F(z, w, p),$$

where F is rational in $p \equiv \mathrm{d}w/\mathrm{d}z$, algebraic in w, and analytic in z, which have all of their critical points fixed; 'fixed' means independent of the initial condition (the constants of integration). Critical points are branch points and essential singularities: thus the problem is to find all second order ODEs of this form whose moveable singularities are poles.

Ince [32] lists 50 such ODEs. Of these the 'most interesting' are the six *irreducible* equations not replaceable by a simpler equation or a combination of these, and which therefore determine new transcendentals. The first three of these ODEs listed by Ince are

$$\begin{array}{ll}\text{(i)} & w'' = 6w^2 + z,\\ \text{(ii)} & w'' = 2w^3 + zw + \alpha,\\ \text{(iii)} & w'' = w^{-1}(w')^2 - z^{-1}w' + z^{-1}(\alpha w^2 + \beta) + \gamma w^3 + \delta w^{-1},\end{array} \tag{37}$$

and were discovered by Painlevé. Ince quotes Gambier as discovering the remaining three which are somewhat more complicated [32]. [38] quotes Gambier for the first two of the remaining three and quotes Fuchs [39] for the discovery of the one remaining of these three. All of the solutions of each of the six equations are now called 'Painlevé Transcendents'. The six equations themselves are Painlevé I, II, ..., VI. Equations (37) are Painlevé I (Equation (i)), ..., III (Equation (iii)). Painlevé VI actually includes the remaining five equations [40].

Painlevé I and Painlevé II in particular are of interest here because they can be derived by similarity reduction from KdV. If KdV Equation (2) in the moving frame is scaled [12] to

$$u_t + 6uu_x + u_{xxx} = 0, \tag{38}$$

then under the self-similar reduction

$$u(x,t) = (3t)^{-2/3}v(z), \qquad z = x/(3t)^{1/3}, \tag{39}$$

$v(z)$ satisfies

$$v''' + 6vv' - (2v + zv') = 0. \tag{40}$$

Then if

$$v = -w' - w^2, \qquad w = (v' + \alpha)/(2v - z), \tag{41}$$

w satisfies Painlevé II, Equation (37(ii)). Moreover, it is no accident, as we shall see, that under the similarity reduction which is Equations (39) but with $u(x,t) = (3t)^{-1/3}w(z)$ and $u(x,t)$ satisfying the *modified* KdV equation

$$u_t - 6u^2u_x + u_{xxx} = 0, \tag{42}$$

$w(z)$ also satisfies Painlevé II. However, and very much more to the point for quantum gravity (see Section 6), if

$$u(x,t) = -2(w(z) + t), \qquad z = (x + 6t^2), \tag{43}$$

w satisfies $w'' - 6w^2 - z = 0$ which Painlevé I, Equation (37(i)).

The requirement that every ODE arising as a similarity reduction of a completely integrable partial differential equation is of Painlevé type (up to suitable transformations) has been conjectured as a necessary condition for 'complete integrability' [38]. What 'complete integrability' means we discover in the theory as it was developed next.

Kruskal in the period 1955–1965 was concerned with the recurrences exhibited by one-dimensional harmonic lattices – the FPU problem [33] as already mentioned. This led Zabusky and Kruskal [5] to study the equation obtained for a cubic nonlinearity in the continuum limit

$$y_{tt} = y_{xx}(1 + \varepsilon y_x), \tag{44}$$

where the parameter ε which scales the nonlinearity $\sim 1/10$. By looking at the Riemann invariants of the linear problem $y_{tt} = y_{xx}$, namely

$$u = \tfrac{1}{2}(y_x + y_t) = \mathrm{O}(1), \qquad v = \tfrac{1}{2}(y_x - y_t) = \mathrm{O}(\varepsilon) \tag{45}$$

they reduced Equation (44) to the 'simple wave'

$$u_\tau + uu_x = 0 \quad (\tau \equiv 4\varepsilon(t - x)). \tag{46}$$

Characteristics of this equation are $\mathrm{d}x/\mathrm{d}\tau = u = \text{const}$. Intersection of these suggests the equation shocks. It may be integrated by the hodograph transformation to yield the solution $u = u(x - ut)$ which shocks (the speed u is large where the disturbance u is large). An element of dispersion was next added by returning to the discrete lattice model: the result was

$$y_{tt} = y_{xx}(1 + \varepsilon y_x) + \tfrac{1}{12}(\Delta x)^2 y_{xxxx}, \tag{47}$$

where Δx is the lattice spacing, and from this

$$u_\tau + uu_x + \Delta^2 u_{xxx} = 0 \quad (\tau \equiv 4\varepsilon(t - x)), \tag{48}$$

where $\Delta^2 = (\Delta x)^2/12\varepsilon$. For $\Delta x \sim 1/64$, $\varepsilon \sim 1/10$, $\Delta \approx 0.022$.

Numerical solutions of Equation (48) showed [5] that, under periodic boundary conditions, harmonic initial data developed a smooth jump approaching the simple wave shock for the chosen small value of Δ^2. Thereafter, the profile broke up into spikes the collision properties of which caused Zabusky and Kruskal to call them solitons (as described above). In order to characterize the solution across the jump Kruskal and co-workers then looked for conservation laws. Equation (48) is already in conservation law form with a conserved density $T^1 = u$. Then

$$u_\tau + \left(\tfrac{1}{2}u^2 + \Delta^2 u_{xx}\right)_x = 0 \quad \text{and} \quad \int u_\tau \, \mathrm{d}x = \frac{\mathrm{d}}{\mathrm{d}\tau}\int u \, \mathrm{d}x = 0$$

when $(1/2)u^2 + \Delta^2 u_{xx}$ vanishes through boundary conditions at $x = a$ and $x = b > a$. However, $T^2 = u^2$, $T^3 = 2u^3 - u_x^2$, are also conserved densities. Miura *et al.* [41] found 10 *polynomial* conserved densities for KdV this way. They then proved there was an infinity of nontrivial T^r. Each term $\prod y_j^{a_j}$ in T^r (where u_j is the jth x-derivative) has in this case rank

$$r = \sum_j \left[1 + \tfrac{1}{2}j\right] a_j;$$

$T^{10} = u^{10}/10 - 36u^7u_1^2 + \cdots + 419904u_8^2/12155$ has rank 10 and contains 32 terms [41]!

FPU also looked at a lattice with quartic rather than cubic anharmonicity. This leads to $y_{tt} = y_{xx}(1 + \varepsilon y_x^2)$ and to the modified KdV Equation (42). This has conservation laws also. Miura [42] found the transformation $u = v^2 + v_x$ which connects modified KdV Equation (42) and KdV Equation (38) taken with $u \to -u$. The Ricatti transformation $v = \partial(\ln\psi)/\partial x$ linearises $v^2 + v_x$ and $u = \psi_{xx}/\psi$. The Galilean transformation

$$x' = x - Vt, \quad t' \to t, \; u \to u + V, \tag{49}$$

leaves KdV in the form $u_t - uu_x + u_{xxx} = 0$ invariant. Hence, $u \to u + \lambda = \psi_{xx}/\psi$ and the *Schrödinger eigenvalue problem*

$$-\psi_{xx} + u\psi = -\lambda\psi \tag{50}$$

has emerged. Furthermore, $u = \psi_{xx}/\psi - \lambda$ into KdV Equation (38) (with $u \to -u$) yields [6]

$$\begin{array}{ll}\text{(a)} & \lambda_t = 0 \text{ for the bound states } (\lambda > 0) \\ & \text{even though } u \text{ depends on } t; \\ \text{(b)} & \psi_t + \psi_{xxx} - 3(u-1)\psi_x = C\psi + D\psi \int^x \mathrm{d}x\psi^{-2};\end{array} \tag{51}$$

and [6] D in Equation (51) vanishes for $\lambda > 0$, while C vanishes through normalisation of the bound state eigenfunctions. From the *boundary conditions* $u \to 0$ as $|x| \to \infty$ the eigenfunction ψ_n belonging to $\lambda_n > 0$ takes the form $\psi_n \sim C_n(t)\exp(-k_n x)$ for $x \to \infty$ ($k_n = \lambda_n^{1/2}$), and from Equation (51)

$$C_n(t) = C_n(0)\exp 4k_n^3 t.$$

For $\lambda = -k^2 < 0$ a solution of Equation (50) for large $|x|$ is a linear combination of $\exp \pm ikx$. If $\psi \sim \exp(-ikx) + b\exp(ikx)$, $x \to \infty$ a reflection coefficient, and $|a|^2 + |b|^2 = 1$. Since the spectrum for $\lambda < 0$ is continuous, λ can be chosen so that $\lambda_t = 0$ here also. The asymptotics for ψ into Equation (50) yields $D = 0$, $C = 4ik^3$, and two equations which together easily give $a(k,t) = a(k,0)$, $b(k,t) = b(k,0)\exp(8ik^3 t)$.

The key point now is that the *scattering data* a, b, c_n and λ_n are sufficient to allow reconstruction of u at any time t. Let $K(x,y)$ for $y \geqslant x$ be the solution of the Gel'fand–Levitan equation [43, 44]

$$K(x,y) + B(x+y) + \int_x^\infty K(x,z)B(y+z)\,\mathrm{d}z = 0 \tag{52}$$

with

$$B(\xi) = \frac{1}{2\pi}\int_{-\infty}^{\infty} b(k)\exp(ik\xi)\,\mathrm{d}k + \sum_n c_n^2 \exp k_n\xi. \tag{53}$$

Then $u(x,t)$ is given by $u(x,t) = 2\partial K(x,x)/\partial x$. Consequently, initial data $u(x,0)$ solving KdV Equation (38) (with $u \to -u$) at $t = 0$ can be mapped to scattering data $S = \{b(k,0);\ c_n(0);\ \lambda_n\ (n = 1,\dots,N)\}$ and $S = \{b(k,t);\ c_n(t);\ \lambda_n(n = 1,\dots,N)\}$ can be inverted by the Gel'fand–Levitan system Equations (46), (47) to determine $u(x,t)$ from Equation (53) thus solving KdV at $t > 0$ – a marvellous result!

All of this was achieved by 1967 [6] and three important further steps soon followed. First Lax [45] rationalised the procedure by identifying what is now called the Lax pair: given the relatively prime differential operators $u \to \hat{L}_u \equiv (-\partial^2/\partial x^2 + u)$, $\widehat{B}_u = \partial^{2n+1}/\partial x^{2n+1} +$ lower degree ($\widehat{B}$ is skew symmetric, $\widehat{B}_u^\dagger = -\widehat{B}_u$) such that there is a one parameter family of unitary operators $\widehat{U}$ satisfying $\widehat{U}_t = \widehat{B}\widehat{U}$; and $\hat{L}_u$ is unitary equivalent under $\widehat{U}$, i.e. $\widehat{U}^{-1}(t)L_u(t)\widehat{U}(t)$ is independent of t, then [12]: (1) the eigenvalues of $\hat{L}_u$ are integers of motion; (2) $\hat{L}_{u,t} = [\widehat{B}, \hat{L}_u]$ (where $\hat{L}_{u,t}$ is the t-derivative of $\hat{L}_u$); and (3) this is an *evolution* equation $u_t = K[u]$, where $K[u]$ is a nonlinear functional of u (in general). Lax [45] found that with

$$\widehat{B} = \partial^3 - \tfrac{3}{4}\, u\partial - \partial \tfrac{3}{4}\, u \quad (\partial \equiv \partial/\partial x)$$

this equation was in fact the KdV equation. But he found an infinite sequence of operators $\widehat{B} = \partial^5 + \cdots + \partial^7 + \cdots$ and thus found an infinite hierarchy of KdV equations of orders $3, 5, \ldots$, now called the 'Lax hierarchy'.

It is a remarkable fact that Lax's operators $\widehat{B}$ for this hierarchy of KdV equations can be found by defining the square root $\widehat{R}$ of the operator $-\hat{L}_u$ by $\widehat{R}^2 = -L_u$ and $\widehat{R} = \partial + c_0 + c_1\partial^{-1} + c_2\partial^{-2} + \cdots$. Then $-\hat{L}_u = \partial^2 - u$ and $\widehat{R}^2 = -\hat{L}_u$ is $\partial^2 + 2c_0\partial + (c_{0,x} + c_0^2 + 2c_1) + (c_{1,x} + 2c_2 + 2c_0c_1)\partial^{-1} + \cdots = \partial^2 - u$ so that $c_0 = 0$, $c_1 = -\frac{1}{2}\, u$, $c_2 = \frac{1}{4}\, u_x, \ldots$. Evidently the principal part of $\widehat{R}$ (that excluding inverse powers of ∂) is simply ∂ and with this principal part $[\widehat{R}, \hat{L}_u] = \hat{L}_{u,t}$ is $u_x = u_t$. However the principal part of $(-\hat{L}_u)^{3/2} = (-L_u)\widehat{R}$ is $\partial^3 - \partial u - \frac{1}{2}\, u\partial + \frac{1}{4}\, u_x = \partial^3 - \frac{3}{4}\, \partial u - \frac{3}{4}\, u\partial$ and this is Lax's operator $\widehat{B}$ for the KdV. Other fractional powers $(-\hat{L}_u)^{5/2}$, etc. generate Lax's KdV hierarchy.

Gel'fand and Dikii [46] found the basis for this structure in the resolvent for the Sturm–Liouville eigenvalue problem, the Schrödinger problem Equation (50). A kernel of the resolvent $R(x, y;\ \lambda)$ has the diagonal $R(x, x;\ \lambda) \equiv R(x;\ \lambda)$ and for this they found the asymptotic expansion in half-odd inverse powers of λ

$$R(x;\ \lambda) = \sum_{\ell=0}^{\infty} R_\ell[u]\lambda^{-\ell-1/2}. \tag{54}$$

The $R_\ell[u]$ are polynomials in $u, u', u'', \ldots$ ($u' \equiv u_x$, etc.). They show that R satisfies a third order linear differential equation from which, by substituting Equation (54), they obtain the recurrence relation.

$$R_{\ell+1} = \tfrac{1}{4}\, R_\ell'' - uR_\ell + \tfrac{1}{2}\left(\frac{\mathrm{d}}{\mathrm{d}x}\right)^{-1} u'R_\ell \tag{55}$$

and so find

$$\begin{aligned} R_0 &= \tfrac{1}{2}, \qquad R_1 = -\tfrac{1}{4}\, u, \qquad R_2 = \tfrac{1}{16}(3u^2 - u''), \\ R_3 &= -\tfrac{1}{64}(10u^3 - 10uu'' - 5(u')^2 + u''''), \\ R_4 &= \tfrac{1}{256}(35u^4 - 70u(u')^2 - 70u^2u'' + 21(u')^2 + \\ &\qquad + 28u'u'' + 14uu'''' - u''''''), \end{aligned} \tag{56}$$

and so on. An alternative method for finding recurrence relation between $R_\ell[u]$ leads to a calculation of $(\hat{L}_u)^{n+1/2}$ and justifies the analysis we gave in terms of $\widehat{R} = (-L_u)^{1/2}$ above. We make use of operators like $\widehat{R}$ in connection with the Kadomtsev–Petviashvili equation in Section 6 (the KP equations are given at Equations (73) below).

From the work on the coefficients $R_\ell[u]$ appearing in Equation (54) Novikov [47] also derives

$$\frac{\delta}{\delta u} R_\ell[u] = -\left(\ell - \tfrac{1}{2}\right) R_{\ell-1}[u] \tag{57}$$

where $\delta/\delta u$ is functional derivative. A discovery is to recognize

$$\frac{\delta}{\delta u} R_\ell[u] = 0, \tag{58}$$

which is also $R_{\ell-1}[u] = 0$, *as a stationary KdV flow*. Novikov [47] introduced the stationary KdV flows in connection with his work on the KdV equation under *periodic* boundary conditions.

Gardner [48] had earlier (1971) suggested that the Poisson bracket of two functionals of u, $K_1[u]$, and $K_2[u]$ is properly taken as

$$\{K_1, K_2\} = \int \frac{\delta K_1}{\delta u} \frac{\partial}{\partial x}\left(\frac{\delta K_2}{\delta u}\right) \mathrm{d}x, \tag{59}$$

and the $\partial/\partial x$ acts as symplectic operator. From this bracket Hamilton's equations would be $u_t = [\delta H/\delta u]_x$, a formula also ascribed to Gardner, by Lax [45] in particular. Gel'fand and Dikii [46] suggest first of all that the $R_\ell[u]$ can be identified as *Lagrangian densities* L so Hamiltonian densities H, such that $\mathrm{d}H/\mathrm{d}x \equiv H' = -u'\delta L/\delta u$ can be found and there are Hamiltonian densities H_k related to the $R_k[u]$ as

$$H_k' = \left(k - \tfrac{1}{2}\right) u' R_{k-1}$$

by Equation (58). However Novikov [47] also recognizes Hamiltonian densities H_k through $H_k[u] = R_k[u]$. Then $u_t = [\delta R_k[u]/\delta u]_x$ is a kth KdV flow which by Equation (57), and after scaling t, is $u_t = (R_{k-1}[u])_x$. Then $R_{k-1}[u] = 0$ are the stationary KdV flows [12, 47, 48] while $u_t = [R_{k-1}[u]]_x$ prove to be Lax's KdV hierarchy.

Surprisingly perhaps, and see Section 6 below for more details, we are already close to a result in the theory of two-dimensional quantum gravity – namely [49, 50] that the differential equation for the 'specific heat' for the 'kth multicritical point' of the 'one matrix model' can be written in the form

$$x = \left[2\Gamma\left(\tfrac{1}{2}\right)\Gamma(k+1)/\Gamma\left(k + \tfrac{1}{2}\right)\right](-1)^{k+1} R_k[u(x)]. \tag{60}$$

Thus upto scaling

$$\left(R_\ell[u]\right)_x = 1 \tag{61}$$

and $(R_k[u])_x = [\hat{L}_u, \hat{A}_{k+1}]$, where $\hat{A}_{K+1}$ is Lax's differential operator for the $(k+1)$th KdV equation and $\hat{L}_u$ is the Schrödinger operator $-\partial^2/\partial x^2 + u$. A particular result from Equation (59) is that for $k = 2$

$$x = \left(u(x)\right)^2 - \tfrac{1}{3} u''(x) \tag{62}$$

which is Painlevé I after scaling. We return to these several equations in Section 6 as mentioned.

The Hamiltonian structure for KdV in [47] could be no surprise because Zakharov and Faddeev [51] had already made in 1971 the *second* of 'the three important steps' we mentioned (recall Lax's work [45] was the first of these three important steps). The remarkable properties of the KdV equation were 'explained' by demonstrating that KdV was the first known nonlinear[12] example of an *infinite dimensional* completely integrable Hamiltonian system. For such a system it is necessary that there be a continuous infinity of independent constants of the motion commuting amongst themselves and with the Hamiltonian H. [51] finds a countable infinity of first integrals

$$I_n[u] = \int_{-\infty}^{\infty} P_n(u, u_x, \ldots)\,\mathrm{d}x$$

(the Pn are the conserved densities T^n described above). They took the Hamiltonian form of the KdV equation as $u_t = (\delta I_3[u]/\delta U(x))_x$ ($I_3[u]$ is $R_3[u]$) and demonstrated that the set of constants I_n was a countable part of precisely the right number (the continuous infinity) of constants by constructing the symplectic form

$$\Omega(\delta_1 u, \delta_2 u) = \int_{-\infty}^{\infty} \mathrm{d}x \int_{-\infty}^{\infty} \mathrm{d}y\,[\delta_1 u(x)\delta_2 u(y) - \delta_1 u(y)\delta_2 u(x)] \tag{63}$$

showing that this was closed under transformation to scattering data, demonstrating thereby that the inverse and direct scattering transforms of the inverse method are *canonical* transforms from and to action-angle variables, constructing the Hamiltonian H in terms of these action variables alone, which are constants, and explicitly integrating the equations of motion in these variables. Their result for H was $H[p]$, where

$$H[p] = \int_{-\infty}^{\infty} \omega(k)P(k)\,\mathrm{d}k - \frac{32}{5}\sum_{\ell=1}^{N_S} p_\ell^{5/2} \tag{64}$$

and $\omega(k) = 8k^3$, the (scaled) linearised dispersion relation and where the phase spaces are

$$\begin{array}{ll} \text{(i)} & 0 \leqslant P(k) < \infty,\ 0 \leqslant Q(k) < 2\pi;\ \{P(k), Q(k')\} = \delta(k-k') \\ \text{(ii)} & -\infty < p_i,\ q_i < \infty;\ \{p_i, q_k\} = \delta_{ik}; \end{array} \tag{65}$$

for which $\{\ldots\}$ is Poisson bracket. The $P(k)$ thus provide the continuous infinity of independent constants $\{P(k),\ P(k')\} = 0$; $k, k' \in \mathbf{R}$ and the coordinates of the phase space, relations (65), preserve the symplectic form Ω.

[12] The linear Klein–Gordon equation was recognized as an infinite-dimensional completely integrable system: linearised KdV $u_t = u_{xxx}$ is likewise.

We now know that the beautiful result Equation (64) for $H[p]$, apparently [52] a feature peculiar to KdV, is generic for all of the integrable models in one space and one time dimension. And it was this realisation which was the third important step originating from KdV mentioned.

Indeed the inverse scattering transform usually now called the spectral transform and inverse spectral transform is a canonical transform and inverse transform applicable to the large number of integrable models now known to exist in $(1+1)$ dimensions.

We can pick this up in the Figure 1: Lax pairs can be found in the 'box' top left, second column, in the form $\psi_x = U\psi$, $\psi_t = V\psi$: now U, V are $n \times n$ matrices and U in particular contains $n(n-1)$ 'potentials' like $u(x,t)$ *linearly*; while ψ is an $n \times n$ matrix of independent column vectors; compatibility $\psi_{xt} = \psi_{tx}$ means $U_t\psi + UV\psi = V_x\psi + VU\psi$, and if this is true for any ψ

$$U_t - V_x + [U,V] = 0. \tag{66}$$

But this is a nonlinear evolution equation which can be solved by the inverse method (method of spectral transforms). One easily checks that the KdV Equation (38) takes on the form Equation (66) with

$$U = \begin{bmatrix} 0 & 1 \\ -u+\lambda & 0 \end{bmatrix}$$

and

$$V = \begin{bmatrix} u_x & -2(u+2\lambda) \\ u_{xx} + 2(u-\lambda)(u+2\lambda) & -u_x \end{bmatrix}.$$

It is shown, e.g., in [12], how Equation (66) is a *vanishing curvature condition* $\mathrm{d}\Omega + \Omega \wedge \Omega = 0$.

5. Further Developments since the Discovery of the Inverse Method for KdV

Figure 1 shows these developments to be considerable! Two important aspects have been first of all the discovery of the algebraic structures underlying the integrable systems in $(1+1)$, the discovery and solution of completely integrable *quantum* models in $(1+1)$, and the connection of these to solvable models like the Ising model and the 8-vertex model [52] in the statistical mechanics of '$(2+9)$' dimensions (two Euclidean dimensions and no time). Then the second aspect has been the extension of the spectral transform mechanisms to $(2+1)$ dimensions (and to $(n+1)$ dimensions [29, 30]). To some extent the two authors have diverged separately into these two aspects, one (RKB) into the former the other (PJC) into the latter. We cannot hope to cover here all of the developments as sketched in Figure 1. Professor Faddeev is talking about the history of the quantum inverse method (QIM) which forms the third column to the right of this 'map' and this history is closely related to the discovery of the quantum groups [7, 8]. We shall therefore use this section briefly to summarize some of the results found during 1980–1995 by RKB and collaborators and some of the results found by PJC and collaborators in the two different directions. The work by RKB and collaborators concerns the statistical mechanics of the integrable systems in $(1+1)$ and as such was developed in parallel with, and is essentially complementary to, the work of the Leningrad (now St. Petersburg) group led by Professor Faddeev on the quantum inverse method. It would also seem to be directly relevant to the work on quantum gravity of Section 6.

This work on quantum gravity is directed at certain 'partition functions': for field theories (as in Equations (75)–(78) below for quantum gravity) this partition function is a functional integral: this functional integral is typically the trace of a 'Wick rotated' quantum mechanical 'propagator' [53, 55]: thus partition functions are closely related to quantum mechanics (and the quantum inverse method). Generally the partition function Z finds its origins in statistical mechanics, classical and quantum: the Helmholtz free energy is given by $F = -\beta^{-1} \ln Z$ where $\beta^{-1} = k_B T$, and k_B is Boltzmann's constant with T, the temperature. The energy E is $-(\partial \ln Z/\partial\beta)$ so that the specific heat $\propto -(\partial E/\partial\beta)$ is essentially $\partial^2(\ln Z)/\partial\beta^2$ and it is this quantity which is evaluated in [49, 50] for quantum gravity. RKB and collaborators [53, 55] have been concerned to develop general methods for calculating partition functions Z for a variety of model systems all of which are completely integrable in $(1+1)$ dimensions in the sense described for KdV in Section 5. Since it is the classical action S which appears in the partition function in both the quantum and classical cases this theory concerns the statistical mechanics of both the classical integrable models and the quantum integrable models [53]. *Moreover, since the classical 2-dimensional gravitational field equations are apparently completely integrable* [12] the methods should also apply to this case also. In practice so far different methods have been developed and these are briefly summarized in the Section 6 below.

The case of the classical Toda lattice has been analysed by RKB and collaborators [54] and the Toda lattice becomes KdV in a suitable continuum limit; but the case of KdV itself is not yet done. In any case the 'simple' model is the sine-Gordon model with classical equations of motion [53, 56]

$$\phi_{xx}(x,t) = \phi_{tt}(x,t) = m^2 \sin\phi, \quad m, \phi \in \mathbf{R},\ m > 0. \tag{67}$$

This equation is Hamilton's equation for

$$H[\Pi,\phi] = \gamma_0^{-1} \int \left\{ \tfrac{1}{2}\gamma_0^2\Pi^2 + \tfrac{1}{2}\phi_x^2 + m^2(1-\cos\phi) \right\} \mathrm{d}x \tag{68}$$

with the Poisson bracket $\{\Pi,\phi\} = \delta(x-x')$; $\gamma_0 > 0$ is a classical coupling constant. For connection of the sine-Gordon to KdV one might go to the countably infinite set of polynomial conserved densities commuting with H Equation (68) written in light-cone coordinates to find the Hamiltonian density for the modified

KdV Equation (mKdV) (42). Under the famous Miura transformation [12, 42] mKdV maps to KdV.

RKB and colleagues look at the *quantum* functional integral Z taken in the form

$$Z = \mathrm{Tr} \int \mathcal{D}\Pi\mathcal{D}\phi \exp \hbar^{-1} S[\Pi, \phi]; \tag{69}$$

$S[\Pi, \phi]$ is the *classical* action

$$S[\Pi, \phi] = \int_0^{\beta\hbar} \mathrm{d}t \left[\int \Pi\phi_{,t} \,\mathrm{d}x - H[\Pi, \phi] \right] \tag{70}$$

and $\hbar = h/2\pi$ with h Planck's constant: $\beta^{-1} = k_B T$ and Tr is 'trace' while $\phi_{,t}$ means ϕ_t. The measure $\mathcal{D}\Pi\mathcal{D}\phi$ is a combination of Wiener measures so that Z is defined on a very large dimensional symplectic manifold M with symplectic differential 2-form

$$w^2 \equiv \int_{-L/2}^{L/2} [\mathrm{d}\Pi(x) \wedge \mathrm{d}\phi(x)] \,\mathrm{d}x$$

at fixed t in the usual wedge product description (cf. (63) for the comparable symplectic form Ω) and w^2 is invariant. The classical statistical mechanics is obtained from the classical limit for Z and this limit is $\hbar \to 0$ so that, without winding number problems [55],

$$S[\Pi, \phi] = -\beta\hbar H[\Pi, \phi] \tag{71}$$

in that (semi) classical limit and $\hbar^{-1} S$ is independent of $\hbar$.

The functional integral Equations (69), (70) for Z is *not* the usual Feynman form of Z for field theories. However, this Feynman form can be obtained from it in the case $H[\Pi, \phi]$ is quadratic in Π by performing the functional integral with respect to the measure $\mathcal{D}\Pi$. Evidently in this connection it would be of some interest to set up comparable theory using the symplectic structure found for KdV in Section 4 but this is not yet carried out.

The interest for the integrable systems in $1+1$ is that $S[\Pi, \phi]$ can be expressed in terms of the action-angle type variables and an appropriate canonical transformation made on the measure $\mathcal{D}\Pi\mathcal{D}\phi$ so that this becomes $\mathcal{D}\mu$, say. The route for doing this goes top left to top right on the Figure 1. There are many pit-falls, not least because Z must be evaluated in *thermodynamic limit* at finite density. To this end we can usefully use periodic boundary conditions of period L, find action-angle variables under these periodic boundary conditions, find the measure $\mathcal{D}\mu$ and evaluate the functional integral.[13] For *classical* sine-Gordon, for

[13] Evaluation in effect requires discretization to a lattice of spacing a, under periodic boundary conditions of period L. Then $a \to 0$ and $L \to \infty$ in a double limit at *finite density*. The double 'scaling' limit in quantum gravity emerges as a very difficult but comparable limit (Section 6).

example, we find [53, 56] that the free energy density is given by

$$\lim_{L\to\infty}(-\beta^{-1}\ln Z)L^{-1} = (2\pi\beta)^{-1}\int_{-\infty}^{\infty}\omega(x)\ln(\beta\varepsilon(x))\,\mathrm{d}x - -2(2\hbar\pi\beta)^{-1}\int_{-\infty}^{\infty}E(x)\mathrm{e}^{-\beta\widetilde{E}(x)}\,\mathrm{d}x \tag{72}$$

with $\omega(x) = m\cosh x$, $E(x) = M\cosh x$ for rapidities x: $M \equiv 8m\gamma_0^{-1}$. The excitation energies $\varepsilon(x)$, $\widetilde{E}(x)$ are solutions of the pair of coupled integral equations

$$\begin{aligned}\varepsilon(x) &= \omega(x) + (2\pi\beta)^{-1}\int_{-\infty}^{\infty}[\mathrm{d}\Delta(x,x')/\mathrm{d}x]\ln(\beta\varepsilon(x'))\,\mathrm{d}x' + \\ &\quad +2(2\pi\beta)^{-1}\int_{-\infty}^{\infty}[\mathrm{d}\Delta_k(x,x')/\mathrm{d}x]\mathrm{e}^{-\beta\widetilde{E}(x')}\,\mathrm{d}x', \\ \widetilde{E}(x) &= E(x) - 2(2\pi\beta)^{-1}\int_{-\infty}^{\infty}[\mathrm{d}\Delta_{kk}(x,x')/\mathrm{d}x]\mathrm{e}^{-\beta\widetilde{E}(x')}\,\mathrm{d}x' - \\ &\quad -(2\pi\beta)^{-1}\int_{-\infty}^{\infty}[\mathrm{d}\Delta_k(x,x')/\mathrm{d}x]\ln(\beta\varepsilon(x'))\,\mathrm{d}x'.\end{aligned} \tag{73}$$

The Δ's are 2-body phase shifts: Δ is 'phonon-phonon', Δ_k is 'phonon-kink', and Δ_{kk} is 'kink-kink'. They have been computed and are given in [56]. That only 2-*body* phase shifts appear is characteristic of the integrability of the model. *Iteration* of this system of integral equations yields low temperature asymptotic expansions which can be checked against asymptotic expansions for the eigenvalues of the Mathieu equation [56, 57]. It seems [58] that this *form* of result, Equations (71), (72) is generic for all of the integrable models in $1+1$ dimensions classical *and* quantum. For example, the expression for the free energy density of *quantum* sine-Gordon takes on the general form of Equations (71), (72) although the result is dominated by quantum *breather* contributions and there are no 'phonon' contributions. These phonon contributions should arise from the contribution $\int_{-\infty}^{\infty}\omega(k)P(k)\,\mathrm{d}k$ to a Hamiltonian $H[p]$ in action-variables like that given for KdV in Equation (63). We have shown nevertheless [57] how this quantum result without phonons becomes Equations (71), (72) in semiclassical and subsequent classical limit, and phonon contributions are now identifiable in Equations (71), (72) in the terms in $\ln[\beta\varepsilon(x)]$.

We have carried out similar calculations for some 16 integrable models classical or quantum [58]. These do not yet include KdV, classical or quantum, but the classical Toda lattice is done as mentioned. The Toda lattice already offers special difficulty since one is apparently obliged to work on a constrained phase space [54]. The expression of Z for KdV in terms of its field $u(x,t)$ is not so far written in a form analogous to Equations (68), (69). But if the phase space for KdV coordinatized in action-angle variables Equation (64) can be used directly this seems to offer greater simplicity than we found for the Toda lattice. A significant result for this lattice is that the number of solitons excited at temperature

$T \propto T^{1/3}$, yet this nonanalytic in T result can be combined with a development in terms of low temperature asymptotic power series for the free energy density [54].

We must refer to [58, 60, 61] for results for free energy densities for the quantum integrable models: the free energy density for quantum Toda is not yet calculated, while the zero temperature quantum theory offers special difficulties [62].

Main results by PJC and colleagues include $n \times n$ spectral problems and their inversion through local Riemann–Hilbert problems, passage to $(2+1)$ dimensions through $n \to \infty$, *nonlocal* Riemann–Hilbert problem methods for $(2+1)$ dimensional models like the Kadomtsev–Petviashvili equations KP-I, and so-called 'D-bar' methods for $(2+1)$ dimensional models like KP-II (KP-I and KP-II are given at Equation (73) below), and the extension of these inverse methods to $(n+1)$ dimensions for all $n > 2$. These results are shown up to $n = 2$ in the Figure 1 where the local Riemann problem (in the double lined 'box') is first of all identified in the usual inverse method for $(1+1)$ dimensions (a realisation due first of all to Zakharov and Shabat [63]) and then generalised to the nonlocal Riemann and D-bar problems to the right and used to solve KP-I and KP-II, for example, as is explicitly marked there. The D-bar methods were first introduced by Beales and Coifman [64], and work on KP-I and KP-II is due to Manakov [64] and Ablowitz and Fokas [66], in particular. We shall need the KP equations for quantum gravity in the next Section 6 so we give them here in the form on the 'map' Figure 1 as

$$(u_t + 6uu_x + u_{xxx})_x = \pm u_{yy}; \tag{74}$$

Plus is KP-I, minus is KP-II. In these KP Equations (74), $u = u(x, y, t)$; but when u is independent of y, Equations (74) both reduce to Equation (38) for the KdV when $u \to 0$ as $x \to \infty$ (so fixing the constant of integration).

Finally, we reach the recent work done on two-dimensional quantum gravity where the KdV hierarchy first found by Lax [45] plays a profoundly important role.

6. KdV and 2-Dimensional Quantum Gravity

In formulating this short summary we have been much influenced by the presentation in [67].

In contrast with the partition functions Equations (68) we follow Polyakov [68] and write $(\hbar = 1)$[14]

$$Z = \sum_p \int \mathcal{D}g \int \mathcal{D}X \exp S[g, X;\ p], \tag{75}$$

[14] We follow convertion and omit the surface elements in the surface integrals.

$$S[g, X;\ p] = -\lambda_1 \int_{\Sigma_p} \sqrt{g} - \left(\frac{\lambda_2}{2\pi}\right) \int_{\Sigma_p} R\sqrt{g} - \int_{\Sigma_p} \sqrt{g} g^{ab} \partial_a X^\mu \partial_b X^\mu. \tag{76}$$

The functional integral $\int \mathcal{D}g$ is integration over all possible metrics on the 2-surface Σ_p of genus p and the summation is over such surfaces; $\int \mathcal{D}X$ is integration over all mappings $X\colon \Sigma_p \to \mathbf{R}^D$, where D denotes dimension (these mappings are the string fields) and $\mathbf{R}$ is the scalar curvature of the metric g. The constants $\lambda_1, \lambda_2 \in \mathbf{R}$ are the cosmological constant (λ_1) and the string coupling (λ_2). Pure two-dimensional quantum gravity then has the partition function

$$Z = \sum_p \int \mathcal{D}g \exp S[g;\ p], \tag{77}$$

$$S[g;\ p] = -\lambda_1 \int_{\Sigma_p} \sqrt{g} - \left(\frac{\lambda_2}{2\pi}\right) \int_{\Sigma_p} R\sqrt{g}. \tag{78}$$

A problem now is to make sense of these two heuristically motivated functional integrals: a route has been [50, 69] to consider the triangulations of 2 surfaces and more generally [67] the covering of surfaces by squares, hexagons, etc. A calculation of the number of ways of doing this for each genus p can be used to approximate Z, Equations (75), (76). This approximation is shown to be valid for each genus p for special values of the dimension D, namely for

$$D = 1 - 6[m(m+1)]^{-1}, \quad m = 2, 3, \ldots. \tag{79}$$

Experts may recognize in formula (79) one of the two conditions, this one on the central charge usually called c but here called D, that there must be for there to be a unitary representation of a Virasoro algebra [70]. The particular case $m = 2$ ($D = 0$) is the case of pure quantum gravity Equations (77), (78). More generally one finds the representation for Z Equations (75), (76) which is of the form [67]

$$Z = \sum_{p=0}^{\infty} \int Z_p(A)\, \mathrm{d}A \simeq \sum_{p=0}^{\infty} n^{2-2p} \sum_{q=0}^{\infty} (-\lambda)^q W_{\{x\}}(p;\ q) \tag{80}$$

with[15]

$$\lambda \longrightarrow \lambda_c, \quad n \to \infty; \qquad n(\lambda - \lambda_c)^{-\gamma} = \mathrm{O}(1), \tag{81}$$

and $\gamma = -1 - (2m)^{-1}$ (so that $\gamma = -5/4$ for pure gravity $D = 0$). But this cannot define the functional integral Equations (75), (76) for Z since the right side only

[15] For each p, $\lambda \equiv -\mathrm{e}^{-\lambda_1 \varepsilon}$ and λ_1 renormalises as $\lambda_1 = c\varepsilon^{-1} + \lambda_1^0$ with $\varepsilon \to 0$: λ_c as used in Equation (83) is thus $\lambda_c = -\mathrm{e}^{-c}$.

has meaning asymptotically. However, from the known asymptotic behaviour of $W_{\{x\}}(p;\ q)$ as $q \to \infty$ one deduces

$$Z \equiv Z(t) = \sum_{p=0}^{\infty} t^{2-2p} b_p \Gamma(\gamma(2-2p)) + \text{reg. terms}, \tag{82}$$

where $b_p\{x\}$, $\gamma\{x\}$ for special choices of $m-2$ of the parameters[16] x such that $\gamma = -1-(2m)^{-1}$ (see above) help to determine the asymptotics of $W_{\{x\}}(p,q)$: parameters t in Equation (81) are given by

$$t = h\left(1 - \frac{\lambda}{\lambda_c}\right)^{-\gamma} \tag{83}$$

and are renormalised string couplings with ‘critical behaviours’ as $\lambda \to \lambda_c$.

The asymptotic series Equation (82) defines the perturbative theory for Z Equations (75), (76). This series does note depend on the parameters $\{x\}$ mentioned (see Footnote 15) but only on the integer m which determines D, Equation (79) (and so determines an mth universality class). For a nonperturbative definition of the functional integral Equations (75), (76) one needs a generating function for Equation (80) and a candidate is

$$\ln Z_n\left(\frac{1}{2}, \frac{\lambda}{4n} x_2, \frac{\lambda^2}{6n^2} x_3, \ldots, \frac{\lambda^{N-1}}{2Nn^{N-1}} x_N\right), \tag{84}$$

where $Z_n(t_1, t_2, \ldots, t_N)$ is the partition function of the *Hermitian matrix model*

$$Z_n = \int D\Phi \exp\{-\operatorname{Tr} U(\Phi)\},$$

$$U(z) = \sum_{j=1}^{N} t_j Z^{2j}. \tag{85}$$

In this, Φ is an $n \times n$ Hermitian matrix and

$$\mathrm{d}\Phi = \prod_{i=1}^{a} \mathrm{d}\Phi_{ii} \prod_{i<j} \mathrm{d}\Phi_{ij}\, \mathrm{d}\Phi_{ij} \tag{86}$$

so that Z_n is a well defined finite dimensional integral. Of course this is a discretization for $n < \infty$ of the functional integral Z, and the matrix model Equation (85) becomes nontrivial because one finds [67] that the dimension n of the integral, Equation (85), is $n \to \infty$ within a certain double scaling limit which is the limit Equation (88). First we rescale as

$$\Phi \longrightarrow \beta^{1/2}\Phi, \qquad \lambda\beta/n = 1, \qquad x_j/2j = q_j \tag{87}$$

[16] The parameters $\{x\}$ play a purely auxiliary role.

(so that action $U(\Phi)$ gains a factor β). Then the problem is to evaluate $\ln Z_n(\frac{1}{2}\beta, \beta q_2, \ldots, \beta q_N)$ under the limit

$$\beta = C_1 h^{-4-2m^{-1}}, \qquad n\beta^{-1} = C_2 + C_1^{-1} h^4 \xi, \quad h \to 0, \tag{88}$$

in which

$$C_2 \equiv \lambda_c, \quad \lambda - \lambda_c \equiv C_1^{-1} h^4 \xi, \quad \xi \equiv -t^{2m(2m+1)-1}(C_1 \lambda_c). \tag{89}$$

The equivalence between Z Equations (75), (76) and matrix model Equation (85) in this double scaling limit has sparked many papers which simply start from a matrix model *per se* (e.g., [50, 51, 71–73]). One can reduce the (finite dimensional) matrix model Equation (85) to the expression [49, 67]

$$Z_n = \text{const} \prod_{i=1}^{n} h_{i-1}(t_1, \ldots, t_N), \tag{90}$$

where the h_n are normalisation constants of corresponding orthogonal polynomials $P_n(z)$ of degree n with leading coefficient unity taken against the weighting function

$$\exp\{-U(z)\}\colon\ h_n \delta_{nm} = \int_{-\infty}^{\infty} P_n(z) P_m(z) \exp\left(-\sum_{j=1}^{N} t_j z^{2j}\right) dz.$$

These h_n allow the introduction of $w_n = 4h_n/h_{n-1}$, $n = 1, 2, \ldots$, and these w_n satisfy a difference equation called the 'discrete string equation' studied in [67]. This system is an integrable system which is compatible with the integrable so-called Volterra hierarchy [67, 74].

For our purposes the connection now is that with

$$\beta = C_1 h^{-5}, \qquad n\beta^{-1} = C_2 + C_1^{-1} h^4 \xi,$$

$$w_n \sim \rho(1 - 2h^2 u(\xi)), \quad h \to 0, \tag{91}$$

which incorporates the limit Equation (88) for the case $m = 2$, then, for an appropriate choice of the constants C_1, C_2, ρ, $u(\xi)$ satisfies the Painlevé I equation in the form

$$u_{\xi\xi} = 6u^2 + \xi. \tag{92}$$

More generally, that is for general m, the discrete string equation must be taken under the limit Equation (88), and if $w_n \simeq \rho(1 - 2h^{4m^{-1}} u(\xi))$ then ρ can be chosen so that, after rescaling to Equation (95) below (under which $\xi \to t$)

$$C_k R_k[u(t)] = t; \qquad C_k = \left[2\Gamma\left(\tfrac{1}{2}\right)\Gamma(k+1)/\Gamma\left(k+\tfrac{1}{2}\right)\right](-1)^{k+1}, \tag{93}$$

where

$$R_k[u(t)] = -\left(k - \tfrac{1}{2}\right)\delta R_{k-1}[u(t)]/\delta u$$

by Equation (57) and defines the kth stationary KdV flow. Equation (93) is Equation (59).

In more detail Gross and Migdal [49] start from the matrix model

$$Z_n(\beta) = \int \mathrm{d}\Phi \exp\left[-\beta \mathrm{Tr}\, U(\Phi)\right] \tag{94}$$

(which scales in the β) and see that

$$\ln Z_n(\beta) = -F(t) + \text{reg. terms}, \tag{95}$$

in which, now, $t = (\beta - n)\beta^{-(2k+1)^{-1}}$ (cf. (82)). Then the 'specific heart' is $f(t) = \ddot{F}(t)$ and for the kth 'multicritical point' this is given by the Equation (59) which is:

$$\begin{aligned}
k &= 1\colon\ t = f;\\
k &= 2\colon\ t = f^2 - \tfrac{1}{3} f'';\\
k &= 3\colon\ t = f^3 - ff'' - \tfrac{1}{2}(f')^2 + \tfrac{1}{30} f'''';\\
k &= 4\colon\ t = f^4 - 2f(f')^2 - 2f^2 f'' + \tfrac{3}{5}(f'')^2 +\\
&\qquad\qquad + \tfrac{4}{5}\left(f'f''' + \tfrac{2}{5} ff'''' - \tfrac{1}{35} f''''''\right);\\
k &= 5\colon\ \cdots .
\end{aligned} \tag{96}$$

Evidently these equations can be put in the form of Equation (60) which is

$$[\hat{L}_n, \hat{A}_{k+1}] = 1, \tag{97}$$

where $\hat{L}_n = -\mathrm{d}^2/\mathrm{d}\xi^2 + u(\xi)$, the Schrödinger operator, while $\hat{A}_{k+1}$ is the A-operator forming the Lax pair for the $(k+1)$th KdV equation. Then Painlevé I, Equation (92), is $k = 1$ which means $u_{\xi\xi\xi} - 12uu_\xi = 1$ and corresponds to the KdV equation with that scaling.

These rather remarkable results place the KdV hierarchy in an exceptional position vis-à-vis quantum gravity. It seems natural to look for the origin of this exceptional position in the conformal invariance of the KdV hierarchy. This raises questions about the representations of the Virasoro algebra in relation to two-dimensional quantum gravity. A preferred candidate might have been the covariant Liouville equation

$$u_{xx} - u_{tt} = \mathrm{e}^{\alpha u} \tag{98}$$

which is conformally invariant and plays a fundamental role already in string theory [75].

Conformally invariant field theories induce representations of two identical copies of the Virasoro algebra

$$[L_n, L_m] = (m-n)L_{m+n} + \tfrac{1}{12}\, cm(m^2-1)\delta_{m,-n} \tag{99}$$

with $m, n \in \mathbf{Z}$ and c the central charge. With c identified as the dimension D at Equation (79), we have already seen one (apparent) connection with the Virasoro algebra for which $0 \leqslant c < 1$: we sketch next an alternative theory of two-dimensional quantum gravity where this algebra becomes immediately evident with however $c = 1$.[17] We refer now predominantly to the exposition by Fukuma, Kawai and Nakayama [72].

By starting from the one matrix ψ^4-model which has action S scaled to $S = n\mathrm{Tr}\, U(\phi)$ with ϕ an $n \times n$ Hermitian matrix and

$$U(z) = \tfrac{1}{2}\, z^2 - \tfrac{1}{4}\, \lambda z^4,$$

as is already contained with the different scaling in Equation (85), these authors find Schwinger–Dyson equations for the partition function $Z = \lim_{n\to\infty} Z_n$ taken in a continuum limit involving $a = n^{-2/5} \to 0$, an analogue of the limit Equations (90). We shall now use n for the integers $n = -1, 0, 1, 2, \ldots$ following the notation of [72] (N is used for our *dimension* n in [72]). The authors find there is a tau-function (τ-function) τ related to Z as $\tau^2 = Z$. This τ-function satisfies, for the integers $n = -1, 0, 1, 2, \ldots,$

$$L_n \tau = 0,$$

$$\begin{aligned} 2L_n &= \frac{1}{2} \sum_{p+q=-2n} pqx_p x_q + \sum_{p-q=2n} px_p \frac{\partial}{\partial q} + \\ &+ \frac{1}{2} \sum_{p+q=2n} \partial_p \partial_q + \frac{1}{8}\, \delta_{n,0}. \end{aligned} \tag{100}$$

The p, q run over positive odd integers and the L_n satisfy a Virasoro algebra, Equation (99), with $c = 1$.

The authors of [72] *conjecture*, but virtually prove by explicit calculation, that τ satisfying Equations (100) is actually a τ-function [76] of the KdV hierarchy satisfying *in addition* the equation

$$L_{-1}\tau = 0. \tag{101}$$

The conjectured equivalence of the two statements that $L_n \tau = 0$, Equations (100), and that τ is a τ-function of the KdV hierarchy[18] is strongly supported by the explicit calculation which reaches

$$x_1 + (3x_3 \partial_1^2 \ln \tau + 5x_5 \partial_1 \partial_3 \ln \tau + \cdots) = 0 \tag{102}$$

[17] The meanings of these different choices for the central charge c remain obscure to us at present.

[18] The τ-function as described in [76] can be traced to the work of Hirota [77] and ultimately to his solution of the KdV equation [78]. Caudrey *et al.* [79, 80] gave the comparable solution of the sine-Gordon equation soon after.

(Equation (3.6) of [72]) for the equation $L_{-1}\tau = 0$ differentiated once with respect to x_1. The many variables $x_1, x_3, x_5, \ldots$ arise in the following way. It is necessary to go to the KP hierarchy [76] of which the first number is (up to scaling) the KP-I equation in $(2+1)$ dimensions, Equation (73) above. The KP-I hierarchy has a τ-function [76] and τ satisfies [72, 76]

$$\partial_1 \partial_m \ln \tau = (L^m)_{-1}, \tag{103}$$

where L is the pseudo-differential operator $L = \partial + u_2 \partial^{-1} + \cdots$ essentially the square root $\widehat{R}$ of $-\hat{L}_n$ introduced in Section 4, and ∂ is $\partial_{x_1} \equiv \partial_1$, while $(\ldots)_{-1}$ stands for the coefficient of ∂^{-1}. The KP-I equations in $(2+1)$ dimensions contain by reduction many of the integrable systems in $(1+1)$ dimensions [76], and in particular the condition of 'two-reduction'[19] which is $(L^2)_- = 0$, where $(\ldots)$ stands for the negative-powers-in-∂-part, restricts to the variables $x_1, x_3, x_5, \ldots$, and yields

$$\left(L^{2k-1}\right)_{-1} = 2R_k[-2u_2], \quad k \geqslant 1, \tag{104}$$

where R_k is again the coefficient R_k of the resolvent of the Schrödinger operator equation, Equation (54) and determines the stationary KdV flows. By combining Equations (101)–(103) the authors of [72] then reach

$$\frac{1}{2} x_1 + \sum_{k=1}^{\infty} (2k+1) x_{2k+1} R_k[-2u_2] = 0 \tag{105}$$

from which one picks off Equation (60), up to scaling, by setting all but one of the x_{2k+1} to zero ($R_k[-2u_2]$ depends only on the variable x_1). Thus, recalling the differentiation with respect to x_1, the equation $L_{-1}\tau = 0$ is a once integrated version of Equation (105) under the assumption that τ is a τ-function of the KdV hierarchy. From this the other equations $L_n\tau = 0$, $n \geqslant 0$, should follow from $L_{-1}\tau = 0$; and as an immediate check, with all x's zero except x_1, x_5, Equation (104) is of course the Painlevé I equation

$$f^2 + \tfrac{1}{3}\partial_1^2 f = x_1 \quad \text{and} \quad f = 2\partial_1^2 \ln \tau = 2u_2.$$

Then using this and Equation (103) one finds [72] that $L_0\tau = 0$, $L_1\tau = 0$, at least.

From this result one can guess (as in [72]) that the '3-reduction' of the KP-hierarchy $(L^3)_- = 0$, which will produce the Boussinesq hierarchy whose first member is Equation (29) up to scaling, is equivalent to the Schwinger–Dyson equations

$$3L_n = \frac{1}{2} \sum_{p+q=-3n} pq x_p x_q + \sum_p p x_p \partial_{p+3n} +$$

[19] The case of 2-reduction is through of as pure quantum gravity but includes all of the multi-critical points of [49, 50]: more generally gravity is coupled to '(p,q) conformal matter' [72, 81, 82].

$$+\frac{1}{2}\sum_{p+q=3n}\partial_p\partial_q+\frac{1}{3}\,\delta_{n,0}, \tag{106}$$

and the L_n again satisfy the Virasoro algebra Equations (99) with $c = 1$.

A new feature is that τ therefore add a large set of further equations which form a representation of the so-called W-algebra [72, 81, 82]. The explicit expression for W_n chosen in [72] is

$$\begin{aligned} 3^{3/2}W_n \;=\; & \sum_{p+q+r=-3n} pqrx_px_qx_r + 3\sum_{p+q-r=-3n} pqx_px_q\partial_r + \\ & +3\sum_{p-q-r=-3n} px_p\partial_q\partial_r + \sum_{-p-q-r=-3n}\partial_p\partial_q\partial_r, \end{aligned} \tag{107}$$

where p, q, and r run over positive integers $\not\equiv 0\,(\mathrm{mod}\,3)$. And the total algebra becomes

$$[L_n, L_m] = (m-n)L_{m+n} + \tfrac{1}{12}m(m^2-1)\delta_{m,-n},$$

$$[L_n, W_m] = (2n-m)W_{n+m},$$

$$\begin{aligned} [W_n, W_m] \;=\; & -\tfrac{1}{10}\,\delta_{n+m,0}n(n^2-1)(n^2-4) + \\ & +(n-m)\Big[\tfrac{3}{2}(n^2+4nm+m^2)+\tfrac{27}{2}(n+m)+21\Big]L_{n+m} - \\ & -9(n-m)U_{n+m}, \end{aligned}$$

$$U_n = \sum_{k\leqslant -2} L_kL_{n-k} + \sum_{k\geqslant -1} L_{n-1}L_k. \tag{108}$$

The equations

$$\begin{aligned} & L_n\tau = 0, \quad n = -1, 0, 1, 2, \ldots, \\ & W_n\tau = 0, \quad n = -2, -1, 0, \ldots, \end{aligned} \tag{109}$$

prove to form a closed and consistent set, and it is conjectured, in analogy with the case of 2-reduction and the KdV hierarchy, that an equivalent statement to τ is a τ-function of the KdV hierarchy with $L_{-1}\tau = 0$ is now τ is a τ-function of the *Boussinesq hierarchy* together with, in formal terms, the same additional condition

$$L_{-}\tau = 0. \tag{110}$$

The whole problem has been transformed [7, 81, 82] to the $W_{1+\infty}$ algebra with representation[20] $\partial\tau/\partial_{-1,\ell} = C_{-1,\ell}\tau$, $\partial\tau/\partial t^*_{n,m} = C_{n,m}\tau = 0$, $n \geqslant 0$.

[20] The Virasoro part $[c_{0,n}, c_{0,m}] = (n-m)C_{0,n+m}$ has central charge $c = 0$. As noted already we have not understood the different choices for central charges c in the analysis of this section.

The C_n satisfy $[C_{-1,u}, C_{-1,m}] = 0$, $[C_{-1,n}, C_{0,m}] = nC_{1,n+m}$, $[C_{0,n}, C_{0,n}] = (n-m)C_{0,n+m}, \dots$. The $t^*_{n,m}$ are introduced as in [81]. In this representation of the $W_{1+\infty}$ algebra the $C_{-1,\ell}$ are a representation of the *isospectral symmetries* K_n of the KP-I equations of the letter by one of us [83]: the $C_{n,m}$ for $n \geqslant 0$ include the nonisospectral symmetries also given in that letter. This connection explains the expression of the $W_{1+\infty}$ algebra in the 'box' on the 'map' Figure 1 which lies above, and is connected to the 'box' "2-D-Quantum Gravity, Partition function is a τ-function of p-reduced KP" there (actually as noted the τ-function is the square root of the partition function). The 'box' "Symmetries of KP-I and KP-II Algebra

$$[K_m, K_n] = 0,$$

$$[K_m, \tau_n] = \tfrac{1}{3}(m+n)K_{m+n-2},$$

$$[\tau_m, \tau_n] = \tfrac{1}{3}(m-n)\tau_{m+n-2}, \tag{111}$$

is the actual algebra realised for KP in [83], and the scaling $1/3$ and the form of Virasoro algebra as $[\tau_m, \tau_n] = \frac{1}{3}(m-n)\tau_{n+m-2}$ arise from the details of the scaling of KP and the details of the consequent symmetries K_m, τ_m as they are treated in that letter.

We have thus seen how the KdV *hierarchy* in $(1+1)$-dimensions is embedded in the KP *hierarchy* in $(2+1)$-dimensions, how the Boussinesq *hierarchy* is embedded likewise, and how the whole is ultimately traceable back to the partition functions Z of 2D-quantum gravity coupled most generally [72] to (p,q)-conformal matter as is briefly summarised again in [81]. Given the starting point in 1834 as described in this paper in Section 2, which then led up to the KdV paper of 1895 as described in this paper in Sections 2, 3, we think these final achievements, as they are rather inadequately sketched in this section, are 'absolutely staggering'!

We can say little more than that (!) – but we would still like to add two notes (!):

Note 1. Recent ideas [84, 85] suggest that quantum gravity may be deeply involved in 'consciousness' and our understanding of this still elusive concept: thus in this way we would see a connection of KdV to these ideas also!

Note 2. It has become important in the present climate in the U.K. firmly to indicate the strictly commercial applications of one's researches! We have no difficulty in this connection for solitons and the KdV equation! Russell [2, 3] was actually employed in 1834 to discover whether the canal in question (the Union Canal) could be navigated to advantage by steam-powered vessels. He discovered the important result [2, 4] that boats (barges) navigating the canal of finite depth h should ride on the solitary wave solution, scaled for example as at Equation (33), of the KdV equation and so travel most efficiently at the speed c given by his formula (6). A remarkable corollary of that observation is that the most efficient way (in terms of conservation of *information*) to send an optical pulse down an optical fibre is to work with optical solitons which though dissipated regain their shape and spectrum on successive amplifications: these particular optical solitons[21] are solutions of the integrable attractive nonlinear Schrödinger (NLS) equation $-iu_t = u_{xx} - 2cu^*u^2$, where u is a complex field and $c < 0$, and Professor Hasegawa is talking about these objects. The NLS soliton relates to KdV through their polynomial conservation laws as explained in Section 5. We understand that some Ł40 billion (Ł4 $\times$ 10^{10}) sterling is being invested in trans-Atlantic optical fibre communication of this kind at the present time!

References

1. Korteweg, D. J. and de Vries, G.: On the change of form of long waves advancing in a rectangular canal, and on a new type of long stationary waves, *The London, Edinburgh, and Dublin Philosophical Magazine and Journal of Science*, Series 5, Vol. 39, No. 241, June 1895, pp. 422–443 [*Phil. Mag.* **39** (1895), 422–443].
2. Russell, J. S.: Experimental researches into the laws of certain hydrodynamical phenomena that accompany the motion of floating bodies and have not previously been reduced into conformity with the laws of resistance of fluids, *Trans. Royal Soc. Edinburgh*, XIV (1840), 47–109.
3. Russell, J. S.: Report on waves, *British Association Reports*, 1844.
4. Bullough, R. K.: The wave 'par excellence', the solitary, progressive great wave of equilibrium of the fluid – an early history of the solitary wave, in M. Lakshmanan (ed.), *Solitons*, Springer Ser. in Nonlinear Dynamics, Springer-Verlag, Heidelberg, 1988, pp. 7–42.
5. Zabusky, N. and Kruskal, M. D.: *Phys. Rev. Lett.* **15** (1965), 240.
6. Gardner, C. S., Greene, J. M., Kruskal, M. D., and Miura, R. M.: *Phys. Rev. Lett.* **19** (1967), 1995.
7. Bullough, R. K. and Bogoliubov, N. M.: in Sultan Catto and Alveney Rocha (eds), *Proc XXth Conference on Differential and Geometric Methods in Theoretical Physics*, World Scientific, Singapore, 1992, pp. 488–504.
8. Bullough, R. K. and Bogoliubov, N. M.: in A. S. Fokas, D. J. Kaup, A. C. Newell and V. E. Zakharov (eds), *Nonlinear Processes in Physics*, Springer Ser. in Nonlinear Dynamics, Springer-Verlag, Berlin, 1993, pp. 232–240.
9. Kadomtsev, B. B. and Petviashvili, V. I.: *Sov. Phys. Dokl.* **15** (1970), 539–541.
10. Ikezi, H., Taylor, R. J., and Baker, R. D.: *Phys. Rev. Lett.* **25** (1970), 11. Also see Bullough, R. K.: in *Interaction with Radiation and Matter*, Vol. 1, International Atomic Agency, Vienna (IAEA–SMR–20/51), 1977.
11. Russell, J. S.: The wave of translation and the work it does as the carrier wave of sound, *Proc. Royal Soc.* **32** (1881), 382–383.
12. Bullough, R. K. and Caudrey, P. J. (eds): *Solitons*, Springer Topics in Current Physics, Vol. 17, Springer-Verlag, Heidelberg, 1980, Chapter 1.
13. Kruskal, M. D.: Private communication at the John Scott Russell meeting, 1982.
14. Stokes, Sir George: *Trans. Camb. Phil. Soc.* **8** (1847), 441.
15. Lagrange, J. L.: *Méchanique Analytique*, pp. XII, 512 Chez La Veuve Desaint, Paris (1788), 1811–1815, Nouvelle Edition Augmenté par L'auteur, 2 Vols., Paris.
16. Laplace, P. S. (Marquis de Laplace): *Traité de méchanique céleste*, Chez J. B. M. Duprat Paris au VII (1799), 1799–1823.

[21] We think an 'optical soliton' first appeared in this phrasing in the paper [86] where it was a soliton of the sine-Gordon equation: the NLS optical soliton is the nonrelativistic limit of the s-G optical soliton.

17. Poisson, S. D.: Mémoires sur la théorie des ondes, *Mem. de l'Acad Royale des Sciences* (i), 1816.
18. Weber, E. H. and Weber, W.: *Wellenlehre auf Experimente gegründet, oder über tropfbaren Flüssigkeiten mit Adwendung auf die Schall und Licht-Wellen*, Leipzig, 1825.
19. Emerson, G. S.: *John Scott Russell. A Great Victorian Engineer and Naval Architect*, John Murray, London, 1977.
20. Airy, Sir. G. B. (Astronomer Royal): Tides and waves, *Encyclopaedia Metropolitana*, 1845.
21. de Boussinesq, J.: Théorie des ondes et des remous qui se propagent le long d'un canal rectangulaire horizontal, en communiquant au liquide contenu dans ce canal des vitesses sensiblement pareilles de la surface au fond, *J. Math. Pures et Appliquées* **2** (1872), 55.
22. Rayleigh, Lord (J. W. Strutt): On waves, *Phil. Mag.* **1** (1876), 257.
23. Russell, J. S.: *The Wave of Translation in the Oceans of Water, Air and Ether*, Trubner, London, 1895.
24. Herschel, Sir John: quoted by Russell in his book *The Modern System of Naval Architecture*, Day and Son, London, 1865.
25. Stokes, Sir George: *British Association Reports*, 1846.
26. We cannot now find the explicit reference by Stokes on this point. But (see [4]) Rayleigh [22] in a footnote added in 1899 (to his collected papers) quotes Stokes's supplement to [14] in Stokes's *Mathematical and Physical Papers*, Vol. 1, Cambridge University Press, Cambridge, 1880, indicating that like the work of KdV [1] Stokes's later work confirms the existence of absolutely permanent waves of finite height. In this supplement to [14] Stokes develops the analysis at finite depth to 3rd order. He finds
$$\frac{mc^2}{y} = \frac{D_1}{S_1} + \frac{1}{S_1 D_1}(S_4 + 2S_2 + 12)b^2$$
where $S_i = \mathrm{e}^{imk/c} + \mathrm{e}^{-imkc}$, $i = 1, 2, \ldots$, $D_i = \mathrm{e}^{imk/c} - \mathrm{e}^{imkc}$, $k = ch$ and $b\ D_1 = a$. In this paper he introduces the stream function Ψ as well as the velocity potential $\phi = -k$ (a different k) at the canal bottom.
27. St. Venant, B. de: *Comptes Rendus*, Vol. CI, 1885, reference taken from KdV [1].
28. McCowan, J.: On the solitary wave, *Phil. Mag.* **31** (1891), 45; On the highest wave of permanent type, *Phil. Mag.* **38** (1894), 351.
29. Caudrey, P. J.: *Physica* **D6** (1982), 51.
30. Caudrey, P. J.: in A. P. Fordy (ed.), *Soliton Theory: a Survey of Results*, Manchester University Press, Manchester, 1990, pp. 25–54; 55–74.
31. Lamb, Sir Horace: *Hydrodynamics* (6th edn.), Cambridge University Press, Cambridge, 1952.
32. Ince, E. L.: *Ordinary Differential Equations*, Dover, New York, 1956.
33. Fermi, E., Pasta, J. R., and Ulam, S. M.: *Studies of Nonlinear Problems*, Vol. 1, Los Alamos Report LA-1940, 1940 (May 1955); *Collected Works of E. Fermi*, Vol. 2, University of Chicago Press, 1965, pp. 978–988.
34. Darboux, G.: *Leçons sur la Théorie des Surfaces* (2nd edn.), Gauthier-Villars, Paris, 1915.
35. Matveev, V. B. and Salle, M. A.: *Darboux Transformations and Solitons*, Springer-Verlag, Berlin, 1991.
36. Darboux, G.: At the Oberwolfach Meeting, July 1993 (for example).
37. Painlevé, P.: *Bull. Soc. Math. France* **28** (1900), 227; *Comptes Rendus Acad. Sc. Paris* **135** (1902), pp. 411, 641, 757, 1020.
38. Ablowitz, M. J. and Clarkson, P. A.: *Solitons, Nonlinear Evolution Equations and Inverse Scattering*, London Math. Soc. Lecture Note Series, Vol. 149, Cambridge University Press, Cambridge, 1991, Chapter 7.
39. Fuchs, R: Sur quelques équations différentielles linéaires du second ordre, *Comptes Rendus Acad. Sc. Paris* **141** (1905), 555–558.
40. See footnote of references in Ince [32].
41. Miura, R. M., Gardner, C. S., and Kruskal, M. D.: *J. Math. Phys.* **9** (1968), 1204.
42. Miura, R. M.: *J. Math. Phys.* **9** (1968), 1202.
43. Gel'fand, I. M. and Levitan, B. M.: *Izv. Akad. Nauk. SSR, Ser. Mat.* **15** (1951), 309; English translation *Amer. Math. Soc. Translations*, Ser. 2, **1** (1955), 253. The extension to $-\infty < x < \infty$ is due to V. A. Marchenko. Also see Chadan, K. and Sabatier, P. C.: *Inverse Problems*

in Quantum Scattering Theory, Texts and Monographs in Physics, Springer-Verlag, Berlin, Heidelberg, New York, 1977.
44. Kay, I.: Report No. EM-74, Courant Inst. of Mathematical Sciences, New York University, 1955.
45. Lax, P. D.: *Comm. Pure Appl. Math.* **21** (1968), 467.
46. Gel'fand, I. M. and Dikii, L.: *Uspekhi Mat. Nauk* **30** (1975), 67; *Russian Math. Surveys* **30** (1975), 77; *Funkt. Anal. Prilozh.* **10** (1976), 18.
47. Novikov, S. P.: *Funkt. Anal. Prilozh.* **8**(3) (1974), 54–66.
48. Gardner, C. S.: *J. Math. Phys.* **12** (1971), 1548.
49. Gross, David J. and Migdal, Alexander A.: *Nuclear Physics* **B340** (1990), 332.
50. Migdal, A. A.: Matrix models, in V. Bazhanov (ed.), *Proc. 7th Physics Summer School in Statistical Mechanics and Field Theory*, Australian National University, Canberra, January 1994, World Scientific, Singapore, 1995, to appear.
51. Zakharov, V. E. and Faddeev, L. D.: *Funkt. Anal. Prilozh.* **5** (1971), 18; *Funct. Anal. Appl.* **5** (1971), 280.
52. Baxter, Rodney, J.: *Exactly Solved Models in Statistical Mechanics*, Academic Press, London, 1982.
53. Bullough, R. K. and Timonen, J. T.: Quantum and classical integrable models and statistical mechanics, in V. V. Bazhanov (ed.), *Proc. 7th Physics Summer School in Statistical Mechanics and Field Theory*, Australian National University, Canberra, January 1994, World Scientific, Singapore, 1995, to appear.
54. Bullough, R. K. and Timonen, J.: in A. R. Bishop, V. L. Pokrovsky and V. Tognetti (eds), *Microscopic Aspects of Nonlinearity in Condensed Matter*, NATO ARW Series B: Physics, Vol. 264, Plenum, New York, 1991, pp. 263–280.
55. Bullough, R. K., Chen, Yu-Zhong, and Timonen, J. T.: *Physica* **D68** (1993), 83.
56. Timonen, J. T., Stirland, M., Pilling, D. J., Cheng Yi, and Bullough, R. K.: *Phys. Rev. Lett.* **56** (1986), 2233.
57. Dingle, R. B. and Müller, H. J. W.: *J. Reine und Angewandte Math.* **211** (1962), 11.
58. Bullough, R. K., Chen, Y-Z., and Timonen, J. T.: in V. G. Bar'yakhtar, V. Chernousenko, N. S. Erokhin, A. G. Sitenko, and V. E. Zakharov (eds), *Nonlinear World*, Vol. 2, World Scientific, Singapore, 1990, pp. 1377–1422.
59. Timonen, J., Bullough, R. K., and Pilling, D. J.: *Phys. Rev.* **B34** (1986), 6525.
60. Bullough, R. K., Pilling, D. J., and Timonen, J. T.: in M. Lakshmanan (ed.), *Solitons*, Springer Series in Nonlinear Dynamics, Springer-Verlag, Heidelberg, 1988, pp. 250–281.
61. Korepin, V. E., Bogoliubov, N. M., and Iczersin, A. C.: *Quantum Inverse Scattering Method and Correlation Functions*, Cambridge University Press, Cambridge, 1993 (and references therein).
62. Sklyanin, E. K.: The quantum Toda chain, in N. Sanchez (ed.), *Nonlinear Equations in Classical and Quantum Field Theory*, Lecture Notes in Physics, Vol. 226, Springer-Verlag, Berlin, 1985, pp. 196–233.
63. Zakharov, V. E. and Shabat, A. B.: *Funct. Anal. Appl.* **8** (1974), 226; *Funct. Anal. Appl.* **13** (1979), 166.
64. Beales, R. and Coifman, R. R.: *Physica* **D18** (1986), 242.
65. Manakov, S. V.: *Physica* **D3** (1981), 420.
66. Ablowitz, M. J., Yaacov, D. Bar, and Fokas, A. S.: *Stud. Appl. Math.* **69** (1983), 135 (and references therein).
67. Fokas, A. S., Its, A. R., and Kitaev, A. V.: *Commun. Math. Phys.* **147** (1992), 395–430.
68. Polyakov, A. M.: *Phys. Lett.* **B103** (1981), 207; *Modern Phys. Lett.* **A2** (1987), 899.
69. Migdal, A. A.: in D. J. Gross, T. Piran and S. Weinberg (eds), *Two Dimensional Quantum Gravity and Random Surfaces*, Jerusalem Winter School for Theoretical Physics, Vol. 8, World Scientific, Singapore, 1992, pp. 41–79 and other papers there.
70. Goddard, Peter and Olive, David: *Int. J. Modern Physics* **A1**(2) (1986), 320–325.
71. David, F.: *Mod. Phys. Lett.* **A5**(13) (1990), 1019–1029.
72. Fukuma, M., Kawai, H., and Ryuichi Nakayama: *Int. J. Modern Physics* **A6**(8) (1991), 1385–1406.

73. Dijkgraaf, R., Verlinde, H., and Verlinde, E.: *Nuclear Phys.* **B348** (1991), 435–456.
74. Alvarez-Gammé, Gomez, C., and Lacki, J.: *Phys. Lett.* **B253** (1991), 56–62.
75. Gross, D. J.: String theory, in V. V. Bazhanov (ed.), *Proc. 7th Physics Summer School in Statistical Mechanics and Field Theory*, Australian National University, Canberra, January 1994, World Scientific, Singapore, 1995, to appear.
76. Jimbo Michio and Miwa Tetsuji: in G. M. d'Ariano, A. Montorsi and M. G. Rasetti (eds), *Integrable Systems in Statistical Mechanics*, World Scientific, Singapore, 1985, pp. 65–127.
77. Hirota, R.: Chapter 5 of Ref. [12], 1980.
78. Hirota, R.: *Phys. Rev. Lett.* **27** (1971), 1192.
79. Caudrey, P. J., Gibbon, J. D., Eilbeck, J. C., and Bullough, R. K.: *Phys. Rev. Lett.* **30** (1973), 237.
80. Caudrey, P. J., Eilbeck, J. C., and Gibbon, J. D.: *J. Inst. Maths. Appl.* **14** (1974), 375.
81. Aoyama, S. and Kodama, Y.: *Phys. Lett.* **B278** (1992), 56–62.
82. Aoyama, S. and Kodama, Y.: *Phys. Lett.* **B295** (1992), 190–198.
83. Cheng, Yi, Li, Yi-Shen and Bullough, R. K.: *J. Phys. A: Math. Gen.* **21** (1988), L443.
84. Penrose, R.: *Shadows of the Mind*, Oxford University Press, Oxford, 1994; also in *Times Higher Educational Supplement*, Times Supplements Ltd, London E1 9XY, p. 15, Oct. 14, 1994.
85. Josephson, Brian: To err is mechanical, in *Times Higher Educational Supplement*, Times Supplements Ltd, London E1 9XY, p. 23, Nov. 4, 1994.
86. Bullough, R. K. and Caudrey, P. J.: Optical solitons and their spin wave analogues in ^{3}He, in L. Mandel and E. Wolf (eds), *Coherence and Quantum Optics IV*, Plenum, New York, 1978, pp. 767–780.

Acta Applicandae Mathematicae **39:** 229–244, 1995.

Integrable Nonlinear Evolution Equations and Dynamical Systems in Multidimensions

FRANCESCO CALOGERO
* *Dipartimento di Fisica, Università di Roma I La Sapienza, 00185 Rome, Italy, and Istituto Nazionale di Fisica Nucleare, Sezione di Roma, Italy*

(Received: 15 September 1994)

Abstract. The investigation of nonlinear evolution equations and dynamical systems integrable in multidimensions constitutes at present our main research interest. Here we survey findings obtained recently as well as over time: solvable equations (both PDEs and ODEs) are reported, philosophical motivations and methodological approaches are outlined. For more detailed treatments, including the display and analysis of solutions, the interested reader is referred to the original papers.

Mathematics Subject Classifications (1991). 35Q20, 58F07.

Key words: integrable PDEs, integrable dynamical systems, integrability in multidimensions.

1. Introduction

Over the last quarter century the investigation of *integrable* Nonlinear Evolution Equations and Dynamical Systems (NEEDS: including both PDEs and ODEs) has constituted a (perhaps the) major development in mathematical physics, as well as in applied and pure mathematics. In the following we outline some of these findings; in so doing, in some cases we do not back our statements with references, since too many would have to be quoted; in other cases, we do provide a selected list of references, even though we are well aware of its inadequacy; in all cases, we apologize for the many omissions.

At the origin of this upsurge of fruitful research – which has occured worldwide, and has created an international community of scientists bonded by the commonality of their research interests in *integrable* NEEDS – stands the Korteweg–de Vries (KdV) equation [1], whose hundredth anniversary we are now celebrating. Indeed, the beginning of this development can be identified with the discovery [2] of the 'inverse scattering' (or 'spectral transform') technique to solve this, as well as several other, important nonlinear evolution equations, such as nonlinear Schrödinger (NLS) [3], sine-Gordon (SG) [4], etc. These are nonlinear PDEs in $1+1$ dimensions (one space dimension and one time dimen-

* On leave while serving as Secretary General, Pugwash Conferences on Science and World Affairs, Geneva, London, Rome.

sion); their importance is due to their integrable character, which underpins many beautiful mathematical properties, as well as to their applicative relevance, which may be traced to a common origin with their integrability (at least for 'universal' equations such as NLS; see below) [5]; and it is of course also due to the remarkable phenomenology displayed by their solutions ('solitons'). Let us also recall that a seminal role in these developments was played by the early introduction of the 'Lax pair' technique [6].

In a parallel evolution, integrable dynamical systems were discovered [7, 8], which could be naturally interpreted as one-dimensional many-body problems; their integrability in the classical case was then related [9, 10] to the 'Lax pair' technique; and again it yielded a rich harvest of results [11]. Remarkable relations among integrable PDEs and dynamical systems were moreover uncovered, such as those emerging from the identification of the poles of special solutions of $(1+1)$-dimensional integrable PDEs with the dependent variables of one-dimensional (complexified) dynamical systems [12]. This also led to the introduction of new techniques to identify and solve new integrable dynamical systems [13].

Most of these developments, with few exceptions [14, 15], were confined to one-dimensional space. Then several integrable nonlinear evolution equations in $2+1$ dimensions (i.e., with two space variables, in addition to time) were discovered and investigated; but the extension of these results to more space dimensions (including the all-important case of 3-dimensional space, where we all live) proved unwieldy. In this paper we instead focus on techniques which are applicable in spaces with more than just two space dimensions: although sometimes we illustrate these techniques below by exhibiting examples in two-dimensional space, too.

Let us now outline two, somewhat heuristic, notions, which are useful to put in proper perspective some of the results for integrable nonlinear PDEs reviewed below.

The first notion is the distinction among 'S-integrable' and 'C-integrable' nonlinear PDEs [16, 5]. We call 'S-integrable' those nonlinear PDEs which can be solved by the 'inverse Scattering' or 'Spectral transform' technique, and 'C-integrable' those that can be solved via an appropriate 'Change of (dependent or independent) variables'. A more precise definition is given in the Addendum to Ref. [5]; it implies that C-integrable equations are also S-integrable.

The second notion we like to review is that of 'universal' equations. We say that an equation is 'universal' if it is associated to a *large* class of equations – for instance, if it can be derived, by a limiting process, from *all* (or *almost all*) the equations of that *large* class. For instance, in the context of integrable dynamical systems ('ODEs'), the (linear) equations which describe (coupled) harmonic oscillators are *universal*, since they characterize the behaviour of almost any dynamical system in the neighborhood of its (stable) equilibrium configurations ('Hooke's law', 'theory of small oscillations'). Likewise, in the $(1+1)$-dimensional PDE context, the NLS equation is *universal*, since it describes, in the limit of weak but not altogether negligible nonlinearity, the amplitude modulation, in appropriately rescaled and shifted variables, of a single dispersive wave, for (almost) any nonlinear PDE whose linear part is dispersive and whose nonlinear part is (in some very weak sense) analytic [5, 16, 17]. Indeed, the remarkable feature of NLS, to be both *widely applicable and integrable*, can be traced precisely to this *universality* (see below) [16, 5].

Universal nonlinear equations are clearly an interesting target for investigation, especially if they are also *integrable* and *widely applicable*. They provide an appealing escape from the quandary that tends to bedevil the study of nonlinear equations: on one side, the investigation of sufficiently general equations cannot hope to yield much detailed understanding of the actual behaviour of their solutions, beyond proofs of existence, uniqueness, perhaps inclusion in some functional class; on the other hand, by concentrating on special equations, which allow a more detailed understanding of the behaviour of their solutions, one runs the risk of focusing on uninteresting flukes.

A technique to identify *universal* equations which are also likely to be *integrable* and *widely applicable* is to start from some equation – possibly an artificially concocted one – which belongs to a *large* class (for instance to the class mentioned above, of the nonlinear evolution equations whose linear part is dispersive and whose nonlinear part is weakly analytic), and then apply to it some limiting process, obtaining thereby a new, generally simpler, equation. If the limiting process is well chosen, it yields the same equation (generally up to some constant parameters), from *all* (or *almost all*) the equations of the *large* class; hence the equation obtained in this manner – namely, by starting from a particular, artificially concocted, equation – stands nevertheless a good chance of being *universal*. Moreover, if the (artificially concocted) starting equation is *integrable*, and the limiting procedure is, in some (generally asymptotic) sense *exact*, and therefore such as to preserve the property of integrability, then the universal equation so obtained is generally itself *integrable*; indeed, whether (and how) this is the case can generally be ascertained by applying the limiting procedure not only to the original integrable equation, but also to the machinery that underpins its integrability, deriving thereby the machinery that underpins the integrability of the universal equation. Finally, if the limiting procedure is also 'reasonable in an applicative context', namely if it reflects circumstances likely to arise in applications, then it is also likely that *the universal integrable equation manufactured in this manner shall itself be relevant to applications*: since the large class, just because it is large, is likely to contain some equations modeling phenomenological applications; and such modeling shall remain relevant, under the hypotheses indicated above, in phenomenologically interesting circumstances, even after the limiting process.

The NLS equation, which can be obtained in this manner, for instance from KdV or SG, is a clear demonstration of the viability of this approach in the $(1+1)$-dimensional context; several other examples are also available in the same context [5]; as well as in the $(2+1)$-dimensional context, where the universal Davey–Stewardson (DS) equations take the role of NLS, and, for instance, the Kadomtsev–Petviashvili (KP) or the Konopelchenko–Rogers (KR) equations take the role of KdV or SG [5, 18].

It is natural to try and exploit this same approach in order to discover *universal integrable nonlinear equations in $N+1$ dimensions*. The difficulty, however, is to identify a suitable starting point for this treatment – since the available inventory of integrable nonlinear PDEs in $N+1$ dimensions is poorly endowed. Hence it has been necessary to begin by manufacturing some integrable equations in $N+1$ dimensions; this turned out to be possible by focusing attention on C-*integrable* equations [19]. It was then possible to derive from these, by a suitable limiting process of the kind mentioned above, a new equation, that has therefore some claim to be considered universal, and which, as expected, is also C-*integrable* [20]. Of course the restriction to C-*integrable* equations implies a narrowing down of the character of universality. This phenomenon was indeed well understood, in the $(1+1)$-dimensional context, some years ago [5, 16, 21]: it entails the replacement of the ('more universal') NLS equation by the ('less universal') Eckhaus equation [22]. It was therefore justified to suggest [20] that the newly derived *universal* C-*integrable equation* be called the '$(N+1)$-dimensional Eckhaus equation'; moreover, this nonlinear PDE in $N+1$ dimensions (whose form is displayed below, see Section 3) does indeed reduce to the standard Eckhaus equation [22] for $N=1$.

Most of our discussion thus far has focused on nonlinear evolution equations ('PDEs'); let us now outline some developments on dynamical systems ('ODEs'). As mentioned above, one technique to identify solvable dynamical systems is to relate them to integrable (linear or nonlinear) PDEs; much has been done in this direction, in the context of one-dimensional space [13]. Recently, techniques have been introduced [23] to extend these approaches to N-dimensional space, deriving thereby several examples of solvable dynamical systems, which can be interpreted as (*nonrelativistic, classical*) n-body problems in N-dimensional space, being characterized by equations of motion of Newtonian type ('mass times acceleration equals force'; albeit with special many-body and velocity-dependent forces) [23, 24]. The *leitmotif* of these methods is to identify nonlinear evolutions that can be solved via appropriate changes of variables. Representative examples of such models are reported below (Section 5).

Let us end this introductory section with a general remark. There are, roughly speaking, two complementary approaches to mathematical physics. In the first one, a (hopefully interesting) physical problem is the starting point of the analysis; and the main task of the mathematical physicist is to provide for it an appropriate mathematical formulation, then obtain some 'rigorous' results. In the second approach, the starting point is rather a (hopefully novel and powerful) mathematical technique; and the main task of the mathematical physicist is then to explore the range of 'physical' problems that can be tackled by exploiting it.

Clearly, provided they are practised *cum grano salis*, both approaches are likely to be fruitful, hence worthy of dedicated investigation; although it is common practice to present, perhaps a bit hypocritically, findings arrived at via the second approach, as if they had been obtained by the first. In most of our investigations, as indeed in much of the research on integrable NEEDS whose upsurge has characterized mathematical physics during the last quarter century, we follow instead, unashamedly, the second approach, as recommended by C. Jacobi [25], who was indeed a pioneer in the study of integrable dynamical systems [26].

This paper is organized as follows. In Section 2 we focus on the S-integrable equations in N+1 dimensions which were introduced [14] and developed [27] long ago; the general idea behind these results is to introduce, in the (by now) standard context of S-integrable equations, several time variables, and then to reinterpret all of them but one as space variables. This approach does indeed produce classes of nonlinear evolution equations in $N+1$ dimensions, whose initial value problems are solvable via the spectral transform technique [14, 27]. A representative example is displayed below; many more, and their solutions, are discussed in the literature quoted above.

In Section 3, the $(N+1)$-dimensional Eckhaus equation is displayed, as well as some of the nonlinear evolution equations in $N+1$ dimensions which were instrumental in its derivation, as explained above.

In Section 4, some (systems of coupled) nonlinear evolution equations in $N+1$ dimensions are displayed, which are C-*integrable* and can be solved [28] by invertible nonlinear transformations such as those that relate the coefficients of a polynomial to its zeros. Appealing features of these equations are the 'physical' nature of their linear parts (which reproduce standard 'physical' equations such as the Klein–Gordon, Schrödinger and diffusion equations in $N+1$ dimensions – for any N), as well as their invariance properties (under Lorentz transformations, space rotations, etc., as the case may be; see below). This suggests a potential for applications, perhaps 'physical', more likely 'theoretical' (possibly in quantum field theory).

Finally, in Section 5, several instances of solvable dynamical systems which can be interpreted as (classical, nonrelativistic) n-body problems in N-dimensional space (with $N = 2, 3$ and $n = 3, 4$) are exihibited; as mentioned above, they are characterized by equations of motion of Newtonian type ('mass times acceleration equals force'), with (appropriately chosen) many-body, velocity-dependent forces.

2. S-Integrable Nonlinear PDEs in Multidimensions

Already in the relatively early days of the development of the spectral transform technique to identify and solve integrable nonlinear evolution equations, a simple but effective trick was noted to extend the standard treatment, originally developed in the context of $1+1$ dimensions (one space and one time variables), firstly to $1+N$ dimensions (one space and N time variables), then, by a suitable exchange of the roles of the time and space variables (which is formally trivial, but is in fact quite significant in the context of solving the initial-value problem), to $N+1$ dimensions (N space and one time variables) [14]. This approach was then fully developed, indeed included in a book-form presentation of the spectral transform technique to solve and investigate nonlinear evolution equations [27]; hence it is justified to limit our report here to the exhibition of just a single example of nonlinear evolution equation of this kind:

$$\begin{aligned} u_t &= u_{xxx} - 6u_x u + \mathbf{v}_1 \cdot \nabla_{\mathbf{y}} u + \\ &\quad + \mathbf{v}_2 \cdot [\nabla_{\mathbf{y}} u_{xx} - 4u \nabla_{\mathbf{y}} u - 2u_x \nabla_{\mathbf{y}} w], \end{aligned} \tag{2.1a}$$

$$w_x = u. \tag{2.1b}$$

Here $u \equiv u(\mathbf{r}, t)$ is the main unknown field (dependent variable) evolving in time, $w \equiv w(\mathbf{r}, t)$ is an auxiliary field, and the independent variable $\mathbf{r} \equiv (x, \mathbf{y})$ is an N-dimensional 'space' vector; accordingly $\mathbf{y}$ is to be interpreted as an $(N-1)$-dimensional 'space' vector, $\nabla_{\mathbf{y}}$ is the corresponding $(N-1)$-dimensional gradient, and $\mathbf{v}_1, \mathbf{v}_2$ are two (arbitrarily assigned) $(N-1)$-vectors.

The solution of this nonlinear evolution equation is uniquely determined by assigning, at the 'initial time' t_0, the field $u(\mathbf{r}, t)$,

$$u(\mathbf{r}, t_0) = u_0(\mathbf{r}),$$

and moreover by giving a boundary condition for the auxiliary field $w(\mathbf{r}, t)$, say

$$w(x_0, \mathbf{y}, t) = w_0(\mathbf{y}, t)$$

(a typical choice is $x_0 = -\infty, w_0(\mathbf{y}, t) = 0$).

Of course, (2.1) is in effect a (3+1)-dimensional equation, since the only space variables that are really effective are x, and the two components of $\mathbf{y}$ along $\mathbf{v}_1$ and $\mathbf{v}_2$; but genuinely $(N+1)$-dimensional S-integrable nonlinear evolution equations can be easily written, although they are more complicated, and of higher order [26].

3. Universal C-Integrable Nonlinear Partial Differential Equation in $N+1$ Dimensions

Notation. Let $\mathbf{r}$ be a vector in N-dimensional space, x its 'first' Cartesian component, and $\mathbf{y}$ the vector in $(N-1)$-dimensional space of the remaining Cartesian components of $\mathbf{r}$,

$$\mathbf{r} \equiv (x, \mathbf{y}). \tag{3.1}$$

Let the N-vector ∇, resp. the $(N-1)$-vector $\nabla_{\mathbf{y}}$, be the (first-order) differential operators in N-space, resp. $(N-1)$-space (namely in $\mathbf{r}$-space, resp. $\mathbf{y}$-space), and Δ, resp. $\Delta_{\mathbf{y}}$ the Laplace operators in N-dimensional, resp. $(N-1)$-dimensional space:

$$\Delta \equiv \nabla \cdot \nabla \equiv \Delta_{\mathbf{y}} + \partial^2/\partial x^2 \equiv \nabla_{\mathbf{y}} \cdot \nabla_{\mathbf{y}} + \partial^2/\partial x^2. \tag{3.2}$$

Let $\psi \equiv \psi(\mathbf{r}, t)$ be a complex scalar field, $f \equiv f(\mathbf{r}, t)$ a real scalar field, and $\mathbf{g} \equiv \mathbf{g}(\mathbf{r}, t)$ a real $(N-1)$-dimensional vector field.

Then the '$(N+1)$-dimensional Eckhaus equation' reads as follows:

$$\begin{aligned} &i\psi_t + \Delta\psi + [|\psi|^4 + 2(|\psi|^2)_x]\psi + if\psi + \\ &\quad +2(\mathbf{g} \cdot \nabla_{\mathbf{y}})\psi + [(\nabla_{\mathbf{y}} \cdot \mathbf{g}) + \mathbf{g}^2]\psi = 0, \end{aligned} \tag{3.3a}$$

$$\mathbf{g}_x = \nabla_{\mathbf{y}}(|\psi|^2), \tag{3.3b}$$

$$f_x + 2|\psi|^2 f = 2i\mathbf{g} \cdot (\psi^* \nabla_{\mathbf{y}} \psi - \psi \nabla_{\mathbf{y}} \psi^*) + i(\psi^* \Delta_{\mathbf{y}} \psi - \psi \Delta_{\mathbf{y}} \psi^*). \tag{3.3c}$$

Note that this equation is not spherically symmetrical (in N-dimensional space); the x direction plays a special role (see below). In this equation the complex scalar field $\psi(\mathbf{r}, t)$ is the main unknown (dependent variable), and the two real scalar, resp. $(N-1)$-vector fields, $f(\mathbf{r}, t)$, resp. $\mathbf{g}(\mathbf{r}, t)$, are auxiliary dependent variables. The solution of this equation is uniquely determined by assigning the field ψ at the 'initial' time t_0,

$$\psi(\mathbf{r}, t_0) = \psi_0(\mathbf{r}), \tag{3.4}$$

and, in addition, boundary conditions for the auxiliary fields f and $\mathbf{g}$, say

$$f(x_0, \mathbf{y}, t) = f_0(\mathbf{y}, t), \tag{3.5}$$

$$\mathbf{g}(x_0, \mathbf{y}, t) = g_0(\mathbf{y}, t). \tag{3.6}$$

The nonlinear evolution equation (3.3) is C-integrable, but only for a subclass of its solutions, namely those restricted by the condition that the $(N-1)$-vector field $\mathbf{g}(\mathbf{r}, t)$ be irrotational in the $(N-1)$-dimensional $\mathbf{y}$-space, namely

$$\nabla_{\mathbf{y}} \wedge \mathbf{g}(x, \mathbf{y}, t) = 0. \tag{3.7}$$

Note that (3.3c) implies that this condition is compatible with the time-evolution implied by (3.3), and that it is sufficient that this property of irrotationality hold at the boundary, say (see (3.6))

$$\nabla_{\mathbf{y}} \wedge \mathbf{g}_0(\mathbf{y}, t) = 0, \tag{3.8}$$

for it to hold everywhere. A class of solutions for which this property is automatically satisfied are those characterized by a vanishing boundary value for $\mathbf{g}$, namely $\mathbf{g}_0(\mathbf{y},t) = 0$; this condition might only hold asymptotically, namely correspond to the choice, say, $x_0 = -\infty$, in (3.6).

For this class of solutions, the initial boundary value problem for (3.3) can be solved by quadratures [20]; moreover, many solutions (including those allowing a 'multisolitonic' interpretation) can be explicitly exhibited [20]. This is the case even for the more general version of (3.3) that is obtained by adding, in the l.h.s. of (3.3a), an additional term $V\psi$, so that the linear part of this equation represent the linear Schrödinger equation in $N+1$ dimensions in the presence of an (arbitrarily assigned) external potential $V(\mathbf{r})$.

For the special subclass of solutions of (3.3) that are independent of the variable $\mathbf{y}$, it is clearly possible to set to zero both auxiliary fields, f and $\mathbf{g}$; then (3.3) reduces to the $(1+1)$-dimensional Eckhaus equation [22].

To explain in which sense (3.3) can be considered a *universal* equation, it is convenient to display another C-integrable equation, which in any case also deserves some mention in its own right. It reads as follows:

$$\begin{aligned} u_{tt} &- \Delta u + u + (2+p)u^p u_t - 2(\mathbf{h}\cdot\nabla)u + \\ &+ \left[u^{2p} - (\nabla\cdot\mathbf{h}) - \mathbf{h}^2\right]u = 0, \end{aligned} \tag{3.9a}$$

$$\mathbf{h}_t = p u^{p-1}\nabla u = \nabla(u^p). \tag{3.9b}$$

Here $u \equiv u(\mathbf{r},t)$ is a (real) scalar, and $\mathbf{h} \equiv \mathbf{h}(\mathbf{r},t)$ a (real) N-vector; both these fields appear now as time-evolving dependent variables. Note that, in contrast to (3.3), this system is spherically symmetrical (no privileged space direction); moreover, in contrast to (3.3a), the evolution equation (3.9a) is second-order in time.

The initial-value problem for this equation requires as input initial data for u, u_t and $\mathbf{h}$, say

$$u(\mathbf{r},t_0) = u_0(\mathbf{r}), \tag{3.10a}$$

$$u_t(\mathbf{r},t_0) = u_1(\mathbf{r}), \tag{3.10b}$$

$$\mathbf{h}(\mathbf{r},t_0) = \mathbf{h}_0(\mathbf{r}); \tag{3.10c}$$

clearly these assignments are sufficient to determine uniquely a solution of (3.9).

This nonlinear evolution equation is C-integrable, but only for a subclass of its solutions, namely those restricted by the condition that the N-vector field $\mathbf{h}(\mathbf{r},t)$ be irrotational (of course, in the N-dimensional $\mathbf{r}$-space), namely

$$\nabla \wedge \mathbf{h}(\mathbf{r},t) = 0. \tag{3.11}$$

Note that (3.9b) implies that this condition is compatible with the time evolution of $\mathbf{h}(\mathbf{r},t)$; hence, it is sufficient that this property of irrotationality hold at any one time, for it to hold for all time. In particular, it is sufficient that the initial datum, see (3.10c), satisfy this condition,

$$\nabla \wedge \mathbf{h}_0(\mathbf{r}) = 0, \tag{3.12}$$

to guarantee that the corresponding solution be in the class characterized by (3.11).

Within this class, the C-integrable character of (3.9) entails the possibility to solve by quadratures its initial-value problem, and moreover to manufacture many explicit solutions (including those admitting a 'multisolitonic' interpretation) [19].

An appealing feature of the nonlinear evolution equation (3.9) (in the case $p = 2$, to which attention is hereafter for simplicity restricted) is displayed by performing the rescaling $u \rightarrow \varepsilon^{1/2}u$, $\mathbf{h} \rightarrow \varepsilon\mathbf{h}$:

$$\begin{aligned} u_{tt} - \Delta u + u &= \varepsilon\big[-4u^2u_t + 2\mathbf{h}\cdot\nabla u + (\nabla\cdot\mathbf{h})u\big] + \\ &\quad +\varepsilon^2\big[-u^4 + \mathbf{h}^2\big]u, \end{aligned} \tag{3.13a}$$

$$\mathbf{h}_t = 2u\nabla u = \nabla(u^2). \tag{3.13b}$$

It is then clear that this equation belongs to the class of nonlinear PDEs whose linear part is dispersive and whose nonlinear part is weakly analytic. Indeed, in the extreme 'weak field' limit ($\varepsilon = 0$), u satisfies the Klein–Gordon equation

$$u_{tt} - \Delta u + u = 0, \tag{3.14}$$

and it admits therefore the solution

$$u(\mathbf{r},t) = A\exp\left[i(kx - \omega t)\right] + c.c., \tag{3.15}$$

representing a single dispersive wave (traveling in the x direction; see (3.1)). Here the 'wave number' k is an (arbitrarily assigned and hereafter held fixed) parameter, ω is the corresponding 'circular frequency',

$$\omega = (k^2 + 1)^{1/2}, \tag{3.16}$$

and the complex constant A is the 'amplitude' of the dispersive wave (3.15). The fact that this solution represent a *dispersive* wave is demonstrated by the difference among the 'group velocity' c associated with (3.16),

$$c = k/\omega, \tag{3.17}$$

and the corresponding 'phase velocity' $v = \omega/k$.

If the parameter ε is small but not quite vanishing, it stands to reason that Equations (3.13) admit a solution in which u is 'close' to the 'single dispersive wave' (3.15); understanding how such a solution looks, corresponds to studying how weak nonlinear effects affect the traveling wave solution (3.15) (this is clearly a question that arises in many phenomenological circumstances, modeled by nonlinear evolution equations whose linear part is dispersive, in which a single 'carrier wave' solution of the linear part of the equation plays a dominant role, in a weakly nonlinear context). The answer is [20], that the main effect is to induce an 'amplitude modulation' of the dispersive wave: the *constant* amplitude A gets replaced by a *function* ψ, depending on *coarse-grained*, *shifted* and *slow* variables:

$$\psi \equiv \psi(x', \mathbf{y}', t'), \tag{3.18a}$$

$$x' = \varepsilon\alpha(x - ct), \tag{3.18b}$$

$$\mathbf{y}' = \varepsilon\beta\mathbf{y}, \tag{3.18c}$$

$$t' = \varepsilon^2\gamma t. \tag{3.18d}$$

Here of course ε is the small parameter, c the group velocity (3.17), and α, β, γ are three rescaling parameters (introduced for cosmetic reasons, see below); and we are of course using the notation (3.1), so that $\mathbf{y}$ and $\mathbf{y}'$ are $(N-1)$-dimensional vectors. It turns moreover out [20] that *the time-evolution of ψ is given precisely by the $(N+1)$-dimensional Eckhaus equation* (3.3), rewritten of course in 'primed' variables (and with an appropriate choice for the three 'cosmetic' constants α, β, γ).

This argument illustrates the origin of the $(N+1)$-dimensional Eckhaus equation (3.3), and provids some justification for the claim that this is a *universal* equation: it is indeed clear that one might also arrive at (3.3) starting from other equations, instead of (3.13). For a more detailed discussion – quite essential for a better understanding of the universality of (3.3), connected with the fact that the starting point of this derivation, namely the nonlinear evolution Equation (3.13), is C-integrable – the interested reader is referred to the literature [5, 19, 20].

Let us end this section by reporting (without any detail; for these, as well as for more general examples, see [19]) two other C-integrable nonlinear evolution equations in $N+1$ dimensions, which are in some sense analogous to (3.9). They read as follows:

$$u_t + \Delta u + 2\mathbf{h} \cdot \nabla u + \left[u^p + (\nabla \cdot \mathbf{h}) + \mathbf{h}^2\right]u = 0, \tag{3.19a}$$

$$\mathbf{h}_t = p u^{p-1} \nabla u = \nabla(u^p); \tag{3.19b}$$

$$i\psi_t + \Delta\psi + V(\mathbf{r})\psi + 2\mathbf{h} \cdot \nabla\psi + \left[i|\psi|^p + (\nabla \cdot \mathbf{h}) + \mathbf{h}^2\right]\psi = 0, \tag{3.20a}$$

$$\mathbf{h}_t = p|\psi|^{p-1}\nabla|\psi| = \nabla(|\psi|^p). \tag{3.20b}$$

4. Coupled C-Integrable PDEs in Multidimensions

In this section we exhibit two examples of (highly) nonlinear evolution equations in $N+1$ dimensions for n scalar fields, which can be linearized by a simple change of dependent variables, corresponding to the invertible transformation among the n coefficients of a (properly normalized) polynomial of degree n, and its n zeros. For a detailed illustration of how such a transformation entails the possibility to solve the nonlinear evolution equations displayed below, as well as an indication of the make-up of the classes of nonlinear evolution equations solvable by such techniques, which include of course many more examples, the interested reader is referred to [28], and to the literature quoted there.

The two examples read as follows:

$$\begin{aligned}&[\gamma\partial/\partial t - \Delta + V(\mathbf{r})]f_j(\mathbf{r},t)\\ &\quad = 2\sum_{k=1,\ k\neq j}^{n}\{A^2 + [\nabla f_j(\mathbf{r},t)]\cdot[\nabla f_k(\mathbf{r},t)]\}/[f_j(\mathbf{r},t) - f_k(\mathbf{r},t)],\end{aligned} \tag{4.1}$$

$$\begin{aligned}&[\partial^2/\partial t^2 - \Delta + M^2]f_j(\mathbf{r},t)\\ &\quad = 2\sum_{k=1,\ k\neq j}^{n}\{A^2 + f_{j,\mu}(\mathbf{r},t)f_{k,\mu}(\mathbf{r},t)\}/[f_j(\mathbf{r},t) - f_k(\mathbf{r},t)].\end{aligned} \tag{4.2}$$

Notation. Here the index j runs from 1 to n, and the n scalar fields $f_j(\mathbf{r},t)$ are the dependent variables; the N-dimensional space vector $\mathbf{r} \equiv (x_1, x_2, \ldots, x_N)$, and the time $t \equiv ix_0$, are the independent variables; ∇ is the N-dimensional gradient, $\Delta \equiv \nabla^2$ the Laplacian, $f_{j,\mu} \equiv \partial f_j/\partial x_\mu$, $\mu = 0, 1, \ldots, N$; $V(\mathbf{r})$ is an arbitrary given function; γ, A^2, M^2 are arbitrary constants (which could be rescaled away, or set to zero). Of course in (4.2) the repeated index μ is summed upon, from 0 to N; note that, for $N = 3$, this is a relativistically invariant equation in our physical space-time. The number of fields n, and the dimensionality N of space, are two arbitrary positive integers; $n > 1$ in order that these two evolution equations be indeed nonlinear.

Clearly by summing over j from 1 to n the r.h.s. of these equations vanish, hence the quantity

$$F(\mathbf{r},t) = \sum_{j=1}^{n} f_j(\mathbf{r},t) \tag{4.3}$$

satisfies, as the case may be, the equations

$$[\gamma\partial/\partial t - \Delta + V(\mathbf{r})]f(\mathbf{r},t) = 0, \tag{4.4}$$

$$[\partial^2/\partial t^2 - \Delta + M^2]F(\mathbf{r},t) = 0. \tag{4.5}$$

Note that, for $N = 3$, (4.5) is the standard Klein–Gordon equation, while for $\gamma = i$ (4.4) is the standard linear Schrödinger equation in the 'external potential' $V(\mathbf{r})$, for $\gamma = 1$ and $V(\mathbf{r}) = 0$ it is the standard diffusion equation.

5. Solvable (Nonrelativistic, Classical) n-Body Problems in Multidimensions

In this Section we report the equations of motion that characterize three solvable many-body problems: a 3-body problem in 2-dimensional space, an analogous 4-body problem in 3-dimensional space, and a somewhat different 3-body problem in 2-dimensional space. The interested reader will find a more complete discussion of the approach that has led to these models in [23], and a remarkably explicit display of their solutions in [24]. In this literature other solvable models of this kind are also displayed, as well, of course, as indications of how to manufacture additional ones.

A solvable 3-body problem in 2-dimensional space is characterized by the equations of motion

$$\dot{\mathbf{p}}_j = \sum_{k=1}^{3} \mathbf{p}_k\left[\mathbf{z}\cdot\dot{\mathbf{r}}_j \wedge (\mathbf{r}_{k+1} - \mathbf{r}_{k-1})\right]/\left[\mathbf{z}\cdot(\mathbf{r}_1\wedge\mathbf{r}_2 + \mathbf{r}_2\wedge\mathbf{r}_3 + \mathbf{r}_3\wedge\mathbf{r}_1)\right],$$
$$j = 1,2,3, \tag{5.1}$$

where

$$\mathbf{p}_j \equiv m_j\dot{\mathbf{r}}_j + \sum_{k=1}^{3} \nu_{jk}\mathbf{r}_k. \tag{5.2}$$

Here $\mathbf{r}_j \equiv \mathbf{r}_j(t)$ are the coordinates of the three particles in the plane, and we use for notational convenience a 3-dimensional notation, namely

$$\mathbf{r}_j \equiv (x_j, y_j, 0), \tag{5.3a}$$

$$\mathbf{z} = (0,0,1); \tag{5.3b}$$

the superimposed dot of course denotes time-differentiation; and in the r.h.s. of (5.1) the index k is defined mod (3). The three 'masses' $\mathbf{m}_j$, and the nine parameters ν_{jk}, are arbitrary; if these latter parameters satisfy the three conditions

$$\sum_{k=1}^{3} \nu_{jk} = 0, \tag{5.4}$$

then the equations of motion (5.1) are translation-invariant; they are generally rotation-invariant, but not Galileo-invariant.

An analogous solvable 4-body problem in 3-dimensional space is characterized by the following equations of motion (in self-evident notation):

$$\dot{\mathbf{p}}_j = \sum_{k=1}^{4}(-)^k \mathbf{p}_k\left[\dot{\mathbf{r}}_j \cdot (\mathbf{r}_{k+1} - \mathbf{r}_{k+2}) \wedge (\mathbf{r}_{k+2} - \mathbf{r}_{k-1})\right]/ \\ \left[(\mathbf{r}_2 - \mathbf{r}_1) \cdot (\mathbf{r}_3 - \mathbf{r}_1) \wedge (\mathbf{r}_4 - \mathbf{r}_1)\right], \quad j = 1,2,3,4, \tag{5.5}$$

$$\mathbf{p}_j \equiv m_j \dot{\mathbf{r}}_j + \sum_{k=1}^{4} \nu_{jk} \mathbf{r}_k. \tag{5.6}$$

The third (hitherto unpublished) model we report is characterized by rather unwieldy equations of motion (but quite interesting solutions; see below). Since it is again a 3-body problem in 2-dimensional space, we use the notation (5.3) to display it. Its equations of motion read:

$$\ddot{\mathbf{r}}_j = \sum_{k=1}^{3}[A_{jk}\mathbf{r}_k + B_{jk}\dot{\mathbf{r}}_k], \quad j = 1,2,3, \tag{5.7}$$

with the 18 scalar quantities A_{jk}, B_{jk} defined by the following formulas:

$$A_{jk} = (A_{jk}^{(0)} + \lambda A_{jk}^{(1)} + \omega^2 A_{jk}^{(2)})/ \\ \left[(\alpha_j - \alpha_{j+1})(\alpha_j - \alpha_{j-1})(\mathbf{z} \cdot \mathbf{r}_{j+1} \wedge \mathbf{r}_{j-1})\right], \tag{5.8a}$$

$$A_{jk}^{(0)} = -2\sum_{i=1}^{3}\left[\mathbf{z} \cdot (\mathbf{r}_{i+1} - \mathbf{r}_{i-1}) \wedge \dot{\mathbf{r}}_j/\Delta\right] \times \\ \times \mathbf{z} \cdot \Big\{\left[(\alpha_j\alpha_k - \alpha_{k+1}\alpha_{k-1})(\mathbf{r}_{k+1} - \mathbf{r}_{k-1}) + \right. \\ \left. + \alpha_i(\alpha_{k+1}\mathbf{r}_{k+1} - \alpha_{k-1}\mathbf{r}_{k-1})\right] \wedge \dot{\mathbf{r}}_i - \\ -(\mathbf{r}_{k+1} - \mathbf{r}_{k-1}) \wedge \dot{\mathbf{r}}_i\left[(\alpha_k - \alpha_{k-1})\alpha_{k+1}(\mathbf{z} \cdot \mathbf{r}_k \wedge \mathbf{r}_{k+1}) - \right. \\ \left. -(\alpha_k - \alpha_{k+1})\alpha_{k-1}(\mathbf{z} \cdot \mathbf{r}_k \wedge \mathbf{r}_{k-1})\right]/\Delta\Big\}, \tag{5.8b}$$

$$A_{jj}^{(1)} = -A_{jj}^{(2)}\left[q + (1-q)(\mathbf{z} \cdot \mathbf{r}_{j+1} \wedge \mathbf{r}_{j-1})/\Delta\right] + \\ +(1-q)\alpha_{j+1}\alpha_{j-1}(\mathbf{z} \cdot \mathbf{r}_{j+1} \wedge \mathbf{r}_{j-1}) \times \\ \times\left[1 - (\mathbf{z} \cdot \mathbf{r}_{j+1} \wedge \mathbf{r}_{j-1})/\Delta\right], \tag{5.8c}$$

$$A_{j,j\pm1}^{(1)} = -A_{j,j\pm1}^{(2)}\Big\{q + (1-q)(\alpha_{j\pm1}/\alpha_{j\mp1}) + \\ +(1-q)\mathbf{z} \cdot \left(\left[1 - (\alpha_{j\pm1}/\alpha_{j\mp1})\right]\mathbf{r}_{j+1} \wedge \mathbf{r}_{j-1} \pm \right. \\ \left. \pm\left[1 - (\alpha_{j\pm1}/\alpha_{j\mp1})\right]\mathbf{r}_j \wedge \mathbf{r}_{j\pm1}\right)/\Delta\Big\}, \tag{5.8d}$$

$$A_{jj}^{(2)} = \alpha_j \mathbf{z} \cdot \left[(\alpha_{j+1} - \alpha_j) \mathbf{r}_j \wedge \mathbf{r}_{j+1} + \right.$$
$$\left. + (\alpha_{j-1} - \alpha_j) \mathbf{r}_j \wedge \mathbf{r}_{j-1} - \alpha_j \mathbf{r}_{j+1} \wedge \mathbf{r}_{j-1} \right], \tag{5.8e}$$

$$A_{j,j\pm 1}^{(2)} = \pm \alpha_j \alpha_{j\mp 1} \mathbf{z} \cdot \mathbf{r}_j \wedge \mathbf{r}_{j\mp 1}, \tag{5.8f}$$

$$B_{jk} = \{ 2\alpha_j \alpha_k / [(\alpha_j - \alpha_{j+1})(\alpha_j - \alpha_{j-1})] \} \times$$
$$\times \left[\mathbf{z} \cdot \dot{\mathbf{r}}_j \wedge (\mathbf{r}_{k+1} - \mathbf{r}_{k-1}) \right] / (\mathbf{z} \cdot \mathbf{r}_{j+1} \wedge \mathbf{r}_{j-1}), \tag{5.9}$$

$$\Delta = \mathbf{z} \cdot (\mathbf{r}_1 \wedge \mathbf{r}_2 + \mathbf{r}_2 \wedge \mathbf{r}_3 + \mathbf{r}_3 \wedge \mathbf{r}_1). \tag{5.10}$$

These equations contain six arbitrary (real) parameters, namely the three constants α_j, $j = 1, 2, 3$, the dimensionless parameter q, and the two constants λ and ω^2, both of which have the dimension of 'inverse time squared' (but for $q = 1$, only the combination $\omega^2 - \lambda$ enters). Note however that the three constants α_j – which are of course assumed to be all different, $\alpha_j \neq \alpha_k$ – in fact introduce only two parameters in the equations of motion (5.7–10), since the quantities A_{jk} and B_{jk} are clearly invariant (see (5.8–10)) under a rescaling of these constants, $\alpha_j \to \alpha'_j = c\alpha_j$.

Since this model has, as yet, not been published, we report its solution here:

$$\mathbf{r}_j(t) = \{ \alpha_j [\mathbf{v}_1 \cos(\omega_1 t) + \mathbf{v}_2 \sin(\omega_1 t)] + \mathbf{r}(t) \} / D_j(t), \tag{5.11a}$$

$$\mathbf{r}(t) = \mathbf{v}_3 \cos\left[(\omega_1 + \omega_2) t \right] + \mathbf{v}_4 \sin\left[(\omega_1 + \omega_2) t \right] +$$
$$+ \mathbf{v}_5 \cos\left[(\omega_1 - \omega_2) t \right] + \mathbf{v}_6 \sin\left[(\omega_1 - \omega_2) t \right], \tag{5.11b}$$

$$D_j(t) = \alpha_j^2 + \alpha_j \left[c_1 \cos(\omega_2 t) + c_2 \sin(\omega_2 t) \right] +$$
$$+ c_3 + c_4 \cos(2\omega_2 t) + c_5 \sin(2\omega_2 t), \tag{5.11c}$$

$$\omega_1 = (\omega^2 - \lambda)^{1/2}, \tag{5.11d}$$

$$\omega_2 = (\omega^2 - q\lambda)^{1/2}. \tag{5.11e}$$

For explicit expressions of the six constant two-vectors $\mathbf{v}_s$, and of the five scalar constants c_s, in terms of the initial data, $\mathbf{r}_j(0)$ and $\dot{\mathbf{r}}_j(0)$, $j = 1, 2, 3$, the interested reader is referred to the literature [24].

This model is rotation-invariant, but not translation-invariant; it would be easy to manufacture a generalized version of it that is also translation-invariant, or analogous 3- or 4-body models in 3-dimensional space.

References

1. Korteweg, D. J. and de Vries, G.: On the change of form of long waves advancing in a rectangular canal, and on a new type of long stationary waves, *Phil. Mag.* **39** (1895), 422–443.
2. Gardner, C. S., Greene, J. M., Kruskal, M. D., and Miura, R. M.: Method for solving the Korteweg–de Vries equation, *Phys. Rev. Lett.* **19** (1967), 1095–1097.
3. Zakharov, V. E. and Shabat, A. B.: Exact theory of two-dimensional self-focusing and one-dimensional self-modulation of waves in nonlinear media, *Soviet Physics JETP* **34** (1972), 62–69.
4. Ablowitz, M. J., Kaup, D. J., Newell, A. C., and Segur, H.: The inverse scattering transform – Fourier analysis for nonlinear problems, *Stud. Appl. Math.* **53** (1974), 249–315.
   Zakharov, V. E., Takhtadzjan, L. A., and Faddeev, L. D.: A complete description of the solution of the sine-Gordon equation, *Soviet Phys. Dokl.* **19** (1975), 824–826.
5. Calogero, F.: Why are certain nonlinear PDEs both widely applicable and integrable?, in: Zakharov, V. E. (Ed.) *What is Integrability*?, Springer, 1991, pp. 1–62.
6. Lax, P. D.: Integrals of nonlinear equations of evolution and solitary waves, *Commun. Pure Appl. Math.* **21** (1968), 467–490.
7. Toda, M.: Vibration of a chain with nonlinear interaction, *J. Phys. Soc. Japan* **29** (1967), 431–436;
   Waves in nonlinear lattice, *Prog. Theor. Phys. Suppl.* **45** (1970), 174–200.
8. Calogero, F.: Solution of the one-dimensional N-body problem with quadratic and/or inversely-quadratic pair potentials, *J. Math. Phys.* **12** (1971), 419–436.
9. Manakov, S. V.: Complete integrability and stochastization of discrete dynamical systems, *Sov. Phys. JETP* (1974), 269–274.
   Flaschka, H.: The Toda lattice. I and II, *Phys. Rev.* **B9** (1974), 1924–1925 and *Prog. Theor. Phys.* **51** (1974), 703–716.
10. Moser, J.: Three integrable Hamiltonian systems connected with isospectral deformations, *Adv. in Math.* **16** (1975), 197–220.
11. Calogero, F.: Exactly solvable one-dimensional many-body problems, *Lett. Nuovo Cimento* **13** (1975), 411–416;
    Integrable dynamical systems and related mathematical results, in: Wolf, K. B. (Ed.) *Nonlinear Phenomena*, Lecture Notes in Physics **189**, Springer, 1983, pp. 47–109;
    Integrable dynamical systems and some other mathematical results (remarkable matrices, identities, basic hypergeometric functions), in: Winternitz, P. (Ed.) *Systèmes Dynamiques Non Linéaires: Integrabilité et Comportement Qualitatif*, Presse de l'Université de Montréal, **SMS 120**, 1986, pp. 40–70;
    Remarks on certain integrable one-dimensional many-body problems, *Physics Letters* **A183** (1993), 85–88.
    Olshanetsky, M. A. and Perelomov, A. M.: Classical integrable finite-dimensional systems related to Lie algebras and Quantal integrable systems related to Lie algebras, *Physics Reports* **71** (1981), 313–400 and **94** (1983), 313–404.
12. Kruskal, M. D.: *Lectures in App. Math.* **15** (1974), 61–83.
    Thickstun, W. R.: *J. Math. Anal. Appl.* **55** (1976), 335.
    Airault, H., McKean, H. P., and Moser, J.: Rational and elliptic solutions of the Korteweg-de Vries equation and a related many-body problem, *Commun. Pure Appl. Math.* **30** (1977), 95–148.
    Choodnovsky, D. V. and Choodnovsky, G. V.: Pole expansions of nonlinear partial differential equations, *Nuovo Cimento* **40B** (1977), 339–353.

13. Calogero, F.: Motion of poles and zeros of special solutions of nonlinear and linear partial differential equations and related 'solvable' many-body problems, *Nuovo Cimento* **43B** (1978), 177–241.
A class of solvabe dynamical systems, *Physica* **D18** (1986), 280–302.
A class of integrable dynamical systems, *Inverse Problems* **1** (1985), L21–L24.
14. Calogero, F.: A method to generate solvable nonlinear evolution equations, *Lett. Nuovo Cimento* **14** (1975), 443–448.
15. Calogero, F.: Integrable many-body problems in more than one dimension, *Rep. Math. Phys.* **24** (1986), 141–143.
16. Calogero, F. and Eckhaus, W.: Nonlinear evolution equations, rescalings, model PDEs and their integrability. I and II, *Inverse Problems* **3** (1987), 229–262 and **4** (1988), 11–33.
17. Papers by Taniuti and others in: *Suppl. Prog. Theor. Phys.* **55**, 1974 (issue devoted to the 'Reductive perturbation method for nonlinear wave propagation').
Eckhaus, W.: The long-time behaviour for perturbed wave-equations and related problems, preprint **404**, Department of Mathematics, University of Utrecht, December 1985, published in part in *Lectures Notes in Physics*, Springer, 1986.
18. Calogero, F.: Universal integrable nonlinear PDEs', in: P. A. Clarkson (ed.) *Applications of Analytical and Geometrical Methods to Nonlinear Differential Equations*, NATO ASI Series C: Mathematical and Physical Sciences, Vol. 413, Kluwer, 1993, pp. 109–114.
19. Calogero, F.: C-integrable nonlinear partial differential equations in $N+1$ dimensions, *J. Math. Phys.* **33** (1992), 1257–1271.
20. Calogero, F.: Universal C-integrable nonlinear partial differential equation in $N+1$ dimensions, *J. Math. Phys.* **34** (1993), 3197–3209.
21. Calogero, F.: The evolution partial differential equation $u_t = u_{xxx} + 3(u_{xx}u^2 + 3u_x^2 u) + 3u_x u^4$, *J. Math. Phys.* **28** (1987), 538–555.
22. Calogero, F. and De Lillo, S.: The Eckhaus PDE $i\psi_t + \psi_{xx} + 2(|\psi|^2)_x\psi + |\psi|^4\psi = 0$, *Inverse Problems* **3** (1987), 633–681; Corrigendum: *ibid.* **4** (1988) 571.
23. Calogero, F.: Interpolation in multidimensions, a convenient finite-dimensional matrix representation of the (partial) differential operators, and some applications, *J. Math. Phys.* **34** (1993), 2704–4724.
24. Calogero, F. and Xiaoda, Ji: Solvable (nonrelativistic, classical) n-body problems in multidimensions. I & II, *J. Math. Phys.* **35** (1994), 710–733 and in M. Costato, A. Degasperis and D. Milani (eds), *Nonlinear Dynamics*, Conference Proceedings, Vol. 48, Pavullo nel Frignano, Italy, 19–22 May 1994, Soc. Italiana di Fisica, Bologna, 1995.
Solvable (nonrelativistic, classical) n-body problems on the line. I, *J. Math. Phys.* **34** (1993), 5659–5670.
25. Arnold, V. I.: *Mathematical Methods of Classical Mechanics*, 2nd edn, Springer, p. 266.
26. Jacobi, C.: Problema trium corporum mutis attractionibus cubus distantiarum inverse proportionalibus recta linea se moventium, in: *Gesammelte Werke,* Bd. IV, Berlin, 1866.
27. Calogero, F. and Degasperis, A.: Nonlinear evolution equations solvable by the inverse spectral transform. I and II, *Nuovo Cimento* **32B** (1976), 201–242 and **39B** (1977), 1–54.
Spectral Transform and Solitons: Tools to Solve and Investigate Nonlinear Evolution Equation, Vol. 1, North-Holland, 1982.
28. Calogero, F.: A class of C-integrable PDEs in multidimensions, *Inverse Problems* **10** (1994), 1231–1234.

Acta Applicandae Mathematicae **39:** 245–276, 1995.

Symmetry Reductions and Exact Solutions of Shallow Water Wave Equations

PETTER A. CLARKSON and ELIZABETH L. MANSFIELD
Department of Mathematics, University of Exeter, Exeter EX4 4QE, U.K.

(Received: 8 February 1995)

Abstract. In this paper we study symmetry reductions and exact solutions of the shallow water wave (SWW) equation

$$u_{xxxt} + \alpha u_x u_{xt} + \beta u_t u_{xx} - u_{xt} - u_{xx} = 0,$$

where α and β are arbitrary, nonzero, constants, which is derivable using the so-called Boussinesq approximation. Two special cases of this equation, or the equivalent nonlocal equation obtained by setting $u_x = U$, have been discussed in the literature. The case $\alpha = 2\beta$ was discussed by Ablowitz, Kaup, Newell and Segur (*Stud. Appl. Math.*, **53** (1974), 249), who showed that this case was solvable by inverse scattering through a second-order linear problem. This case and the case $\alpha = \beta$ were studied by Hirota and Satsuma (*J. Phys. Soc. Japan*, **40** (1976), 611) using Hirota's bi-linear technique. Further, the case $\alpha = \beta$ is solvable by inverse scattering through a third-order linear problem.

In this paper, a catalogue of symmetry reductions is obtained using the classical Lie method and the nonclassical method due to Bluman and Cole (*J. Math. Mech.*, **18** (1969), 1025). The classical Lie method yields symmetry reductions of (1) expressible in terms of the first, third and fifth Painlevé transcendents and Weierstrass elliptic functions. The nonclassical method yields a plethora of exact solutions of (1) with $\alpha = \beta$ which possess a rich variety of qualitative behaviours. These solutions all like a two-soliton solution for $t < 0$ but differ radically for $t > 0$ and may be viewed as a nonlinear superposition of two solitons, one travelling to the left with arbitrary speed and the other to the right with equal and opposite speed. These families of solutions have important implications with regard to the numerical analysis of SWW and suggests that solving (1) numerically could pose some fundamental difficulties. In particular, one would not be able to distinguish the solutions in an initial-value problem since an exponentially small change in the initial conditions can result in completely different qualitative behaviours.

We compare the two-soliton solutions obtained using the nonclassical method to those obtained using the singular manifold method and Hirota's bi-linear method.

Further, we show that there is an analogous nonlinear superposition of solutions for two (2+1)-dimensional generalisations of the SWW Equation (1) with $\alpha = \beta$. This yields solutions expressible as the sum of two solutions of the Korteweg–de Vries equation.

Mathematics Subject Classification (1991): 58F07.

Key words: symmetry reductions, exact solutions, shallow water wave equation.

1. Introduction

In this paper we discuss symmetry reductions and exact solutions for the shallow water wave (SWW) equation

$$\Delta \equiv u_{xxxt} + \alpha u_x u_{xt} + \beta u_t u_{xx} - u_{xt} - u_{xx} = 0, \tag{1.1}$$

where α and β are arbitrary, nonzero, constants, which can be derived from the classical shallow water theory in the Boussinesq approximation (cf. [26]). This equation may be written in the nonlocal form (set $u_x = U$)

$$U_{xxt} + \alpha U U_t - \beta U_x \partial_x^{-1} U_t - U_t - U_x = 0, \tag{1.2}$$

where

$$\left(\partial_x^{-1} f\right)(x) = \int_x^\infty f(y)\,\mathrm{d}y;$$

in fact this is the form of the physical model derived in [26].

Two special cases of (1.2) have attracted some attention in the literature, namely $\alpha = 2\beta$ and $\alpha = \beta$. In their seminal paper on soliton theory, Ablowitz, Kaup, Newell and Segur [2] show that

$$U_{xxt} + 2\beta U U_t - \beta U_x \partial_x^{-1} U_t - U_t - U_x = 0, \tag{1.3}$$

which is (1.2) with $\alpha = 2\beta$ is solvable by inverse scattering. Further, Ablowitz *et al.* [2] remark that (1.3) reduces in the long wave, small amplitude limit to the celebrated Korteweg–de Vries (KdV) equation

$$u_t + u_{xxx} + 6uu_x = 0, \tag{1.4}$$

which was the first equation to be solved by inverse scattering [27], and they also comment that (1.3) also has the desirable properties of the regularized long wave (RLW) equation [8, 47]

$$u_{xxt} + uu_x - u_t - u_x = 0, \tag{1.5}$$

sometimes known as the Benjamin–Bona–Mahoney equation, in that it responds feebly to short waves. Additionally, we note that (1.5) and (1.3) have the same linear dispersion relation $\omega(k) = -k/(1+k^2)$ for the complex exponential $u(x,t) \sim \exp\{\mathrm{i}[kx + \omega(k)t]\}$. However, in contrast to (1.3), the RLW Equation (1.5) is thought *not* to be solvable by inverse scattering (cf. [42]). Hirota and Satsuma [34] studied both (1.3) and

$$U_{xxt} + \beta U U_t - \beta U_x \partial_x^{-1} U_t - U_t - U_x = 0, \tag{1.6}$$

i.e. (1.2) with $\alpha = 2\beta$, using Hirota's bi-linear method [32] and obtained N-soliton solutions for both equations (see also [43]).

In the sequel, we shall refer to (1.1) with $\alpha = 2\beta$

$$u_{xxxt} + 2\beta u_x u_{xt} + \beta u_t u_{xx} - u_{xt} - u_{xx} = 0, \tag{1.7}$$

as the SWW–AKNS equation and (1.1) with $\alpha = \beta$

$$u_{xxxt} + \beta u_x u_{xt} + \beta u_t u_{xx} - u_{xt} - u_{xx} = 0, \tag{1.8}$$

as the SWW–HS equation.

The scattering problem for the SWW–AKNS Equation (1.7) is the second-order problem [2]

$$\psi_{xx} + \tfrac{1}{2}\beta u_x \psi = \lambda\psi, \tag{1.9}$$

with associated time-dependence

$$(4\lambda - 1)\psi_t = (1 - \beta u_t)\psi_x + \tfrac{1}{2}\beta u_{xt}\psi, \tag{1.10}$$

where λ is the constant eigenvalue, and $\psi_{xxt} = \psi_{txx}$ if and only if u satisfies (1.7). We note that (1.9) is the time-independent Schrödinger equation which is also the scattering problem for the KdV Equation (1.4) [27]. In contrast, the scattering problem for the SWW–HS Equation (1.8) is the third-order problem [23]

$$\psi_{xxx} + \left(\tfrac{1}{2}\beta u_x - 1\right)\psi_x = \lambda\psi, \tag{1.11}$$

with associated time-dependence

$$3\lambda\psi_t = (1 - \beta u_t)\psi_{xx} + \beta u_{xt}\psi_x. \tag{1.12}$$

We remark that (1.11) is similar to the scattering problem

$$\psi_{xxx} + \tfrac{1}{4}(1 + 6u)\psi_x + \tfrac{3}{4}\left[u_x - \mathrm{i}\sqrt{3}\,\partial_x^{-1}(v_t)\right]\psi = \lambda\psi \tag{1.13}$$

which is the scattering problem for the Boussinesq equation

$$u_{xxxx} + 3(u^2)_{xx} + u_{xx} = u_{tt}, \tag{1.14}$$

and which has been comprehensively studied by Deift, Tomei and Trubowitz [24].

The SWW Equation (1.1) was also discussed by Hietarinta [31] who shows that it can be expressed in Hirota's bi-linear form [32] if and only if either (i), $\alpha = \beta$, when it reduces to (1.8), or (ii), $\alpha = 2\beta$, when it reduces to (1.7). Further, in [19] it is shown that the SWW Equation (1.1) satisfies the necessary conditions of the Painlevé tests due to Ablowitz, Ramani and Segur [3, 4] and Weiss, Tabor and Carnevale [55] to be completely integrable if and only if either $\alpha = \beta$ or $\alpha = 2\beta$. These results strongly suggest that the SWW Equation (1.1) is completely integrable if and only if it has one of the two special forms (1.7) or (1.8), which are both known to be solvable by inverse scattering.

The classical method for finding symmetry reductions of partial differential equations (PDEs) is the Lie group method of infinitesimal transformations (cf. [10, 44]). Though this method is entirely algorithmic, it often involves a large amount of tedious algebra and auxiliary calculations which can become virtually unmanageable if attempted manually, and so symbolic manipulation programs have been developed, for example in MACSYMA, MAPLE, MATHEMATICA,

MUMATH and REDUCE, to facilitate the calculations. An excellent survey of the different packages presently available and a discussion of their strengths and applications is given by Hereman [30].

In recent years the nonclassical method due Bluman and Cole [9] (in the sequel referred to as the *nonclassical method*) and the direct method of Clarkson and Kruskal [17] have been used to generate many new symmetry reductions and exact solutions for several physically significant PDEs that are *not* obtainable using the classical Lie method, which represents important progress (see [16, 29] and references therein).

Symmetry groups and associated reductions and exact solutions have several different important applications in the context of differential equations (see, for example, [10, 16, 44] for further details and references):

- *Derive new solutions from old solutions.* Applying the symmetry group of a differential equation to a known solution yields a family of new solutions. Quite often interesting solutions can be obtained from trivial ones.
- *Integration of ODEs.* Symmetry groups of ODEs can be used to reduce the order of the equation, for example to reduce a second-order equation to first order.
- *Reductions of PDEs.* Symmetry groups of PDEs are used to reduce the total number of dependent and independent variables, for example from a PDE with two independent and one dependent variables to an ODE.
- *Linearisation of PDEs.* Symmetries groups can be used to discover whether or not a PDE can be linearised and to construct an explicit linearisation when one exists.
- *Classification of equations.* Symmetry groups can be used to classify differential equations into equivalence classes and to choose simple representatives of such classes.
- *Asymptotics of solutions of PDEs.* It is known that as solutions of PDEs asymptotically tend to solutions of lower-dimensional equations obtained by symmetry reduction, some of these special solutions will illustrate important physical phenomena. In particular, exact solutions arising from symmetry methods can often be effectively used to study properties such as asymptotics and 'blow-up'.
- *Numerical methods and testing computer coding.* Symmetry groups and exact solutions of physically relevant PDEs are used in the design, testing and evaluation of numerical algorithms; these solutions provide an important practical check on the accuracy and reliability of such integrators.
- *Conservation laws.* The application of symmetries to conservation laws dates back to the work of Noether who proved the remarkable result that for systems arising from a variational principle, every conservation law of the system comes from a corresponding symmetry property.

- *Further applications.* There are several other important applications of symmetry groups including bifurcation theory, control theory, special function theory, boundary value problems and free boundary problems.

The method used to find solutions of the determining equations for the infinitesimals in both the classical and nonclassical case is that of Differential Gröbner Bases (DGBs), defined to be a basis $\mathcal{B}$ of the differential ideal generated by the system such that every member of the ideal pseudo-reduces to zero with respect to $\mathcal{B}$. This method provides a systematic framework for finding integrability and compatibility conditions of an overdetermined system of PDEs. It avoids the problems of infinite loops in reduction processes, and yields, as far as is currently possible, a 'triangulation' of the system from which the solution setcan be derived more easily (cf. [18, 41, 48, 49]). In a sense, a Differential Gröbner Bases provides the maximum amount of information possible using elementary differential and algebraic processes in a finite time.

The triangulations of the systems of determining equations for infinitesimals arising in the classical and nonclassical methods in this article were all performed using the MAPLE package DIFFGROB2 [40]. This package was written specifically to handle fully nonlinear equations of polynomial type; packages such as those in [50, 51, 53] have been developed for the study of linear equations. All calculations are strictly 'polynomial', that is, there is no division. Implemented there are the Kolchin–Ritt algorithm, the differential analogue of Buchberger's algorithm [14] using pseudo-reduction instead of reduction, and extra algorithms needed to calculate a Differential Gröbner Bases (as far as possible using the current theory), for those cases where the Kolchin–Ritt algorithm is not sufficient [41]. Designed to be used interactively as well as algorithmically, the package has proved useful for solving some fully nonlinear systems. As yet, however, algorithmic methods for finding the most efficient orderings, the best method of choosing the sequence of pairs to be cross-differentiated, for deciding when to integrate and read off coefficients of independent functions in one of the variables, for finding the best change of coordinates, and so on, are still the subject of much investigation.

In Section 2 we find the classical Lie group of symmetries and associated reductions of (1.1), which are expressible in terms of the first, third and fifth Painlevé transcendents and Weierstrass elliptic functions. Then in Section 3 we discuss the nonclassical symmetries and reductions of (1.1). In particular, the nonclassical symmetry reductions obtained for (1.8) generate a wide variety of interesting exact analytical solutions of the equations which we plot using MAPLE. In Section 4 we compare the two-soliton solutions generated in Section 3 with those obtained using the singular manifold method and Hirota's bi-linear method. In Section 5 we show that there are analogous symmetry reductions and exact solutions also occurs for two $(2+1)$-dimensional generalisations of the SWW Equation (1) with $\alpha = \beta$. We discuss our results in Section 6.

2. Classical Symmetry Reductions of the SWW Equation

To apply the classical method to the SWW Equation (1.1) we consider the one-parameter Lie group of infinitesimal transformations in (x, t, u) given by

$$x = x + \varepsilon\xi(x,t,u) + \mathrm{O}(\varepsilon^2), \tag{2.1a}$$

$$t = t + \varepsilon\tau(x,t,u) + \mathrm{O}(\varepsilon^2), \tag{2.1b}$$

$$u = u + \varepsilon\phi(x,t,u) + \mathrm{O}(\varepsilon^2), \tag{2.1c}$$

where ε is the group parameter. Then one requires that this transformation leaves invariant the set

$$\mathcal{S}_\Delta \equiv \{u(x,t)\colon\ \Delta = 0\}, \tag{2.2}$$

of solutions of (1.1). This yields an overdetermined, linear system of equations for the infinitesimals $\xi(x,t,u)$, $\tau(x,t,u)$ and $\phi(x,t,u)$. The associated Lie algebra of infinitesimal symmetries is the set of vector fields of the form

$$\mathbf{v} = \xi(x,t,u)\partial_x + \tau(x,t,u)\partial_t + \phi(x,t,u)\partial_u. \tag{2.3}$$

Having determined the infinitesimals, the symmetry variables are found by solving the characteristic equations

$$\frac{\mathrm{d}x}{\xi(x,t,u)} = \frac{\mathrm{d}t}{\tau(x,t,u)} = \frac{\mathrm{d}u}{\phi(x,t,u)}, \tag{2.4}$$

which is equivalent to solving the invariant surface condition

$$\psi \equiv \xi(x,t,u)u_x + \tau(x,t,u)u_t - \phi(x,t,u) = 0. \tag{2.5}$$

The set $\mathcal{S}_\Delta$ is invariant under the transformation (2.1) provided that

$$\mathrm{pr}^{(4)}\mathbf{v}(\Delta)|_{\Delta=0} = 0, \tag{2.6}$$

where $\mathrm{pr}^{(4)}\mathbf{v}$ is the fourth prolongation of the vector field (2.3), which is given explicitly in terms of ξ, τ and ϕ (cf. [44]). This yields a system of fourteen determining equations, as calculated using the MACSYMA package SYMMGRP.MAX [15]. A triangulation or standard form [18, 48] of these determining equations is the following system of eight equations,

$$\xi_u = 0, \qquad \xi_t = 0, \qquad \xi_{xx} = 0, \qquad \tau_u = 0, \qquad \tau_x = 0,$$

$$\alpha\phi_x - 2\xi_x = 0, \qquad \beta\phi_t - \tau_t - \phi_x = 0, \qquad \phi_u + \xi_x = 0,$$

from which we easily obtain the following infinitesimals,

$$\xi = \kappa_1 x + \kappa_2, \qquad \tau = g(t),$$

$$\phi = -\kappa_1 \left(u - \frac{2x}{\alpha} - \frac{t}{\beta} \right) + \frac{g(t)}{\beta} + \kappa_3, \tag{2.7}$$

where $g(t)$ is an arbitrary differentiable function and κ_1, κ_2 and κ_3 are arbitrary constants. The associated Lie algebra is spanned by the vector fields

$$\mathbf{v}_1 = x\partial_x - \left(u - \frac{2x}{\alpha} - \frac{t}{\beta} \right) \partial_u, \qquad \mathbf{v}_2 = \partial_x,$$

$$\mathbf{v}_3 = \partial_u, \qquad \mathbf{v}_4(g) = g(t) \left(\partial_t + \beta^{-1}\partial_u \right).$$

Solving (2.4) with ξ, τ and ϕ given by (2.7), or equivalently solving (2.5), we obtain the following two canonical symmetry reductions.

Case 2.1. $\kappa_1 \neq 0$. In this case we set $g(t) = f(t)/\dot{f}(t)$, $\kappa_1 = 1$ and $\kappa_2 = \kappa_3 = 0$ (with $\dot{f} \equiv \mathrm{d}f/\mathrm{d}t$). Hence we obtain the symmetry reduction

$$u(x,t) = f(t)w(z) + \frac{x}{\alpha} + \frac{t}{\beta}, \quad z = xf(t), \tag{2.8}$$

where $w(z)$ satisfies

$$zw'''' + 4w''' + (\alpha+\beta)zw'w'' + \beta ww'' + 2\alpha \left(w'\right)^2 = 0, \tag{2.9}$$

where $' = \mathrm{d}/\mathrm{d}z$. It is straightforward to show using the algorithm of Ablowitz *et al.* [4] that this equation is of Painlevé-type, i.e., its solutions have no movable singularities other than poles, only if either (i) $\alpha = \beta$ or (ii) $\alpha = 2\beta$. These two special cases (2.9) are solvable in terms of solutions of the third Painlevé equation [36]

$$\frac{\mathrm{d}^2 y}{\mathrm{d}x^2} = \frac{1}{y}\left(\frac{\mathrm{d}y}{\mathrm{d}x}\right)^2 - \frac{1}{x}\frac{\mathrm{d}y}{\mathrm{d}x} + ay^3 + \frac{by^2+c}{x} + \frac{d}{y}, \tag{2.10}$$

and the fifth Painlevé equation,

$$\frac{\mathrm{d}^2 y}{\mathrm{d}x^2} = \left(\frac{1}{2y} + \frac{1}{y-1}\right)\left(\frac{\mathrm{d}y}{\mathrm{d}x}\right)^2 - \frac{1}{x}\frac{\mathrm{d}y}{\mathrm{d}x} + \frac{(y-1)^2}{x^2}\left(ay + \frac{b}{y}\right) + \frac{cy}{x} + \frac{dy(y+1)}{y-1}, \tag{2.11}$$

with *a*, *b*, *c* and *d* constants (see [19] for details). Hence, the Painlevé Conjecture [3, 4] predicts that a necessary condition for (1.1) to be completely integrable is that $(\alpha-\beta)(\alpha-2\beta) = 0$, i.e., only if (1.1) has one of the two special forms (1.8) or (1.7).

Case 2.2. $\kappa_1 = 0$. In this case we set $g(t) = 1/\dot{f}(t)$, $\kappa_2 = 1$ and $\kappa_3 = -1/\beta$ and obtain the symmetry reduction

$$u(x,t) = w(z) + t/\beta, \qquad z = x - f(t), \tag{2.12}$$

where $W(z) = w'(z)$ satisfies

$$(W')^2 + \tfrac{1}{3}(\alpha+\beta)W^3 = AW + B, \tag{2.13}$$

with A and B arbitrary constants. This equation is equivalent to the Weierstrass elliptic function equation

$$(\wp')^2 = 4\wp^3 - g_2\wp - g_3, \tag{2.14}$$

where g_2 and g_3 are arbitrary constants [56].

We remark that the vector field $\mathbf{v}_4(g)$ shows that (1.1) is invariant under the following variable coefficient transformation

$$\tilde{x} = x, \qquad \tilde{t} = g(t), \qquad \tilde{u} = u + [g(t) - t]/\beta, \tag{2.15}$$

i.e., if $u(x,t)$ is a solution of (1.1), then so is $\tilde{u}(\tilde{x},\tilde{t})$. Further, the associated Lie algebra $\mathcal{G} = \{\mathbf{v}_4(g)\}$, with $g(t) \in C^\infty(\mathbb{R})$, is a Virasoro algebra since

$$[\mathbf{v}_4(g_1), \mathbf{v}_4(g_2)] = \mathbf{v}_4(g_1\dot{g}_2 - g_2\dot{g}_1),$$

where $[A,B] = AB - BA$ is the standard commutator. Virasoro algebras frequently arise in the study of symmetry reductions for $(2+1)$-dimensional equations such as the Kadomtsev–Petviashvili and Davey–Stewartson equations (see, for example, [57] and the references therein), but not $(1+1)$-dimensional equations such as (1.1).

It might be thought that the presence of the arbitrary function $g(t)$ in the infinitesimals associated with the SWW Equation (1.1) is an artefact of the fact that we are considering (1.1) rather than (1.2), which is the physical equation. In view of the nonlocal term, we write the nonlocal Equation (1.2) as the system

$$U_{xxt} + \alpha UU_t + \beta VU_x - U_t - U_x = 0, \quad V_x = U_t, \tag{2.16}$$

where α and β are arbitrary, nonzero constants. To apply the classical method to (2.16), we consider the one-parameter Lie group of infinitesimal transformations in (x,y,U,V) given by

$$\begin{aligned}
\tilde{x} &= x + \varepsilon\xi_1(x,y,U,V) + O(\varepsilon^2),\\
\tilde{y} &= y + \varepsilon\xi_2(x,y,U,V) + O(\varepsilon^2),\\
\widetilde{U} &= U + \varepsilon\phi_1(x,y,U,V) + O(\varepsilon^2),\\
\widetilde{V} &= V + \varepsilon\phi_2(x,y,U,V) + O(\varepsilon^2),
\end{aligned}$$

where ε is the group parameter. Solving the resulting determining equations yields the infinitesimals

$$\xi = \kappa_1 x + \kappa_2, \qquad \tau = g(t), \qquad \phi_1 = -2\kappa_1(U - 1/\alpha),$$

$$\phi_2 = -\left(\kappa_1 + \frac{\mathrm{d}g}{\mathrm{d}t}\right)(V - 1/\beta),$$

where κ_1 and κ_2 are arbitrary constants and $g(t)$ is an arbitrary differentiable function and the associated vector fields are

$$\mathbf{w}_1 = x\partial_x - 2(U - 1/\alpha)\partial_U + (v - 1/\beta)\partial_V, \qquad \mathbf{w}_2 = \partial_x,$$

$$\mathbf{w}_3(g) = g(t)\partial_t + \frac{\mathrm{d}g}{\mathrm{d}t}(V - 1/\beta)\partial_V.$$

This means that the system (2.16) is invariant under the transformation

$$\tilde{x} = x, \qquad \tilde{t} = g(t), \qquad \tilde{U} = U, \qquad \tilde{V} = \frac{\beta V + \dot{g}(t) - t}{\beta \dot{g}(t)}.$$

Hence if $\{U(x,t), V(x,t)\}$ are solutions of (2.16), then so are $\{\tilde{U}(\tilde{x},\tilde{t}), \tilde{V}(\tilde{x},\tilde{t})\}$.

3. Nonclassical Symmetry Reductions of the SWW Equation

There have been several generalisations of the classical Lie group method for symmetry reductions. Bluman and Cole [9], in their study of symmetry reductions of the linear heat equation, proposed the so-called nonclassical method of group-invariant solutions. This method involves considerably more algebra and associated calculations than the classical Lie method. In fact, it has been suggested that for some PDEs, the calculation of these nonclassical reductions might be too difficult to do explicitly [45], especially if attempted manually since the associated determining equations are now an overdetermined, *nonlinear* system. For some equations such as the KdV Equation (1.4), the nonclassical method does not yield any additional symmetry reductions to those obtained using the classical Lie method, while there are PDEs which do possess symmetry reductions *not* obtainable using the classical Lie group method. It should be emphasised that the associated vector fields arising from the nonclassical method do not form a vector space, still less a Lie algebra, since the invariant surface condition (2.5) depends upon the particular reduction. Subsequently, these methods were further generalised by Olver and Rosenau [45, 46] to include 'weak symmetries' and, even more generally, 'side conditions' or 'differential constraints', and they concluded that "the unifying theme behind finding special solutions of PDEs is not, as is commonly supposed, group theory, but rather the more analytic subject of overdetermined systems of PDEs".

In the nonclassical method one requires only the subset of $\mathcal{S}_\Delta$ given by

$$\mathcal{S}_{\Delta,\psi} = \{u(x,t)\colon\ \Delta(u) = 0,\ \psi(u) = 0\}, \tag{3.1}$$

where $\mathcal{S}_\Delta$ is as defined in (2.2) and $\psi = 0$ is the invariant surface condition (2.5), is invariant under the transformation (2.1). The usual method of applying the nonclassical method (e.g., as described in [39]), to the SWW Equation (1.1) involves applying the prolongation $\mathrm{pr}^{(4)}\,\mathbf{v}$ to the system of equations given by

(1.1) and the invariant surface condition (2.5) and requiring that the resulting expressions vanish for $u \in \mathcal{S}_{\Delta,\psi}$, i.e.,

$$\mathrm{pr}^{(4)}\,\mathbf{v}(\Delta)\,|_{\Delta=0,\psi=0} = 0, \qquad \mathrm{pr}^{(1)}\,\mathbf{v}(\psi)|_{\Delta=0,\psi=0} = 0. \tag{3.2}$$

It is easily shown that

$$\mathrm{pr}^{(1)}\,\mathbf{v}(\psi) = -\left(\xi_u u_x + \tau_u u_t - \phi_u\right)\psi,$$

which vanishes identically when $\psi = 0$ without imposing any conditions upon ξ, τ and ϕ. However, as shown in [20], this procedure for applying the nonclassical method can create difficulties, particularly when implemented in symbolic manipulation programs. These difficulties often arise for equations such as (1.1) which require the use of differential consequences of the invariant surface condition (2.5). In [20] we proposed an algorithm for calculating the determining equations associated with the nonclassical method which avoids many of the difficulties commonly encountered; we use that algorithm here.

In the canonical case when $\tau \not\equiv 0$ we set $\tau = 1$. (We shall omit the special case $\tau \equiv 0$; in that case one obtains a single condition for ϕ with 424 summands, and even considering the subcase $\phi_u = 0$ leads to an equation more complex than the one we are studying.) Eliminating u_t, u_{xt} and u_{xxxt}, in (1.1) using the invariant surface condition (2.5) yields

$$\begin{aligned}\tilde{\Delta} \equiv\ & \phi_u u_{xxx} + 3\phi_{uu}u_x u_{xx} + 3\phi_{ux}u_{xx} + \phi_{uuu}u_x^3 + 3\phi_{xuu}u_x^2 + \\ & +3\phi_{xxu}u_x + \phi_{xxx} - \xi_{xxx}u_x - 3\xi_{xx}u_{xx} - 3\xi_x u_{xxx} - \xi_{uuu}u_x^4 - \\ & -3\xi_{xuu}u_x^3 - 6\xi_{uu}u_x^2 u_{xx} - 3\xi_{xxu}u_x^2 - 9\xi_{xu}u_x u_{xx} - \\ & -\xi_u\left(4u_x u_{xxx} + 3u_{xx}^2\right) - \xi u_{xxxx} + \\ & +(\alpha u_x - 1)\left[\phi_x + \phi_u u_x - \xi_x u_x - \xi_u u_x^2 - \xi u_{xx}\right] + \\ & +u_{xx}\left[\beta(\phi - \xi u_x) - 1\right] = 0,\end{aligned} \tag{3.3}$$

with t a parameter, which involves the infinitesimals ξ and ϕ that are to be determined. Now we apply the classical Lie algorithm to this equation using the fourth prolongation $\mathrm{pr}^{(4)}\,\mathbf{v}$ and eliminate u_{xxxx} using (3.3). This yields the following overdetermined, nonlinear system of equations for ξ and ϕ (contrast the classical case discussed in Section 2 above where they are linear).

$$\xi_u = 0, \tag{3.4i}$$

$$\phi_{uu} = 0, \tag{3.4ii}$$

$$(\alpha + \beta)(\phi_u + \xi_x) = 0, \tag{3.4iii}$$

$$\xi\phi_{tu} + 3\xi^2\xi_{xx} + 3\xi_t\xi_x - 3\xi\xi_{xt} - 3\xi^2\phi_{xu} - \xi_t\phi_u = 0, \tag{3.4iv}$$

$$\alpha\xi\phi_u^2 - \alpha\xi_t\phi_u + \alpha\xi\phi_{tu} - 2\beta\xi^2\phi_{xu} - \alpha\xi^2\phi_{xu} + \\ + \beta\xi^2\xi_{xx} - \alpha\xi\xi_x^2 + \alpha\xi_t\xi_x - \alpha\xi\xi_{xt} = 0, \tag{3.4v}$$

$$3\xi\phi_{xu}\phi_{xx} + \beta\xi\phi\phi_{xx} - \xi_t\phi_{xxx} - 3\xi\xi_{xx}\phi_{xx} - \xi\phi_{xx} + \\ + \alpha\xi\phi_x^2 + 3\xi\phi_{xxu}\phi_x - \xi\xi_{xxx}\phi_x - 2\xi\xi_x\phi_x + \\ + \xi_t\phi_x + \xi\phi\phi_{xxxu} - \xi\phi\phi_{xu} + \xi\phi_{xxxt} - \xi\phi_{xt} = 0, \tag{3.4vi}$$

$$3\xi\phi_u\phi_{xu} - 6\xi\xi_x\phi_{xu} - 3\xi_t\phi_{xu} - \alpha\xi^2\phi_x - 3\xi^2\phi_{xxu} + \beta\xi\phi\phi_u - \\ -3\xi\xi_{xx}\phi_u + 3\xi\phi_{xtu} + \beta\xi\phi_t + \beta\xi\xi_x\phi - \beta\xi_t\phi + \xi^2\xi_{xxx} + \\ + 6\xi\xi_x\xi_{xx} + 3\xi_t\xi_{xx} + 2\xi^2\xi_x - \xi\xi_x - 3\xi\xi_{xxt} + \xi_t = 0, \tag{3.4vii}$$

$$9\xi\xi_{xx}\phi_{xu} + \beta\xi^2\phi_{xx} + \alpha\xi_t\phi_x + 3\xi\xi_x\phi_{xxu} - 2\alpha\xi\phi_u\phi_x - 6\xi\phi_{xu}^2 - \\ -\alpha\xi\phi_{xt} - \xi\xi_x\xi_{xxx} + 2\xi\phi_{xu} + \xi\xi_{xxx}\phi_u - 3\xi\phi_u\phi_{xxu} - 3\xi\xi_{xx}^2 - \\ -2\beta\xi\phi\phi_{xu} - 3\xi\phi_{xxtu} + 2\xi\xi_x\phi_u - 2\xi\xi_x^2 - \xi^2\phi_{xu} - \xi_t\phi_u + \\ + \xi\phi_{tu} + \xi_t\xi_x - \xi\xi_{xt} + \beta\xi\xi_{xx}\phi + \xi^2\phi_{xxxu} - \\ -\alpha\xi\phi\phi_{xu} - \xi_t\xi_{xxx} - \xi\xi_{xx} + \xi\xi_{xxxt} + 3\xi_t\phi_{xxu} = 0. \tag{3.4viii}$$

These equations were calculated using the MACSYMA package SYMMGRP.MAX [15]. We then used the method of DGBs as outlined in [18, 20] to solve this system.

Case 3.1. $\alpha + \beta \neq 0$. In this case it is straightforward to obtain from (3.4i–v), the condition

$$\xi_{xx}\xi^2(\alpha + \beta)(3\beta - 2\alpha) = 0.$$

The case $3\beta - 2\alpha = 0$ leads to no solutions different from those obtained using the classical method.

Subcase 3.1.1. $\xi_x \neq 0$. This is the generic case which has the solution

$$\xi = (\kappa_1 x + \kappa_2) f(t), \qquad \phi = -\kappa_1 f(t)\left(u - \frac{2x}{\alpha} + \frac{\kappa_2 - t}{\beta}\right) + \frac{1}{\beta},$$

where $f(t)$ is an arbitrary differentiable function and $\kappa_1 \neq 0$ and κ_2 are arbitrary constants. These are equivalent to the infinitesimals (2.7) obtained using the classical method.

Subcase 3.1.2. $\xi_x = 0$. In this case it is easy to obtain the condition

$$\phi_{xx}\xi^3(\beta - \alpha)(\alpha + \beta) = 0.$$

There are two subcases to consider.

(i) $\alpha \neq \beta$, $\xi_x = 0$. In this case the solution is

$$\xi = f(t), \qquad \phi = \kappa_3 f(t) + 1/\beta,$$

where $f(t)$ is an arbitrary differentiable function and κ_3 is an arbitrary constant, which is equivalent to the infinitesimals (2.7) obtained using the classical method in the case when $\kappa_1 = 0$.

(ii) $\alpha = \beta$, $\xi_x = 0$. In this case, we obtain the following Differential Gröbner Bases for ξ, ϕ

$$\xi_u = 0, \qquad \xi_x = 0, \qquad \phi_u = 0,$$

$$\xi\phi_{xxxx} - (\xi + 1)\phi_{xx} + \beta\phi\phi_{xx} + \beta\phi_x^2 = 0,$$

$$\beta\xi^2\phi_x - \beta\xi\phi_t + \beta\xi_t\phi - \xi_t = 0.$$

Solving these yields

$$\xi = \frac{\mathrm{d}f}{\mathrm{d}t}, \qquad \phi = 2V(\zeta)\frac{\mathrm{d}f}{\mathrm{d}t} + \frac{1}{\beta}, \tag{3.5}$$

where $\zeta = x + f(t)$, $f(t)$ is an arbitrary differentiable function and $V(\zeta)$ satisfies

$$V_{\zeta\zeta} + \beta V^2 - V = \kappa_4\zeta + \kappa_5, \tag{3.6}$$

with κ_4 and κ_5 arbitrary constants. If $\kappa_4 \neq 0$, then this equation is equivalent to the first Painlevé equation [36]

$$\frac{\mathrm{d}^2 y}{\mathrm{d}x^2} = 6y^2 + x, \tag{3.7}$$

otherwise it is equivalent to the Weierstrass elliptic function Equation (2.14).

Thus solving the characteristic Equations (2.4) yields the nonclassical reduction

$$u(x,t) = v(\zeta) + w(z) + t/\beta, \qquad \zeta = x + f(t), \quad z = x - f(t), \tag{3.8}$$

where $f(t)$ is an arbitrary function and $V(\zeta) = v_\zeta$ and $W(z) = w_z$ satisfy

$$V_{\zeta\zeta} + \beta V^2 - V = -\lambda\zeta + \mu_1, \tag{3.9a}$$

and

$$W_{zz} + \beta W^2 - W = \lambda z + \mu_2, \tag{3.9b}$$

respectively, where μ_1 and μ_2 are arbitrary constants and λ is (effectively) a 'separation' constant. If $\lambda \neq 0$ then these equations are equivalent to the first Painlevé Equation (3.7), whilst if $\lambda = 0$ then they are equivalent to the Weierstrass elliptic function Equation (2.14).

In particular, if $\lambda = 0$ (we set $\beta = 1$ without loss of generality) then Equations (3.9) possess the special solutions

$$V(\zeta) = \left\{6\kappa_1^2 \operatorname{sech}^2(\kappa_1\zeta) + \tfrac{1}{2} - 2\kappa_1^2\right\},$$

$$W(z) = \left\{6\kappa_2^2 \operatorname{sech}^2(\kappa_2 z) + \tfrac{1}{2} - 2\kappa_2^2\right\},$$

where

$$\kappa_1 = \tfrac{1}{2}(1 + 4\mu_1)^{1/4} \quad \text{and} \quad \kappa_2 = \tfrac{1}{2}(1 + 4\mu_2)^{1/4}.$$

Hence, we obtain the exact solution of (1.8) given by

$$\begin{aligned} u(x,t) &= \Big\{6\kappa_1 \tanh\{\kappa_1[x + f(t)]\} + 6\kappa_2 \tanh\{\kappa_2[x - f(t)]\} + \\ &\quad + x(1 - 2\kappa_1^2 - 2\kappa_2^2) + 2f(t)(\kappa_2^2 - \kappa_1^2) + t\Big\}, \end{aligned} \tag{3.10}$$

where $f(t)$ is an arbitrary differentiable function.

If $\mu_1 = \mu_2 = 0$ then $\kappa_1 = \kappa_2 = \frac{1}{2}$ (with $\beta = 1$) and (3.10) simplifies to

$$u(x,t) = \left\{3\tanh\left\{\tfrac{1}{2}[x + f(t)]\right\} + 3\tanh\left\{\tfrac{1}{2}[x - f(t)]\right\} + t\right\}. \tag{3.11}$$

In Figure 1 we plot u_x with u given by (3.11) for various choices of the arbitrary function $f(t)$. This is one of the simplest, nontrivial family of solutions of (1.1) with $\alpha = \beta (= 1)$, using this reduction. In Figure 1, $f(t)$ is chosen so that $f(t) \sim t + t_0$, as $t \to -\infty$, where t_0 is a constant. Consequently all the solutions plotted in Figure 1 have a similar asymptotic behaviour as $t \to -\infty$. However, the asymptotic behaviours as $t \to \infty$ are radically different.

In the special case when $f(t) = ct$, then choosing

$$\kappa_1 = \tfrac{1}{2}(1 + 1/c)^{1/2} \quad \text{and} \quad \kappa_2 = \tfrac{1}{2}(1 - 1/c)^{1/2}$$

in (3.10) yields the two-soliton solution for (1.8) given by

$$\begin{aligned} u(x,t) &= \frac{3}{\sqrt{c}}\left\{\sqrt{c+1}\tanh\left[\tfrac{1}{2}\sqrt{(1+1/c)}\,(x+ct)\right] + \right. \\ &\quad \left. + \sqrt{c-1}\tanh\left[\tfrac{1}{2}\sqrt{(1-1/c)}\,(x-ct)\right]\right\}. \end{aligned} \tag{3.12}$$

This solution is of special interest since such two-soliton solutions are normally associated with so-called Lie–Bäcklund transformations (cf. [7]), whereas (3.12) has arisen from a Lie point symmetry, albeit nonclassical. Plots of (3.12) and

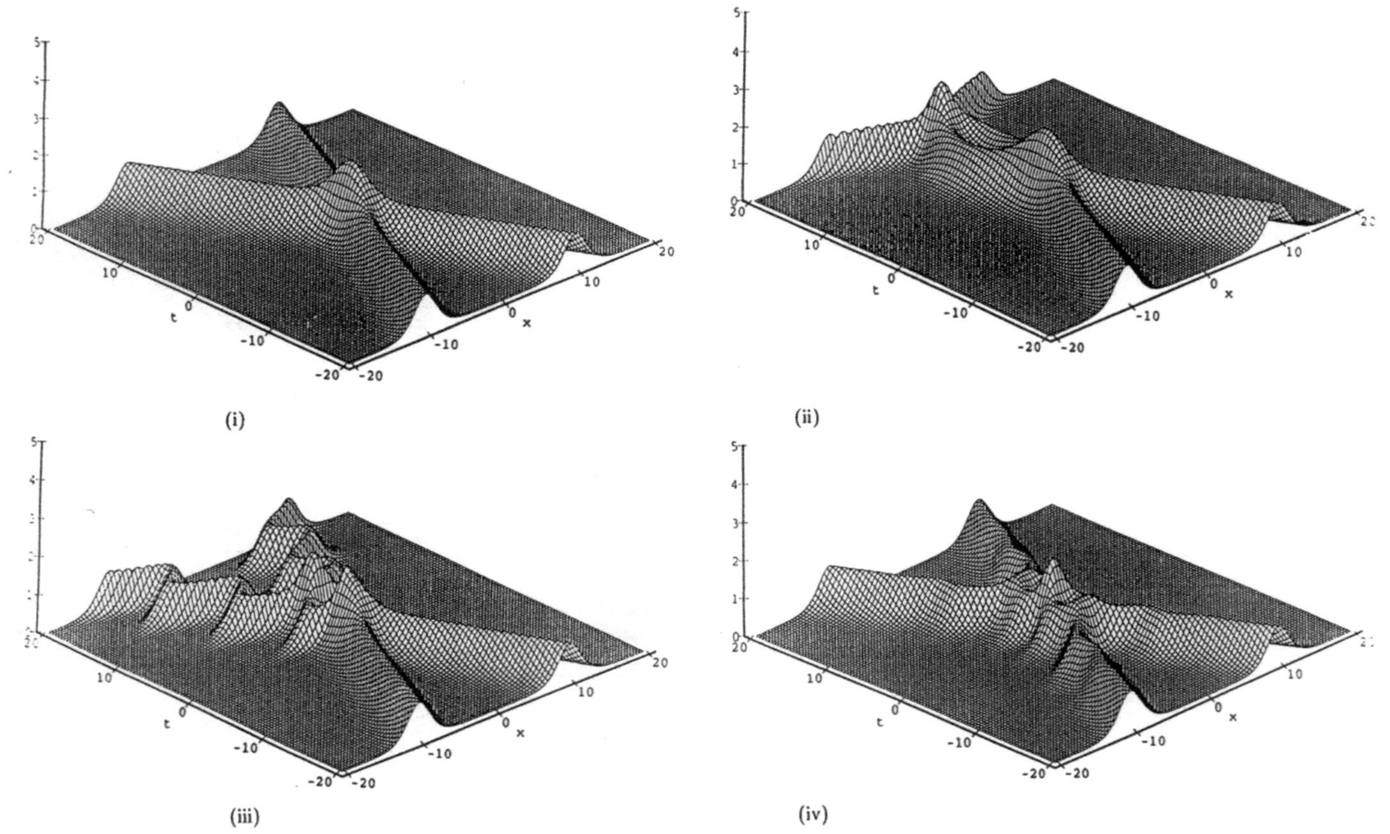

Fig. 1. The solution (3.11) where (i) $f(t) = \frac{1}{2}t$, (ii) $f(t) = \frac{1}{2}t - \exp\left(\frac{1}{5}t - 1\right)$, (iii) $f(t) = \frac{1}{2}t + [1 - \tanh t]\sin t$, (iv) $\frac{1}{2}t - \exp\left(-t^2/100\right)\sin t$.

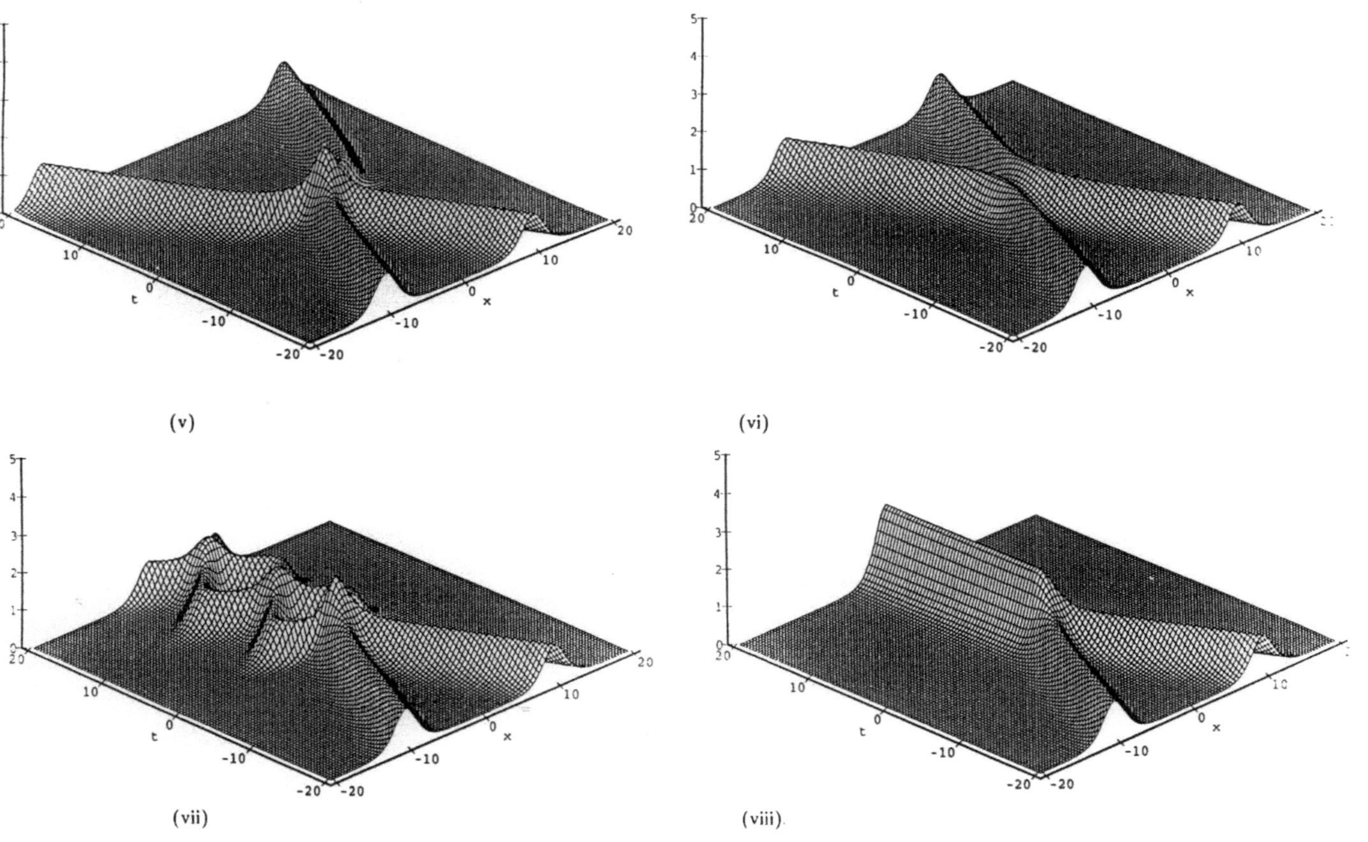

Fig. 1. (Continued). (v) $f(t) = \frac{1}{2}t + 6\pi^{-1}\tan^{-1}t$, (vi) $f(t) = \frac{1}{2}\sqrt{t^2+20}$, (vii) $f(t) = f(t) = \frac{1}{4}t(1-\tanh t) + \frac{1}{2}[2\sin(\frac{2}{3}t) + 3](1+\tanh t)$, (viii) $f(t) = \frac{1}{4}t(1-\tanh t)$.

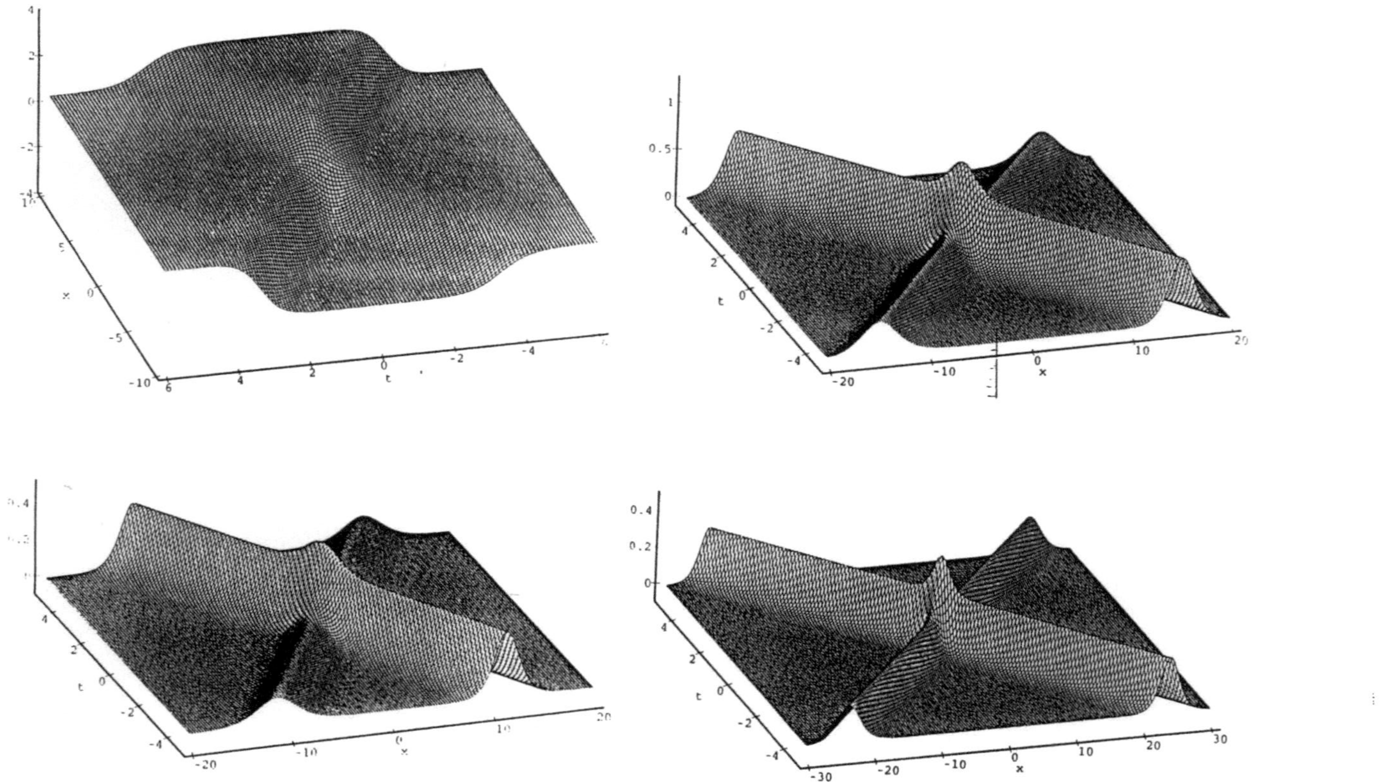

Fig. 2. (a) The solution (3.12) of the SWW–HS Equation (1.8) and its x-derivative for $c = 3$; (b) The x-derivative of solution (4.6) of the SWW–HS Equation (1.8) for $\kappa_1 = 2$, $\kappa_2 = 1.7$ and $\kappa_1 = \frac{3}{4}$, $\kappa_2 = \frac{2}{3}$.

its x-derivative are given in Figure 2(a) for $c = 3$. These should be compared with the plots in Figure 2(b) of the x-derivative of solution (4.6) below for the SWW–HS equation with $\kappa_1 = 2$, $\kappa_2 = 1.7$ and $\kappa_1 = \frac{3}{4}$, $\kappa_2 = \frac{2}{3}$ (so that $A_{12} = 1$ in both cases).

We stress that the 'decoupling' of the nonclassical reduction (3.8) into a function of $\zeta = x + f(t)$ and a function of $z = x - f(t)$ occurs for the SWW Equation (1.1) *only* in this special case when $\alpha = \beta$.

Case 3.2. $\alpha + \beta = 0$. This case also leads to no solutions different from those obtained using the classical method in Section 2.

To conclude this section we briefly consider the equations

$$u_{xxxx} + \alpha u_x u_{xt} + \beta u_t u_{xx} - u_{xt} - u_{xx} = 0, \tag{3.13}$$

$$u_{xxtt} + \alpha u_x u_{xt} + \beta u_t u_{xx} - u_{xt} - u_{xx} = 0, \tag{3.14}$$

where α and β are arbitrary, nonzero, constants, which are variants of the SWW Equation (1.1). Both these equations are thought to be *nonintegrable*, i.e. not solvable by inverse scattering, for all values of α and β since it is straightforward to show that neither satisfies the Painlevé PDE test due to Weiss *et al.* [55].

Applying the classical Lie method to Equations (3.13) and (3.14) yields the infinitesimals

$$\xi = \kappa_1 x + \kappa_2, \qquad \tau = (3\kappa_1 + \kappa_3)t + \kappa_4,$$

$$\phi = (\kappa_1 + \kappa_3)u - \frac{\kappa_3 x}{\alpha} + \frac{2\kappa_1 t}{\beta} + \kappa_5,$$

$$\xi = \kappa_1 x + \kappa_2, \qquad \tau = -(\kappa_1 + \alpha\kappa_3)t + \kappa_4,$$

$$\phi = (\kappa_1 + \alpha\kappa_3)u - \kappa_3 x - 2(\kappa_1 + \alpha\kappa_3)t/\beta,$$

respectively, where $\kappa_1, \kappa_2, \ldots, \kappa_5$ are arbitrary constants. In contrast to the SWW Equation (1.1), these symmetry groups are finite-dimensional.

If $\alpha = \beta$, then Equations (3.13) and (3.14) both possess nonclassical reductions of the form

$$u(x,t) = v(\zeta) + w(z), \qquad \zeta = x + ct, \quad z = x - ct,$$

where for (3.13), $V(\zeta) = v_\zeta$ and $W(z) = w_z$ satisfy

$$V_{\zeta\zeta} - (1 + c)V + \beta c V^2 = -\lambda\zeta + \lambda_1, \tag{3.15i}$$

$$W_{zz} + (c - 1)W - \beta c W^2 = \lambda z + \lambda_2, \tag{3.15ii}$$

and for (3.14), $V(\zeta) = v_\zeta$ and $W(z) = w_z$ satisfy

$$c^2 V_{\zeta\zeta} - (1 + c)V + \beta c V^2 = -\lambda\zeta + \lambda_1, \tag{3.16i}$$

$$c^2W_{zz} + (c-1)W - \beta cW^2 = \lambda z + \lambda_2, \tag{3.16ii}$$

with λ, λ_1, and λ_2 arbitrary constants. If $\lambda \neq 0$, then Equations (3.15), (3.16) are solvable in terms of the first Painlevé Equation (3.7), whilst if $\lambda = 0$ then they are equivalent to the Weierstrass elliptic function Equation (2.14).

Setting $\lambda = \lambda_1 = \lambda_2 = 0$, and solving Equations (3.15) and (3.16) yields the exact solutions of (3.13) and (3.14) with $\alpha = \beta$ given by

$$\begin{aligned} u(x,t) &= \frac{3}{\beta c}\left\{\sqrt{1+c}\tanh\left[\tfrac{1}{2}\sqrt{1+c}\,(x+ct)\right] - \right. \\ &\qquad \left. -\sqrt{1-c}\tanh\left[\tfrac{1}{2}\sqrt{1-c}\,(x-ct)\right]\right\}, \end{aligned} \tag{3.17}$$

$$\begin{aligned} u(x,t) &= \frac{3}{\beta}\left\{\sqrt{1+c}\tanh\left[\frac{\sqrt{1+c}}{2c}(x+ct)\right] - \right. \\ &\qquad \left. -\sqrt{1-c}\tanh\left[\frac{\sqrt{1-c}}{2c}(x-ct)\right]\right\}, \end{aligned} \tag{3.18}$$

respectively, which are analogues of (3.12). Plots of the solutions (3.17) and (3.18) and their derivatives with respect to x are given in Figures 3 and 4, respectively. The solutions (3.17) and (3.18) are plotted in Figures 3(i) and 4(i) and look like the elastic interaction of a 'kink' and an 'anti-kink' solution. The x-derivatives of (3.17) and (3.18) are plotted in Figures 3(ii) and 4(ii) and look like the elastic interaction of two 'soliton' solutions. These plots are very similar to the plots of (3.12) and its x-derivative given in Figure 2, however whereas (1.8) is completely integrable, both (3.13) and (3.14) are seemingly non-integrable. These solutions (3.17) and (3.18) are of particular interest since such solutions are normally associated with integrable equations, whereas they arise here for nonintegrable equations. Furthermore, like (3.12), these 'two-soliton' solutions have also arisen from nonclassical reductions. As mentioned above, such solutions are normally associated with Lie–Bäcklund transformations (cf. [7]).

4. Soliton Solutions of the SWW–HS and SWW–AKNS Equations

4.1. The SWW–HS Equation

Exact solutions of the SWW–HS Equation (1.8) can be obtained using the so-called singularity manifold method which uses truncated Painlevé expansions [54, 55]. If we seek a solution of (1.8) in the form

$$u(x,t) = \frac{6}{\beta}\frac{\phi_x(x,t)}{\phi(x,t)}, \tag{4.1}$$

and then equate coefficients of powers of ϕ to zero, we find that $\phi(x,t)$ satisfies the overdetermined system

$$\phi_{xxxt} - \phi_{xx} - \phi_{xt} = 0, \tag{4.2a}$$

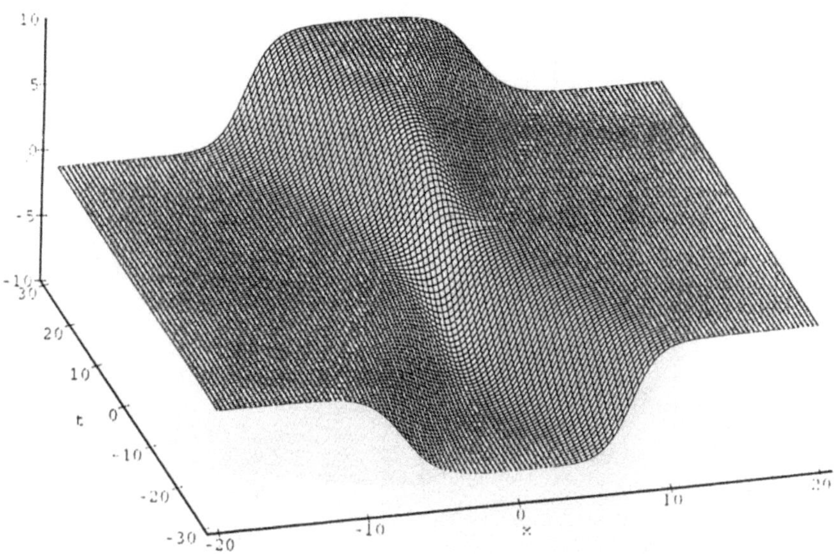

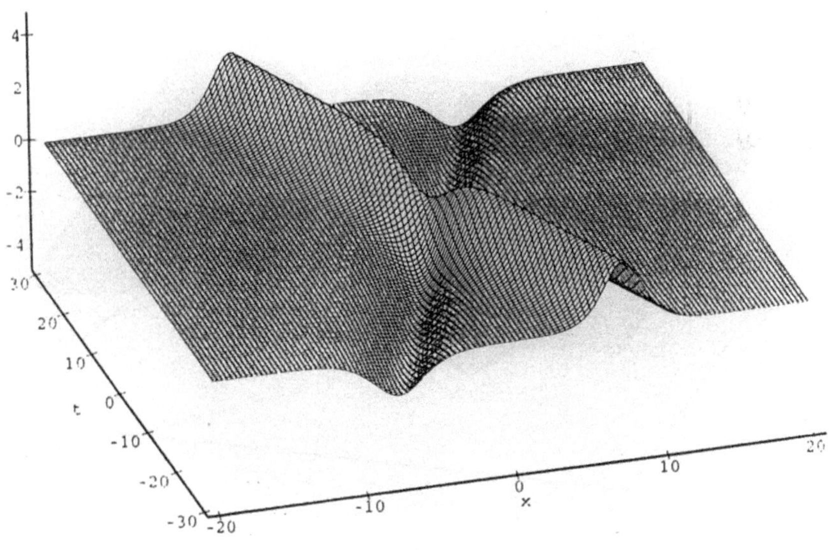

Fig. 3. The solution (3.17) of Equations (3.13) and its x-derivative for $c = \frac{1}{4}$.

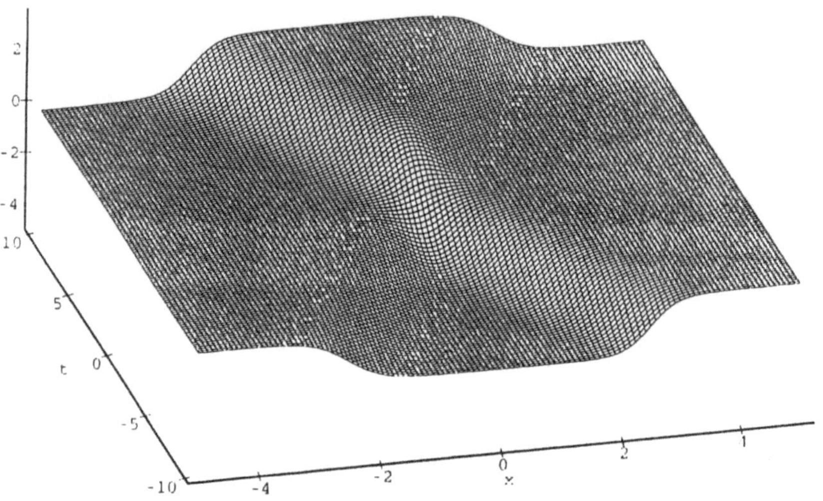

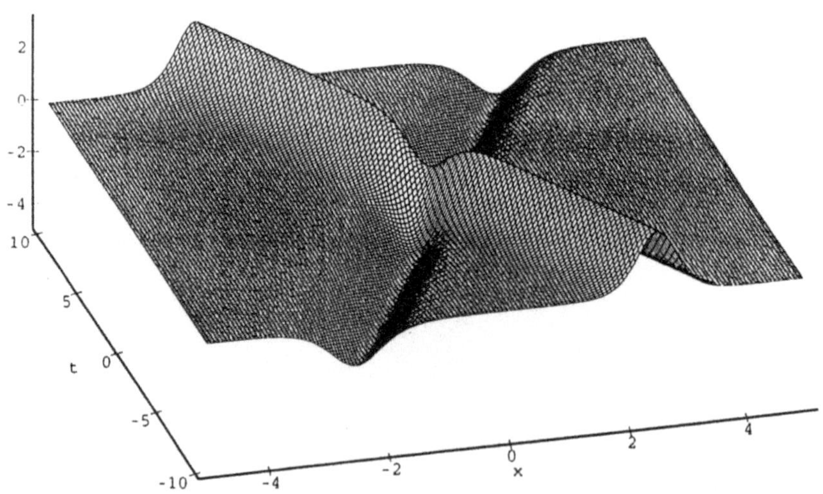

Fig. 4. The solution (3.18) of Equations (3.14) and its x-derivative for $c = \frac{1}{4}$.

$$\phi_t\phi_{xxx} - 3\phi_{xt}\phi_{xx} + 3\phi_x\phi_{xxt} - \phi_x(\phi_x + \phi_t) = 0. \tag{4.2b}$$

(A DGB analysis of this system leads to some very complex expressions. Although it does yield some ODEs in x for ϕ in the various subcases they appear difficult to solve.) Now suppose we seek a solution of these equations in the form

$$\phi(x,t) = \alpha_1 \exp\{\kappa_1 x + \mu_1 t\} + \alpha_2 \exp\{\kappa_2 x + \mu_2 t\} + \alpha_0, \tag{4.3}$$

where α_0, α_1, α_2, κ_1, κ_2, μ_1 and μ_2 are constants such that $\alpha_0\alpha_1\alpha_2 \neq 0$. It is straightforward to show that Equations (4.2) have a solution of the form (4.3) provided that $\mu_1 = \kappa_1/(\kappa_1^2 - 1)$, $\mu_2 = \kappa_2/(\kappa_2^2 - 1)$ and κ_1 and κ_2 satisfy the constraint

$$\kappa_1^2 - \kappa_1\kappa_2 + \kappa_2^2 = 3. \tag{4.4}$$

Thus we obtain the following exact solution of the SWW–HS Equation (1.8) given by

$$u(x,t) = \frac{6}{\beta}\,\frac{\alpha_1\kappa_1 \exp\left\{\kappa_1 x + \kappa_1 t/(\kappa_1^2 - 1)\right\} + \alpha_2\kappa_2 \exp\left\{\kappa_2 x + \kappa_2 t/(\kappa_2^2 - 1)\right\}}{\alpha_0 + \alpha_1 \exp\left\{\kappa_1 x + \kappa_1 t/(\kappa_1^2 - 1)\right\} + \alpha_2 \exp\left\{\kappa_2 x + \kappa_2 t/(\kappa_2^2 - 1)\right\}}, \tag{4.5}$$

provided κ_1 and κ_2 satisfy (4.4). It should be noted that (4.5) and (3.12) are fundamentally different solutions of the SWW–HS Equation (1.8) as we shall now demonstrate.

The general two-soliton solution of (1.8) is given by

$$u(x,t) = \frac{6}{\beta}\,\frac{\kappa_1\alpha_1 \exp(\eta_1) + \kappa_2\alpha_2 \exp(\eta_2) + (\kappa_1 + \kappa_2)A_{12} \exp(\eta_1 + \eta_2)}{1 + \alpha_1 \exp(\eta_1) + \alpha_2 \exp(\eta_2) + A_{12} \exp(\eta_1 + \eta_2)}, \tag{4.6a}$$

where

$$\eta_1 = \kappa_1 x + \kappa_1 t/(\kappa_1^2 - 1), \qquad \eta_2 = \kappa_2 x + \kappa_2 t/(\kappa_2^2 - 1), \tag{4.6b}$$

$$A_{12} = \alpha_1\alpha_2 \frac{(\kappa_1 - \kappa_2)^2(\kappa_1^2 - \kappa_1\kappa_2 + \kappa_2^2 - 3)}{(\kappa_1 + \kappa_2)^2(\kappa_1^2 + \kappa_1\kappa_2 + \kappa_2^2 - 3)}, \tag{4.6c}$$

with α_1, α_2, κ_1 and κ_2 arbitrary constants [34] (see also [43]). A straightforward method of obtaining this solution is to substitute (4.1) into SWW–HS Equation (1.8) and then integrate once. This yields the bi-linear equation,

$$\begin{aligned}&\phi\phi_{xxxt} - \phi_t\phi_{xxx} + 3(\phi_{xx}\phi_{xt} - \phi_x\phi_{xxt}) - \\ &\quad -\phi(\phi_{xx} + \phi_{xt}) + \phi_x(\phi_x + \phi_t) = 0,\end{aligned} \tag{4.7}$$

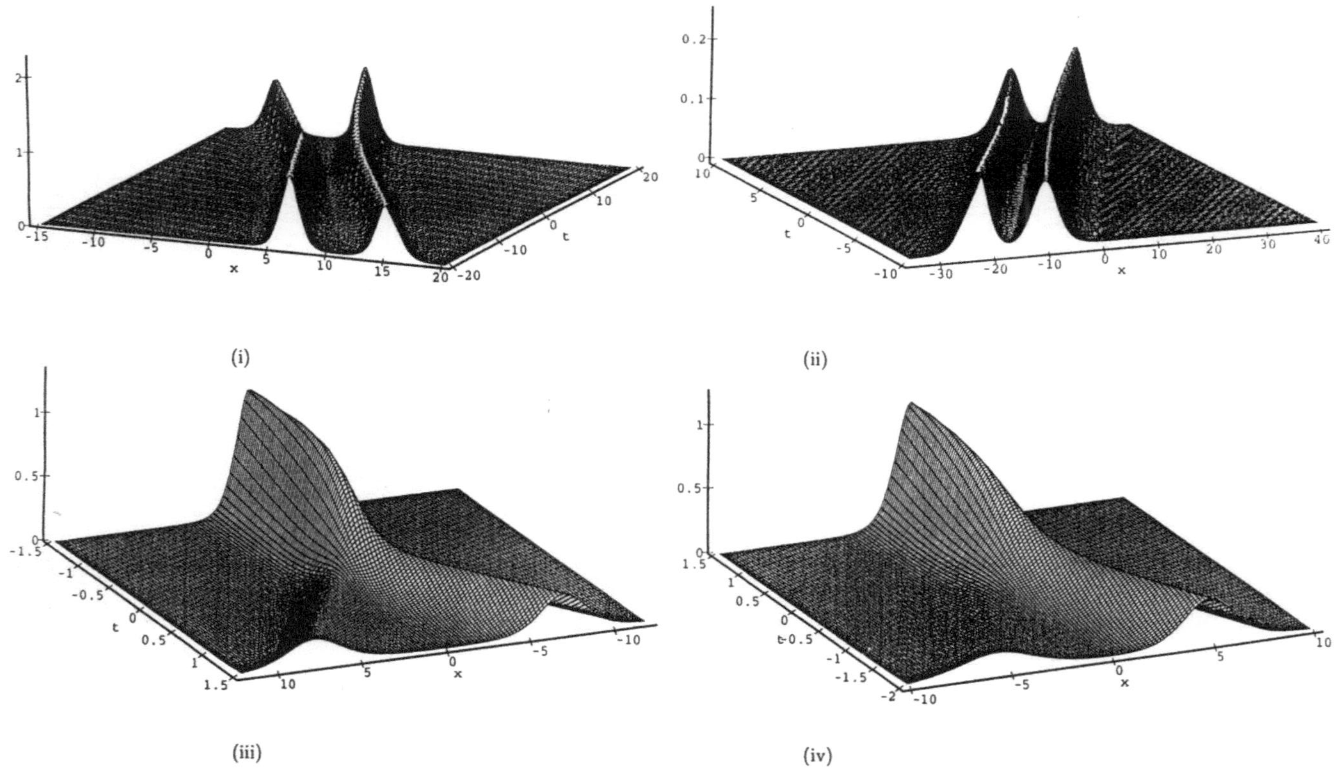

Fig. 5. The solution (4.6) of the SWW–HS Equation (1.8). (i) $\kappa_1 = 2$, $\kappa_2 = 1.7$, (ii) $\kappa_1 = \frac{3}{4}$, $\kappa_2 = \frac{2}{3}$, (iii) $\kappa_1 = 1$, $\kappa_2 = (11 + 3\sqrt{93})/20$ $(A_{12} = 0)$, (iv) $\kappa_1 = 0.8$, $\kappa_2 = \frac{1}{5}(2 + 3\sqrt{7})$ $(A_{12} = 0)$.

where we have set the function of integration to zero. Now, by seeking a solution of this equation, in the form

$$\phi(x,t) = 1 + \alpha_1 \exp(\eta_1) + \alpha_2 \exp(\eta_2) + A_{12} \exp(\eta_1 + \eta_2),$$

with $\eta_1 = \kappa_1 x + \mu_1 t$ and $\eta_2 = \kappa_2 x + \mu_2 t$, it is easy to show that necessarily $\mu_1 = \kappa_1/(\kappa_1^2 - 1)$, $\mu_2 = \kappa_2/(\kappa_2^2 - 1)$ and A_{12} is given by (4.6c); this technique may be viewed as a variant of Hirota's bi-linear method [32]. Four plots of the x-derivative of (4.6), i.e. solutions of (1.6), are given in Figure 5 for

(i) $\kappa_1 = 2$, $\kappa_2 = 1.7$,
(ii) $\kappa_1 = \frac{3}{4}$, $\kappa_2 = \frac{2}{3}$,
(iii) $\kappa_1 = 1.1$, $\kappa_2 = (11 + 3\sqrt{93})/20$

(so that $A_{12} = 0$),

(iv) $\kappa_1 = 0.8$, $\kappa_2 = \frac{1}{5}(2 + 3\sqrt{7})$

(so that $A_{12} = 0$).

Plots 5 (i) and (ii) illustrate 'standard' two-soliton interaction whilst plots 5 (iii) and (iv) are in the special case when $A_{12} = 0$. Plots of the x-derivative of (4.6) for Figure 2(b) of the x-derivative of solution (4.6) below for the SWW–HS equation, i.e. solutions of (1.6), for $\kappa_1 = 2$, $\kappa_2 = 1.7$ and $\kappa_1 = \frac{3}{4}$, $\kappa_2 = \frac{2}{3}$ (so that $A_{12} = 1$ in both cases).

The solution (4.5) is the special case of (4.6) with $A_{12} = 0$; Hirota and Ito [33] refer to this as being the 'resonant state' where either a single soliton splits into two solitons, see Figure 5 (iii), or two solitons fuse together after colliding with each other, see Figure 5 (iv). On the other hand, (3.12) is the special case of (4.6) with $A_{12} = 1$ where two solitons pass through each other with no phase shift as a consequence of the interaction. Thus whereas both solutions are asymptotically equivalent as $t \to -\infty$, they are qualitatively very different as $t \to \infty$. This shows that the nonclassical method and the singular manifold method do not, in general, yield the same solution set.

We remark that the SWW–HS Equation (1.8) also possesses the solution

$$u(x,t) = \frac{6}{\beta} \frac{\kappa_1[\alpha_1 \exp(\eta_1) - \alpha_2 \exp(-\eta_1)] - \kappa_2 B_{12} \sin(\eta_2 + \delta_0)}{\alpha_1 \exp(\eta_1) + \alpha_2 \exp(-\eta_1) + B_{12} \cos(\eta_2 + \delta_0)}, \tag{4.8a}$$

where

$$\eta_1 = \kappa_1 x + \frac{(\kappa_1^2 + \kappa_2^2 - 1)t}{(\kappa_1^2 - 2\kappa_1 + \kappa_2^2 + 1)(\kappa_1^2 + 2\kappa_1 + \kappa_2^2 + 1)}, \tag{4.8b}$$

$$\eta_2 = \kappa_2 x - \frac{(\kappa_1^2 + \kappa_2^2 + 1)t}{(\kappa_1^2 - 2\kappa_1 + \kappa_2^2 + 1)(\kappa_1^2 + 2\kappa_1 + \kappa_2^2 + 1)}, \tag{4.8c}$$

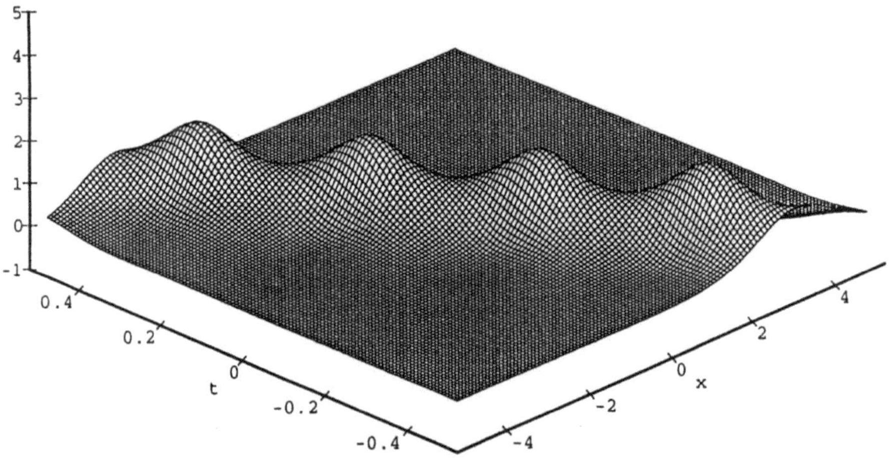

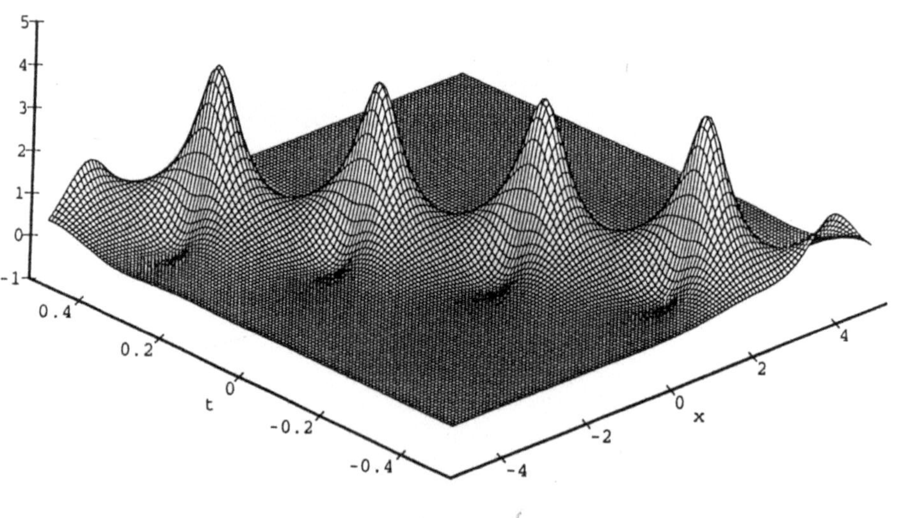

Fig. 6. The solution (4.9) of the SWW–HS Equation (1.8).

$$B_{12} = 2(\alpha_1\alpha_2)^{1/2}\frac{\kappa_1}{\kappa_2}\left(\frac{3\kappa_1^2-\kappa_2^2-3}{3\kappa_2^2-\kappa_1^2+3}\right)^{1/2}, \tag{4.8d}$$

with α_1, α_2, κ_1, κ_2 and δ_0 arbitrary constants. Leble and Ustinov [38] show that solutions of the form (4.8) exist for several of PDEs that are solvable by inverse scattering through third order linear problems. Two plots of the x-derivative of (4.8), i.e. solutions of (1.6), are given in Figure 6 for (i) $\kappa_1 = 1$, $\kappa_2 = 1.17$ and (ii) $\kappa_1 = 1$, $\kappa_2 = 1.24$.

4.2. THE SWW–AKNS EQUATION

The general two-soliton solution of the SWW–AKNS Equation (1.7) is given by

$$u(x,t) = \frac{6}{\beta}\,\frac{\kappa_1\alpha_1\exp(\eta_1)+\kappa_2\alpha_2\exp(\eta_2)+(\kappa_1+\kappa_2)A_{12}\exp(\eta_1+\eta_2)}{1+\alpha_1\exp(\eta_1)+\alpha_2\exp(\eta_2)+A_{12}\exp(\eta_1+\eta_2)}, \tag{4.9a}$$

where

$$\begin{aligned}\eta_1 &= \kappa_1 x + \kappa_1 t/(\kappa_1^2-1), \qquad \eta_2 = \kappa_2 x + \kappa_2 t/(\kappa_2^2-1),\\ A_{12} &= \alpha_1\alpha_2\left(\frac{\kappa_1-\kappa_2}{\kappa_1+\kappa_2}\right)^2,\end{aligned} \tag{4.9b}$$

with with α_1, α_2, κ_1 and κ_2 arbitrary constants [34] (see also [43]). Two plots of the x-derivative of (4.9), i.e. solutions of (1.3), are given in Figure 7 for (i) $\kappa_1 = 2$, $\kappa_2 = 1.7$ and (ii) $\kappa_1 = \frac{3}{4}$, $\kappa_2 = \frac{4}{5}$.

To apply the singularity manifold method to the SWW–AKNS Equation (1.7), we seek a solution in the form

$$u(x,t) = \frac{4}{\beta}\,\frac{\phi_x(x,t)}{\phi(x,t)}. \tag{4.10}$$

Equating coefficients of powers of ϕ to zero, yields the overdetermined system

$$\phi_{xxxxt} - \phi_{xxx} - \phi_{xxt} = 0, \tag{4.11a}$$

$$\phi_t\phi_{xxxx} + 4\phi_x\phi_{xxxt} - 2\phi_{xx}\phi_{xxt} - 3\phi_x\phi_{xx} - \phi_t\phi_{xx} - 2\phi_x\phi_{xt} = 0, \tag{4.11b}$$

$$2\phi_t\phi_x\phi_{xxx} - 2\phi_x\phi_{xx}\phi_{xt} + 2\phi_x^2\phi_{xxt} - \phi_t(\phi_x^2+\phi_{xx}^2) - \phi_x^3 = 0. \tag{4.11c}$$

We note that this is a system of three equations in contrast to that for the SWW–HS Equation (1.8) where $\phi(x,t)$ satisfies a system of two Equations (4.2). It is easily shown that there exist *no* solutions of (4.11) in the form (4.3) if $\alpha_0\alpha_1\alpha_2 \neq 0$. Additionally substituting (4.10) into (1.7) yields a *tri-linear* equation for $\phi(x,t)$, whereas the analogous equation for (1.8), i.e. (4.7), is bi-linear.

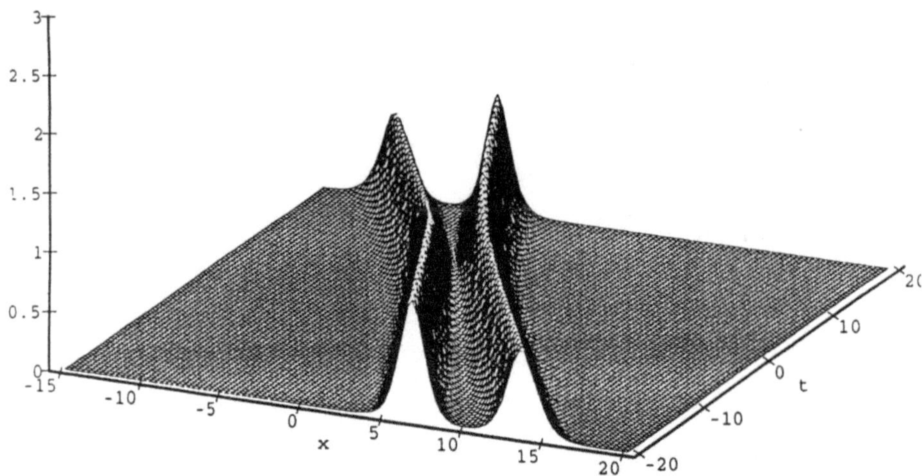

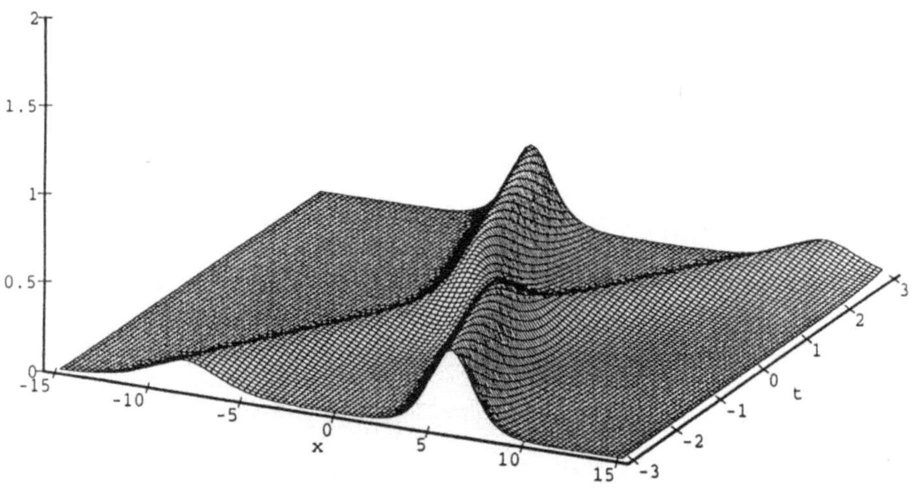

Fig. 7. Two soliton solutions (4.10) of the SWW–AKNS Equation (1.7).

5. Nonlinear Superposition for $(2+1)$-Dimensional Equations

In this section we show that nonlinear superposition of solutions can also occur for two $(2+1)$-dimensional equations of the SWW–HS Equation (1.8). Two $(2+1)$-dimensional generalisations of the SWW Equation (1.1) are the following equations

$$u_{yt} + \alpha u_x u_{xy} + \beta u_y u_{xx} - u_{xxxy} = 0, \tag{5.1}$$

$$u_{xt} + \alpha u_x u_{xy} + \beta u_y u_{xx} - u_{xxxy} = 0. \tag{5.2}$$

We remark that these two equations differ only in the first term, which belongs to the linear part, yet their symmetries differ remarkably. Further both (5.1) and (5.2) reduce to the KdV Equation (1.4) if $y = x$.

Boiti, Leon, Manna and Pempinelli [13] developed an inverse scattering scheme to solve the Cauchy problem for (5.1) with $\alpha = \beta$, i.e.

$$u_{yt} + u_{xxxy} + \beta u_{xx} u_y + \beta u_x u_{xy} = 0, \tag{5.3}$$

which is a $(2+1)$-dimensional generalisation of the SWW–HS Equation (1.8), for initial data decaying sufficiently rapidly at infinity. This inverse scattering scheme is formulated as a nonlocal Riemann–Hilbert problem and involves a so-called 'weak-Lax pair'. We note that both (5.3) and SWW–HS (1.8) are reductions of the $(3+1)$-dimensional equation

$$u_{yt} + u_{xxxy} + \beta u_{xx} u_y + \beta u_x u_{xy} - u_{xz} = 0, \tag{5.4}$$

which was introduced by Jimbo and Miwa [37] as the second equation in the so-called Kadomtsev–Petviashvili hierarchy, though (5.4) is not completely integrable in the usual sense (cf. [25]). Bogoyaviemskii [11, 12] discusses the inverse scattering method of solution for (5.2) with $\alpha = 2\beta$, i.e.,

$$u_{xt} + u_{xxxy} + \beta u_{xx} u_y + 2\beta u_x u_{xy} = 0, \tag{5.5}$$

which is a $(2+1)$-dimensional generalisation of the SWW–AKNS Equation (1.7); it should be noted that the $(2+1)$-dimensional generalisation of (1.7) discussed by Gilson *et al.* [28] is different to (5.2).

It is routine to show that (5.1) satisfies the necessary conditions of the Painlevé PDE test due to Weiss *et al.* [55] to be completely integrable if and only if $\alpha = \beta$, i.e. when it reduces to (5.3), and (5.2) satisfies these necessary conditions if and only if $\alpha = 2\beta$, i.e. when it reduces to (5.5). This strongly suggests that (5.1) and (5.2) are solvable by inverse scattering only in these two special cases, both of which are known to be completely integrable [13] and [11, 12], respectively).

5.1. Classical Symmetries

To apply the classical method to (5.1), we consider the one-parameter Lie group of infinitesimal transformations in (x, y, t, u) given by

$$\tilde{x} = x + \varepsilon \xi_1(x, y, t, u) + O(\varepsilon^2),$$

$$\tilde{y} = y + \varepsilon\xi_2(x,y,t,u) + O(\varepsilon^2),$$
$$\tilde{t} = t + \varepsilon\xi_3(x,y,t,u) + O(\varepsilon^2),$$
$$\tilde{u} = u + \varepsilon\phi(x,y,t,u) + O(\varepsilon^2),$$

where ε is the group parameter. Solving the resulting determining equations yields the infinitesimals

$$\xi_1 = \kappa_1 x + f_1(t), \qquad \xi_2 = g(y), \qquad \xi_3 = 3\kappa_1 t + \kappa_2,$$
$$\phi = -\kappa_1 u + \frac{x}{\alpha}\frac{\mathrm{d}f_1}{\mathrm{d}t} + f_2(t),$$

if $\alpha \neq \beta$ and

$$\xi_1 = x\frac{\mathrm{d}f_1}{\mathrm{d}t} + f_2(t), \qquad \xi_2 = g(y), \qquad \xi_3 = 3f_1(t),$$
$$\phi = -u\frac{\mathrm{d}f_1}{\mathrm{d}t} + \frac{x^2}{2\alpha}\frac{\mathrm{d}^2 f_1}{\mathrm{d}t^2} + \frac{x}{\alpha}\frac{\mathrm{d}f_2}{\mathrm{d}t} + f_3(t),$$

if $\alpha = \beta$, where κ_1, κ_2 and κ_3 are arbitrary constants and $f_1(t)$, $f_2(t)$, $f_3(t)$ and $g(y)$ are arbitrary differentiable functions [21] (classical symmetries in the case $\alpha = \beta$ are also discussed in [52]).

Applying the classical Lie method to (5.2) yields the infinitesimals

$$\xi_1 = \kappa_1 x + f_1(t), \qquad \xi_2 = \kappa_2 t + \kappa_3 y + \kappa_4, \qquad \xi_3 = (2\kappa_1 + \kappa_3)t + \kappa_5,$$
$$\phi = -\kappa_1 u + \frac{y}{\beta}\frac{\mathrm{d}f_1}{\mathrm{d}t} + \frac{\kappa_2 x}{\alpha} + f_2(t),$$

if $\alpha \neq 2\beta$ and

$$\xi_1 = (\kappa_1 t + \kappa_2)x + f_1(t), \qquad \xi_2 = (2\kappa_1 y + \kappa_3)t + \kappa_4 y + \kappa_5,$$
$$\xi_3 = 2\kappa_1 t^2 + (2\kappa_2 + \kappa_4)t + \kappa_6,$$
$$\phi = -(\kappa_1 t + \kappa_2)u + \frac{y}{\beta}\left(\kappa_1 x + \frac{\mathrm{d}f_1}{\mathrm{d}t}\right) + \frac{\kappa_3 x}{2\beta} + f_2(t),$$

if $\alpha = 2\beta$, where $\kappa_1, \kappa_2, \ldots, \kappa_6$ are arbitrary constants and $f_1(t)$ and $f_2(t)$ are arbitrary differentiable functions [21].

5.2. Nonclassical Reductions

Applying the nonclassical method to (5.1) yields an additional reduction in the case when $\alpha = \beta$, when it reduces to (5.3), This nonclassical reduction is given by

$$u(x,y,t) = v(\eta,\tau) + w(\zeta,\tau) - \frac{f(t)g^2(y)}{2\beta}\frac{\mathrm{d}f}{\mathrm{d}t},$$

$$\eta = x + f(t)g(y), \qquad \zeta = x - f(t)g(y), \qquad \tau = t, \tag{5.6}$$

where $V(\eta,\tau) = v_\eta$ and $W(\zeta,\tau) = w_\zeta$ satisfy the variable-coefficient KdV equations

$$\begin{aligned} &V_{\eta\eta\eta} - 2\beta V V_\eta - V_\tau - F(\tau)V + \\ &\quad + \frac{\eta}{2\beta}\left\{\frac{\mathrm{d}F}{\mathrm{d}\tau} + 2F^2(\tau)\right\} + Q(\tau) = 0, \end{aligned} \tag{5.7}$$

$$\begin{aligned} &W_{\zeta\zeta\zeta} - 2\beta W W_\zeta - W_\tau - F(\tau)W + \\ &\quad + \frac{\zeta}{2\beta}\left\{\frac{\mathrm{d}F}{\mathrm{d}\tau} + 2F^2(\tau)\right\} + Q(\tau) = 0, \end{aligned} \tag{5.8}$$

with $F(\tau) = \dot{f}(t)/f(t)$ and $Q(\tau)$ an arbitrary function. It is easily shown that (5.7) and (5.8) satisfy the necessary conditions of the Painlevé test due to Weiss *et al.* [55] to be completely integrable and both equations can be transformed into the usual KdV Equation (1.4). In particular, if in (5.7) and (5.8) we set $f(t) = \kappa$, a constant, then we obtain solutions to (5.1) with $\alpha = \beta$ in terms of sums of solutions of the KdV Equation (1.4), with arguments $x \pm g(y)$, where $g(y)$ is an arbitrary function; this is perhaps the simplest family of solutions to (5.3) found using this nonclassical reduction.

Applying the nonclassical method to (5.2) yields an additional reduction in the case when $\alpha = \beta$, given by

$$u(x,y,t) = v(\eta,\tau) + w(\zeta,\tau),$$

$$\eta = x + \kappa y, \qquad \zeta = x - \kappa y, \qquad \tau = t, \tag{5.9}$$

where $V(\eta,\tau) = v_\eta$ and $W(\zeta,\tau) = w_\zeta$ satisfy

$$\kappa V_{\eta\eta\eta} - \beta\kappa V V_\eta - V_\tau + Q(\tau) = 0, \tag{5.10}$$

$$\beta\kappa W_{\zeta\zeta\zeta} - \beta\kappa W W_\zeta - W_\tau + Q(\tau) = 0, \tag{5.11}$$

with $Q(\tau)$ an arbitrary function.

6. Discussion

In this paper we have discussed symmetry reductions and exact solutions for the shallow water wave Equation (1.1). In particular, for the special case of (1.1) given by the SWW–HS Equation (1.8), using the nonclassical symmetry reduction method originally proposed by Bluman and Cole [8], we obtained a family of solutions (3.10) which have a rich variety of qualitative behaviours. This is due to the freedom in the choice of the arbitrary function $f(t)$. One can choose $f_1(t)$ and $f_2(t)$ such $|f_1(t) - f_2(t)|$ is exponentially small as $t \to -\infty$, yet $f_1(t)$ and $f_2(t)$ are quite different as $t \to -\infty$, so that as $t \to -\infty$ the two solutions are essentially the same, yet as $t \to \infty$ they are radically different. In Figure 1 we show that by a judicious choice of $f(t)$ we can exhibit a plethora of different solutions. We believe that these results suggest that solving the SWW–HS Equation (1.8) numerically for initial conditions such as those in the solutions plotted in Figure 1 could pose some fundamental difficulties. An exponentially small change in the initial data yields a fundamentally different solution as $t \to \infty$. How can any numerical scheme in current use cope with such behaviour?

The solution (3.10) appears to be a nonlinear superposition of solutions suggesting that the SWW–HS Equation (1.8) may be linearisable through a transformation to a linear PDE, analogous to the linearisation of Burgers' equation

$$u_t = u_{xx} + 2uu_x, \tag{6.1}$$

which is mapped to the linear heat equation through the Cole–Hopf transformation [22, 35]. If so then the solution (3.10) could be viewed as an artefact of the fact that the SWW–HS Equation (1.8) is linearisable. However the SWW–HS Equation (1.8) can be expressed as the compatibility condition of the third-order spectral problem (1.11), (1.12). Further the associated scattering problem (1.11) is very similar to that for the Boussinesq equation which has been thoroughly studied by Deift *et al.* [24]. This strongly suggests the SWW–HS Equation (1.8) is solvable by inverse scattering. The spatial part of the inverse scattering formalism (1.11) only defines u up to an arbitrary additive function of t; this arbitrary function may be incorporated into u using the variable-coefficient transformation (2.15). Indeed the initial value problem for the SWW–HS Equation (1.8) is *not* well-posed without the imposition of an additional constraint since if $u(x,t)$ is a solution of (1.8) satisfying the inital condition $u(x,0) = \phi(x)$, then so is

$$\tilde{u}(\tilde{x},\tilde{t}) = u(x,t) + [g(t) - t]/\beta, \qquad \tilde{x} = x, \quad \tilde{t} = g(t), \tag{6.2}$$

where $g(t)$ is *any* differentiable function such that $g(0) = 0$. It appears likely that the inverse scattering formalism for the SWW–HS Equation (1.8) will require that $u(x,t)$ satisfies a constraint of the form

$$\int_{-\infty}^{\infty} |u(x,t)|\,\mathrm{d}x < \infty, \tag{6.3}$$

for all t. It is well-known that such constraints are required in the inverse scattering formalism for $(2+1)$-dimensional equations such as the Kadomtsev–Petviashvili equation (cf. [6], see also [1] and the references therein).

Since the shallow water wave Equation (1.1) is invariant under the variable-coefficient transformation (2.15) for all nonzero α and β, one can take any solution of (1.1) and using (2.15) generate some interesting solutions. For example, one can apply the transformation (2.15) to the two-soliton solution (4.9) and thus generate some exotic solutions for the SWW–AKNS Equation (1.7) analogous to the solution (3.10) of the SWW–HS Equation (1.8). Consequently the above remarks about difficulties in solving (1.8) numerically apply to (1.1) in general. Additionally, the inverse scattering formalism for the SWW–AKNS Equation (1.8) will probably require that $u(x,t)$ satisfies a constraint such as (6.3).

Recently, Ablowitz, Schober and Herbst [5] have shown that the focusing nonlinear Schrödinger equation

$$iu_t + u_{xx} + |u|^2 u = 0, \tag{6.4}$$

exhibits numerical chaos created by small errors on the order of roundoff. The results of Ablowitz *et al.* [5] together with those given in this paper suggest that numerical analysts need to take care to ensure the accuracy of their programs. Numerical predictions of chaos may not always be what they seem!

Acknowledgements

It is a pleasure to thank Mark Ablowitz, Chris Cosgrove, Jim Curry, Pilar Estévez, Willy Hereman and Colin Rogers for several illuminating discussions. We also thank the Program in Applied Mathematics, University of Colorado at Boulder and the Department of Mathematics and Statistics, University of Pittsburgh for their hospitality during our visits whilst some of this work was done. The support of EPSRC (grant GR/H39420) is gratefully acknowledged.

References

1. Ablowitz, M. J. and Clarkson, P. A.: *Solitons, Nonlinear Evolution Equations and Inverse Scattering*, Lect. Notes Math., Vol. 149, C.U.P., Cambridge, 1991.
2. Ablowitz, M. J., Kaup, D. J., Newell, A. C., and Segur, H.: *Stud. Appl. Math.* **53** (1974), 249–315.
3. Ablowitz, M. J., Ramani, A., and Segur, H.: *Phys. Rev. Lett.* **23** (1978), 333–338.
4. Ablowitz, M. J., Ramani, A., and Segur, H.: *J. Math. Phys.* **21** (1980), 715–721.
5. Ablowitz, M. J., Schober, C., and Herbst, B. M.: *Phys. Rev. Lett.* **71** (1993), 2683–2686.
6. Ablowitz, M. J. and Villarroel, J.: *Stud. Appl. Math.* **85** (1991), 195–213.
7. Anderson, R. L. and Ibragimov, N. H.: *Lie–Bäcklund Transformations in Applications*, SIAM, Philadelphia, 1979.
8. Benjamin, T. B., Bona, J. L., and Mahoney, J.: *Phil. Trans. R. Soc. Lond. Ser. A* **272** (1972), 47–78.
9. Bluman, G. W. and Cole, J. D.: *J. Math. Mech.* **18** (1969), 1025–1042.
10. Bluman, G. W. and Kumei, S.: *Symmetries and Differential Equations*, in Appl. Math. Sci., Vol. 81, Springer-Verlag, Berlin, 1989.
11. Bogoyavlenskii, O. I.: *Math. USSR Izves.* **34** (1990), 245–259.
12. Bogoyavlenskii, O. I.: *Russ. Math. Surv.* **45** (1990), 1–86.
13. Boiti, M., Leon, J. J-P., Manna, M., and Pempinelli, F.: *Inverse Problems* **2** (1986), 271–279.
14. Buchberger, B.: in J. Rice (ed.), *Mathematical Aspects of Scientific Software*, Springer-Verlag, 1988, pp. 59–87.
15. Champagne, B., Hereman, W., and Winternitz, P.: *Comp. Phys. Comm.* **66** (1991), 319–340.
16. Clarkson, P. A.: Nonclassical symmetry reductions for the Boussinesq equation, in *Chaos, Solitons and Fractals*, 1994, to appear.
17. Clarkson, P. A. and Kruskal, M. D.: *J. Math. Phys.* **30** (1989), 2201–2213.
18. Clarkson, P. A. and Mansfield, E. L.: *Physica D* **70** (1994), 250–288.

19. Clarkson, P. A. and Mansfield, E. L.: *Nonlinearity* **7** (1994), 975–1000.
20. Clarkson, P. A. and Mansfield, E. L.: Algorithms for the nonclassical method of symmetry reductions, *SIAM J. Appl. Math.*, 1994, to appear.
21. Clarkson, P. A. and Mansfield, E. L.: Exact solutions for some $(2+1)$-dimensional shallow water wave equations, Preprint, Department of Mathematics, University of Exeter, 1994.
22. Cole, J. D.: *Quart. Appl. Math.* **9** (1951), 225–236.
23. Conte, R. and Musette, M.: *J. Math. Phys.* **32** (1991), 1450–1457.
24. Deift, P.,Tomei, C., and Trubowitz, E.: *Comm. Pure Appl. Math.* **35** (1982), 567–628.
25. Dorizzi, B., Grammaticos, B., Ramani, A., and Winternitz, P.: *J. Math. Phys.* **27** (1986), 2848–2852.
26. Espinosa, A. and Fujioka, J.: *J. Phys. Soc. Japan* **63** (1994), 1289–1294.
27. Gardner, C. S., Greene, J. M., Kruskal, M. D., and Miura, R.: *Phys. Rev. Lett.* **19** (1967), 1095–1097.
28. Gilson, C. R., Nimmo, J. J. C., and Willox, R.: *Phys. Lett.* **180A** (1993), 337–345.
29. Fushchich, W. I.: *Ukrain. Math. J.* **43** (1991), 1456–1470.
30. Hereman, W.: *Euromath Bull.* **1**(2) (1994), 45–79.
31. Hietarinta, J.: in R. Conte and N. Boccara (eds), *Partially Integrable Evolution Equations in Physics*, NATO ASI Series C: Mathematical and Physical Sciences, Vol. 310, Kluwer, Dordrecht, 1990, pp. 459–478.
32. Hirota, R.: in R. K. Bullough and P. J. Caudrey (eds), *Solitons, Topics in Current Physics*, Vol. 17, Springer-Verlag, Berlin, 1980, pp. 157–176.
33. Hirota, R. and Itô, M.: *J. Phys. Soc. Japan* **52** (1983), 744–748.
34. Hirota, R. and Satsuma, J.: *J. Phys. Soc. Japan* **40** (1976), 611–612.
35. Hopf, E.: *Comm. Pure Appl. Math.* **3** (1950), 201–250.
36. Ince, E. L.: *Ordinary Differential Equations*, Dover, New York, 1956.
37. Jimbo, M. and Miwa, T.: *Publ. R.I.M.S.* **19** (1983), 943–1001.
38. Leble, S. B. and Ustinov, N. V.: *Inverse Problems* **210** (1994), 617–633.
39. Levi, D. and Winternitz, P.: *J. Phys. A: Math. Gen.* **22** (1989), 2915–2924.
40. Mansfield, E. L.: DIFFGROB2: A symbolic algebra package for analysing systems of PDE using Maple, ftp euclid.exeter.ac.uk, login: anonymous, password: your email address, directory: pub/liz, 1993.
41. Mansfield, E. L. and Fackerell, E. D.: Differential Gröbner Bases, Preprint 92/108, Macquarie University, Sydney, Australia, 1992.
42. McLeod, J. B. and Olver, P. J.: *SIAM J. Math. Anal.* **14** (1983), 488–506.
43. Musette, M., Lambert, F., and Decuyper, J. C.: *J. Phys. A: Math. Gen.* **20** (1987), 6223–6235.
44. Olver, P. J.: *Applications of Lie Groups to Differential Equations*, 2nd edn, Graduate Texts Math., Vol. 107, Springer-Verlag, New York, 1993.
45. Olver, P. J. and Rosenau, P.: *Phys. Lett.* **114A** (1986), 107–112.
46. Olver, P. J. and Rosenau, P.: *SIAM J. Appl. Math.* **47** (1987), 263–275.
47. Peregrine, H.: *J. Fluid Mech.* **25** (1966), 321–330.
48. Reid, G. J.: *J. Phys. A: Math. Gen.* **23** (1990), L853–L859.
49. Reid, G. J.: *Europ. J. Appl. Math.* **2** (1991), 293–318.
50. Reid, G. J. and Wittkopf, A.: A Differential Algebra Package for Maple, ftp 137.82.36.21 login: anonymous, password: your email address, directory: pub/standardform, 1993.
51. Schwarz, F.: *Computing* **49** (1992), 95–115.
52. Tamizhmani, K. M. and Punithavathi, P.: *J. Phys. Soc. Japan* **59** (1990), 843–847.
53. Topunov, V. L.: *Acta Appl. Math.* **16** (1989), 191–206.
54. Weiss, J.: *J. Math. Phys.* **24** (1983), 1405–1413.
55. Weiss, J., Tabor, M., and Carnevale, G.: *J. Math. Phys.* **24** (1983), 522–526.
56. Whittaker, E. E. and Watson, G. M.: *Modern Analysis*, 4th edn, C.U.P., Cambridge, 1927.
57. Winternitz, P.: Lie groups and solutions of nonlinear partial differential equations, in L. A. Ibort and M. A. Rodriguez (eds), *Integrable Systems, Quantum Groups, and Quantum Field Theories*, NATO ASI Series C, Vol. 409, Kluwer, Dordrecht, 1993, pp. 429–495.

Acta Applicandae Mathematicae **39:** 277–294, 1995.

A KdV Equation in 2 + 1 Dimensions: Painlevé Analysis, Solutions and Similarity Reductions

P. G. ESTÉVEZ
Area de Física Teórica, Facultad de Ciencias, Universidad de Salamanca, 37008 Salamanca, Spain

and

S. B. LEBLE
Theoretical Physics Department, Kaliningrad State University, Aleksandr Nevsky Street 14, 236041 Kaliningrad, Russia

(Received: 3 August 1994)

Abstract. The nonlinear equation $m_{ty} = (m_{yxx} + m_x m_y)_x$ is throughly analyzed. The Painlevé test yields a positive result. The Bäcklund transformations are found and the Darboux–Moutard–Matveev formalism arises in the context of this analysis. Some solutions and their interactions are also analyzed. The singular manifold equations are also used to determine symmetry reductions. This procedure can be related with the direct method of Clarkson and Kruskal.

Mathematics Subject Classifications (1991): 35Q53, 58F07.

Key words: Painlevé analysis, Bäcklund transformation, Kadomtsev–Petviashvili equation, singular manifold methods.

1. Introduction

The problem unde consideration in the present paper involves the following system of equations

$$u_x + v_y = 0 \tag{1.1a}$$

$$u_t = u_{xxx} - (uv)_x. \tag{1.1b}$$

The spectral transform for this system has been investigated in [1] and more details can be found in [2]. This system has been considered in [3] as a generalization to 2 + 1 of the results of Hirota and Satsuma [4]. The nonclassical symmetries, Painlevé property and solutions of the same 1 + 1 reduction of (1.1) considered in [4] have been studied by Clarkson and Mansfield in [5]. It is also interesting to notice that for $x = y$ the system (1.1) reduces to the KdV equation [1].

An alternative form of (1.1) can be obtained by first equating:

$$u = m_y. \tag{1.2}$$

The Equations (1.1) now provide the PDE

$$m_{ty} = (m_{xxy} + m_y m_x)_x. \tag{1.3}$$

Investigations of the singular manifold method (SMM) in $2+1$ are not excesive in number. Nevertheless, we refer the reader to [7, 8] for the Kadomtsev–Petviashvili equation and [9, 10] for the Davey–Stewartson equation. Furthermore, in [8] the author has obtained some $2+1$ equations similar to (1.3). We also make reference to [24], where the authors show an equation with the same linear part but with a different nonlinear one. An analysis of this equation is planned by the authors of [25].

The plan of the present paper is as follows: Section 2 is devoted to the SMM and to the construction of a Lax pair for (1.3). In Section 3 we apply the SMM to the Lax pair itself [8, 11]. The result leads us to an iterative procedure for the construction of solutions. This procedure corresponds exactly to the Moutard–Darboux–Matveev transformations [12, 13].

In Section 4 we deal with the symmetry reductions (SR) of Equation (1.3). We can relate our results with the direct method of Clarkson and Kruskal [15] and with those revealed in [5]. We end this paper with a Conclusions Section and an Appendix where some particular solutions will be described.

2. Painlevé Test and the Singular Manifold Method

Acting in correspondence with the main direction of [6], let us present the solutions of Equation (1.3) as the power series expansion

$$m = \sum_{j=0}^{\infty} m_j \phi^{j-a} \tag{2.1}$$

and list the standard results below.

The leading index is $a = 1$ and the resonances are located in $j = 1, 4, 6$. It can be proved by using direct calculations (see Acknowledgements) that the conditions at the resonances are identically satisfied. This means that Equation (1.3) has the Painlevé property and its solutions can be locally expanded in terms of four arbitrary functions ϕ, m_1, m_4 and m_6.

The next step consists in the use of the singular manifold method [6, 7] by truncating the expansion (2.1) at the constant level. In such a case the truncation procedure provides an auto-Bäcklund transformation between two different solutions m and m' of (1.3). In our case this transformation is:

$$m' = m + 6\frac{\phi_x}{\phi} \tag{2.2}$$

and the substitution of (2.2) in (1.3) requires ϕ to be the 'singular manifold' that satisfies:

$$0 = \{v_{yx} + 2vv_y + vm_y - w_y + (z_x + vz)(v_x + v^2 - w + m_x)\}\phi -$$

$$-\{m_y + 3v_y + z(v_x + v^2 - w + m_x)\}\phi_x \tag{2.3}$$

where we have used the definition (see [16]):

$$v = \frac{\phi_{xx}}{\phi_x} \tag{2.4a}$$

$$w = \frac{\phi_t}{\phi_x} \tag{2.4b}$$

$$z = \frac{\phi_y}{\phi_x} \tag{2.4c}$$

The following identities are nothing other than the compatibility conditions between the definitions (2.4):

$$v_y = (z_x + zv)_x \tag{2.5a}$$

$$v_t = (w_x + wv)_x \tag{2.5b}$$

$$w_y = z_t + zw_x - wz_x. \tag{2.5c}$$

The usual way to proceed in the SMM [6, 7] is to assume that all coefficients for each power of ϕ in Equation (2.3) vanish independently. Nevertheless, in our case, it is easy to see that in so doing the resulting m_x and m_y are rational functions of v, w and z. These expressions cannot be linearized in order to provide a Lax pair. Actually the solution to this problem can be found by taking into account that the assumption that the coefficients have to vanish independently for each power of ϕ is too restrictive. In fact this point has been noticed many times in the literature [17–19]. The authors of [18] have suggested that an expression such as (2.3) can be satisfied in a more general way by doing:

$$\{v_{yx} + 2vv_yvm_y - w_y + (z_x + vz)(v_x + v^2 - w + m_x)\}\phi = A\phi_x \tag{2.6a}$$

and

$$m_y + 3v_y + z(v_x + v^2 - w + m_x) = A \tag{2.6b}$$

where, in the words of these authors, A is a function that must be 'judiciously chosen'. From (2.6) m_x and m_y can be obtained as:

$$\begin{aligned} m_x &= w - v_x - v^2 + \\ &\quad + \left(\frac{1}{z_x}\right)\left\{w_y - v_{xy} + vv_y + A\left(\frac{\phi_x}{\phi}\right) - Av\right\} \end{aligned} \tag{2.7a}$$

$$\begin{aligned} m_y &= A - 3v_y - \\ &\quad - \left(\frac{z}{z_x}\right)\left\{w_y - v_{xy} + vv_y + A\left(\frac{\phi_x}{\phi}\right) - Av\right\}. \end{aligned} \tag{2.7b}$$

The form of the Equations (2.7) suggests that the most 'judicious' selection for A is not zero, as usual, but:

$$0 = w_y - v_{xy} + vv_y + A\left(\frac{\phi_x}{\phi}\right) - Av. \tag{2.8a}$$

In such a case, the compatibility condition $m_{xy} = m_{yx}$ in (2.7) provides:

$$0 = A_x - 2v_{xy} + 2v_x v_y - w_y \tag{2.8b}$$

and the addition of both Equations (2.8a) and (2.8b) leads to the following equation for A.

$$0 = A_x - 3v_{xy} + 3v_x v_y + A\left(\frac{\phi_x}{\phi}\right) - Av \tag{2.9}$$

that can be integrated as:

$$A = 3v_y - 3\left(\frac{\phi_{xy}}{\phi}\right) + 3\lambda(y,t)\left(\frac{\phi_x}{\phi}\right) \tag{2.10}$$

where $\lambda(y,t)$ is an arbitrary function. Equations (2.7) can now be written as:

$$0 = \phi_{xxx} - \phi_t + m_x \phi_x \tag{2.11a}$$

$$0 = 3\phi_{xy} - 3\lambda\phi_x + m_y \phi \tag{2.11b}$$

and the compatibility between (2.11a) and (2.11b) requires λ to be a function of y only. Nevertheless the function λ can be easily removed from Equation (2.11b) by the gauge transformation:

$$\phi' = \phi \exp\left\{\int \lambda(y)\,\mathrm{d}y\right\}.$$

In consequence, there is no restriction in setting $\lambda = 0$ in (2.11). So the Equations (2.11) can be interpreted as the following Lax pair [3] for (1.3)

$$0 = \phi_{xxx} - \phi_t + m_x \phi_x \tag{2.12a}$$

$$0 = 3\phi_{xy} + m_y \phi \tag{2.12b}$$

and the equation for the singular manifold ϕ is obtained from the compatibility condition $m_{xy} = m_{yx}$ in (2.12). The result is:

$$\left(\frac{\phi_{xxx} - \phi_t}{\phi_x}\right)_y = 3\left(\frac{\phi_{xy}}{\phi}\right)_x. \tag{2.13}$$

3. Iteration Procedure for the Solutions

We have proved in the last section that (2.13) is a Lax pair for the solution m that appears in (2.3). In the same way we can write a Lax pair for the solution m' in the form:

$$0 = \psi'_{xxx} - \psi'_t + m'_x\psi'_x \tag{3.1a}$$

$$0 = 3\psi'_{xy} + m'_y\psi' \tag{3.1b}$$

and now (see [7] and [11]), we can consider (3.1) as a nonlinear coupled system for ψ and m. So if m' is expanded as (2.2),

$$m' = m + 6\frac{\phi_x}{\phi} \tag{3.2a}$$

ψ' must be expanded as:

$$\psi' = \psi + \frac{p}{\phi} \tag{3.2b}$$

where ψ is an eigenfunction for m and therefore it satisfies:

$$0 = \psi_{xxx} - \psi_t + m_x\psi_x \tag{3.3a}$$

$$0 = 3\psi_{xy} + m_y\psi \tag{3.3b}$$

and, as far as ϕ is a singular manifold for m, it is a solution of the linear system:

$$0 = \phi_{xxx} - \phi_t + m_x\phi_x \tag{3.4a}$$

$$0 = 3\phi_{xy} + m_y\phi \tag{3.4b}$$

and p can be obtained by direct substitution of (3.2) into (3.1). The resulting equations for p are:

$$\begin{aligned} 0 \;=\; & \phi\{p_{xxx} - p_t + m_x p_x + 6\phi_{xx}\psi_x\} + \\ & +3\{-p_{xx}\phi_x + p_x\phi_{xx} - 2\phi_x^2\psi_x\} \end{aligned} \tag{3.5a}$$

$$\begin{aligned} 0 \;=\; & \phi\left\{p_{xy} + 2\psi\phi_{xy} + \frac{p}{3}m_y\right\} + \\ & +\{-p_x\phi_y - p_y\phi_x + p\phi_{xy} - 2\phi_x\phi_y\psi\}. \end{aligned} \tag{3.5b}$$

From (3.5a), we can obtain:

$$0 = p_{xxx} - p_t + m_x p_x + 6\phi_{xx}\psi_x \tag{3.6a}$$

and

$$p_x = -2(\psi + M)\phi_x \tag{3.6b}$$

where M is an arbitrary function of y and t. The introduction of (3.6b) in to (3.5b) provides:

$$p_y = -2[\psi_y + M_y]\phi - 2M\frac{\phi\phi_{xy}}{\phi_x}. \tag{3.6c}$$

It can be easily proved that the compatibility condition $p_{xy} = p_{yx}$ between (3.6b) and (3.6c) requires $M = 0$. In consequence, Equations (3.6) allow us to write the derivatives of p as:

$$p_x = -2\psi\phi_x \tag{3.7a}$$

$$p_y = -2\phi\psi_y \tag{3.7b}$$

$$p_t = 2\psi_x\phi_{xx} - 2\phi_x\psi_{xx} - 2\psi\phi_t. \tag{3.7c}$$

Equations (3.7) are an important point of this paper because they represent the connexion between the singular manifold method [6, 7] and the generalized Darboux convariance [3, 12, 13]. In fact, it can be done by taking into account that

$$p = p(\psi, \phi) \tag{3.8}$$

where both, ψ and ϕ are two solutions of the same linear system (3.3)–(3.4) for the solution m. This means that, starting with the trivial solution $m = 0$ the solutions of the Lax pair (3.1) can be written according to (3.2) and (3.8) in the form:

$$m' = 6\frac{\phi_x}{\phi} \tag{3.9a}$$

and,

$$\psi' = \psi + \frac{p(\psi, \phi)}{\phi} \tag{3.9b}$$

where both ψ and ϕ verify:

$$0 = \psi_{xxx} - \psi_t \tag{3.10a}$$

$$0 = 3\psi_{xy} \tag{3.10b}$$

and,

$$0 = \phi_{xxx} - \phi_t \tag{3.11a}$$

$$0 = 3\phi_{xy}. \tag{3.11b}$$

The solution of (3.10) and (3.11) can be trivially obtained as:

$$\psi = R_1(y) + Q_1(x,t) \tag{3.12a}$$

$$\phi = R_2(y) + Q_2(x,t) \tag{3.12b}$$

where the $R_i(y)$ are arbitrary functions of y and the $Q_i(x,y)$ are solutions of the linear KdV equation:

$$(Q_i)_{xxx} = (Q_i)_t. \tag{3.13}$$

Furthermore, as far as ψ' verifies (3.1), it means that ψ' is a singular manifold for m' and therefore we can construct a new solution as:

$$m'' = m' + 6\frac{\psi'_x}{\psi'} \tag{3.14}$$

and by using (3.9a), we can write:

$$m'' = 6\frac{\tau_x}{\tau} \tag{3.15}$$

where we have defined τ as:

$$\tau = \psi'\phi \tag{3.16}$$

and with the aid of (3.2b)

$$\tau = \psi\phi + p(\psi,\phi). \tag{3.17}$$

The equivalence between the solution (3.15) and those obtained in the Darboux formalism [3] can be established if we define:

$$\Delta = \tau^2 \tag{3.18}$$

in such a way that (3.15) can be written as:

$$m'' = 3\frac{\Delta_x}{\Delta} \tag{3.19}$$

and, combining (3.17) and (3.18), Δ can be expressed as a determinant:

$$\Delta = \det|\Delta_{ij}| \tag{3.20}$$

where,

$$\Delta_{11} = \psi^2 = p(\psi,\psi) \tag{3.21a}$$

$$\Delta_{22} = \phi^2 = p(\phi,\phi) \tag{3.21b}$$

$$\Delta_{12} = p(\psi, \phi) \tag{3.21c}$$

$$\Delta_{21} = -2\psi\phi + p(\psi, \phi) = -p(\phi, \psi) \tag{3.21d}$$

and we have used (3.7) and (3.8) in order to stablish the second part of the identifies (3.21). Equations (3.9) and (3.19)–(3.21) constitute the Darboux transformations for the Lax pair (3.1). In consequence we have proved the equivalence between both formalisms. Furthermore, the procedure developed above allows us to construct one- and two-soliton solutions with the elementary solutions (3.12). Particular solutions of this kind will be analyzed in the Appendix.

At this point one can establish a comparison between our results and those of [5]. In this interesting paper, the authors claim (for a $1+1$ reduction of (1.3)) that only some special cases of the two-soliton solution can be achieved by using the singular manifold. We think that this is true only if we restrict ourselves to the first iteration. Actually, in the Appendix we will prove that the second iteration provides the complete two-soliton solution.

4. Similarity Reductions and the Singular Manifold Method

The relationship between the Painlevé property and the similarity reductions (SR) of a given PDE was at the core of the celebrated AR (Ablowitz–Ramani–Segur) conjecture [20]. More recently, one of us (P. G. E.) has proved [14, 21] that the singular manifold method can be used as a useful tool in order to identify the SR of a PDE. The equivalence between this procedure and the usual ones (nonclassical method [22], and the direct method [5, 15, 23]) have been discussed in [14, 21, 24]. We will adress ouselves in this section to obtaining reductions of Equation (1.3) to an ODE by using the singular manifold equations.

From the results of Section 2 we have concluded that the singular manifold ϕ verifies Equation (2.13). We write this equation again

$$\left(\frac{\phi_{xxx} - \phi_t}{\phi_x}\right)_y = 3\left(\frac{\phi_{xy}}{\phi}\right)_x. \tag{4.1}$$

Let us suppose $\phi(x, y, t)$ is a function of just one reduced variable $z = z(x, y, t)$. In such a case Equation (4.1) is:

$$\begin{aligned}
0 = {} & -z_x^2 z_y\left(\frac{F_{zz} + 3FF_z + F^3}{F}\right)_z - 2z_x z_{xy}\left(\frac{F_{zz} + 3FF_z + F^3}{F}\right) - \\
& -3z_{xx}z_y\left(\frac{F_z + F^2}{F}\right)_z - 3z_{xxy}\left(\frac{F_z + F^2}{F}\right) + 3z_x z_{xy}F_z + \\
& +3z_{xxy}F + 3z_x^2 z_y(F_z + F^2)_z + 3(z_x z_y)_x(F_z + F^2) + \\
& +\left(\frac{z_t}{z_x} - \frac{z_{xxx}}{z_x}\right)_y
\end{aligned} \tag{4.2}$$

where we have defined:

$$F(z) = \frac{\phi_z}{\phi}. \tag{4.3}$$

Now, we can proceed as in the direct method of Clarkson and Kruskal [15]. If Equation (4.2) were an ODE in the reduced variable z, the following quantities:

$$\frac{z_{xy}}{z_x z_y} \tag{4.4a}$$

$$\frac{z_{xx}}{z_x^2} \tag{4.4b}$$

$$\frac{z_{xxy}}{z_x^2 z_y} \tag{4.4c}$$

$$\left(\frac{1}{z_x^2 z_y}\right)\left(\frac{z_t}{z_x}\right)_y \tag{4.4d}$$

$$\left(\frac{1}{z_x^2 z_y}\right)\left(\frac{z_{xxx}}{z_x}\right)_y \tag{4.4e}$$

need to be just functions on z. From (4.4) it is not difficult to see that this goal can be achieved only if z is of the form:

$$z = \theta(y,t)\left[x + g(t)\right] \tag{4.5}$$

where $g(t)$ is an arbitrary function of t and $\theta(y,t)$ has to verify:

$$\left(\frac{\theta_t}{\theta}\right)_y = C_0 \theta^2 \theta_y \tag{4.6}$$

with C_0 an arbitrary constant.

With this prescription, Equation (4.2) is the ODE:

$$\left(\frac{F_{zz}}{F} - 2F^2\right)_z + \left(\frac{1}{z}\right)\left(2\frac{F_{zz}}{F} - F^2\right) = C_0. \tag{4.7}$$

Now, the question is: how to relate the reduction (4.5) and Equation (4.7) with the similarity reductions of Equation (1.3)? This is the question that we are going to solve now.

From (2.12) we have:

$$m_x = \frac{\phi_t - \phi_{xxx}}{\phi_x} \tag{4.8a}$$

$$m_y = -3\frac{\phi_{xy}}{\phi_x} \tag{4.8b}$$

and with reduction $\phi = \phi(z)$, the expressions for m_x and m_y are:

$$m_x = \frac{z_t}{z_x} - z_x^2\left(\frac{F_{zz}}{F} + 3F_z + F^2\right) \tag{4.9a}$$

$$m_y = -3z_{xy}F - 3z_x z_y(F_z + F^2). \tag{4.9b}$$

Expression (4.9a) can be integrated as:

$$m = \alpha(x, y, t) + z_x M(z) \tag{4.10}$$

where

$$\alpha = \left(\frac{\theta_t}{\theta}\right)x^2 + \left(\frac{\theta_t}{\theta}\,g + g_t\right)x + h(y, t) \tag{4.11}$$

and

$$M_z = -\frac{F_{zz}}{F} - 3F_z - F^2, \tag{4.12}$$

$h(y, t)$ is an arbitrary function of y and t.

From the comparison between (4.12) and (4.7) it is not difficult to see

$$M = z\left(\frac{F_{zz}}{F - 2F^2}\right) - 3F - \frac{C_0}{2}\,z^2. \tag{4.13}$$

On the other hand, taking the derivative of (4.10) with respect to y and by identifying it with (4.9b), it is possible, with the aid of (4.7) and (4.13), to obtain:

$$h(y, t) = \frac{C_0}{6}\,g^2\theta^3 + f(t) \tag{4.14}$$

and $f(t)$ is a totally arbitrary function of t.

To summarize, the ansatz (4.10) together with:

$$\alpha = \left(\frac{\theta_t}{\theta}\right)x^2 + \left(\frac{\theta_t}{\theta}\,g + g_t\right)x + \frac{C_0}{6}\,g^2\theta^3 + f(t), \tag{4.15}$$

where M is expressed by (4.13), reduces Equation (1.3) to the ODE (4.7). Furthermore, the combination of (4.7) and (4.13) allows us to obtain an ODE just for M. Such an ODE is:

$$\begin{aligned} 0 = {} & M_{zzzz} + 2M_z M_{zz} + \frac{1}{z}(4M_{zzz} + MM_{zz} + 2M_z^2) + \\ & + \frac{C_0}{2}\,zM_{zz} - \frac{C_0}{2}\,M - \frac{C_0^2}{2}\,z. \end{aligned} \tag{4.16}$$

At this point it could be interesting to notice that if $C_0 = 0$, Equation (4.16) is exactly Equation (2.12) of [5]. In this reference Clarkson and Mansfield use the nonclassical method in order to get reductions of a $(1+1)$ version of (1.3). As they proved in that paper, Equation (4.16) (for the C_0 case) can be integrated in terms of the third Painlevé transcendent.

5. Conclusions

We have proved the Painlevé property for (1.3). The singular manifold method for this equation needs to include some modifications in the direction suggested in [18]. In so doing, we can obtain a Lax pair for the equation and an iterative procedure for the generation of solutions. This method includes a linearization of the equation in terms of the linear KdV equation. The equivalence between the singular manifold method and the Darboux transformation has been described in Section 3. The singular manifold also provides a useful tool for the derivation of similarity reductions of (1.3), as has been done in Section 4. Some particular solutions are described in the Appendix.

Acknowledgements

The Painlevé property of (1.3) has been tested using a computer program by O. Zurrón of the University of Exeter. We thank P. Clarkson for such a possibility and for providing many interesting references. One of us (S. Leble) wishes to express his thanks the kind hospitality of the Theoretical Physics Department of the University of Salamanca.

Finally, we want to thank Professor J. M. Cerveró for his suggestions about the final form of this paper and J. Villarroel for enlightening discussions.

Appendix

In this Appendix we obtain some particular solutions for Equation (1.3) by using the general method developed in Section 3.

(1) If we choose the following solutions for Equations (3.12), (3.13),

$$R_i = \mathrm{e}^{b_i y} \tag{A.1a}$$

$$Q_i = \mathrm{e}^{a_i x + a_i^3 t + c_i} \tag{A.1b}$$

where a_i, b_i and c_i are constants, the expressions for ϕ and ψ (see (3.12)) are:

$$\psi = R_1\left(1 + \mathrm{e}^{a_1 x + a_1^3 t - b_1 y + c_1}\right) \tag{A.2a}$$

$$\phi = R_2\left(1 + \mathrm{e}^{a_2 x + a_2^3 t - b_2 y + c_2}\right) \tag{A.2b}$$

and the direct integration of (3.7), (3.8) provides:

$$\Omega = R_1 R_2 \left[\frac{b_1}{b_1 + b_2} + \frac{Q_2}{R_2} + \left(\frac{a_2}{a_1 + a_2}\right)\frac{Q_1}{R_1}\frac{Q_2}{R_2}\right]. \tag{A.3}$$

With the aid of (A.2) and (A.3), we can write (3.17) as:

$$\tau = \left(\frac{b_2 - b_1}{b_2 + b_1}\right) R_1 R_2 (1 + f_1 + f_2 + A f_1 f_2) \tag{A.4a}$$

where

$$f_i = \mathrm{e}^{a_i x + a_i^3 t - b_i y + \theta_i} \tag{A.4b}$$

$$A = \left(\frac{a_1 - a_2}{a_1 + a_2}\right)\left(\frac{b_1 - b_2}{b_1 + b_2}\right) \tag{A.4c}$$

$$\mathrm{e}^{\theta_1} = \mathrm{e}^{c_1} \frac{b_2 + b_1}{b_2 - b_1} \tag{A.4d}$$

$$\mathrm{e}^{\theta_2} = -\mathrm{e}^{c_2} \frac{b_2 + b_1}{b_2 - b_1}. \tag{A.4e}$$

The corresponding solutions for the first and second iterations are respectively given by (3.9a) and (3.15). In our case they are:

$$m' = 6a_2 \left(\frac{\mathrm{e}^{a_2 x + a_2^3 t - b_2 y + c_2}}{1 + \mathrm{e}^{a_2 x + a_2^3 t - b_2 y + c_2}}\right) \tag{A.5}$$

and

$$m'' = 6\, \frac{a_1 f_1 + a_2 f_2 + (a_1 + a_2) A f_1 f_2}{1 + f_1 + f_2 + A f_1 f_2} \tag{A.6}$$

m' is a two-dimensional kink solution that propagates with a velocity $(-a_2^2, a_2^3/b)$ (see Figure 1). On the other hand, m'' represents the interaction between two such kinks (Figure 2). An interesting particular case can be obtained by choosing $b_1 = a_1 c$ and $b_2 = -a_2 c$; in such a case (A.4c) provides $A = 1$ and the solution (A.6) factorizes as a linear superposition of two solutions as (A.5):

$$m'' = 6a_1 \left(\frac{\mathrm{e}^{a_1(x+cy) + a_1^3 t}}{1 + \mathrm{e}^{a_1(x+cy) + a_1^3 t}}\right) + 6a_2 \left(\frac{\mathrm{e}^{a_2(x-cy) + a_2^3 t}}{1 + \mathrm{e}^{a_2(x-cy) + a_2^3 t}}\right).$$

(2) Different solutions can be obtained if we choose the following solutions of (3.12), (3.13)

$$R_i = B_i \cosh(b_i y) \tag{A.7a}$$

$$Q_i = A_i \cosh(a_i x + a_i^3 t) \tag{A.7b}$$

where A_i, B_i, a_i, b_i are arbitrary constants. The expressions (3.12) are:

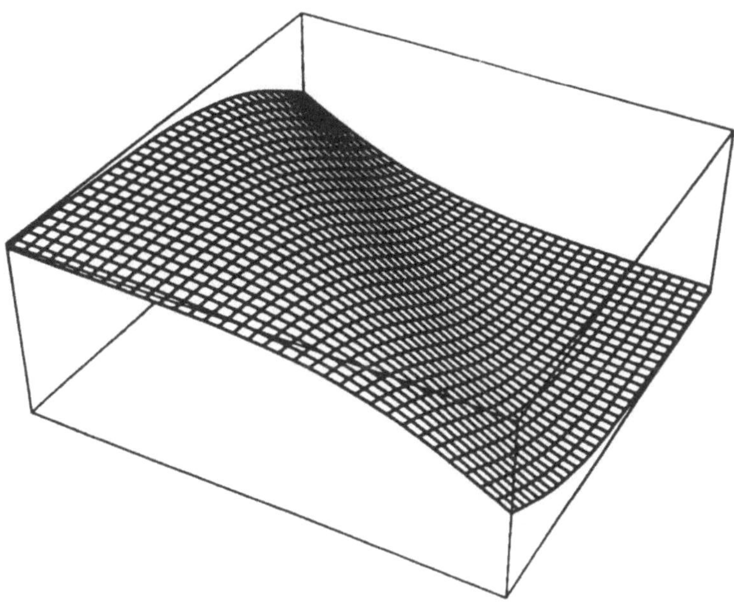

Fig. 1. Two-dimensional kink $m' = 6(\ln\, \phi)_x$.

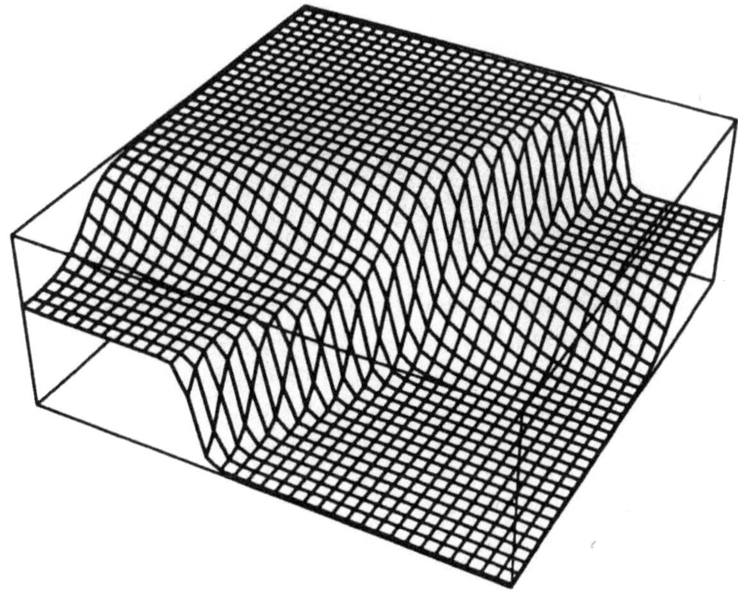

Fig. 2. Interaction of two kinks $m'' = 6(\ln\, \tau)_x$.

$$\psi = A_1 \cosh(a_1 x + a_1^3 t) + B_1 \cosh(b_1 y) \tag{A.8a}$$

$$\phi = A_2 \cosh(a_2 x + a_2^3 t) + B_2 \cosh(b_2 y). \tag{A.8b}$$

With these expressions for ϕ and ψ it is possible to integrate (3.7), (3.8)

$$\begin{aligned}
\Omega = {} & B_1 A_2 \cosh(b_1 y)\cosh(a_2 x + a_2^3 t) + \\
& + A_1 A_2 \left(\frac{a_2}{a_2^2 - a_1^2}\right)\{a_2 \cosh(a_1 x + a_1^3 t)\cosh(a_2 x + a_2^3 t) - \\
& - a_1 \sinh(a_1 x + a_1^3 t)\sinh(a_2 x + a_2^3 t)\} + \\
& + B_1 B_2 \left(\frac{b_1}{b_2^2 - b_1^2}\right)\{b_2 \sinh(b_1 y)\sinh(b_2 y) - \\
& - b_1 \cosh(b_1 y)\cosh(b_2 y)\}.
\end{aligned} \tag{A.9}$$

(A.8), (A.9) can be used in (3.17) in order to obtain τ as:

$$\begin{aligned}
\tau = {} & A_1 B_2 \cosh(a_1 x + a_1^3 t)\cosh(b_2 y) - A_2 B_2 \cosh(a_2 x + a_2^3 t)\cosh(b_1 y) + \\
& + \left(\frac{A_1 A_2}{a_2^2 - a_1^2}\right)\{2a_1 a_2 \sinh(a_1 x + a_1^3 t)\sinh(a_2 x + a_2^3 t) - \\
& - (a_2^2 + a_1^2)\cosh(a_1 x + a_1^3 t)\cosh(a_2 x + a_2^3 t)\} - \\
& - \left(\frac{B_1 B_2}{b_2^2 - b_1^2}\right)\{2b_1 b_2 \sinh(b_1 y)\sinh(b_2 y) - \\
& - (b_2^2 + b_1^2)\cosh(b_1 y)\cosh(b_2 y)\}.
\end{aligned} \tag{A.10}$$

The solutions for the first and second iteration can now be easily obtained. In fact, (3.9a) and (A.8b) provide the following expression for m':

$$m' = 6A_2 \left\{\frac{a_2 \sinh(a_2 x + a_2^3 t)}{A_2 \cosh(a_2 x + a_2^3 t) + B_2 \cosh(b_2 y)}\right\}. \tag{A.11}$$

In Figures 3 and 4 we have represented $m' = 6(\ln\, \phi)_x$ and $v' = 6(\ln\, \phi)_{xx}$. Figure 5 corresponds to the function $\beta' = 6(\ln\, \phi)_{pq}$ where $p = x + y$ and $q = x - y$. It is interesting to notice that m' and v' are not localized; nevertheless, β' is a localized function with a dromionic behaviour. The solution m'' for the second iteration can be obtained from (3.14) with τ given by (A.10). In Figures (6.a–e) we have represented $\beta'' = 6(\ln\, \tau)_{pq}$. These figures represent the elastic interaction between two dromions.

(3) We can conclude that (3.6) can be interpreted as a transformation that relates (1.3) with the linear equation (3.12). This relation is similar to the Cole–Hopf transformation between the Burgers and the heat equation [26]. Due to this linearization, there is another relationship between Burgers and (1.3) because they both have solutions that can be interpreted as fusion of solitons, like [19,

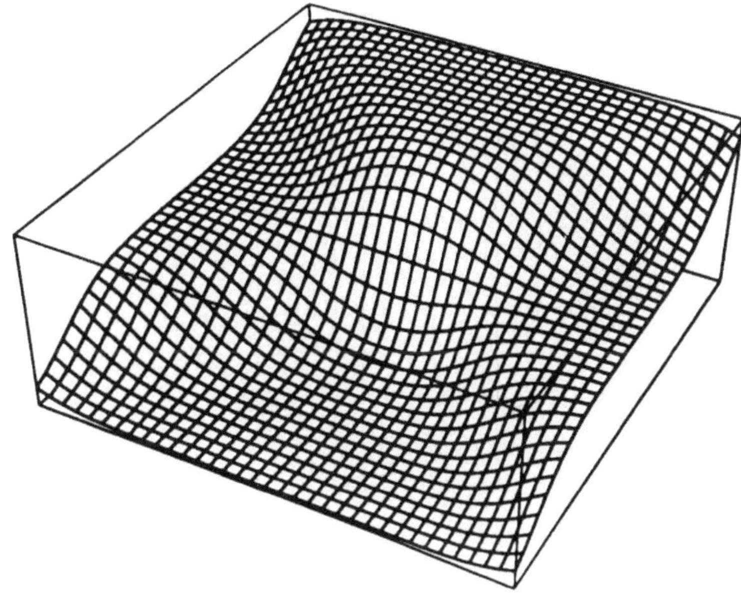

Fig. 3. $m' = 6(\ln\phi)_x$ for the dromion solution.

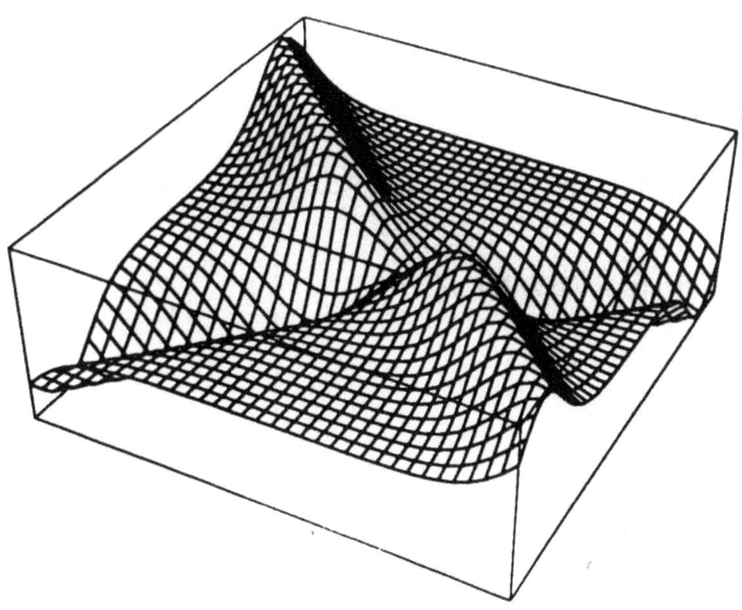

Fig. 4. $v' = 6(\ln\phi)_{xx}$ for the dromion solution.

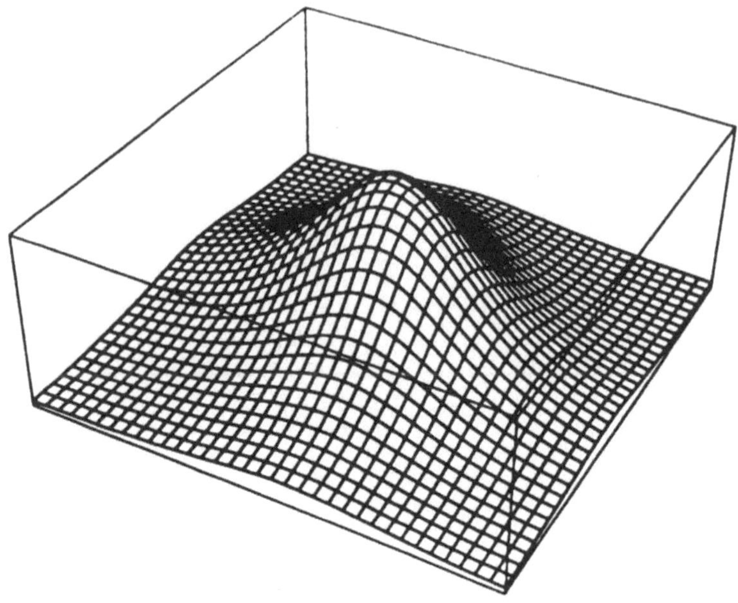

Fig. 5. $\beta' = 6(\ln \phi)_{pq}$ for the dromion solution.

27], but two-dimensional. This solution can be obtained in our framework by choosing a superposition of modes as the solution of (3.13). For

$$R_2 = e^{by} \tag{A.12a}$$

$$Q_2 = e^{a_1 x + a_1^3 t + c_1} + e^{a_2 x + a_2^3 t + c_3} \tag{A.12b}$$

we obtain the following solution for m':

$$m' = 6\,\frac{a_1 f_1 + a_2 f_2}{1 + f_1 + f_2} \tag{A.13}$$

where

$$f_i = e^{a_1 x + a_1^3 t - by + c_1}. \tag{A.14}$$

This solution can be interpreted as a particular case of (A.6) with $A = 0$. Nevertheless, the main difference is that (A.13) has been obtained in the first iteration while (A.6) needs to go to the second iteration.

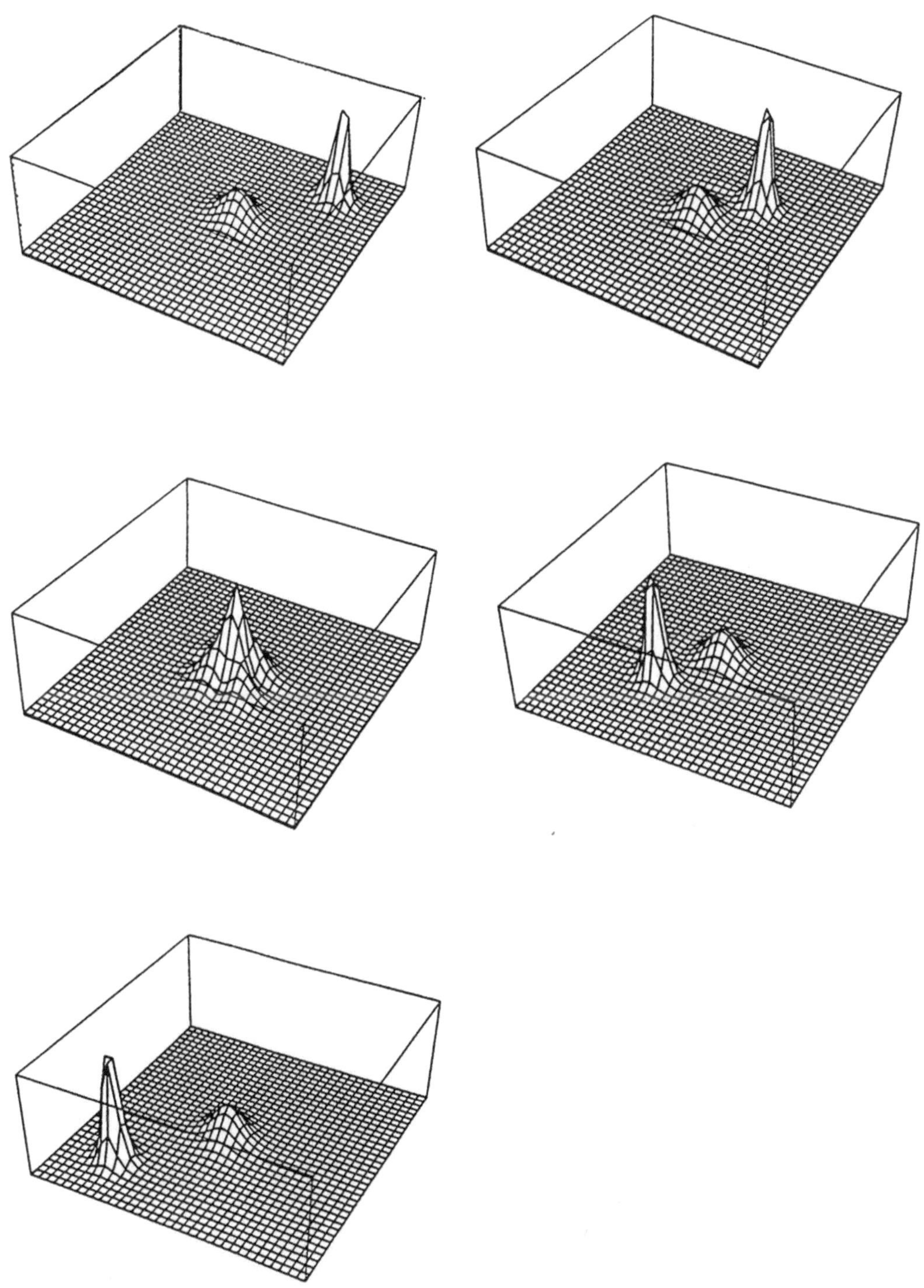

Fig. 6. Interaction of two dromions $\beta'' = 6(\ln\,\tau)_{pq}$.

References

1. Boiti, M., Leon, J. P., Manna, M., and Pempinelli, F.: *Inverse Problems* **2** (1986), 271.
2. Ablowitz, M. J. and Clarkson, P.: *Solitons, Nonlinear Evolution Equations and Inverse Scattering*, Lecture Notes Series, Vol. 149, pp. 269, Cambridge University Press, 1991.
3. Leble, S. B. and Ustinov, N. V.: *Inverse Problems*, to appear, 1994.
4. Hirota, R. and Satsuma, J.: *J. Phys. Soc. Japan* **40** (1976), 611.
5. Clarkson, P. A. and Mansfield, E. L.: *Nonlinearity* **7** (1994), 975.
6. Weiss, J., Tabor, M., and Carnevale, G.: *J. Math. Physics* **24** (1983), 522.
7. Weiss, J.: *Bäcklund Transformations and the Painlevé Property*, in P. J. Olver and D. Sattinger (eds), *Solitons in Physics, Mathematics and Nonlinear Optics*, IMA Volumes in Mathematics, Springer-Verlag, Berlin, 1990.
8. Kudryashov, N.: *Phys. Lett. A* **182** (1993), 356.
   Clarkson, P. A.: *Physica D* **19** (1986), 447.
9. Ganesan, S. and Lakshmanan, M.: *J. Phys. A* **20** (1987), L1143.
10. Mancarella, R. A., Soliani, G., and Solombrino, L.: *J. Math. Phys.* **29** (1988), 2666.
11. Konopelchenko, B. G. and Stramp, W.: *J. Math. Phys.* **32** (1991), 40.
12. Matveev, V. B.: *Lett. Math. Phys.* **3** (1979), 213.
13. Athorne, C.: *Inverse Problems* **9** (1993), 217.
14. Estévez, P. G.: *Phys. Lett A* **171** (1992), 259.
    Estévez, P. G.: *J. Phys. A* **27** (1994), 2113.
15. Clarkson, P. and Kruskal, M.: *J. Math. Phys. A* **30** (1989), 2201.
16. Cariello, F. and Tabor, M.: *Physica D* **39** (1989), 77.
    Conte, R.: *Phys. Lett. A* **134** (1988), 100.
17. Newell, A. C., Tabor, M., and Zeng, Y. B.: *Physica D* **29** (1987), 1.
18. Flaschka, H., Newell, A. C., and Tabor, M.: What is integrability? in V. E. Zakharov (ed.), *Nonlinear Dynamics*, Springer Series, 1991.
19. Estévez, P. G., Gordoa, P. R., Alonso, L. M., and Reus, E. M.: *J. Phys. A* **26** (1993), 1915.
20. Ablowitz, M. J., Ramani, A., and Segur, H.: *Lett. Nuov. Cim.* **23** (1978), 333.
21. Estévez, P. G. and Gordoa, P. R.: *Nonclassical Symmetries and the Singular Manifold Method: Theory and Six Examples*, Preprint of the University of Salamanca, 1993.
22. Olver, P. J.: *Applications of Lie Groups to Differential Equations*, Graduate Texts Math., Vol. 107, Springer-Verlag, 1986.
    Bluman, G. W. and Cole, J. D.: *Simmilarity Methods for Differential Equations*, Appl. Math. Sci. Vol., 81, Springer-Verlag, 1989.
    Levi, D. and Winternitz, P. J.: *J. Phys. A* **22** (1989), 2915.
23. Clarkson, P. A. and Hood, S. J.: *J. Math. Phys.* **35** (1994), 255.
    Nucci, M. C. and Clarkson, P. A.: *Phys. Lett.* **164 A** (1992), 49.
24. Bogoyavlenskii, O. I.: *Uspekhi Mat. Nauk.* **45** (1990), 17.
25. Li, Y. S. and Zhang, Y. J.: *J. Phys. A* **26** (1993), 7487.
26. Whitham, G. B.: *Linear and Nonlinear Waves*, Wiley, New York, 1974.
27. Estévez, P. G., Gordoa, P. R., Alonso, L. M., and Reus, E. M.: *Inverse Problems*, to appear, 1994.

Acta Applicandae Mathematicae **39**: 295–305, 1995.

The Korteweg–de Vries Equation and Beyond

A. S. FOKAS
Department of Mathematics and Computer Science and Institute for Nonlinear Studies, Clarkson University, Potsdam, NY 13699-5815, U.S.A.

(Received: 15 December 1994)

Abstract. We review a new method for linearizing the initial-boundary value problem of the KdV on the semi-infinite line for decaying initial and boundary data. We also present a novel class of physically important integrable equations. These equations, which include generalizations of the KdV, of the modified KdV, of the nonlinear Schrödinger and of the N-wave interactions, are as generic as their celebrated counterparts and, furthermore it appears that they describe certain physical situations more accurately.

Mathematics Subject Classification (1991): 58F07.

Key words: initial-boundary value problem, semi-infinite line, integrable equations.

1. Introduction

The KdV equation arises in a great number of physical situations. The reason for the ubiquitous occurrence of KdV equation is at least twofold. First, it combines simple dispersion with weak nonlinearity. Second, using the asymptotic method of multiple scales it can be shown that it describes (on suitable scales) the Riemann invariants of any hyperbolic system with weak nonlinearity and dispersion.

The KdV equation is integrable [1], i.e. it can be written as the compatibility condition of a pair of linear eigenvalue equations, called the Lax pair [2]. Indeed, the KdV equation

$$u_t + u_{xxx} + 6uu_x = 0, \tag{1.1}$$

is the compatibility condition of

$$\mu_{xx} + (u + k^2)\mu = 0, \tag{1.2}$$

and of

$$\mu_t = u_x\mu + (4k^2 - 2u)\mu_x. \tag{1.3}$$

Equation (1.2) is the celebrated time-independent Schrödinger equation. The spectral analysis of this equation, with u on the infinite line and decaying sufficiently fast as $|x| \to \infty$, yields a certain nonlinear Fourier transform. The derivation of this nonlinear transform involves the solution of a certain classical mathematical problem called the Riemann–Hilbert (RH) problem. This problem, which can be solved via a linear integral equation, is formulated in the complex k-plane; in particular the relevant 'jump' occurs on $\text{Im}(k) = 0$. Using this nonlinear Fourier transform it is possible to solve the Cauchy problem of KdV, $u(x,0) = u_0(x)$, for $u_0(x)$ decaying as $|x| \to \infty$. Equation (1.3) plays only an auxiliary role in this analysis, namely it can be used to show that the nonlinear inverse Fourier data evolve linearly in time.

In many physical situations, KdV arises in the form of an initial-boundary value problem on the semi-infinite line. This is for example the case of a certain laboratory study of water waves [3]. However, the solution of initial-boundary value problems for integrable equations was until recently open.

We have developed a method [4] for solving initial-boundary value problems on the half-infinite line for decaying initial and boundary data. This method, which also reduces the solution of the initial-boundary value problem to the solution of a single RH problem, exhibits the following novelties: (a) The formulation of the relevant RH problem is based on the analysis of both the x- and the t-parts of the Lax pair. Actually, it is the t-part which determines where, in the complex k-plane, the jumps occur. For the KdV equation the jumps of the RH problem occur on $\text{Im}(k^3) = 0$, which is a reflection of the fact that the t-part of the Lax pair contains k^3, which in turn is a consequence of the fact that the KdV involves a third derivative in x. (b) The analysis of initial-boundary value problems for integrable equations such as the KdV equation, in addition to techniques from exact integrability it also requires the essential use of general PDE techniques. In the case of initial value problems, the exact methods could be used to establish existence of global solutions as well as to study the properties of the solutions. In contrast, for initial-boundary value problems exact methods are used only to study the properties of solutions.

A summary of our linearization scheme for the KdV [5] is given in Section 2.

In addition to the KdV equation there exist several other integrable nonlinear PDEs which are also generic, i.e. they can also be derived, using physically meaningful asymptotic techniques, from very large classes of nonlinear PDEs [6, 7]. The most well known such equations are the nonlinear Schrödinger (NLS) and the N-wave interactions. The NLS describes the amplitude $A(\xi, \tau)$ of the modulation of a monochromatic wave of *any* dispersive system, where $\xi = \varepsilon x$ and $\tau = \varepsilon^2 t$. Similarly, the N-wave interactions describe the modulation of N simple dispersive waves.

We have recently shown [8, 9] that there exist integrable generalizations of the KdV, of the modified KdV, of the NLS, and of the N-wave interactions, which are as generic, and furthermore they describe physics more accurately than their celebrated counterparts. For example, if one looks for an equation describing the modulation of a monochromatic wave which is valid to $\text{O}(\varepsilon^3 t)$ one obtains our generalized NLS [9].

The integrable generalization of the KdV is the equation

$$u_t + u_x + \nu u_{xxt} + \gamma u_{xxx} + \alpha u u_x + \frac{\alpha\nu}{3}(u u_{xxx} + 2u_x u_{xx}) = 0, \qquad (1.4)$$

where α, ν, γ are constants. As a concrete physical application of this equation we consider the motion of a 2-dimensional, inviscid, incompressible, irrotational, fluid above a horizontal flat bottom located at $y = -h_0$ (h_0 constant). This system is characterized by two parameters $\alpha = a/h_0$ and $\beta = h_0^2/l^2$, where a and l are typical values of the amplitude and of the wavelength of the waves. Assuming that the waves are unidirectional and neglecting terms of $O(\alpha^3, \alpha^2\beta, \beta^2)$, the location $\zeta(x,t)$ of the free surface satisfies [8, 10]

$$\eta_t + \eta_x + \alpha\eta\eta_x + \beta\eta_{xxx} + \rho_1\alpha^2\eta^2\eta_x + \alpha\beta(\rho_2\eta\eta_{xxx} + \rho_3\eta_x\eta_{xx}) = 0, \qquad (1.5)$$

where $\rho_1 = -1/6$, $\rho_2 = 5/3$, $\rho_3 = 23/6$. This equation, under the local transformation

$$\eta = u + \lambda_1\alpha u^2 + \lambda_2\beta u_{xx}, \qquad (1.6)$$

where $\lambda_1 = -\rho_1$, $\lambda_2 = -3\rho_1 - 7\rho_2/4 + \rho_3/2$, is equivalent to $O(\alpha^2, \alpha\beta, \beta)$ to Equation (1.4) with $\nu = -3\rho_2\beta/2$, $\gamma = (1 - 3\rho_2/2)\beta$.

The integrable generalizations of modified KdV and of NLS are given in Section 3.

2. KdV on the Semi-Infinite Line

We consider the KdV Equation (1.1) with $x, t \in [0, \infty)$ and with $u(x,0) = q(x)$, $u(0,t) = v(x)$ given, where $q(x)$ and $v(t)$ decay for large x and large t, respectively, and such that $q(0) = v(0)$.

We have developed the following linearization scheme for this initial-boundary value problem. Given $u(x,0)$ construct

$$s_1^+(k) = \frac{1}{2}w(0,k) - \frac{1}{2ik}w_x(0,k)$$

and

$$s_2^+(k) = \frac{1}{2}w(0,k) + \frac{1}{2ik}w_x(0,k),$$

where $w(x,k)$ is the solution of Equation (1.2) with $u(x,t)$ replaced by $u(x,0)$ satisfying the boundary condition

$$\lim_{x\to\infty} w(x,k)\exp(-ikx) = 1.$$

Define $b(k)$ by $b = s_1^+/\bar{s}_2^+$. Let II denote the domain of the complex k-plane bounded by the rays $\arg k = \pi/3$ and $\arg k = 2\pi/3$. Let $c(k)$, $\arg k = \pi/3$ and $\arg k = 2\pi/3$, be the boundary value of some function meromorphic for $k \in II$,

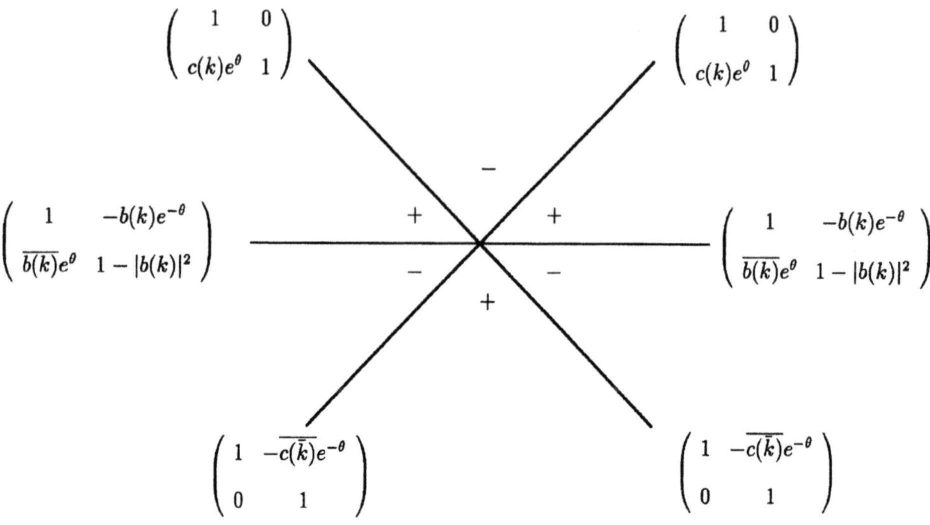

Fig. 1. The RH problem associated with the initial-boundary value problem of the KdV. The x, t dependence enters only through $\theta(x,t) = 2i(kx + 4k^3 t)$. The domains formed by the rays imaginary $k^3 = 0$ will be denoted by I,...,VI.

with poles at the zeros of $s_2^+(k)$ and at the points $\{k_j\}_1^N$, $k_j \in i\mathbb{R}^+$, and such that $c(k) \to 0$ as $k \to \infty$. Having $s_1^+(k)$, $s_2^+(k)$, and $c(k)$, solve a RH problem for the sectionally meromorphic function $\widetilde{Z}_p(x,t,k)$ with the jumps shown in Figure 1. This function can have poles only at $\{k_j\}_1^N$. Define $\widehat{Z}_p(x,t,k)$ by

$$\widehat{Z}_p(x,t,k) = I + \frac{iy(x,t)}{2k}\begin{pmatrix} 1 & 1 \\ -1 & -1 \end{pmatrix}\widetilde{Z}_p(x,t,k);$$

$$y(x,t) \doteqdot \lim_{k\to\infty}(k\widetilde{Z}_p)_{12}. \tag{2.1}$$

Then determine $u(x,t)$ by

$$u(x,t) = 2i\partial_x\left[\lim_{k\to\infty}(k\widehat{Z}_p(x,t,k))_{11}\right].$$

The points $\{k_j\}_1^N$ give rise to solitons which move away from the origin.

The points $\{k_j\}_1^N$ have the following spectral characterization. Let $(\widehat{\psi}_1(t,k), \widehat{\psi}_2(t,k))^T$ be the solution of the vector equation

$$\widehat{\psi}_t(t,k) + \\ + \left(4ik^3\sigma_3 - 2ku(0,t)\sigma_2 - u_x(0,t)\sigma_1 - \frac{q(0,t)}{k}(i\sigma_3 - \sigma_2)\right)\widehat{\psi}(t,k) = 0, \tag{2.2}$$

where $q(0,t) \doteqdot \frac{1}{2}u_{xx}(0,t) + u^2(0,t)$, and σ_1, σ_2, σ_3 are the three Pauli matrices. This equation is the vector version of Equation (1.3). Then $\{k_j\}_1^N$ is the

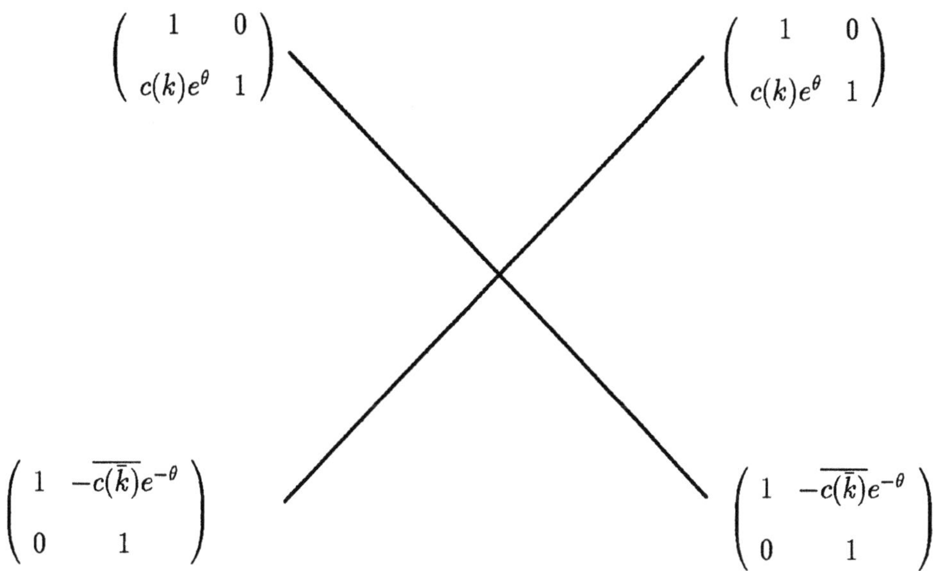

Fig. 2. The RH problem associated with the initial-boundary value problem of the KdV with $u(x,0)=0$. The x and t dependents enter only through $\theta(x,t)=2i(kx+4k^3t)$.

discrete spectrum of this equation supplemented with the boundary condition $\widehat{\psi}_1(0,k)s_2^+(k)-\widehat{\psi}_2(0,k)s_1^+(k)=0$.

In the particular case of $u(x,0)=0$, the RH problem reduces to the one depicted in Figure 2; it is specified by the boundary values of $c(k)$ for $\arg k=\pi/3$ and $\arg k=2\pi/3$, by $\{k_j\}_1^N$ and by $\{c_j\}_1^N$, where c_j denotes the residue of $c(k)$ at k_j.

Unfortunately, although we have a complete characterization of the analytic properties of $c(k)$, we have not found an effective way of computing $c(k)$ in terms of $u(0,t)$. Let $c(k)$ be *any* suitably decaying function meromorphic for $k \in II$. Then the solution of the RH problem of Figure 1 generates the solution $u(x,t)$ corresponding to initial data $u(x,0)$ and *some* boundary data $u(0,t)$. The main limitation of our result is that for given $u(x,0)$ and $u(0,t)$ we cannot construct $c(k)$ by solving a linear problem. Nevertheless, for *any* given $u(x,0)$ and $u(0,t)$ the corresponding function $c(k)$ exists; in other words, the RH problem of Figure 1.1 solves the initial-boundary value problem (1.1) for *general* initial-boundary data.

The above RH problem is quite natural. To appreciate this we first recall the RH problem that corresponds to the KdV with $x \in (-\infty,\infty)$. This RH problem is depicted in Figure 3.

Comparing the RH problems of Figure 1 and Figure 3 we see that the jumps for $k \in \mathbb{R}$ are identical. The jumps for $\arg k=\pi/3$ and $\arg k=2\pi/3$ cannot have a nonzero entry in the 12 position since $e^{-\theta}$ is unbounded for $k \in \mathbb{C}^+$. The jumps for $k \in \mathbb{C}^-$ follow by symmetry considerations. The fact that $c(k)$ has

$$+ \qquad \begin{pmatrix} 1 & -b(k)e^{-\theta} \\ \overline{b(k)}e^{\theta} & 1-|b(k)|^2 \end{pmatrix}. \qquad -$$

Fig. 3.

analytic continuation for $k \in II$ can also be easily understood. At $t = 0$, the RH problem of Figure 1 must be reduced to the one that defines $u(x, 0)$. At $t = 0$, the term e^{θ} has analytic continuation in II. In this way, at $t = 0$ one finds the RH problem of Figure 3 with θ replaced by $2ikx$. This RH problem corresponds precisely to $u(x, 0)$.

The fact that $u(x, 0)$, $u(0, t)$ and $c(k)$ are related in a nonlinear way is a reflection of the fact that $u_x(0, t)$ and $u_{xx}(0, t)$ depend nonlinearly on $u(x, 0)$ and $u(0, t)$. To appreciate this we first recall the solution of the linearized problem

$$u_t + u_{xxx} = 0, \quad x, t \in [0, \infty), \tag{2.3}$$

where $u(0, t)$ and $u(x, 0)$ are given and decaying for large t and large x. In order to draw comparisons with the nonlinear problem we use a Fourier transform to solve Equation (2.3),

$$\widehat{u}(k, t) = \int_0^{\infty} \mathrm{d}x\, \mathrm{e}^{ikx} u(x, t). \tag{2.4}$$

The evolution of the Fourier data $\widehat{u}(k, t)$ is given by

$$\widehat{u}_t + ik^3 \widehat{u} = u_{xx}(0, t) - iku_x(0, t) - k^2 u(0, t). \tag{2.5}$$

In this equation $u(0, t)$ is known but $u_x(0, t)$ and $u_{xx}(0, t)$ are unknown. These unknown functions can be found by using the fact that the solution $\widehat{u}(k, t)$ of (2.5) is analytic in the upper half of the k complex plane. It turns out that this requirement implies

$$\widehat{u}(k, 0) = \int_0^{\infty} \mathrm{d}t\, \mathrm{e}^{ik^3 t} [-u_{xx}(0, t) + iku_x(0, t) + k^2 u(0, t)]. \tag{2.6}$$

Letting $k = \mathrm{e}^{i\pi/6}\sqrt{\rho}$, $\rho > 0$, and $k = \mathrm{e}^{5i\pi/6}\sqrt{\rho}$, $\rho > 0$, this equation yields the equations

$$\begin{aligned} &\widehat{u}(\mathrm{e}^{i\pi/6}\sqrt{\rho}, 0) \\ &\quad = \int_0^{\infty} \mathrm{d}t\, \mathrm{e}^{-\rho t} \left[-u_{xx}(0, t) + i\mathrm{e}^{i\pi/6}\sqrt{\rho} u_x(0, t) + \mathrm{e}^{i\pi/3} \rho u(0, t)\right], \end{aligned} \tag{2.7}$$

and

$$\widehat{u}(-\mathrm{e}^{-i\pi/6}\sqrt{\rho},0) = \int_0^{\infty} \mathrm{d}t\,\mathrm{e}^{-\rho t}\left[-u_{xx}(0,t) - i\mathrm{e}^{-i\pi/6}\sqrt{\rho}u_x(0,t) + \mathrm{e}^{-i\pi/3}\rho u(0,t)\right]. \quad (2.8)$$

If $u \in \mathbb{R}$, Equation (2.8) is the complex conjugate of Equation (2.7). Given $u(x,0)$ and $u(0,t)$, Equations (2.7) and (2.8) can be solved for $u_x(0,t)$ and $u_{xx}(0,t)$.

It is important to notice that if $u(0,t)$, $u_x(0,t)$, and $u_{xx}(0,t)$ are arbitrary decaying functions, then the rhs of Equation (2.6) will be analytic for $k \in I \cup III \cup V$. However, in order for these functions to be the boundary values of the solution of Equation (2.3), it is necessary and sufficient that the rhs of Equation (2.6) has analytic continuation across the rays $\arg k = \pi/3$ and $\arg k = 2\pi/3$ ($\widehat{u}(k,0)$ is analytic for k in the upper half of the complex plane).

We now discuss the nonlinear problem. Let $(\widehat{\psi}_1(t,k), \widehat{\psi}_2(t,k))^T$ be the solution of the vector Equation (2.2) satisfying the boundary condition

$$\lim_{t\to\infty} [(\widehat{\psi}_1, \widehat{\psi}_2)\exp(-4ik^3t)] = (0,1)^T.$$

Let $r(k) \doteqdot \widehat{\psi}_1(0,k)/\widehat{\psi}_2(0,k)$. It turns out that for arbitrary decaying functions $u(0,t)$, $u_x(0,t)$, and $u_{xx}(0,t)$, the function $r(k)$ is a meromorphic function for $k \in I \cup III \cup V$ and $r(k) \to 0$ as $k \to \infty$. However, if $u(0,t)$, $u_x(0,t)$, and $u_{xx}(0,t)$ are the boundary values of the KdV equation then in addition $r(k)$ satisfies

$$r(k) = \frac{s_1^+(k)}{s_2^+(k)}, \quad k \in I \cup III, \quad (2.9)$$

where s_1^+ and s_2^+ are determined from $u(x,0)$ and are analytic for k in the upper half complex plane. Equation (2.9) is the analogue of Equation (2.6). It shows that although the relationship between $u(x,0)$, $u(0,t)$, $u_x(0,t)$, and $u_{xx}(0,t)$ is highly nonlinear, its reflection in the k-plane (scattering space) is rather simple: $r(k)$ has analytic continuation across the rays $\arg k = \pi/3$ and $\arg k = 2\pi/3$.

There exists an invertible correspondence between the 'potential' $Q(t) \doteqdot \{u(0,t), u_x(0,t), u_{xx}(0,t)\}$ and the scattering data $r(k)$: Given $Q(t)$, Equation (2.2) implies $r(k)$. Conversely given a meromorphic function $r(k)$, one can find $Q(t)$ by solving a RH problem with the jump $r(k)$ for $k^3 \in \mathbb{R}$. Given $u(0,t)$ and $u(x,0)$, we cannot compute $r(k)$ for $k \in V$ by solving a linear problem. This is a consequence of the fact that now we have a 'mixed' problem where one gives part of the potential, i.e. $u(0,t)$, and part of the scattering data, i.e. $r(k)$, for $k \in I \cup III$. It turns out that this problem can be formulated as a *nonlinear* RH problem and will be discussed elsewhere.

In spite of the fact that we have no effective way of computing $c(k)$ in terms of $u(0,t)$ we can give an effective description of the long time asymptotics.

The study of the large t behavior of $u(x,t)$ reduces to the study of the large t behavior of the RH problem of Figure 1. Because the x,t dependence of this RH problem is rather simple, it is possible to give an effective asymptotic description of $u(x,t)$ as $t \to \infty$. It turns out that the asymptotics of $u(x,t)$ is given by the usual N-soliton solution of the KdV equation.

It was mentioned in the introduction that the analysis of initial-boundary value problems requires an essential use of general PDE techniques. This follows from the fact that in order to study the map between $Q(t)$ and $r(k)$ one needs apriori estimates for $u_x(0,t)$ and $u_{xx}(0,t)$. The uniqueness and existence of a global solution for the KdV on the quarter plane is established in [11, 12]. This result is based on L_2 estimates. However, the methods of exact integrability are based on L_1 estimates. It is therefore crucial for us to extend these results from L_2 to L_1. This poses significant technical difficulties which for the case of the nonlinear Schrödinger equation were discussed in [13, 14]. The situation of the KdV equation is more complicated. This and other rigorous aspects of the present formalism are presently under consideration.

3. Novel Generic Integrable Equations

Equation (1.4) was derived in the late seventies from mathematical considerations (see Equations (26e) and (30) of [15]). It has recently been rederived from physical considerations [16, 8]. Equation (1.4) is actually a member of the infinite hierarchy of integrable equations,

$$q_t + \sum_{j=0}^{\infty} c_j \Phi^j q_x = 0,$$

$$\Phi \doteqdot \left[\gamma\partial^3 + \frac{\alpha}{3}(q\partial + \partial q)\right](\partial + \nu\partial^3)^{-1}, \quad q = u + \nu u_{xx}, \tag{3.1}$$

where $\partial \doteqdot \partial_x$, and c, ν, c_j are arbitrary constants. Equation (3.1) with $c_0 = c = c_1 = 1$, $c_j = 0$ for $j \neq 0, 1$, is Equation (1.4). The integrability of Equation (3.1) is based on the following general result proven independently in [17] and [18]. Assume that the operator $\theta_1 + \lambda\theta_2$ is a Hamiltonian operator for all values of the constant parameter λ. Then each member of the hierarchy of equations $q_t = (\theta_2\theta_1^{-1})^n q_x$, $n \in \mathbb{Z}$, is an integrable equation (it is a bi-Hamiltonian system possessing infinitely many conserved quantities). The operator $\partial^3 + \lambda_1\partial + \lambda_2(q\partial + \partial q)$ is Hamiltonian for all values of the constants λ_1 and λ_2. Let

$$\theta_2 = \gamma\partial^3 + \frac{\alpha}{3}(q\partial + \partial q);$$

the choice of $\theta_1 = \partial$ yields the KdV hierarchy, while the choice of $\theta_1 = \partial + \nu\partial^3$ yields Equation (3.1).

The Lax pair of Equation (1.4) (with $\alpha = 1$) is given by

$$\psi_{xx} + \left[k^2(1-v) + \frac{v}{4\nu}\right]\psi = 0, \quad \sigma v = u + \nu u_{xx}, \tag{3.2}$$

and

$$\psi_t + \left(-\tfrac{1}{2}u_x + c\right)\psi + \left(u + \frac{\gamma}{\nu} + \frac{2\sigma}{4\nu k^2 - 1}\right)\psi_x = 0, \tag{3.3}$$

where $\sigma = (\gamma - \nu)/2\nu$, and c is an arbitrary constant.

Equation (1.4) exhibits a much richer phenomenology than the KdV equation. In particular it possesses peakons [16] (if both u_x and u_{xxx} are absent) and compactons [19] (if $\nu > 0$). The 1-compacton solution of (1.4) (with $\alpha = 1$) is

$$\begin{aligned} u = {} & H\left[\frac{x-\gamma t/\nu}{\sqrt{\nu}} + \pi\right] H\left[\pi - \frac{x - \gamma t/\nu}{\sqrt{\nu}}\right] \times \\ & \times \left(\frac{\gamma - 2\nu}{2\nu}\right)\left[1 + \cos\left(\frac{x-\gamma t/\nu}{\sqrt{\nu}}\right)\right], \end{aligned} \tag{3.4}$$

where $H(x)$ is the Heaviside function, i.e. $H(x) = 0$ for $x < 0$ and $H(x) = 1$ for $x > 0$.

The linearization of Equation (1.4) using the acoustic spectral problem (3.2) is given in [20]. Among the novel features of the inverse spectral method for this case we note that: (a) while for $v < 1$ the relevant RH problem has jumps on $\mathrm{Im}(k) = 0$, for $v > 1$ the jumps occur on $\mathrm{Im}(k^2) = 0$; (b) If $v > 1$ it is possible to have infinitely many solitons.

We now discuss the 1-soliton solution of (1.4) for $v < 1$. This solution is given by

$$\begin{aligned} u(X) &= \frac{c_3 \exp\left\{-(2p + 1/\sqrt{|\nu|})U - 2pX\right\}}{\left[c_2\exp\{-2p(U+X)\} + 1\right]^2}, \quad X = x - \frac{c}{p}t, \\ c &= -p\left(\frac{\gamma}{\nu} - \frac{2\sigma}{1 + 4\nu p^2}\right), \end{aligned} \tag{3.5}$$

where

$$\begin{aligned} c_1 &= \left(\lambda - \sqrt{\lambda^2 - 1}\right)e^{2pU_1}, \qquad c_2 = \left(\lambda + \sqrt{\lambda^2-1}\right)e^{2pU_1}, \\ c_3 &= 4p\sqrt{|\nu|}\sqrt{\lambda^2 - 1}\,e^{2pU_1}, \qquad \lambda = \frac{1 - 4\nu p^2}{1 + 4\nu p^2}, \end{aligned} \tag{3.6}$$

p is an arbitrary positive constant, U_1 is an arbitrary constant, and $U(X)$ is the solution of the transcendental equation

$$\exp\left\{\frac{U(X)}{\sqrt{|\nu|}}\right\} = \frac{c_1\exp\{-2p(U+X)\} + 1}{c_2\exp\{-2p(U+X)\} + 1}. \tag{3.7}$$

It turns out that the zeros of $\mathrm{d}u/\mathrm{d}X$ are given by the cubic equation

$$E^3 + \frac{E^2}{c_2} - \frac{E}{c_1c_2} - \frac{1}{c_1c_2^2} = 0. \tag{3.8}$$

Thus depending on the constants c_1 and c_2 the 1-soliton consists of one or two humps.

The integrable generalization of the modified KdV equation is

$$\begin{aligned} &u_t + u_x + \nu u_{xxt} + \gamma u_{xxx} + \alpha u u_x + \frac{\nu\alpha}{3}(u u_{xxx} + 2u_x u_{xx}) + \\ &\quad +3\mu\alpha^2 u^2 u_x + \nu\mu\alpha^2(u^2 u_{xxx} + u_x^3 + 4u u_x u_{xx}) + \\ &\quad +\nu^2\mu\alpha^2(u_x^2 u_{xxx} + 2u_x u_{xx}^2) = 0, \end{aligned} \tag{3.9}$$

where α, γ, ν, μ are constants. This equation can be derived using the bi-Hamiltonian approach outlined above with

$$\theta_2 = \partial + \gamma\partial^3 + 2\left(\frac{\alpha}{3}q + \mu\alpha^2 q^2\right)\partial + q_x\partial^{-1}\left(\frac{\alpha}{3} + 2\mu\alpha^2 q\right)\partial. \tag{3.10}$$

The choice $\theta_1 = \partial$ yields the Gardner equation, i.e. a combination of KdV and of modified KdV. The choice $\theta_1 = \partial + \nu\partial^3$ yields (3.9).

Equation (3.9) also arises in water waves: if one keeps terms of $\mathrm{O}(\alpha^3, \alpha^2\beta)$ one obtains, instead of Equation (1.5), an equation which is equivalent to Equation (3.9).

The integrable generalization of the NLS is [9],

$$iQ_t + iQ_x + \nu Q_{xt} + \gamma Q_{xx} + \rho Q|Q|^2 - i\rho\nu|Q|^2 Q_x = 0. \tag{3.11}$$

The integral generalization of the N-wave interaction is given in [9].

We note that using our approach it is possible to obtain similar generalizations starting with any integrable equation in one or two spatial dimensions. For example the sine-Gordon equation generalizes to

$$u_{xt} = \sin(u + \nu u_{xx}). \tag{3.12}$$

It is quite interesting that one can also find a similar integrable generalization of Burger's equation.

It appears that the results summarized in this section together with the notion of equivalence (see also the important results of [21]) will enhance significantly our understanding of several physical systems.

Acknowledgements

This work was partially supported by the National Science Foundation under Grant Number DMS–9111611, by the Air Force Office of Scientific Research under Grant Number AFOSR–89–0510, and by SERC.

References

1. Gardner, C. S., Green, J. M., Kruskal, M. D., and Miura, R. M.: *Phys. Rev. Lett.* **19** (1967), 1095.
   For recent developments, see A. S. Fokas and V. E. Zakharov (eds), *Important Developments in Soliton Theory*, Springer-Verlag, New York, 1993.
2. Lax, P.: *Comm. Pure Appl. Math.* **21** (1968), 467.
3. Bona, J., Pritchard, W. G., and Scott, L. R.: *Philos. Trans. Royal Soc. London, A* **302** (1981), 457–510.
4. Fokas, A. S.: *Proceedings of the III Potsdam–V Kiev International Workshop 1991*, Springer-Verlag, Berlin, 1992;
   Fokas, A. S. and Its, A. R.: *Phys. Rev. Lett.* **68** (1992), 3117.
5. Fokas, A. S. and Its, A. R.: An initial-boundary value problem for the KdV equation, Preprint, 1993.
6. Taniuti, T. and Wei, C. C.: *J. Phys. Soc. Japan* **24** (1968), 941;
   Taniuti, T.: *Progr. Theor. Phys.* **55** (1974), 1;
   Kodama, Y. and Taniuti, T.: *J. Phys. Soc. Japan* **45** (1978), 298.
7. Calogero, F.: Why are certain nonlinear PDEs both widely applicable and integrable, in V. E. Zakharov (ed.), *What is Integrability*, Springer-Verlag, New York, 1992.
8. Fokas, A. S.: Moderately long waves of moderate amplitude and integrable, Preprint, 1994.
9. Fokas, A. S.: Asymptotic integrability, Preprint, 1994.
10. Whitham, G. B.: *Linear and Nonlinear Waves*, Wiley–Intersci., New York, 1974.
11. Bona, J. and Winther, R.: *SIAM J. Math. Anal.* **14** (1983), 1056–1106.
12. Bona, J. and Winther, R.: *Differ. Integral Equations* **2** (1989), 228–250.
13. Fokas, A. S. and Its, A. R.: The linearization of the initial-boundary value problem of the nonlinear Schrödinger equation, to appear in *SIAM J. Math. Anal.* (1994).
14. Sung, L.: Solution of the initial-boundary value problem of NLS using PDE techniques, Preprint, Clarkson University, 1994.
15. Fuchssteiner, B. and Fokas, A. S.: *Physica* **4D** (1981), 47.
16. Holm, D. and Camassa, R.: *Phys. Rev. Lett.* **71** (1993), 1671.
17. Fokas, A. S. and Fuchssteiner, B.: *Lett. Nuovo Cimento* **28** (1980), 299.
18. Gel'fand, I. M. and Dorfman, I.: *Funct. Anal. Appl.* **13** (1979), 13; **14** (1980), 71.
19. Rosenau, P.: *Phys. Rev. Lett.* (1994).
20. Fokas, A. S. and Santini, P. M.: An inverse acoustic problem and linearization of moderate amplitude dispersive waves, Preprint, 1994.
21. Kodama, Y.: *Phys. Lett. A* **112** (1985), 193.

Acta Applicandae Mathematicae **39:** 307–314, 1995.

On the Background of Limit Pass for Korteweg–de Vries Equation as the Dispersion Vanishes

V. A. GALKIN and V. V. RUSSKIKH
Department of Applied Mathematics, Institute of Atomic Energetics, Obninsk 249020, Russia

(Received: 2 September 1994)

Abstract. We present the new approach to the background of approximate methods of convergence based on the theory of functional solutions and solutions in the mean one for conservation laws. The applications to the Cauchy problem to KdV equation, when dispersion tends to zero are considered. Also the Galerkin method for a periodic problem for the KdV equation is considered.

Mathematics Subject Classification (1991): 58F07.

Key words: functionals, limit pass, dispersion, Cauchy problem, Galerkin method.

1. Functional Solutions Theory for Conservation Laws

Below we consider functional solutions of the Cauchy problem for the system of conservation laws [1, 2]

$$J(u) \stackrel{\text{def}}{=} \partial_t u^{(w)}(x,t) + \sum_{j=1}^{n} \partial_{x_j} f_j^{(w)}(u,x,t) - f_{n+1}^{(w)}(u,x,t) = 0,$$

$$u|_{t=0} = u_0, \quad w \in W, \; x \in \mathbb{R}^n, \; t > 0, \tag{1}$$

for the unknown variable u, where W is the given set of parameters, x are spatial variables and t is the time. Let W be a locally compact metric space with a Borel measure μ, which is bounded on compact sets in W, strictly positive on open sets and the property of density for μ is supposed; $\mathrm{d}x$ and $\mathrm{d}t$ are the Lebesgue measures on $\mathbb{R}^n$ and $\mathbb{R}^1$, respectively; $\dot{B}$ is a set of bounded compactly supported Borel functions on the topological product $Q = W \times \mathbb{R}^n \times \mathbb{R}^1$; $Q_0 = W \times \mathbb{R}^n$; $\dot{B}^\infty$ are functions infinitely differentiable with respect to (x,t), whose derivatives belong to $\dot{B}$; $L(A,q)$ is the set of Borel functions on the set A, which are locally summable with respect to measure q. Suppose $\mathcal{D} \subset L_1^{\mathrm{loc}}(W,\nu)$, f_j: $\mathcal{D} \times \mathbb{R}^n \times \mathbb{R}_+^1 \to L_1^{\mathrm{loc}}(Q,\nu)$, $1 \leqslant j \leqslant n$, are defined mappings, where $\nu = \mu \otimes \mathrm{d}x \otimes \mathrm{d}t$. On the set of functions

$$M = \Big\{ u \in L_1(Q,\nu)\colon \; u|_{(x,t)\in\mathbb{R}^n\times\mathbb{R}_+^1} \subset \mathcal{D}, \\ (f_j \circ u) \in L_1^{\mathrm{loc}}(Q,\nu), \; 1 \leqslant j \leqslant n+1 \Big\},$$

we consider the system of equations for the unknown variable $u \in M$

$$\int_Q \left[u\partial_t g + \sum_{j=1}^{n}(f_j \circ u)\partial_{x_j} g + (f_{n+1} \circ u)\right]\nu(\mathrm{d}Q) + \\ + \int_{Q_0} g|_{t=0} u_0\, \nu_0(\mathrm{d}Q_0) = 0 \quad \forall g \in \dot{B}^{\infty}, \tag{2}$$

which are used as a definition of the generalized solution of the Cauchy problem for conservation laws system (1) with initial data u_0. For the case $\operatorname{card} W > 1$, the concept of a generalized solution is restrictive. The extension of the concept of a solution proposed below (functional solutions) makes it possible to justify the convergence of approximate methods in the presence of an *a-priori* estimate of an approximation in $L_1^{\mathrm{loc}}(Q, \nu)$, which is uniform in the parameter.

We denote $\mathcal{E}$ as the vector space which consists of linear combinations

$$F_{g,g_1} = \sum_{j=1}^{n} f_j \partial_{x_j} g + f_{n+1} g + u g_1, \quad u \in \mathcal{D},$$

with arbitrary $g \in \dot{B}^{\infty}$ and $g_1 \in \dot{B}$. For each vector in $\mathcal{E}$, we define the operators π, π_0 by means of relations

$$\pi(F_{g,g_1}) = F_{g,\partial g/\partial t}, \qquad \pi_0(F_{g,g_1}) = F_{0,g|_{t=0}}.$$

Let $\mathcal{E}^+ = \{l\}$ be the algebraic dual space to $\mathcal{E}$. On $\mathcal{E}$ we define the topology $\sigma(\mathcal{E}^+, \mathcal{E})$ by the system of seminorms $p_F(l) = |l(F)|$, $F \in \mathcal{E}$, and consider an imbedding of the set M into $\mathcal{E}$ by the formula

$$\forall u \in M\colon\ u \mapsto l_u \in \mathcal{E}^+, \qquad l_u(F) = \int_Q (F \circ u)\, \nu(\mathrm{d}Q), \quad F \in \mathcal{E}.$$

This imbedding is monomorphic on the equivalence classes of functions in M. Denote by $[M]$ the closure of the range of imbeddings of the set M in $\mathcal{E}$; on the $[M]$ induce topology $\sigma(\mathcal{E}^+, \mathcal{E})$. A similar imbedding was made with the set of initial data M_0. If function $u \in M$ is a solution of problem (2) (i.e. generalized solution), then relations (2) are equivalent to the relations

$$l_u(\pi(F)) + l_{u_0}^{(0)}(\pi_0(F)) = 0 \quad \forall F \in \mathcal{E}.$$

DEFINITION 1. We call an element $l \in [M]$ a functional solution of Cauchy problem (2) if for each $F \in \mathcal{E}$ the following relation holds

$$l(\pi(F)) + l_{u_0}^{(0)}(\pi_0(F)) = 0.$$

It can be said that we have specified a regular approximate method of solving problem (2), denoted $\mathfrak{RAM}$, if the collection of parameteric families of elements in M is indicated as solutions of the equations $J_\alpha(u_\alpha) = 0$. It is supposed that operators J_α are defined on the set $V \subset M$, the closure of which in L_1^{loc} contains M and there exists subsequent parameters α_n, $n \in \mathbb{N}$, when J_{α_n} converges to J defined by (1) in every point of V. The functional solution which is the limit point of approximations obtained by means of the regular method $\mathfrak{RAM}$ is called a regular functional solution.

DEFINITION 2. $\mathfrak{RAM}$ weakly approximates problem (2) if it is possible to find a sequence $u_{\alpha_n} \in M$ for which the values of the error

$$\begin{aligned}\delta_n(F) = {} & \left| l_{u_{\alpha n}}(\pi(F)) + l_{u_{\alpha n}|_{t=0}}(\pi_0(F)) \right| + \\ & + \left| \left(l^{(0)}_{u_0} - l^{(0)}_{u_{\alpha n}|_{t=0}} \right)(\pi_0(F)) \right|, \quad F \in \mathcal{E},\ n \in \mathbb{N},\end{aligned}$$

vanish for each F in $\mathcal{E}$ as $n \to \infty$. It is supposed that respective operators J_{α_n} converge as mentioned above for this sequence of parameters.

DEFINITION 3. $\mathfrak{RAM}$ converges if there exists a sequence $l_{u_{\alpha n}}$ convergent in $[M]$, whose limit is a functional solution of problem (2).

DEFINITION 4. $\mathfrak{RAM}$ is stable if on each compact set $K \subset Q$ the following relation holds

$$\sup_{\alpha \in A} \int_K |u_\alpha| \, \nu(\mathrm{d}Q) < \infty.$$

THEOREM 1. *Suppose the method $\mathfrak{RAM}$ is stable and there is the property of weak approximation. Then this method converges to a regular functional solution of Cauchy problem* (2).

THEOREM 2. *Every regular functional solution is defined by the unique element of space $L_1^{\text{loc}}(Q, \nu)$.*

THEOREM 3. *Suppose the relations* (2) *are produced by the Cauchy problem for the ordinary system of differential equations*

$$\mathrm{d}u/\mathrm{d}t = f(u),$$

(*system* (1) *without spatial variables terms*) *with measurable bounded vector function $f = f_{n+1} \in \mathbb{R}^m$. Then the Euler finite difference method uniformly converges to the regular functional solution of the Cauchy problem which is unique Filippov's global solution* [4].

EXAMPLE 1. Consider the ordinary differential equation $\mathrm{d}u/\mathrm{d}t = f(u)$, where $f(u) = 1$ for $u < 0$ and $f(u) = -1$ on contrary. The Euler method uniformly converges to $u(t) = 0$, $t \geqslant 0$ (if $u(0) = 0$), which is not a classical or generalized solution, but it is a regular functional (Filippov's) solution.

The same situation is connected with the Cauchy problem for the scalar quasilinear equation

$$\partial_t u(x,t) + \partial_x (xu(x,t)/|u(x,t)|) = 0.$$

The continuous regular global functional solution for initial date u_0 is given by the formula $u(x,t) = u_0(x) - \text{sgn}(u_0(x))t$, $0 \leqslant t \leqslant |u_0(x)|$, and $u(x,t) = 0$, $t \geqslant |u_0(x)|$.

THEOREM 4. *Suppose the compact set K consists of regular functional solutions which are produced by the compact set Y of initial data and there exists a unique element in K for every initial data in Y. Then the functional solutions in K are continuously dependent on the initial data in Y if K and Y are equipped by the topology of weak convergence in L_1^{loc}.*

Remark. The choice of K is usually defined by regularization $\mathfrak{RAM}$ because of (2), as rule is a ill posed problem. The type of $\mathfrak{RAM}$ is strictly connected with the physical nature of system (1).

There exists an example of nonempty above-mentioned sets K which consists of an infinite number of elements.

Below we describe some applications of functional solutions to quasilinear and semilinear systems (1) connected with unsolved problems of mathematical physics in classes of generalized solutions.

EXAMPLE 2. (Vanishing viscosity method for quasilinear system of conservation laws.) We consider the Cauchy problem for system (1) where $\text{card}\, W$ is a finite number, $f_{n+1} = 0$. (The Euler equations of incompressible fluid dynamics are included in this case [3].) To solve this problem approximately, we use the regularization of (1) by substitution of the right-hand side of (1) to spatially elliptic terms (viscosity forces) $\alpha(w)\Delta u$

$$\partial_t u_\alpha^{(w)}(x,t) + \sum_{j=1}^{n} \partial_{x_j} f_j^{(w)}(u_\alpha, x, t) = \alpha(w)\Delta u_\alpha^{(w)}.$$

A sufficient condition for convergence of this method is the condition of uniform local summability of the approximations. This estimation is usual for gradient systems, where $f_j = \nabla_u F_j(u)$. Then *a priori* L_2 estimation is available. The *a priori* estimations of this type take place for Navier–Stokes equations which are used as regularization for Euler equations of incompressible fluid dynamics if the initial global kinetic energy of the fluid is bounded [3, 5, 6].

2. On Background of Limit Pass for Korteweg–de Vries (KdV) Equation as the Dispersion Vanishes

Consider the following problem

$$u_t + uu_x = \alpha u_{xxx}, \tag{3}$$

$$u(x,0) = u^0(x), \quad x, \alpha \in \mathbb{R},\ t \in [0,T],\ 0 < T \leqslant \infty. \tag{4}$$

Equation (3) is the KdV one. (If $\alpha = 0$, we have the Hopf equation which describes the infinite stream of noninteracting particles.) The background of the vanishing dispersion method for Equation (3) in the limits of the classic theory is a very difficult problem. Owing to the physical features of the considered phenomena [11], one may consider it reasonable to find a measure $\mu_{x,t}(\mathrm{d}u)$ (which is defined by respective distribution function) as the limit as $\alpha \to 0$ of approximations (3). In this part, such analysis is based on the theory of solutions in the mean (SM) [8], which is closely connected with the theory of measure-value (MV) solutions [9].

Following [8], we consider the imbedding of the space of Borel locally summable functions u: $[0,T] \times \mathbb{R} \to \mathbb{R}$ with respect to Lebesgue measure $\mathrm{d}x\,\mathrm{d}t$ into the space of regular Borel measures bounded on compact sets $\mathfrak{M}^+ = \mathfrak{M}^+(\mathbb{R} \times [0,T] \times \mathbb{R})$ by means of the formula

$$u \longmapsto \mu_u \in \mathfrak{M}^+\colon\ \mu_u(A) = \int_{\pi(A\cap u)} \mathrm{d}t\,\mathrm{d}x, \quad \pi(\lambda,t,x) \in \mathbb{R} \times [0,T] \times \mathbb{R},$$

where the set A belongs to the Borel σ-algebra on $\mathbb{R} \times [0,T] \times \mathbb{R}$.

DEFINITION 5. The measure $\mu \in \mathfrak{M}^+$ is a SM of the problem (3) corresponding to the initial data $\mu_0 \in \mathfrak{M}_0^+$ if the equality

$$\left\langle \lambda\phi_t + \frac{\lambda^2}{2}\phi_x, \mu \right\rangle + \langle \phi(x,0)\lambda, \mu_0\rangle_0 = 0$$

is satisfied with arbitrary compactly supported, smooth function ϕ defined on $[0,T] \times \mathbb{R}$. Here

$$\langle f, \mu\rangle \int_{\mathbb{R}\times[0,T]\times\mathbb{R}} f(\lambda,t,x)\,\mu(\mathrm{d}\lambda\,\mathrm{d}t\,\mathrm{d}x),$$

$$\langle f, \mu_0\rangle_0 = \int_{\mathbb{R}\times\mathbb{R}} f(\lambda,x)\,\mu_0(\mathrm{d}\lambda\,\mathrm{d}x), \quad \mu_0 \in \mathfrak{M}_0^+ = \mathfrak{M}^+(\mathbb{R}\times\mathbb{R}).$$

The principal result of [8] conformably written for our problem is

THEOREM 5. *Suppose there exists a continuous positive function* η: $\mathbb{R} \to \mathbb{R}^+$ *such that for each* $0 < T_1 \leqslant T$ *the following* a priori *estimate holds uniformly with respect to* α:

$$\int_0^{T_1} \int_{\mathbb{R}} \eta(u^\alpha(t,x))\,\mathrm{d}x\,\mathrm{d}t \leqslant C(T_1) < \infty,$$

where u^α *is the weak solution of the problem* (3), (4) *by fixed* α. *If*

$$\lim_{u\to\infty} [\eta(u)]^{-1} \sup_{(x,t)\in K} u^2 = 0 \quad \forall K \Subset [0,T]\times\mathbb{R},$$

$$\lim_{u\to\infty} [\eta(u)]^{-1}|u| < \infty,$$

then the problem Cauchy for the Hopf equation has a SM.

Below on the basis of this theorem, we will prove the following assertion.

PROPOSITION 1. *Suppose that*
(1) *the problem* (3), (4) *has a compactly supported weak solution* u^α ($\forall t \in [0,T]$: $\operatorname{supp} u^\alpha \Subset Q$, *where* Q *is the bound domain in* $\mathbb{R}$);
(2) *the following a-priori estimates uniformly hold with respect to* α:

$$\int_{\mathbb{R}} (u^\alpha)^2\,\mathrm{d}x \leqslant C_1 < \infty, \tag{5}$$

$$\left|\int_{\mathbb{R}} \left[\alpha(u_x^\alpha)^2 - \frac{(u^\alpha)^3}{3}\right]\mathrm{d}x\right| \leqslant C_2 < \infty. \tag{6}$$

Then the vanishing dispersion method converges to the SM of the problem Cauchy for the Hopf equation.

Remark. Here we do not consider conditions (1) and (2) of the Proposition, as these are discussed in detail in the respective literature (see, for example, [10, 11]).

Proof of Proposition. Owing to (5), (6) and Sobolev theorems of imbedding, the sequence $|\alpha|^{1/2}u_x^\alpha$ is bound in the space $L_2(\mathbb{R})$. Then from (6) we have that the sequence u^α is bound in the space $L_3(\mathbb{R})$ uniformly with respect to α. Clearly, the function $\eta(\lambda) = \lambda^3$ satisfies all conditions of Theorem 4. Proposition 1 is proved. □

PROPOSITION 2. *Suppose the estimate* (5) *holds uniformly with respect to* α. *Then the vanishing dispersion method converges to the functional solution of the problem* (3), (4) *by* $\alpha = 0$.

3. Galerkin Method for the Periodic Problem for the KdV Equation

We consider the following problem

$$u_t + uu_x = \alpha u_{xxx}, \tag{7}$$

$$u(x,0) = u^0(x), \tag{8}$$

$$\frac{\partial^j u(0,t)}{\partial x^j} = \frac{\partial^j u(1,t)}{\partial x^j}, \quad j = 0,1,2, \ x \in [0,1], \ t \in \mathbb{R}^+. \tag{9}$$

Before the existence of solutions of the problem, (7)–(9) was proved on the basis of the parabolic regularization method under the condition that yet one periodic condition is given ($\partial^3 u(0,t)/\partial x^3 = \partial^3 u(1,t)/\partial x^3$). Here we prove the solvability of the problem (7)–(9) by the Galerkin method where this condition is not used.

So, let the functions system $\{\phi_j\}_{j=1}^{\infty}$ be the basis in the space

$$W = \left\{ v \in W_2^4[0,1]\colon \ \frac{\partial^j v(0,t)}{\partial x^j} = \frac{\partial^j v(1,t)}{\partial x^j}, \ j = 0,1,2 \right\}.$$

We define the approximate method for the problem (7)–(9) by the following relationships

$$u^m = \sum_{j=1}^{m} C_j^m(t)\phi_j, \quad m \geqslant 1, \tag{10}$$

$$\int_0^1 \left[u_t^m \phi_k - (u^m)^2/2\phi_{kx} - \alpha u_x^m \phi_{kxx} \right] \mathrm{d}x = 0, \tag{11}$$

$$u^m|_{t=0} = u_0^m, \tag{12}$$

where u_0^m is a function's sequence from $W_2^1[0,1]$ such that $u_0^m \to u^0$ in $W_2^1[0,1]$ as $m \to \infty$. On the basis of this method, the following assertion was proved

PROPOSITION 3. *There exists a general, in the Sobolev sense, a solution of the problem* (7)–(9).

References

1. Galkin, V. A.: Functional solutions of conservation laws, *Doklady Akad. Nauk SSSR* **310** (1990), 834–839; English transl. in *Sov. Phys. Dokl.* **35**(2) (1990), 133–135.
2. Galkin, V. A.: in *Proc. Int. Meeting on Ordinary Diff. Equations and Their Applications*, Italy, Florence, Sept. 1993, p. 52.
3. Galkin, V. A.: Numerical stability and convergence of approximate methods for conservation laws, *J. Modern Phys. C* (to appear).

4. Filippov, A. F.: Ordinary differential equations with discontinuity in right-hand side, *Mat. Sbornik SSSR* **51** (1960), (in Russian).
5. Galkin, V. A. and Russkikh, V. V.: *Proc. Inst. Atomic Energetics* **2** (1992).
6. Galkin, V. A. and Russkikh, V. V.: Convergence of approximate methods for equations of incompressible fluid dynamics, *Mathematical Modeling* **6** (1994), 101–113 (in Russian).
7. Galkin, V. A.: Generalized solution of the Smoluchowski kinetic equation for spatially inhomogeneous systems, *Sov. Phys. Dokl.* **32**(3) (1987), 200–201.
8. Galkin, V. A. and Tupchiev, V. A.: On solvability in the mean of a system of quasilinear conservation laws, *DAN SSSR* **300** (1988).
9. Di Perna, R. J.: Measure-value solutions of conservation laws, *Arch. Ration. Mech. Anal.* **88** (1985), 223–270.
10. Lax, P. D.: Integrals of nonlinear equations of evolution and solitary waves, *Commun. Pure Appl. Math.* **21** (1968).
11. Kadomtzev, B. B.: *Collective Phenomena in Plasma*, Nauka, Moscow, 1976.

Acta Applicandae Mathematicae **39:** 315–333, 1995.

On New Trace Formulae for Schrödinger Operators

F. GESZTESY
Department of Mathematics, University of Missouri, Columbia, MO 65211, U.S.A.
e-mail: mathfg@mizzou1.missouri.edu

and

H. HOLDEN
Department of Mathematical Sciences, Norwegian Institute of Technology, University of Trondheim, N–7034 Trondheim, Norway, e-mail: holden@imf.unit.no

(Received: 28 November 1994)

Abstract. We review a variety of recently obtained trace formulae for (multidimensional) Schrödinger operators and indicate their connections with the KdV hierarchy in one dimension. Our principal new result in this paper concerns a set of trace formulae in $1 \leqslant d \leqslant 3$ dimensions related to point interactions.

Mathematics Subject Classifications (1991): 35Q53, 34L05, 35P05, 47A10.

Key words: trace formulas, Schrödinger operators, Krein's spectral shift function, KdV hierarchy.

1. Introduction

We give a brief account of some of our recent work on new trace formulae for Schrödinger operators obtained in various joint collaborations with B. Simon and Z. Zhao [14–16, 19–21].

While [14] is devoted to a detailed study of trace formulae and their relation to the Korteweg–de Vries (KdV) hierarchy in connection with KdV solutions which have sufficiently fast decay at spatial infinity, the main results in [15, 16, and 20] are trace formulae for Schrödinger operators $H = -\mathrm{d}^2/\mathrm{d}x^2 + V$ in $L^2(\mathbb{R}; \mathrm{d}x)$ for a large class of potentials which satisfy a minimal set of *a priori* restrictions. In particular, we were interested in finding a general trace formula for any continuous potential V bounded from below. In [21] we derived trace formulae for multi-dimensional Schrödinger operators, and we review these results here.

As shown in great detail in [5] in connection with short-range potentials, the trace formula (2.65) for $V(x)$ is a central object in inverse scattering theory. Similarly, the trace formula (2.31) studied in [4, 6–9, 11, 22, 23, 26–29, 32] in connection with periodic potentials, algebro-geometric quasi-periodic finite-gap potentials, and certain classes of almost-periodic potentials has been vital in solving the associated inverse spectral problem and the corresponding Cauchy problem for the KdV equation. Moreover, the description of isospectral manifolds of quasi-periodic finite-gap potentials (as well as solutions of the KdV hierarchy), obtained by algebro-geometric methods, is based on trace formulae. Hence it is natural to look for extensions of these special trace formulae to larger classes of potentials in order to be able to study inverse spectral problems and the KdV initial value problem in a more general setting. Our main strategy in this endeavor is to compare $H = -\mathrm{d}^2/\mathrm{d}x^2 + V$ and H_y^D, the operator H plus an additional Dirichlet boundary condition at the point $y \in \mathbb{R}$. The spectral characteristics of H and H_x^D, especially the corresponding Krein spectral shift function $\xi(\lambda, x)$ associated with the pair (H_x^D, H) [24] then enables one to recover the potential $V(x)$. As has been demonstrated in [15], the function $\xi(\lambda, x)$ provides a powerful new tool in the spectral and inverse spectral analysis of H.

We close our paper with a new set of multi-dimensional trace formulae based on point interactions. For a general reference on point interactions, see [1]. Here we compare the Schrödinger operators $H = -\Delta + V$ (on $L^2(\mathbb{R}^\nu)$, $\nu = 1, 2, 3$) and H with an additional boundary condition which corresponds to adding a point interaction at a point $x \in \mathbb{R}^\nu$ with strength α. The difference of the resolvents of these two operators is a rank-one operator, and one can use an analysis similar to the one-dimensional case.

2. One-Dimensional Trace Formulae and KdV Invariants

Throughout this section we shall assume that $V \in C^\infty(\mathbb{R})$ is real-valued and bounded from below. We then define the self-adjoint Schrödinger operator

$$Hf = \tau f, \qquad \tau = -\frac{\mathrm{d}^2}{\mathrm{d}x^2} + V(x),$$

$$\mathcal{D}(H) = \{g \in L^2(\mathbb{R}) \mid g,\ g' \in AC_{\mathrm{loc}}(\mathbb{R}),\ \tau g \in L^2(\mathbb{R})\} \tag{2.1}$$

and the corresponding self-adjoint Dirichlet Schrödinger operator

$$H_y^D f = \tau f, \qquad y \in \mathbb{R},$$

$$\mathcal{D}(H_y^D) = \Big\{ g \in L^2(\mathbb{R}) \mid g,\ g' \in AC_{\mathrm{loc}}(\mathbb{R}\backslash\{y\}),$$
$$\lim_{\varepsilon \downarrow 0} g(y \pm \varepsilon) = 0,\ \tau g \in L^2(\mathbb{R}) \Big\} \tag{2.2}$$

in $L^2(\mathbb{R})$. In order to gain information about $V(x)$ we shall compare H and H_x^D which are 'close' in the sense that their resolvents differ only by a rank-one perturbation. More precisely, if $G(z,x,x')$ denotes the Green's function of H (i.e., the integral kernel of $(H-z)^{-1}$) then

$$(H_y^D - z)^{-1} = (H-z)^{-1} - G(z,y,y)^{-1}(\overline{G(z,y,\cdot)},\cdot)G(z,\cdot,y),$$
$$y \in \mathbb{R},\ z \in \mathbb{C}\backslash\{\sigma(H_y^D)\cup\sigma(H)\}, \tag{2.3}$$

where $(\cdot,\cdot)$ denotes the scalar product in $L^2(\mathbb{R})$. Thus one computes for the trace of (2.3)

$$\mathrm{Tr}\left[(H_x^D - z)^{-1} - (H-z)^{-1}\right] = -\frac{\mathrm{d}}{\mathrm{d}z}\ln\left[G(z,x,x)\right], \quad x\in\mathbb{R}. \tag{2.4}$$

Since we shall need a high-energy asymptotic expansion of (2.4) we recall the following result

$$\mathrm{Tr}\left[\mathrm{e}^{-\tau H_x^D} - \mathrm{e}^{-\tau H}\right]\widetilde{_{s\downarrow 0}}\sum_{\ell=0}^{\infty} s_\ell(x)\tau^\ell, \tag{2.5}$$

$$\mathrm{Tr}\left[(H_x^D - z)^{-1} - (H-z)^{-1}\right]\widetilde{_{z\in\mathbb{C}\backslash C_\varepsilon,\ |z|\to\infty}}\sum_{j=0}^{\infty} r_j(x)z^{-j-1}, \tag{2.6}$$

where C_ε is a cone with apex at $E_0 := \inf[\sigma(H)]$ and opening angle $\varepsilon > 0$ (arbitrarily small) along the semi-axis $[E_0,\infty)$ and

$$s_\ell(x) = (-1)^{\ell+1}(\ell!)^{-1}r_\ell(x), \quad \ell\in\mathbb{N}_0 := \mathbb{N}\cup\{0\}, \tag{2.7}$$

$$r_0(x) = \tfrac{1}{2}, \qquad r_1(x) = \tfrac{1}{2}V(x),$$

$$\begin{aligned} r_j(x) &= (-1)^{j+1}2^{1-2j}j\phi_{2j-1}(x) + \\ &\quad + \sum_{\ell=1}^{j-1}(-1)^{j-\ell+1}2^{1-2(j-\ell)}\phi_{2(j-\ell)-1}(x)r_\ell(x), \quad j\geqslant 2, \end{aligned} \tag{2.8}$$

$$\begin{aligned} \phi_1(x) &= V(x), \qquad \phi_2(x) = -V'(x), \\ \phi_{j+1}(x) &= -\phi_j'(x) - \sum_{\ell=1}^{j-1}\phi_\ell(x)\phi_{j-\ell}(x), \quad j\geqslant 2. \end{aligned} \tag{2.9}$$

The asymptotic expansion (2.5) is proved in [20] by means of the Feynman–Kac formula, (2.6) is then a consequence of (2.5). Moreover, one can prove that

$$r_j'(x) = (-1)^j 2^{1-2j} j\phi_{2j}(x), \quad j\in\mathbb{N}. \tag{2.10}$$

Explicitly, one computes

$$\begin{aligned} r_0(x) &= \tfrac{1}{2}, \qquad r_1(x) = \tfrac{1}{2}V(x), \\ r_2(x) &= \tfrac{1}{2}V(x)^2 - \tfrac{1}{4}V''(x), \qquad \text{etc.,} \end{aligned} \tag{2.11}$$

$$\begin{aligned} s_0(x) &= -\tfrac{1}{2}, \qquad s_1(x) = \tfrac{1}{2}V(x), \\ s_2(x) &= \tfrac{1}{8}V''(x) - \tfrac{1}{4}V(x)^2, \qquad \text{etc.} \end{aligned} \tag{2.12}$$

Next we employ the fact that for all $x \in \mathbb{R}$, $G(z,x,x)$ is a Herglotz function with respect to z, i.e., $G(\cdot,x,x)$ is analytic in $\mathbb{C}_+ := \{z \in \mathbb{C} \mid \mathrm{Im}(z) > 0\}$ and $\mathrm{Im}[G(z,x,x)] > 0$ for all $z \in \mathbb{C}_+$. Hence $G(\cdot,x,x)$ allows the representation (see, e.g., [2])

$$G(z,x,x) = \exp\left\{c + \int_{\mathbb{R}} \mathrm{d}\lambda \left[\frac{1}{\lambda - z} - \frac{\lambda}{1+\lambda^2}\right]\xi(\lambda,x)\right\}, \tag{2.13}$$

where $c \in \mathbb{R}$, $\xi(\cdot,x) \in L^1_{\mathrm{loc}}(\mathbb{R}; \mathrm{d}\lambda)$ is real-valued, and

$$\int_{\mathbb{R}} \mathrm{d}\lambda (1+\lambda^2)^{-1}\xi(\lambda,x) < \infty.$$

Fatou's lemma then implies the existence of

$$\begin{aligned} \xi(\lambda,x) &= \pi^{-1} \lim_{\varepsilon \downarrow 0} \mathrm{Im}\left\{\ln\left[G(\lambda + i\varepsilon, x, x)\right]\right\} \\ &= \pi^{-1} \arg\left[G(\lambda + i0, x, x)\right] \qquad \text{for a.e. } \lambda \in \mathbb{R} \end{aligned} \tag{2.14}$$

for all $x \in \mathbb{R}$. Moreover, $0 \leqslant \xi(\lambda,x) \leqslant 1$ for all $x \in \mathbb{R}$ and for a.e. $\lambda \in \mathbb{R}$ due to the Herglotz property of $G(z,x,x)$. In particular, $\xi(\lambda,x)$ is Krein's spectral shift function [24] for the pair (H_x^D, H). Before we continue it might be appropriate to give the following simple example.

EXAMPLE 2.1.

$$V^{(0)}(x) \equiv 0, \qquad H^{(0)} = -\frac{\mathrm{d}^2}{\mathrm{d}x^2},$$

$$G^{(0)}(z,x,x') = \frac{i}{2z^{1/2}}\,\mathrm{e}^{iz^{1/2}|x-x'|}, \qquad \mathrm{Im}(z^{1/2}) \geqslant 0.$$

Then

$$\xi^{(0)}(\lambda) = \pi^{-1}\arg\left[G^{(0)}(\lambda + i0, x, x)\right] = \begin{cases} 1/2, & \lambda > 0 \\ 0, & \lambda < 0. \end{cases} \tag{2.15}$$

From now on we normalize $\xi(\lambda, x)$ to be zero below the spectrum of H, i.e.,

$$\xi(\lambda, x) = 0, \qquad \lambda < E_0 = \inf\left[\sigma(H)\right]. \tag{2.16}$$

This is consistent with the fact that $H_x^D \geqslant H$ and hence $\inf[\sigma(H_x^D)] \geqslant \inf[\sigma(H)] = E_0$ and that $G(\lambda + i0, x, x) > 0$, $\lambda < E_0$, $x \in \mathbb{R}$. Inserting (2.13) into (2.4) yields

$$\operatorname{Tr}\left[(H_x^D - z)^{-1} - (H - z)^{-1}\right] = -\int_{E_0}^{\infty} \frac{\mathrm{d}\lambda\, \xi(\lambda, x)}{(\lambda - z)^2}, \tag{2.17}$$

$$z \in \mathbb{C}\backslash\{\sigma(H_x^D) \cup \sigma(H)\}.$$

More generally, one can show that

$$\operatorname{Tr}\left[F(H_x^D) - F(H)\right] = \int_{E_0}^{\infty} \mathrm{d}\lambda\, F'(\lambda)\xi(\lambda, x) \tag{2.18}$$

whenever $F \in C^2(\mathbb{R})$, $(1 + \lambda^2)F^{(j)} \in L^2((0, \infty))$, $j = 1, 2$. In particular,

$$\operatorname{Tr}\left[\mathrm{e}^{-\tau H_x^D} - \mathrm{e}^{-\tau H}\right] = -\tau \int_{E_0}^{\infty} \mathrm{d}\lambda\, \mathrm{e}^{-\tau\lambda}\xi(\lambda, x), \quad \tau > 0. \tag{2.19}$$

Hence combining (2.5), (2.12), and (2.19) yields

$$\begin{aligned} \operatorname{Tr}\left[\mathrm{e}^{-\tau H_x^D} - \mathrm{e}^{-\tau H}\right] &\underset{\tau\downarrow 0}{=} -\frac{1}{2} + \frac{1}{2}V(x)\tau + 0(\tau^2) \\ &= -\tau \int_{E_0}^{\infty} \mathrm{d}\lambda\, \mathrm{e}^{-\tau\lambda}\xi(\lambda, x) \\ &= -\frac{1}{2} + \frac{\tau}{2}\int_{E_0}^{\infty} \mathrm{d}\lambda\, \mathrm{e}^{-\tau\lambda}\left[\mathrm{e}^{\tau E_0} - 2\xi(\lambda, x)\right] \end{aligned} \tag{2.20}$$

implying the general trace formula for $V(x)$

$$V(x) = \lim_{\tau\downarrow 0} \int_{E_0}^{\infty} \mathrm{d}\lambda\, \mathrm{e}^{-\tau\lambda}\left[\mathrm{e}^{\tau E_0} - 2\xi(\lambda, x)\right]. \tag{2.21}$$

We emphasize the Abelian limit in (2.21) which cannot be removed in general. In particular, for potentials satisfying $V(x) \to \infty$ as $|x| \to \infty$ (such as the harmonic oscillator) one verifies $|1 - 2\xi(\lambda, x)| = 1$ and hence the Abelian limit in (2.21) cannot be removed. The computation (2.20) easily generalizes to all higher KdV invariants:

THEOREM 2.2.

$$s_0(x) = -\tfrac{1}{2},$$

$$s_\ell(x) = \frac{(-1)^{\ell-1}}{\ell!} \times \\ \times\left\{\frac{E_0^\ell}{2} + \ell \lim_{t\downarrow 0} \int_{E_0}^{\infty} d\lambda\, e^{-t\lambda}\lambda^{\ell-1}[\tfrac{1}{2} - \xi(\lambda, x)]\right\}, \quad \ell \in \mathbb{N}. \tag{2.22}$$

Explicitly, this yields

$$\begin{aligned} s_0(x) &= -\tfrac{1}{2}, \qquad s_1(x) = \tfrac{1}{2}V(x) = \frac{E_0}{2} + \lim_{t\downarrow 0}\int_{E_0}^{\infty} d\lambda\, e^{-t\lambda}\left[\tfrac{1}{2} - \xi(\lambda, x)\right], \\ s_2(x) &= \tfrac{1}{8}V''(x) - \tfrac{1}{4}V(x)^2 \\ &= -\frac{E_0^2}{4} - \lim_{t\downarrow 0}\int_{E_0}^{\infty} d\lambda\, e^{-t\lambda}\lambda\left[\tfrac{1}{2} - \xi(\lambda, x)\right], \quad \text{etc.} \end{aligned} \tag{2.23}$$

The behavior of $\xi(\lambda, x)$ for λ in spectral gaps of H is particularly simple. In fact, since V is real-valued, $G(\lambda + i0, x, x)$ is real-valued for $\lambda \in \mathbb{R}\backslash\sigma(H)$ and hence

$$\xi(\lambda, x) = \pi^{-1} \arg\left[G(\lambda + i0, x, x)\right] = 1 \text{ or } 0 \text{ for a.e. } \lambda \in \mathbb{R}\backslash\sigma(H). \tag{2.24}$$

More precisely, suppose $(\lambda_1, \lambda_2) \subset \mathbb{R}\backslash\sigma(H)$ is a spectral gap of H and $\mu(x) \in (\lambda_1, \lambda_2)$ is a (Dirichlet) eigenvalue of H_x^D, i.e., $\mu(x) \in \sigma_p(H_x^D)$. Then one easily verifies

$$\xi(\lambda, x) = \begin{cases} 1, & \lambda_1 < \lambda < \mu(x) \\ 0, & \mu(x) < \lambda < \lambda_2. \end{cases} \tag{2.25}$$

We note that since $(H_x^D - z)^{-1} - (H - z)^{-1}$ is of rank one, there is at most one Dirichlet eigenvalue (i.e., an eigenvalue of H_x^D) in each spectral gap of H. One then has in general

$$\sigma(H_x^D) = \sigma_{\text{ess}}(H) \cup \{\mu_j(x)\}_{j\in J},$$

where J is a countable index set (possibly empty).

A particularly interesting special case is the one where $V(x)$ is periodic of period $a > 0$

$$V(x + a) = V(x), \quad x \in \mathbb{R}. \tag{2.26}$$

Standard Floquet theory then yields

$$\sigma(H) = \bigcup_{n=1}^{\infty} \left[E_{2(n-1)}, E_{2n-1}\right], \quad E_0 < E_1 \leqslant E_2 < E_3 \leqslant \cdots \tag{2.27}$$

and

$$\begin{aligned} \sigma(H_x^D) &= \sigma(H) \cup \{\mu_n(x)\}_{n\in\mathbb{N}}, \\ E_{2n-1} &\leqslant \mu_n(x) \leqslant E_{2n}, \quad x \in \mathbb{R},\ n \in \mathbb{N}. \end{aligned} \tag{2.28}$$

We then obtain a new proof of the famous trace formula for periodic potentials $V(x)$ originally due to Gelfand and Levitan [11] (see also [4, 6–9, 22, 23, 26–29, 32]).

THEOREM 2.3. *Let* $V \in C^\infty(\mathbb{R})$, $V(x+a) = V(x)$, $x \in \mathbb{R}$ *for some* $a > 0$. *Then*

$$\begin{aligned} &2(-1)^{\ell+1}\ell! s_\ell(x) \\ &\quad = E_0^\ell + \sum_{n=1}^{\infty} \left[E_{2n-1}^\ell + E_{2n}^\ell - 2\mu_n(x)^\ell\right], \quad \ell \in \mathbb{N},\ x \in \mathbb{R}. \end{aligned} \tag{2.29}$$

Since in the periodic case $G(\lambda + i0, x, x)$ is purely imaginary for $\lambda \in \sigma(H)$, one infers

$$\xi(\lambda, x) = \begin{cases} 0, & \lambda < E_0,\ \mu_n(x) < \lambda < E_{2n},\ n \in \mathbb{N} \\ 1, & E_{2n-1} < \lambda < \mu_n(x),\ n \in \mathbb{N} \\ \frac{1}{2}, & E_{2(n-1)} < \lambda < E_{2n-1},\ n \in \mathbb{N} \end{cases} \tag{2.30}$$

(and similarly for limiting cases where $\mu_n(x) \in \{E_{2n-1}, E_{2n}\}$ for some $n \in \mathbb{N}$). Insertion of the step function (2.30) into (2.22) proves (2.29). In particular, for $\ell = 1$ this yields the trace formula for periodic potentials $V(x)$,

$$V(x) = E_0 + \sum_{n=1}^{\infty} \left[E_{2n-1} + E_{2n} - 2\mu_n(x)\right]. \tag{2.31}$$

We note in passing that (2.31) extends to periodic potentials satisfying $V \in H^{1,2}([0,a])$ ($H^{m,p}(\cdot)$ the usual Sobolev spaces whose elements have up to m distributional derivatives in $L^p(\mathbb{R})$) since then the total gap length is finite, i.e.,

$$\sum_{n=1}^{\infty} |E_{2n} - E_{2n-1}| < \infty.$$

Moreover, as discussed in [4, 15, 20, 26, 27], the trace formulae (2.29) and especially (2.31) extend to certain classes of almost periodic potentials including, in particular, all quasi-periodic finite-gap potentials with purely absolutely continuous spectra obtained by algebro-geometric methods in connection with hyperelliptic curves. In addition, $\xi(\lambda, x)$ is a pure step function in the case of soliton potentials and solitons relative to the almost periodic potentials mentioned above (constructed, e.g., by the double commutation method [12]) and in the case of purely discrete spectra of H, where $V(x) \to \infty$ as $|x| \to \infty$ (in the latter case $\xi(\lambda, x) = 1$ or 0 for all $x \in \mathbb{R}$ and a.e. $\lambda \in \mathbb{R}$).

We have used the Dirichlet boundary condition at $x \in \mathbb{R}$ mainly for convenience. In the remainder of this section we shall briefly indicate how to replace it by other self-adjoint boundary conditions. Hence we replace (2.2) by

$$\begin{aligned} H_y^\beta f &= \tau f, \quad y \in \mathbb{R},\ \beta \in \mathbb{R}, \\ \mathcal{D}(H_y^\beta) &= \Big\{ g \in L^2(\mathbb{R}) \mid g, g' \in AC([y, \pm R]),\ R > 0, \\ &\qquad \lim_{\varepsilon \downarrow 0} \left[g'(y \pm \varepsilon) + \beta g(y \pm \varepsilon)\right] = 0,\ \tau g \in L^2(\mathbb{R}) \Big\}. \end{aligned} \tag{2.32}$$

(The particular case $\beta = 0$ represents a Neumann boundary condition at $y \in \mathbb{R}$.) The Green's function $G_y^\beta(z, x, x')$ of H_y^β is now given by

$$G_y^\beta(z,x,x') = G(z,x,x') - [(\beta+\partial_1)(\beta+\partial_2)G(z,y,y)]^{-1} \cdot \\ \cdot[(\beta+\partial_2)G(z,x,y)]\,[(\beta+\partial_1)G(z,y,x')], \tag{2.33}$$

where we abbreviate

$$\partial_1 G(z,y,x') := \partial_x G(z,x,x')|_{x=y}, \quad \partial_2 G(z,x,y) := \partial_{x'} G(z,x,x')|_{x'=y},$$

etc. One verifies that $\partial_1 G(z,y,x) = \partial_2 G(z,x,y)$, $x \neq y$, which proves that the rank-one integral kernel on the right-hand-side of (2.33) generates a self-adjoint operator for $z < \inf[\sigma(H_y^\beta)]$. The role of the Herglotz function $G(z,x,x)$ in the Dirichlet case is now played by $(\beta+\partial_1)(\beta+\partial_2)G(z,x,x)$. Since

$$(\beta+\partial_1)(\beta+\partial_2)G(\lambda+i0,x,x) < 0 \quad \text{for } -\lambda > 0 \text{ large enough} \tag{2.34}$$

and $\zeta_{\beta,0} := \inf[\sigma(H_x^\beta)] \leqslant \inf[\sigma(H)] = E_0$, we now have $-1 \leqslant \xi_\beta(\lambda,x) \leqslant 0$ for all $x \in \mathbb{R}$ and a.e. $\lambda \in \mathbb{R}$, and the analogs of (2.14), (2.16), and (2.18) then read

$$\xi_\beta(\lambda,x) = \pi^{-1} \lim_{\varepsilon\downarrow 0} \operatorname{Im}\{\ln[(\beta+\partial_1)(\beta+\partial_2)G(\lambda+i\varepsilon,x,x)]\} - 1, \tag{2.35}$$

$$\beta \in \mathbb{R}, \quad x \in \mathbb{R}, \quad \text{a.e. } \lambda \in \mathbb{R},$$

$$\xi_\beta(\lambda,x) = 0, \quad \lambda < \zeta_{\beta,0}(x) = \inf[\sigma(H_x^\beta)], \tag{2.36}$$

$$\operatorname{Tr}[F(H_x^\beta) - F(H)] = \int_{\zeta_{\beta,0}(x)}^{\infty} d\lambda\, F'(\lambda)\xi_\beta(\lambda,x). \tag{2.37}$$

Example 2.1 now turns into

EXAMPLE 2.4.

$$V^{(0)}(x) \equiv 0, \qquad H^{(0)} = -d^2/dx^2,$$

$$(\beta+\partial_1)(\beta+\partial_2)G^{(0)}(z,x,x) = \frac{i}{2}\left[\frac{\beta^2}{z^{1/2}} + z^{1/2}\right], \quad \operatorname{Im}(z^{1/2}) \geqslant 0, \tag{2.38}$$

$$\zeta_{\beta,0}(x) = -\beta^2, \quad x \in \mathbb{R}, \tag{2.39}$$

$$\xi_\beta^{(0)}(\lambda,x) = \begin{cases} 0, & \lambda < -\beta^2 \\ -1, & -\beta^2 < \lambda < 0, \\ -\frac{1}{2}, & \lambda > 0 \end{cases} \quad \beta \in \mathbb{R}\backslash\{0\}, \ x \in \mathbb{R}, \tag{2.40}$$

$$\xi_0^{(0)}(\lambda,x)=\begin{cases}0, & \lambda<0\\ -\frac{1}{2}, & \lambda>0\end{cases}\quad \beta=0,\quad x\in\mathbb{R}, \tag{2.41}$$

$$\mathrm{Tr}\left[e^{-\tau H_x^{(0),\beta}}-e^{-\tau H^{(0)}}\right]=-\tfrac{1}{2}+e^{\tau\beta^2},\quad \tau>0, \tag{2.42}$$

$$\mathrm{Tr}\left[(H_x^{(0),\beta}-z)^{-1}-(H^{(0)}-z)^{-1}\right]=\frac{\beta^2-z}{2z(z+\beta^2)},\quad z\in\mathbb{C}\backslash\sigma(H_x^{(0),\beta}). \tag{2.43}$$

In analogy to (2.5) one obtains

$$\mathrm{Tr}\left[e^{-\tau H_x^{\beta}}-e^{-tH}\right]\underset{\tau\downarrow 0}{\widetilde{}}\sum_{\ell=0}^{\infty}s_{\beta,\ell}(x)\tau^{\ell}, \tag{2.44}$$

where the $s_{\beta,\ell}(x)$ can be computed recursively as follows.

$$s_{\beta,\ell}(x)=(-1)^{\ell+1}(\ell!)^{-1}r_{\beta,\ell}(x),\quad \ell\in\mathbb{N}_0, \tag{2.45}$$

where

$$r_{\beta,0}(x)=-\tfrac{1}{2},\qquad r_{\beta,j}=jc_{\beta,j}(x)-\sum_{\ell=1}^{j-1}c_{\beta,j-\ell}(x)r_{\beta,\ell}(x),\quad j\in\mathbb{N}, \tag{2.46}$$

and

$$\begin{aligned}c_{\beta,0}(x)&=1,\\ c_{\beta,j}(x)&=(-1)^{j-1}2^{-2(j-1)}\Big[\beta^2\omega_{2(j-1)}(x)+\beta\omega_{2(j-1)}'(x)+\\ &\quad+2^{-2}\omega_{2j}(x)+\int^{x}dx'\,V(x')\omega_{2(j-1)}'(x')\Big],\quad j\in\mathbb{N}.\end{aligned} \tag{2.47}$$

Here the $\omega_{2\ell}$ are the homogeneous differential polynomials in V of degree $\deg(\omega_{2\ell})=4\ell$, defining $\deg(V^{(m)})=m+2, m\in\mathbb{N}_0$. Moreover $V\omega_{2\ell}'=h_\ell'$ is also a homogeneous differential polynomial in V and $\int V\omega_{2\ell}'=h_\ell$ in (2.47) denotes homogeneous integration, that is, all integration constants are put zero. Explicitly one obtains

$$\begin{aligned}s_{\beta,0}(x)&=\tfrac{1}{2},\qquad s_{\beta,1}(x)=\beta^2-\tfrac{1}{2}V(x),\\ s_{\beta,2}(x)&=\tfrac{1}{2}\beta^4-\beta^2V(x)-\tfrac{1}{2}\beta V'(x)+\tfrac{1}{4}V(x)^2-\tfrac{1}{8}V''(x),\quad \text{etc.}\end{aligned} \tag{2.48}$$

The analog of Theorem 2.2 then reads

$$\begin{aligned}s_{\beta,\ell}(x)&=\frac{(-1)^{\ell-1}}{\ell!}\Big\{-\frac{\zeta_{\beta,0}(x)^{\ell}}{2}+\\ &\quad+\ell\lim_{t\downarrow 0}\int_{\zeta_{\beta,0}(x)}^{\infty}d\lambda\, e^{-t\lambda}\lambda^{\ell-1}\left[-\tfrac{1}{2}-\xi_\beta(\lambda,x)\right]\Big\},\quad \ell\in\mathbb{N}.\end{aligned} \tag{2.49}$$

Explicitly,

$$\begin{aligned} s_{\beta,0}(x) &= \tfrac{1}{2}, \\ s_{\beta,1}(x) &= \beta^2 - \tfrac{1}{2}V(x) \\ &= \lim_{\tau\downarrow 0} \int_{\zeta_{\beta,0}(x)}^{\infty} \mathrm{d}\lambda\, \mathrm{e}^{-\tau\lambda}\big[-\tfrac{1}{2}\mathrm{e}^{\tau\zeta_{\beta,0}(x)} - \xi_\beta(\lambda,x)\big], \quad \text{etc.} \end{aligned} \tag{2.50}$$

Finally, if $V(x)$ is periodic of period a by (2.26), one infers in analogy to (2.28) that

$$\begin{aligned} \sigma(H_x^\beta) &= \sigma(H) \cup \{\zeta_{\beta,n}(x)\}_{n\in\mathbb{N}_0}, \\ \zeta_{\beta,0}(x) &\leqslant E_0, \quad E_{2n-1} \leqslant \zeta_{\beta,n}(x) \leqslant E_{2n}, \quad x\in\mathbb{R},\ n\in\mathbb{N}, \end{aligned} \tag{2.51}$$

with $\sigma(H)$ given as in (2.27). Equation (2.30) then turns into

$$\xi_\beta(\lambda,x) = \begin{cases} 0, & \lambda < \zeta_{\beta,0}(x),\ E_{2n-1} < \lambda < \zeta_{\beta,n}(x),\ n\in\mathbb{N} \\ -1, & \zeta_{\beta,n}(x) < \lambda < E_{2n},\ n\in\mathbb{N}_0 \\ -\tfrac{1}{2}, & E_{2(n-1)} < \lambda < E_{2n-1},\ n\in\mathbb{N} \end{cases} \tag{2.52}$$

and the periodic trace formulae (2.29) become

$$\begin{aligned} 2(-1)^\ell \ell!\, s_{\beta,\ell}(x) &= 2\zeta_{\beta,0}(x)^\ell - E_0^\ell + \\ &\quad + \sum_{n=1}^{\infty} \big[2\zeta_{\beta,n}(x)^\ell - E_{2n-1}^\ell - E_{2n}^\ell\big], \quad \ell\in\mathbb{N},\ x\in\mathbb{R}. \end{aligned} \tag{2.53}$$

In particular, for $\ell = 1$ the trace formula for $V(x)$ reads

$$V(x) = 2\beta^2 + 2\zeta_{\beta,0}(x) - E_0 + \sum_{n=1}^{\infty} \big[2\zeta_{\beta,n}(x) - E_{2n-1} - E_{2n}\big]. \tag{2.54}$$

For a detailed treatment of these results see [20].

Returning to the trace formula (2.21) we shall now consider the special case where V is a real-valued short-range potential satisfying $V \in L^1(\mathbb{R}; (1+|x|)\,\mathrm{d}x) \cap H^{2,1}(\mathbb{R})$. In this case one can show that [16]

$$\xi(\lambda,x) = \begin{cases} \tfrac{1}{2} + \pi^{-1}\operatorname{Im}\{\ln[1 + R^{\ell}_{r}(\lambda) f_\mp(\lambda,x)^2 |f_\mp(\lambda,x)|^{-2}]\}, & \lambda > 0 \\ 0, \quad \lambda < E_0 = \inf[\sigma(H)], \end{cases} \tag{2.55}$$

where $f_\pm$ are the Jost solutions of H ($\mathrm{e}^{\mp i\lambda^{1/2}x} f_\pm(\lambda,x) \to 1$ as $x \to \pm\infty$), and $R^{r(\ell)}(\lambda)$ denote the reflection coefficients from right (left) incidence. Moreover,

$$|1 - 2\xi(\lambda,x)| \leqslant |R^{r(\ell)}(\lambda)|, \qquad R^{r(\ell)}(\lambda) = o(\lambda^{-3/2}) \quad \text{as } \lambda\to\infty \tag{2.56}$$

then yields $[1-2\xi(.\,,x)] \in L^1((0,\infty); d\lambda)$ and hence on can remove the Abelian limit in (2.21) to arrive at [16]

$$V(x) = E_0 + \int_{E_0}^{\infty} d\lambda [1-2\xi(\lambda,x)] \tag{2.57}$$

$$= 2\Big\{ e_0 + \sum_{j\in J\setminus\{0\}} [e_j - \mu_j(x)] \Big\} - \\ -\frac{2}{\pi}\int_0^\infty d\lambda\, \mathrm{Im}\{ \ln [1 + R^{\ell}_{r}(\lambda) f_{\mp}(\lambda,x)^2 |f_{\mp}(\lambda,x)|^{-2}]\}. \tag{2.58}$$

Here $\sigma_p(H) = \{e_j\}_{j\in J}$ denotes the (possibly empty) set of (necessarily simple and negative) eigenvalues of H ordered $e_0 < e_1 < \cdots$. While (2.58), in the case $\sigma_p(H) = \emptyset$, is due to Deift and Trubowitz [5], (2.57) under the additional assumption $V \geqslant 0$ of compact support was first derived by Venakides [33] (without identifying ξ as the argument of $G(\lambda + i0, x, x)$ or as Krein's spectral shift function for (H_x^D, H)). Equation (2.58) extends in a straightforward manner to all higher KdV invariants $s_\ell(x)$. Moreover, (2.58) extends to potentials which approach different spatial asymptotics sufficiently fast and to appropriate impurity (defect) interactions in one-dimensional solids as detailed in [16].

Next we briefly mention the standard trace formulae for short-range potentials satisfying $V \in L^1(\mathbb{R}; (1+|x|)dx)$, $V^{(m)} \in L^1(\mathbb{R})$, $m \in \mathbb{N}$. Expanding $G(z,x,x)$ as $|z| \to \infty$ one infers

$$G(z,x,x) \underset{\mathrm{Im}(z^{1/2})\geqslant 0,\ |z|\to\infty}{\widetilde{\ \ }} - \sum_{n=0}^{\infty} \omega_n(x)(2iz^{1/2})^{-n-1}, \tag{2.59}$$

where

$$\omega_0(x) = 1, \qquad \omega_{2\ell+1} = 0, \quad \ell \in \mathbb{N}_0, \tag{2.60}$$

$$\omega_{2\ell}(x) = -2\phi_{2\ell-1}(x) - 2\sum_{j=1}^{\ell-1} \phi_{2(\ell-j)-1}(x)\omega_{2j}(x), \quad \ell \in \mathbb{N}. \tag{2.61}$$

Moreover, one can prove that [10]

$$\omega_{2\ell+2}(x) = -2(2\ell+1)\phi_{2\ell+1}(x) + \zeta_\ell'(x), \quad \ell \in \mathbb{N}_0, \tag{2.62}$$

for appropriate differential polynomials ζ_ℓ in $V(x)$ with $\zeta_\ell^{(m)} \in L^1(\mathbb{R})$, $m \in \mathbb{N}_0$. Explicitly, one obtains

$$\omega_0(x) = 1, \quad \omega_2(x) = -2V(x),\ \omega_4(x) = 6V(x)^2 - 2V''(x), \quad \text{etc.} \tag{2.63}$$

A standard contour integration then yields

THEOREM 2.5. *Let ψ_j be the normalized eigenvectors of H associated with e_j, i.e., $H\psi_j = e_j\psi_j$, $\|\psi_j\|_2 = 1$, $j \in J$. Then*

$$\begin{aligned}\omega_{2\ell+2}(x) &= 2^{2\ell+3}\sum_{j\in J}(-e_j)^{(2\ell+1)/2}\psi_j(x)^2 + \\ &\quad +(-1)^\ell 2^{2\ell+2}\pi^{-1}\int_0^\infty d\lambda\,\lambda^\ell\,\mathrm{Im}\,[R_\ell^r(\lambda)f_\pm(\lambda,x)^2],\end{aligned} \tag{2.64}$$

$\ell \in \mathbb{N}_0,\ x \in \mathbb{R}$.

In particular, for $\ell = 0$ one obtains from (2.64) the well-known trace formula for $V(x)$ (see [5])

$$\begin{aligned}V(x) &= -4\sum_{j\in J}(-e_j)^{1/2}\psi_j(x)^2 - \\ &\quad -2\pi^{-1}\int_0^\infty d\lambda\,\mathrm{Im}\,[R_\ell^r(\lambda)f_\pm(\lambda,x)^2].\end{aligned} \tag{2.65}$$

The trace formula (2.64) first appeared in [3] in connection with the classical moment problem. For various extensions of (2.64) (including applications to the off-diagonal Green's function $G(z,x,x')$) see [13, 14]. In [14] the reader can also find extensions to certain cases where H has infinitely many negative eigenvalues accumulating at zero (including the infinite soliton limits described in [17, 18]).

Finally we briefly turn to the KdV conservation laws. Integrating both sides of (2.64) over $x \in \mathbb{R}$ results, after some manipulations (see [14]), in

$$\begin{aligned}\int_{\mathbb{R}} dx\,\omega_{2\ell+2}(x) &= 2^{2\ell+3}\sum_{j\in J}(-e_j)^{(2\ell+1)/2} + (-1)^\ell 2^{2\ell+2}\pi^{-1}\times \\ &\quad \times(2\ell+1)\int_0^\infty d\lambda\,\lambda^{\ell-1/2}\ln[|T(\lambda)|], \quad \ell\in\mathbb{N}_0,\end{aligned} \tag{2.66}$$

where $T(\lambda)$ denotes the transmission coefficient. Equation (2.66) represents a celebrated result due to Faddeev and Zakharov [35]. Introducing

$$\Omega_{2\ell+2} := \int_{\mathbb{R}} dx\,\omega_{2\ell+2}(x), \qquad I_{2\ell+1} := \int_{\mathbb{R}} dx\,\phi_{2\ell+1}, \quad \ell\in\mathbb{N}_0, \tag{2.67}$$

the identity (2.62) then yields for the KdV invariants $I_{2\ell+1}$,

$$I_{2\ell+1} = -(4\ell+2)^{-1}\Omega_{2\ell+2}, \quad \ell\in\mathbb{N}_0. \tag{2.68}$$

Let $V \in C^\infty(\mathbb{R}^2)$ be a real-valued solution of the KdV equation

$$V_t - 6VV_x + V_{xxx} = 0, \quad (x,t)\in\mathbb{R}^2 \tag{2.69}$$

which results from the usual Lax pair $(H(t), B(t))$ by means of the relation

$$\frac{d}{dt}H(t) - [B(t), H(t)] = 0 \tag{2.70}$$

($[\cdot,\cdot]$ the commutator), where $H(t)$ and $B(t)$ denote the differential expressions

$$H(t) = -\partial_x^2 + V(x,t), \quad B(t) = -4\partial_x^3 + 6V(x,t)\partial_x + 3V_x(x,t). \quad (2.71)$$

If $V(x,t)$ satisfies the KdV equation (2.69) and the short-range conditions for each fixed $t \in \mathbb{R}$ then the substitution $V(x) \equiv V(x,0) \to V(x,t)$ is well known to imply the corresponding substitutions

$$T(\lambda) \longrightarrow T(\lambda,t) = T(\lambda),$$

$$R_{\ell}^{r}(\lambda) \longrightarrow R_{\ell}^{r}(\lambda,t) = R_{\ell}^{r}(\lambda)\mathrm{e}^{\pm i8\lambda^{3/2}t},$$

$$e_j \longrightarrow e_j(t) = e_j, \qquad c_{j,\pm} \longrightarrow c_{j,\pm}(t) = c_{j,\pm}\mathrm{e}^{\pm 4(-e_j)^{3/2}t}, \quad j \in J. \qquad (2.72)$$

Replacing accordingly $\omega_{2\ell+2}(x) \to \omega_{2\ell+2}(x,t)$, $\phi_{2\ell+1}(x) \to \phi_{2\ell+1}(x,t)$, $\ell \in \mathbb{N}_0$, we arrive at the infinite sequence of (polynomial) conservation laws for the KdV hierarchy

$$\frac{\mathrm{d}}{\mathrm{d}t} I_{2\ell+1} = \int_{\mathbb{R}} \mathrm{d}x \, \phi_{2\ell+1,t}(x,t) = 0, \quad \ell \in \mathbb{N}_0, \qquad (2.73)$$

since both e_j and $T(\lambda)$ in (2.66) are t-independent.

3. Multi-Dimensional Trace Formulae

First, reporting on recent work in [21], we attempt to extend the leading behavior in (2.5),

$$2\,\mathrm{Tr}\left[\mathrm{e}^{-\tau H} - \mathrm{e}^{-\tau H_x^D}\right] = 1 - \tau V(x) + o(\tau) \quad \text{as } \tau \downarrow 0 \qquad (3.1)$$

to arbitrary space dimensions $\nu \in \mathbb{N}$. The key to such an extension is an appropriate combination of Dirichlet and Neumann boundary conditions on various hyperplanes through the point $x \in \mathbb{R}^\nu$ taking into account that (3.1) is equivalent to

$$\mathrm{Tr}\left[\mathrm{e}^{-\tau H_x^N} - \mathrm{e}^{-\tau H_x^D}\right] = 1 - \tau V(x) + o(\tau) \quad \text{as } \tau \downarrow 0, \qquad (3.2)$$

where $H_x^N = H_x^0$ denotes the operator (2.32) with a Neumann boundary condition at $x \in \mathbb{R}$. We start by introducing proper notations. In the following let V be a real-valued continuous function on $\mathbb{R}^\nu$ bounded from below and define the self-adjoint operator

$$H = -\Delta \dotplus V \qquad (3.3)$$

as a form sum in $L^2(\mathbb{R}^\nu)$. Next, let $A \subseteq \{1,\ldots,\nu\}$ and denote by $|A|$ the number of elements of A. Moreover, let $B_\alpha, \alpha \subseteq \{1,\ldots,\nu\}$ be the 2^ν blocks obtained by removing the hyperplanes $\mathcal{P}_j^{(x)} = \{y \in \mathbb{R}^\nu \mid y_j = x_j\}$ from $\mathbb{R}^\nu$, that is, $B_\alpha = \{y \in \mathbb{R}^\nu \mid y_\ell > 0 \text{ if } \ell \in \alpha,\ y_\ell < 0 \text{ if } \ell \notin \alpha\}$ and denote by $\mathcal{P}_\nu$ the power set of $\{1, \ldots, \nu\}$. The operator $H_{A;x}$ is then defined to be $-\Delta + V$ on $\bigoplus_{\alpha \in \mathcal{P}_\nu} L^2(B_\alpha)$ with Dirichlet boundary conditions on $\{P_j^{(x)}\}_{j \in A}$ and Neumann boundary conditions on $\{P_j^{(x)}\}_{j \notin A}$.

THEOREM 3.1 ([21]). *Define*

$$C_\tau = \sum_{A \in \mathcal{B}_\nu} (-1)^{|A|} \mathrm{e}^{-\tau H_{A;0}}, \quad \tau > 0.$$

Then the integral kernel of C_τ is given by

$$C_\tau(x, x') = \begin{cases} 2^\nu \mathrm{e}^{-\tau H}(x, -x'), & x, x' \text{ in the same orthant} \\ 0, & \text{otherwise.} \end{cases} \tag{3.4}$$

Moreover, $C_\tau, \tau > 0$ is a trace class operator in $L^2(\mathbb{R}^\nu)$ and

$$\mathrm{Tr}(C_\tau) = 2^\nu \int_{\mathbb{R}^\nu} \mathrm{d}^\nu x \, \mathrm{e}^{-tH}(x, -x), \quad \tau > 0. \tag{3.5}$$

The proof of (3.4) in [21] is based on the method of images while the trace class property of C_τ and (3.5) follow from the direct sum decomposition of C_τ in $\bigoplus_{\alpha \in \mathcal{P}_\nu} L^2(B_\alpha)$.

Applying a Feynman–Kac-type analysis then yields the following ν-dimensional generalization of (3.2).

THEOREM 3.2 ([21]).

$$\mathrm{Tr}\left(\sum_{A \in \mathcal{P}_\nu} (-1)^{|A|} \mathrm{e}^{-\tau H_{A;x}} \right) = 1 - \tau V(x) + o(\tau) \quad \text{as } \tau \downarrow 0. \tag{3.6}$$

While Theorem 3.2 represents a multidimensional trace formula for Schrödinger operators associated with unbounded regions in $\mathbb{R}^\nu$, one can also prove new trace formulae for Schrödinger operators defined in boxes. One obtains, e.g.,

THEOREM 3.3 ([21]). *Let V be continuous on $[0,1]^\nu$. For $A \subseteq \{1, \ldots, \nu\}$, let H_A be $-\Delta + V$ on $L^2([0,1]^\nu)$ with Dirichlet boundary conditions on the hyperplanes with $x_j = 0$ or 1 and $j \in A$ and Neumann boundary conditions on the hyperplanes with $x_j = 0$ or 1 and $j \notin A$. Let $\langle V \rangle$ be the average of V at the 2^ν corners of $[0,1]^\nu$. Then*

$$\sum_{A \in \mathcal{P}_\nu} (-1)^{|A|} \, \mathrm{Tr}\, (\mathrm{e}^{-\tau H_A}) = 1 - \tau \langle V \rangle + o(\tau) \quad \text{as } \tau \downarrow 0. \tag{3.7}$$

This result also holds for rectangular boxes $\times_{j=1}^{\nu} [a_j, b_j]$ but the rectangular symmetry is crucial. Similarly one can prove

THEOREM 3.4 ([21]). *Let V be continuous on $[0,1]^\nu$. For $A \subseteq \{1,\ldots,\nu\}$ let $\tilde{H}_A$ be $-\Delta + V$ on $L^2([0,1]^\nu)$ with Dirichlet boundary conditions on the hyperplanes with $x_j = 0$ for $j \in A$ and Neumann boundary conditions on the hyperplanes with $x_j = 0$ for $j \notin A$ or $x_k = 1$ for all $k \in \{1,\ldots,\nu\}$. Then*

$$\sum_{A\in\mathcal{P}_\nu} (-1)^{|A|}\operatorname{Tr}(e^{-\tau\tilde{H}_A}) = 2^{-\nu}[1-\tau V(0)+o(\tau)] \quad \text{as } \tau\downarrow 0. \tag{3.8}$$

Finally, we mention an Abelianized version of a trace formula that Lax [25] derived formally in two dimensions.

THEOREM 3.5 ([21]). *Let V be a continuous periodic function on $\mathbb{R}^2$ with $V(x_1+n_1,x_2+n_2) = V(x_1,x_2)$ for all $(x_1,x_2,n_1,n_2) \in \mathbb{R}^2\times\mathbb{Z}^2$. Let H_P, H_A, H_{AP}, H_{PA}, H_N, and H_D be the operators $-\Delta+V$ on $L^2([0,1]^2)$ with periodic, antiperiodic, AP, PA, Neumann, and Dirichlet boundary conditions, respectively, where AP (resp., PA) means antiperiodic in the x_1 (resp., x_2) direction and periodic in the x_2 (resp., x_1) direction. Then*

$$\begin{aligned}&\operatorname{Tr}\left[e^{-\tau H_P}+e^{-\tau H_A}+e^{-\tau H_{AP}}+e^{-\tau H_{PA}}-2e^{-\tau H_N}-2e^{-\tau H_D}\right]\\&\quad = -1+\tau V(0)+o(\tau) \quad \text{as } \tau\downarrow 0.\end{aligned} \tag{3.9}$$

For a different kind of two-dimensional trace formula for $V(x)$ comparing the heat kernels for $H = -\Delta+V$ and $H_0 = -\Delta$ with Dirichlet boundary conditions on a rectangular box, see [30].

In our last part we mention a different kind of multi-dimensional trace formulae based on point interactions [1] and hence rank-one perturbations of resolvents. Since point interactions (also called contact interactions or δ-interactions) are limited to $1 \leqslant \nu \leqslant 3$ space dimensions, so will be our approach below.

Assuming V to be real-valued, continuous and bounded from below on $\mathbb{R}^\nu$, $1 \leqslant \nu \leqslant 3$, we introduce $H = -\Delta\dot{+}V$ as in (3.3). The resolvent of the self-adjoint Hamiltonian $H_{\alpha,x}$, modeling H plus a point interaction centered at $x \in \mathbb{R}^\nu$ (whose strength is parameterized in terms of $\alpha \in \mathbb{R}$), is defined as follows (see, e.g., [1, 36])

$$\begin{aligned}&(H_{\alpha,x}-z)^{-1} = (H-z)^{-1} + D_{\alpha,x}(z)^{-1}(\overline{G(z,x,\cdot)},\cdot)G(z,\cdot,x),\\&\quad z\in\mathbb{C}\backslash\{\sigma(H_{\alpha,x})\cup\sigma(H)\},\end{aligned} \tag{3.10}$$

where

$$D_{\alpha,x}(z) = \begin{cases} -\alpha^{-1}-\Gamma_\nu(z,x), & \nu=1\\ \alpha-\Gamma_\nu(z,x), & \nu=2,3,\end{cases} \tag{3.11}$$

$$\begin{aligned}\Gamma_1(z,x) &= G(z,x,x),\\ \Gamma_2(z,x) &= \lim_{|\varepsilon|\downarrow 0}\left[G(z,x,x+\varepsilon)-(2\pi)^{-1}\ln(|\varepsilon|)\right],\\ \Gamma_3(z,x) &= \lim_{|\varepsilon|\downarrow 0}\left[G(z,x,x+\varepsilon)-(4\pi|\varepsilon|)^{-1}\right],\end{aligned} \tag{3.12}$$

and $G(z,x,x')$ denotes the Green function of H. In analogy to (2.4) one then computes

$$\operatorname{Tr}\left[(H_{\alpha,x}-z)^{-1}-(H-z)^{-1}\right]=-\frac{\mathrm{d}}{\mathrm{d}z}\ln\left[D_{\alpha,x}(z)\right]. \tag{3.13}$$

In the special case where $V^{(0)}\equiv 0$, one obtains

$$D_{\alpha}^{(0)}(z)=\begin{cases}-\alpha^{-1}-(-4z)^{-1/2}, & \nu=1\\ \tilde{\alpha}+(2\pi)^{-1}\ln((-z)^{1/2}), & \nu=2\\ \alpha+(4\pi)^{-1}(-z)^{1/2}, & \nu=3,\end{cases} \tag{3.14}$$

where $\tilde{\alpha}=\alpha+(2\pi)^{-1}\gamma-(2\pi)^{-1}\ln(2)$ (with $\gamma=0.5772\ldots$ being Euler constant) which leads to

$$\operatorname{Tr}\left[(H_{\alpha,x}^{(0)}-z)^{-1}-(H^{(0)}-z)^{-1}\right]=-\int_{E_{\alpha,0}^{(0)}}^{\infty}\mathrm{d}\lambda\,\frac{\xi_{\alpha}^{(0)}(\lambda)}{(\lambda-z)^{2}}. \tag{3.15}$$

Here, for $\nu=1$,

$$\xi_{\alpha}^{(0)}(\lambda)=\underset{\alpha<0}{\begin{cases}0, & \lambda<-\alpha^2/4\\ -1, & -\alpha^2/4<\lambda<0\\ a_{\alpha}(\lambda)/\pi, & \lambda>0\end{cases}}\quad\underset{\alpha=0}{\begin{cases}0, & \lambda<0\\ \frac{1}{2}, & \lambda>0\end{cases}}\quad\underset{\alpha>0,}{\begin{cases}0,\\ 1-\pi^{-1}a_{\alpha}(\lambda),\end{cases}} \tag{3.16}$$

writing $a_{\alpha}(\lambda)=\arctan(\alpha/2\lambda^{1/2})$, and, for $\nu=2$,

$$\xi_{\alpha}^{(0)}(\lambda)=\begin{cases}0, & \lambda<-\mathrm{e}^{-4\pi\tilde{\alpha}}\\ -1, & -\mathrm{e}^{-4\pi\tilde{\alpha}}<\lambda\leqslant 0\\ -\pi^{-1}\arctan[\pi/(4\pi\tilde{\alpha}+\ln(\lambda))]-1, & 0\leqslant\lambda\leqslant \mathrm{e}^{-4\pi\tilde{\alpha}}\\ -\pi^{-1}\arctan[\pi/(4\pi\tilde{\alpha}+\ln(\lambda))], & \lambda\geqslant \mathrm{e}^{-4\pi\tilde{\alpha}}\end{cases} \tag{3.17}$$

and, finally, for $\nu=3$,

$$\xi_{\alpha}^{(0)}(\lambda)=\underset{\alpha<0}{\begin{cases}0, & \lambda<-(4\pi\alpha)^2\\ -1, & -(4\pi\alpha)^2<\lambda<0\\ -A_{\alpha}(\lambda)/\pi, & \lambda>0\end{cases}}\quad\underset{\alpha=0}{\begin{cases}0, & \lambda<0\\ -\frac{1}{2}, & \lambda>0\end{cases}}\quad\underset{\alpha>0,}{\begin{cases}0,\\ -A_{\alpha}(\lambda)/\pi,\end{cases}} \tag{3.18}$$

writing $A_{\alpha}(\lambda)=\arctan(\lambda^{1/2}/4|\pi\alpha|)$, and

$$E_{\alpha,0}^{(0)}=\underset{\nu=1}{\begin{cases}-\alpha^2/4, & \alpha<0\\ 0, & \alpha\in[0,\infty]\end{cases}}\quad\underset{\nu=2}{\left\{-\mathrm{e}^{-4\pi\tilde{\alpha}}\right.}\quad\underset{\nu=3.}{\begin{cases}-(4\pi\alpha)^2, & \alpha<0\\ 0, & \alpha\geqslant 0\end{cases}} \tag{3.19}$$

For $\nu = 1$ we have $\xi_{\infty}^{(0)}(\lambda) = \xi^{(0)}(\lambda)$ in (2.15).

Krein's spectral shift function for the pair $(H_{\alpha,x}, H)$ is then introduced via

$$\operatorname{Tr}\left[(H_{\alpha,x} - z)^{-1} - (H - z)^{-1}\right] = -\int_{E_{\alpha,x,0}}^{\infty} \mathrm{d}\lambda\, \frac{\xi_{\alpha,x}(\lambda)}{(\lambda - z)^2} \tag{3.20}$$

with $E_{\alpha,x,0} = \inf\{\sigma(H_{\alpha,x}) \cup \sigma(H)\}$ and the normalization

$$\xi_{\alpha,x}(\lambda) = 0, \quad \lambda < E_{\alpha,x,0}. \tag{3.21}$$

The high-energy expansion (see, e.g., [31, 34])

$$\begin{aligned}&\lim_{|\varepsilon|\downarrow 0}\left[G(z,x,x+\varepsilon) - G^{(0)}(z,x,x+\varepsilon)\right]\\ &\quad = -V(x)\begin{cases} 1/4(-z)^{3/2} + o(z^{-3/2}), & \nu = 1\\ -1/4\pi z + o(z^{-1}), & \nu = 2\\ 1/8\pi(-z)^{1/2} + o(z^{-1/2}), & \nu = 3\end{cases}\end{aligned} \tag{3.22}$$

then yields

$$\begin{aligned}&\int_{e_{\alpha,x,0}}^{\infty} \mathrm{d}\lambda\, \frac{[\xi_{\alpha,x}(\lambda) - \xi_{\alpha}^{(0)}(\lambda)]}{(\lambda - z)^2}\\ &\quad = \begin{cases} -\frac{3\alpha V(x)}{8(-z)^{5/2}} + o(z^{-5/2}), & \nu = 1\\ -\frac{V(x)}{z^2 \ln(-z)} + o(z^{-2}(\ln z)^{-1}), & \nu = 2\\ \frac{V(x)}{2z^2} + o(z^{-2}), & \nu = 3,\end{cases}\end{aligned} \tag{3.23}$$

with $e_{\alpha,x,0} = \min\{E_{\alpha,x,0}, E_{\alpha,0}^{(0)}\}$. Summarizing, we have the following new trace formulae.

THEOREM 3.6.

$$V(x) = \lim_{z\downarrow -\infty} \gamma_{\nu}(z) \int_{e_{\alpha,x,0}}^{\infty} \mathrm{d}\lambda\, \frac{[\xi_{\alpha,x}(\lambda) - \xi_{\alpha}^{(0)}(\lambda)]}{(\lambda - z)^2}, \quad 1 \leqslant \nu \leqslant 3, \tag{3.24}$$

where

$$\gamma_{\nu}(z) = \underset{\nu = 1}{\frac{8(-z)^{5/2}}{3\alpha}}, \quad \alpha \neq 0, \quad \underset{\nu = 2}{-z^2 \ln(-z)}, \quad \underset{\nu = 3.}{2z^2} \tag{3.25}$$

Using the systematic high-energy expansion of

$$\lim_{|\varepsilon|\downarrow 0}\left[G(z,x,x+\varepsilon) - G^{(0)}(z,x,x+\varepsilon)\right]$$

in terms of (multi-dimensional) KdV invariants (see, e.g., [31, 34]) one can extend Theorem 3.6 to higher-order trace relations in analogy to (2.22) and (2.49). A detailed account of these results will appear elsewhere.

Acknowledgements

We are indebted to B. Simon and Z. Zhao for joint collaborations which led to most of the results presented in this contribution. Support from the Norwegian Research Council is gratefully acknowledged. F.G. would like to thank the Department of Mathematical Sciences, NTH, University of Trondheim, Norway, for the great hospitality extended to him during a month long stay in the summer of 1994.

References

1. Albeverio, S., Gesztesy, F., Høegh-Krohn, R., and Holden, H.: *Solvable Models in Quantum Mechanics*, Springer, New York, 1988.
2. Aronszajn, N. and Donoghue, W. F.: On exponential representations of analytic functions in the upper half-plane with positive imaginary part, *J. Anal. Math.* **5** (1957), 321–388.
3. Chudnovsky, D. V.: One and multidimensional completely integrable systems arising from the isospectral deformation, in: D. Iagolnitzer (ed.), *Complex Analysis, Microlocal Calculus and Relativistic Quantum Theory*, Lecture Notes in Physics, Vol. 126, Springer, Berlin, 1980, pp. 352–416.
4. Craig, W.: The trace formula for Schrödinger operators on the line, *Commun. Math. Phys.* **126** (1989), 379–407.
5. Deift, P. and Trubowitz, E.: Inverse scattering on the line, *Commun. Pure Appl. Math.* **32** (1979), 121–251.
6. Dikii, L. A.: Trace formulas for Sturm–Liouville differential operators, *Amer. Math. Soc. Transl. Ser.* (2)**18** (1961), 81–115.
7. Dubrovin, B. A.: Periodic problems for the Korteweg–de Vries equation in the class of finite band potentials, *Funct. Anal. Appl.* **9** (1975), 215–223.
8. Flaschka, H.: On the inverse problem for Hill's operator, *Arch. Rat. Mech. Anal.* **59** (1975), 293–309.
9. Gelfand, I. M.: On identities for the eigenvalues of a second-order differential operator, *Uspehi Mat. Nauk* **11**(1) (1956), 191–198 (in Russian).
10. Gelfand, I. M. and Dikii, L. A.: Asymptotic behaviour of the resolvent of Sturm–Liouville equations and the algebra of the Korteweg–de Vries equations, *Russ. Math. Surv.* **30**(5) (1975), 77–113.
11. Gelfand, I. M. and Levitan, B. M.: On a simple identity for the eigenvalues of a second-order differential operator, *Dokl. Akad. Nauk SSSR* **88** (1953), 593–596 (in Russian).
12. Gesztesy, F.: A complete spectral characterization of the double commutation method, *J. Funct. Anal.* **117** (1993), 401–446.
13. Gesztesy, F.: New trace formulas for Schrödinger operators, in: G. Ferreyra, G. Goldstein, F. Neubrander (eds.), *Evolution Equations*, Marcel Dekker, New York, 1995, pp. 201–221.
14. Gesztesy, F. and Holden, H.: Trace formulas and conservation laws for nonlinear evolution equations, *Rev. Math. Phys.* **6** (1994), 51–95.
15. Gesztesy, F. and Simon, B.: The ξ function, *Acta Math.* (to appear).
16. Gesztesy, F., Holden, H., and Simon, B.: Absolute summability of the trace relation for certain Schrödinger operators, *Commun. Math. Phys.* (to appear).
17. Gesztesy, F., Karwowski, W., and Zhao, Z.: New types of soliton solutions, *Bull. Amer. Math. Soc. (New Series)* **27** (1992), 266–272.
18. Gesztesy, F., Karwowski, W., and Zhao, Z.: Limits of soliton solutions, *Duke Math. J.* **68** (1992), 101–150.
19. Gesztesy, F., Holden, H., Simon, B., and Zhao, Z.: Trace formulae and inverse spectral theory for Schrödinger operators, *Bull. Amer. Math. Soc. (New Series)* **29** (1993), 250–255.
20. Gesztesy, F., Holden, H., Simon, B., and Zhao, Z.: Higher order trace relations for Schrödinger operators, *Rev. Math. Phys.* (to appear).

21. Gesztesy, F., Holden, H., Simon, B., and Zhao, Z.: A trace formula for multidimensional Schrödinger operators, preprint.
22. Hochstadt, H.: On the determination of a Hill's equation from its spectrum, *Arch. Rat. Mech. Anal.* **19** (1965), 353–362.
23. Kotani, S. and Krishna, M.: Almost periodicity of some random potentials, *J. Funct. Anal.* **78** (1988), 390–405.
24. Krein, M. G.: Perturbation determinants and a formula for the traces of unitary and self-adjoint operators, *Sov. Math. Dokl.* **3** (1962), 707–710.
25. Lax, P. D.: Trace formulas for the Schroedinger operator, *Commun. Pure Appl. Math.* **47** (1994), 503–512.
26. Levitan, B. M.: On the closure of the set of finite-zone potentials, *Math. USSR Sbornik* **51** (1985), 67–89.
27. Levitan, B. M.: *Inverse Sturm–Liouville Problems*, VNU Science Press, Utrecht, 1987.
28. Marchenko, V. A.: *Sturm–Liouville Operators and Applications*, Birkhäuser, Basel, 1986.
29. McKean, H. P. and van Moerbeke, P.: The spectrum of Hill's equation, *Invent. Math.* **30** (1975), 217–274.
30. Papanicolaou, V.: Trace formulas and the behavior of large eigenvalues, *SIAM J. Math. Anal.*, to appear.
31. Perelomov, A. M.: Schrödinger equation spectrum and Korteweg–de Vries type invariants, *Ann. Inst. H. Poincaré, Sect. A* **24** (1976), 161–164.
32. Trubowitz, E.: The inverse problem for periodic potentials, *Commun. Pure Appl. Math.* **30** (1977), 321–337.
33. Venakides, S.: The infinite period limit of the inverse formalism for periodic potentials, *Commun. Pure Appl. Math.* **41** (1988), 3–17.
34. Wilk, S. F. J., Fujiwara, S., and Osborn, T. A.: N-body Green's function and their semiclassical expansion, *Phys. Rev.* **24A** (1981), 2187–2202.
35. Zakharov, V. E. and Faddeev, L. D.: Korteweg–de Vries equation: a completely integrable Hamiltonian system, *Funct. Anal. Appl.* **5** (1971), 280–287.
36. Zorbas, J.: Perturbation of self-adjoint operators by Dirac distributions, *J. Math. Phys.* **21** (1980), 840–847.

Acta Applicandae Mathematicae **39:** 335–348, 1995.

KdV Equations and Integrability Detectors

B. GRAMMATICOS, V. PAPAGEORGIOU
LPN, Université Paris VII, Tour 24-14, 5ème étage, 75251 Paris, France

and

A. RAMANI
CPT, Ecole Polytechnique, CNRS, UPR 14, 91128 Palaiseau, France

(Received: 3 October 1994)

Abstract. We present a review of the various integrability detectors that have been developed based on the study of the singularities of the solutions of a given equation: the Painlevé method for continuous systems, and the singularity confinement approach for discrete ones. In each case the KdV equation was instrumental in the formulation of the conjectures relating the singularity structure to integrability.

Mathematics Subject Classifications (1991): 35Q53, 39A12, 58F07.

Key words: evolution equations, integrability, singularity analysis, bilinear formalism, Painlevé equations, singularity confinement, discrete Painlevé equations.

1. Introduction

The Korteweg–de Vries equation is the archetype of integrable equations. Initially proposed as an equation with a solitary wave-type solution [1], it turned out (much) later [2] to possess solutions with an arbitrary number of elastically scattering solitary waves. The latter were observed numerically by Kruskal and Zabusky, who decided to call this type of solitary waves 'solitons'. In the years that followed this discovery, Kruskal and his collaborators went on to show that the KdV equation possessed an infinite number of conservation laws and, as the ultimate explanation of these properties, they produced a linear differential system the compatibility of which is just the KdV equation. This linear system is traditionally called the Lax pair and allows the effective linearization of the nonlinear equation. Using techniques developed in the theory of the inverse scattering problems in quantum mechanics (reconstruction of the potential from the scattering data) one can reduce the solution of the KdV equation to that of a *linear* integrodifferential one. This was the final proof of the integrability of KdV. The discovery of an integrable PDE and the techniques for its integration [3] opened a whole new domain that is still the centre of intense activity a quarter-century later.

The final proof of integrability of a given PDE is, roughly stated, its transformation into a linear integrodifferential equation [4]. Although the solution of the linear problem can be particularly delicate, the current consensus is that one does not have to wait for its solution before proclaiming the integrability of a given equation: the existence of the Lax pair suffices. Still, obtaining the Lax pair is a highly nontrivial operation and no systematic, *general* approach exists for that. Thus, in order to assess the integrability of a given PDE one needs a more convenient tool. The existence of a solution with an arbitrary number of solitons could be (and is, in fact) an indication of integrability. However, the difficulties in investigating this situation are considerable. The numerical study of interacting solitary waves necessitates particular care and is of limited applicability. An alternative, *analytical* approach has been proposed by Hirota [5] that allows the calculation of the general, N-soliton solution. It is based on the bilinear formalism that makes possible the transcription of the PDE into a bilinear equation for the related τ-function. A soliton solution corresponds then simply to a τ-function expressed as a sum of exponentials. As expected, the first application of this approach was that concerning KdV: Hirota himself has shown that KdV possesses solutions with an arbitrary number of solitons. Still, the application of the multisoliton solution criterion of integrability meets with considerable difficulties: first the bilinear form of the equation must be obtained and then the proof of the existence of the N-soliton solution must be given by induction. An additional difficulty comes from the fact that several families of solitons, i.e. solutions with different dispersion relations, may exist for the same equation and one must consider not only each of them but also all their possible interactions. So, in practice, the application of the multisoliton integrability criterion reduces to the consideration of the first few nontrivial soliton solutions [6] and it is surmised that once the latter exist the N-soliton solution will exist as well.

Clearly, what is needed in the domain of the prediction of integrability is a flexible criterion that can be algorithmically applied and yield a yes/no answer for as large as possible a class of systems. Such a criterion does exist and is based on the singularity analysis method. This paper is devoted to a review of the singularity analysis approach, also known in the case of continuous systems as the Painlevé method. In the next section we present this approach for continuous systems while, in the remaining sections, we focus on discrete and discrete/continuous systems. Quite naturally, the KdV equation has been instrumental in the formulation of conjectures on integrability, also for discrete systems. This is one more manifestation of the ubiquity of KdV. Since its proposal one century ago the Korteweg–de Vries equation has appeared in innumerable physical models: it is the most widely applicable evolution equation that ever existed.

2. Singularity Analysis: the Painlevé Approach for Continuous Systems

The relation between singularities and integrability can be traced back to Fuchs and Kowalevskaya, but it was Painlevé and his school who first systematized this approach. The starting point is the fact that the solutions of nonlinear differential equations may have singularities that are movable, i.e. their location depends on the initial conditions. The claim of Painlevé was that, for integrability, these singularities should not exhibit branching [7]. Thus integrable systems should have only pole-like singularities. This fact was dubbed the Painlevé property and was considered as a prerequisite for integrability [8]. The most important result of Painlevé and his school was the obtention of the equations known today as the Painlevé equations. The latter possessed the Painlevé property but were not integrable in terms of known 'classical' functions. Painlevé's intuition was correct: he considered that the solutions of these equations defined new transcendents. The confirmation of this point and the actual linearisation of the Painlevé equations, through IST techniques, came much later [9]. This is where the KdV (and related integrable equations) come into play. While examining reductions of integrable PDE's (KdV and mKdV being the most prominent), Ablowitz and Segur [10] made the following interesting remark: "the reductions of an integrable PDE have the Painlevé property" [11]. Let us illustrate this point with the help of KdV:

$$u_t + 6uu_x + u_{xxx} = 0. \tag{2.1}$$

Let us first consider a time-independent solution $u = u(x)$. Equation (2.1) then becomes $6uu_x + u_{xxx} = 0$, which can be integrated twice to yield ($u' \equiv du/dx$):

$$u'^2 = \alpha + \beta u - 2u^3. \tag{2.2}$$

This last equation defines an elliptic function and has the Painlevé property since its singularities are just poles. Next we consider a special solution obtained through the substitution

$$u = 2(t - w(z)) \quad \text{where} \quad z = x - 6t^2. \tag{2.3}$$

We then obtain $w''' - 12ww' - 1 = 0$ which can be integrated to:

$$w'' = 6w^2 + z. \tag{2.4}$$

This is the first of the transcendental equations obtained by Painlevé (on the sole requirement that it possess the Painlevé property). Finally, we consider the self-similar solution to the KdV obtained through:

$$u = (3t)^{-2/3} v(z) \quad \text{where} \quad z = x(3t)^{-1/3}. \tag{2.5}$$

Equation (2.1) reduces to:

$$v''' - 6vv' - 2v - zv' = 0. \tag{2.6}$$

Next we translate $v = w + z/2$ to obtain:

$$w''' + 6ww' + w + 2zw' = 0 \tag{2.7}$$

which can be integrated to:

$$w'' - \frac{w'^2}{2w} + 2w^2 + zw - \frac{\alpha}{w} = 0. \tag{2.8}$$

This is one of the 50 integrable equations, in the Painlevé–Gambier classsification [12]. It is known as P_{34} and is related to the second Painlevé equation through a simple Miura transformation.

These, and similar, examples led Ablowitz and Segur, in collaboration with one of us (AR), to formulate the following conjecture: “Every ODE obtained by an exact reduction of a nonlinear PDE solvable by IST has the Painlevé property” [13]. Moreover they proposed an efficient algorithm for the investigation of the Painlevé property of a given ODE. Since these developments occured long after the revival of KdV they were not useful for the solution of the latter. Rather this is a case where the solution of a seemingly more difficult problem (that of KdV) can be used in order to solve another, *a priori*, easier one: the integration of the Painlevé equations [9]. We are not going to present here any details on the applications of the ARS method on the detection of integrability [14]. We will, rather, proceed to the next important development in integrability detection.

One of the drawbacks of the ARS approach is the fact that one must first reduce the PDE before applying the algorithm. In some cases the reductions are just too trivial to yield interesting information. Weiss and collaborators [15] have done away with this problem by introducing the Painlevé property for the PDE's themselves. In fact according to Weiss *et al.* [15], a PDE will possess the Painlevé property if its solutions are single-valued about a movable singularity manifold (which must be noncharacteristic [16]). To illustrate this approach let us present the example of the KdV equation. The singularity manifold is given by $\phi(x,t) = 0$ and a solution to the PDE is of Painlevé type when

$$u(x,t) = \phi^\alpha \sum_{n=0}^{\infty} u_n \phi^n$$

where α is integer and u_n are analytic in the neighbourhood of the singularity manifold. Saying that the singularity manifold is movable means that it is only determined by the initial conditions without any other constraint besides being noncharacteristic. Substituting in the KdV we find that $\alpha = -2$ and thus the singular expansion is of the form:

$$u = \phi^{-2} \sum_{n=0}^{\infty} u_n \phi^n. \tag{2.9}$$

Next, we substitute (2.9) into (2.1) and establish a recursion relation for the u_n. We readily obtain:

$$(n+1)(n-4)(n-6)u_n = F(u_0,\ldots,u_{n-1},\phi_t,\phi_x,\ldots). \tag{2.10}$$

This means that at $n=-1,4$ and 6, traditionally called 'the resonances', a free u_n appears in the expansion (with -1 being related to the arbitrariness of ϕ). The expansion coefficients can be easily computed: $u_0 = -2\phi_x^2$, $u_1 = 2\phi_{xx},\ldots$ and one finds that the resonance compatibility conditions are satisfied i.e. the r.h.s. of (2.10) vanishes at $n=4$ and 6. This indicates that the singular expansion is devoid of logarithmic terms and thus possesses the Painlevé property. The Weiss approach to the investigation of the Painlevé property of PDEs has been particularly useful [17]. Not only did it allow the investigation of the singularity structure of PDEs in a straightforward way, it also made it possible to obtain the Bäcklund transforms and the Lax pairs of integrable equations [18].

We will conclude this section with another approach to the singularity analysis of PDEs. It is based on the bilinear formalism of Hirota. As explained in the Introduction, Hirota [5] introduced his formalism because it provides a convenient way to compute multisoliton solutions. It is based on a dependent variable transformation which, in the case of the KdV, is $u=-2\partial_x^2\log F$. In terms of the new function F the KdV equation is written:

$$FF_{xxxx} - 4F_xF_{xxx} + 3F_{xx}^2 + FF_{xt} - F_xF_t = 0. \tag{2.11}$$

Hirota introduced the operator D, an antisymmetric derivative acting on the 'dot-product' in the following way:

$$D_xF\cdot G \equiv (\partial_x - \partial_{x'})F(x)G(x')|_{x'=x} = F'G - FG'.$$

Using D we can rewrite (2.11) as:

$$(D_x^4 + D_xD_t)F\cdot F = 0. \tag{2.12}$$

In [19], Hietarinta presented a classification of integrable PDEs in bilinear formalism. Here again the KdV family is the simplest one since it involves only one dependent variable. The singularity analysis of this family of equations has been presented in [20]. It is based on the fact that the τ-function F cannot have any singularities: it can only have zeros. Thus, we look for vanishing solutions of the form $F\sim\phi^n$. One can show that the operator D^{2q} gives leading behaviours with $n=0,1,\ldots,q-1$. In the case of the KdV this means that F is either regular ($n=0$) or vanishes like ϕ. The computation of the resonances in the case $F\sim\phi$ leads to $-1,0,\ldots,2q-3$ and $4q-2$ and there is no surprise that the resonance conditions are satisfied for KdV. In fact, this approach has been used in the case of a general operator D^{2q}, showing that only the cases $q=2$ (KdV) and $q=3$ (Sawada–Kotera) do satisfy the integrability criterion.

The Korteweg–de Vries equation has been the paradigm case for the development of the continuous integrability prognosticator. But that was not the end of the story. The KdV equation was foreordained to play a major role in obtaining another, equally important, integrability detector.

3. Singularity Confinement: The Integrability Detector for Discrete Systems

While an integrability criterion for continuous systems has existed for years and, in fact, can be traced back to the work of the 19th century mathematicians, this was not the case for discrete systems. Examples of integrable discrete systems have been known for years. Ablowitz and Ladik [21] derived integrable integration schemes in order to simulate numerically the behaviour of integrable evolution equations. Hirota [22] obtained the discrete equivalents of most well-known integrable PDEs, starting with the proper discrete formulation of his bilinear formalism. However, the breakthrough in the domain was the obtaining of an integrability criterion. The KdV equation once more played a crucial role here.

As explained in [23] the clue was provided by the *lattice* KdV equation:

$$x_j^{i+1} = x_{j+1}^{i-1} + \frac{1}{x_j^i} - \frac{1}{x_{j+1}^i}. \tag{3.1}$$

The structure of the lattice associated to Equation (3.1) can be easily assessed from Figure 1. Initial data are given on the staircase which joins the points i and $(i+1)$ (thick line), while evolution takes place towards increasing is. Since the mapping variable appears at the denominator the natural question is 'what if x vanishes accidentally?' This is not at all impossible and the point where this occurs depends on the initial data (i.e. the singularity is movable). So, let us assume that $x = 0$ at (i, j). This leads to divergent xs at both $(i+1, j-1)$ and $(i+1, j)$ and a vanishing x at $(i+2, j-1)$ (see Figure 1). The crucial question is what happens at the sites $(i+3, j-2)$ and $(i+3, j-1)$? It turns out that due to the precise form of (3.1) a fine cancellation occurs and one obtains finite values (denoted by 'f' in the figure): $x_{j-1}^{i+3} = x_j^{i-1} + 1/x_{j-1}^i - 1/x_j^{i+2}$ and a similar one for x_{j-2}^{i+3}. Thus the singularity does not propagate beyond a few lattice points and is confined to a small region (shaded region in Figure 1).

A more stringent test to the confinement of singularities can be obtained if one chooses several adjacent zeros leading to more extended divergences (see Figure 2 for an illustration of the case with two zeros).

The pattern of singularities is now more complex but in every case the answer is the same: the singularities are confined. This would not have been true if the evolution did not have the form (3.1). In fact, modifications to the relative coefficients of the various terms in (3.1), that presumably destroy integrability, lead to singularities radiating all the way from the site (i, j) to infinity. The

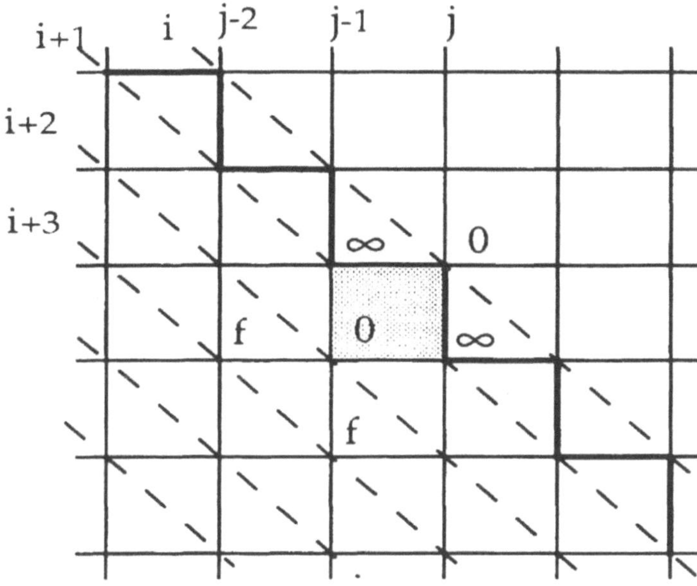

Fig. 1. Lattice KdV with a simple singularity.

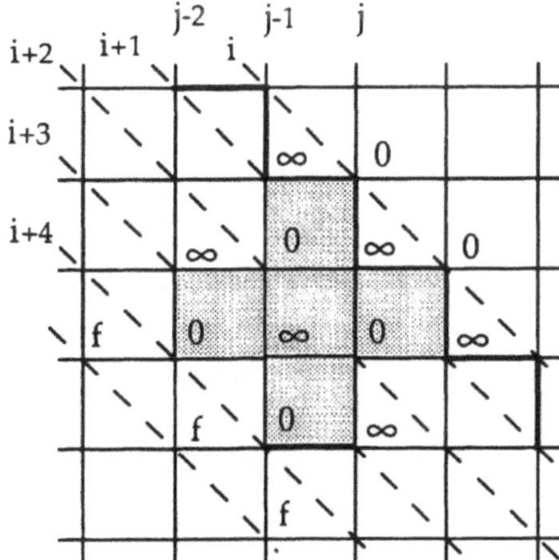

Fig. 2. Lattice KdV with a double singularity

confinement of singularities is reminiscent of the continuous Painlevé property: absence of natural boundaries in integrable 2-D Hamiltonians [14].

This property (singularities that disappear after a few iteration steps) has indeed turned out to be the equivalent of the Painlevé property for discrete systems: it is a necessary condition for integrability [24]. One important application of the singularity confinement method is the derivation of the discrete Painlevé equations [25]. The latter are integrable mappings which at the continuous limit go over to the discrete Painlevé equations. Thus the discrete equivalent to P_I (2.4) is:

$$x_{n+1} + x_n + x_{n-1} = \frac{z_n}{x_n} + c \tag{3.2}$$

where $z_n = an + b$, while P_{34} has the discrete analog:

$$(x_{n+1} + x_n)(x_n + x_{n-1}) = \frac{x_n^2 + c}{x_n + z_n}. \tag{3.3}$$

It would be interesting to point out here that discrete Painlevé equations can indeed be obtained from the similarity reductions of the lattice equations, just like their continuous analogs. In [26] we have shown (among other things) that the similarity reduction of the discrete KdV leads precisely to d-P_{34}, Equation (3.3) above.

We will conclude this section just as we did in the case of continuous equations by showing how the singularity confinement can be applied in the bilinear framework. As was shown by Hirota [22] the bilinear expression of the discrete KdV equation is:

$$\begin{aligned} &\alpha f(m+1,n)f(m-1,n) + \beta f(m,n+1)f(m,n-1)+ \\ &\quad +\gamma f(m+1,n+1)f(m-1,n-1) = 0. \end{aligned} \tag{3.4}$$

The continuous limit of (3.4) is obtained through

$$\gamma = -(\alpha+\beta), \qquad \alpha = 1 + 2\lambda, \qquad \beta = -\lambda(\lambda+2),$$

$D_m = \varepsilon\partial_x + \mu\varepsilon^3\partial_t$ and $D_n = \lambda\varepsilon\partial_x$ with $\mu = (\lambda+1/2)(\lambda+2)(\lambda^2-1)/12(\lambda^2+\lambda+1)$. As $\varepsilon \to 0$ it leads precisely to (2.11), i.e. the bilinear form of the continuous KdV. How can one apply the singularity confinement algorithm to an equation such as (3.4)? The lattice defines an evolution where one of the fs is obtained from the knowledge of the five others. From the bilinear structure it is clear that this recursion will break down when one of the fs becomes zero or infinite. In fact, a diverging f can only be the result of an f vanishing at a previous step, thus only zeros of f need be considered to start with. Indeed, the singularity confinement will precisely consist in demanding that the vanishing of a given f *never* induces a divergence at the next stage. Let us denote by $K(n,m)$ the quantity appearing is the l.h.s. of (3.4). We start with the equation $K(m-1,n-1) = 0$ and assume that $f(m,n) = 0$. This would lead, at the level of the equation $K(m+1,n+1) = 0$, to a divergent $f(m+2,n+2)$. However, when we compute the objects appearing in $K(m+1,n+1)$, i.e. $f(m,n+1)$, $f(m+1,n)$, $f(m+1,n+2)$, $f(m+2,n+1)$ with the help of the equations involving $K(m,n-1)$, $K(m-1,n)$, $K(m,n+1)$, $K(m+1,n)$ we find that the condition for the non divergence of $f(m+2,n+2)$ is satisfied. In fact, this condition is just $\alpha f(m,n-1)f(m-2,n-1)+\beta f(m-1,n)f(m-1,n-2)=0$, i.e. $K(m-1,n-1)=0$ with $f(m,n)=0$. Thus, no singularity ever appears for the bilinear d-KdV. Starting from this simple example of d-KdV, the singularity confinement method can be applied systematically to multidimensional discrete systems [27]. The bilinear formalism (and its multilinear extension [28]) provides a most convenient frame for this investigation and, once again, the KdV equation is the simplest paradigm in this setting.

4. Confined Painlevé Singularities: Blending the Continuous and Discrete Approaches

In the preceding sections we have presented the singularity analysis approach for the detection of integrability of continuous and discrete systems. Although the key word is singularity for both methods, their formulation (in fact their very essence) was such that neither could be adapted to systems that are genuinely *both* discrete and continuous. Clearly, a synthesis is needed. In what follows we propose such a method, incorporating features of both approaches, which will allow us to treat a large variety of systems. Our conjecture will thus provide an integrability criterion for discrete–continuous systems. In order to combine the two methods, assuming that we have a differential-difference system, we will consider the effect of a singularity in the continuous variable when it is iterated following the discrete evolution. The Painlevé property requires that this singularity be pole-like and the singularity confinement criterion consists in demanding that this singularity disappears after some iterations (in the discrete variable). The procedure will become clearer with the examples we give below.

Let us start with the auto-Bäcklund relation for the potential-KdV equation [29], that allows us to relate two solutions with numbers of kinks differing by one unit:

$$u'_{n+1}+u'_n+(u_{n+1}-u_n)^2+K_n=0. \tag{4.1}$$

where the prime denotes the derivative with respect to the continuous variable x, and K_n is related to the energy of the kink which is added (or removed) and should depend on n but *not* on x. We will indeed see that this fact, essentially, emerges from the analysis.

In the spirit of the singularity confinement method, we assume that u_n is regular and that a singularity first appears for u_{n+1}. Given the Riccati form of (4.1) this singularity can only be a simple pole: $u_{n+1}=1/(x-x_0)$ with resonance $\nu=-1$ due to the arbitrariness of x_0. Given u_{n+1} we can compute u_{n+2}. The Riccati now has divergent coefficients and they will induce a singularity that, for u_{n+2}, will have a position x_0 *fixed* by u_{n+1}. We can easily prove that if u_k is of the form $u_k=m(m-1)/2(x-x_0)$ then u_{k+1} will diverge as $A/(x-x_0)$ with $A=\frac{1}{2}m(m+1)$ or $A=\frac{1}{2}(m-1)(m-2)$. Thus starting from $u_{n+1}=1/(x-x_0)$

we get u_{n+2}, which either diverges as $3/(x-x_0)$ or has a finite value. In the latter case the singularity should be confined while a resonance $\nu = 3$ exists. Thus we must expand u_{n+2} in power series and evaluate the resonance compatibility condition. It turns out to be just

$$\frac{\mathrm{d}K_n}{\mathrm{d}x} = \frac{\mathrm{d}K_{n-1}}{\mathrm{d}x},$$

independent of n. This would lead to $K_n = \kappa_n + 2\phi'(x)$. However by redefining u: $u \to u + \phi(x)$, we can absorb ϕ, entirely and without loss of generality we can assume that K_n is indeed x-independent. When this condition is satisfied u_{n+2} is regular, i.e. the singularity is indeed confined. When u_{n+2} diverges we pursue the analysis one step further: u_{n+3} diverges either as $6/(x-x_0)$ or as $1/(x-x_0)$. The first case takes us deeper and deeper in the singularity, while, if we follow the second we see readily that we can emerge at the very next step. We have checked that under the sole condition '$\mathrm{d}K_n/\mathrm{d}x$ independent of n' the singularity is also confined in this case. By now the pattern has become clear: a singularity appears at a certain iteration and, when iterated, it becomes confined. The process may necessitate an arbitrarily large number of steps, but from the structure of the singularity it is clear that we can always emerge. Moreover, the simplest case, i.e. confinement in one step, is the generic one. It is precisely this confinement pattern that gives the constraint on K_n: once it is satisfied, Equation (4.1) becomes integrable. Systems such as (4.1) were previously treated following the 'classical' Painlevé approach, usually through some periodicity assumption. One assumed that for some N, $u_n = u_{n+N}$, and explored the integrability of the finite N-order system for increasing (but still typically small) values of N. This method has two major drawbacks. The first is that small-N cases are far from being generic. The second is more serious for non-autonomous systems. If the n-dependent coefficients (such as K_n in Equation (4.1)) are *not* periodic, there is simply no way to perform the Painlevé analysis through periodification. By contrast, the singularity confinement approach deals with such cases in a straightforward way. The above case has shown clearly the interplay between the processes of (continuous) singularity analysis and confinement.

The situation becomes even simpler when one considers differential-difference equations where the evolution does not enter through derivatives. We will illustrate this through the equation:

$$q_n' = (aq_n^2 + bq_n + c)(q_{n+1} - q_{n-1}) \tag{4.2}$$

that contains both the semidiscrete mKdV [30] ($-a = c = 1, b = 0$) and the semi-discrete KdV or Kac–Van Moerbeke system [31] ($a = c = 0, b = 1$), two well-known integrable equations. We start by rewriting (4.2) as:

$$q_{n+1} = q_{n-1} + \frac{q_n'}{aq_n^2 + bq_n + c} \tag{4.3}$$

and assume that q_n is regular:

$$q_n = q_0 + q_0'(x - x_0) + \frac{1}{2}q_0''(x - x_0)^2 + \cdots,$$

such that $aq_0^2 + bq_0 + c = 0$. Thus q_{n+1} develops a singularity at x_0:

$$q_{n+1} = \frac{1}{(2aq_0 + b)(x - x_0)} + q_{n-1} + \cdots.$$

Iterating we find

$$q_{n+2} = -q_0 - \frac{b}{a} + (x - x_0)A + \cdots$$

where A depends on q_0, q_0' and q_0'' as well as on q_{n-1}. Finally, one further iteration leads to a finite value for q_{n+3}, and the singularity is confined. Two remarks are in order at this point.The KdV/Kac–Van Moerbeke equation is not included in the above analysis since $a = 0$. However, its analysis follows the same steps. We start with

$$q_n = q_0'(x - x_0) + \frac{1}{2}q_0''(x - x_0)^2 + \cdots,$$

leading to

$$q_{n+2} = -\frac{1}{(x - x_0)} + (x - x_0)A + \cdots, \qquad q_{n+3} = (x - x_0)B + \cdots$$

and finally a nonsingular q_{n+4}, that depends on q_{n-1} and q_0'. The second remark concerns the assumption that the first derivative q_0' does not vanish. Although this is the simplest case (the one we consider as 'generic') it is not the only one. For instance in the discrete mKdV case, if $q_0' = 0$ then A in the expression of q_{n+2} diverges and the singularity will need extra steps to be confined. Namely

$$q_{n+1} = \frac{2}{(2aq_0 + b)(x - x_0)} + \mathcal{O}(1), \qquad q_{n+2} = -\frac{b}{2a} + \mathcal{O}(x - x_0),$$

$$q_{n+3} = \frac{2}{(2aq_0 + b)(x - x_0)} + \mathcal{O}(1), \qquad q_{n+4} = -q_0 - \frac{b}{a} + \mathcal{O}((x - x_0)^2)$$

and the singularity is finally confined as q_{n+5} is regular. In principle, one must consider *all* the possible singular behaviors: vanishing derivatives of q_0 up to arbitrary order. Our claim is that the essential constraints already arise at the lowest-order singularity.

This combination of the two integrability detectors is particularly powerful. Its domain of application exceeds that of the simple differential-difference equations. For example, in [32] we have shown how integrodifferential equations of the intermediate long wave/Benjamin–Ono class can be treated with our method, while [33] was devoted to a study of differential-delay systems and led to the discovery of the delay-Painlevé equations.

5. Conclusion (and an Unsusual Application of KdV)

In the preceding sections we presented our results on integrability detection where the KdV equation has played a capital role. The discovery of solitons and the subsequent proof of the complete integrability of KdV was at the origin of the revival of interest is integrable equations. The proposal of efficient integrability detectors was based on extensive 'experimentation' and in every case the KdV equation was the testing ground for each integrability conjecture.

The ubiquity of the KdV equation has been established beyond any doubt. The present conference proceedings constitute the best testimony to that. Nevertheless, we wish to present here a rather unusual application of the KdV equation in the domain of numerical analysis. In [34] we presented a study of a family of algorithms that accelerate the convergence of sequences (and are most often used to estimate the sum of infinite series). The well-known ε-algorithm [35] has the form:

$$x_n^{k+1} = x_n^{k-1} + \frac{1}{x_{n+1}^k - x_{n-1}^k} \tag{5.1}$$

where x_n is the sequence under consideration and x_n^k indicates the kth iteration in the expansion. The initial data are the x_n^1 while the x_n^0 are set to zero. The interesting observation is that (5.1), considered as a two-variable difference equation, is just the lattice (potential) KdV equation [22, 36]. Thus an integrable scheme appears in this context. What is more interesting is that if one tries to generalize this algorithm on the sole requirement that integrability be preserved one obtains the equation:

$$x_n^{k+1} = x_n^{k-1} + \frac{z_n^k}{x_{n+1}^k - x_{n-1}^k} \tag{5.2}$$

with $z_n^k = f(n+k) + g(n-k)$ and f, g free functions. This turns out, unsurprisingly, to be the nonautonomous extension of KdV [26] but at the same time a well-known accelerator, the so-called ρ-algorithm [35]. Once more the KdV equation and its extensions play an important role.

The Korteweg–de Vries equations is 100 years old. Since its revival in the '70s it has been the centre of intense activity. More than that, it is now the axis around which the world of integrability revolves. So, after a famous past, the KdV equation is promised a brilliant future [37].

Acknowledgements

The ideas presented above have matured over the years thanks to discussions, exchange of correspondence or direct collaboration with M. Ablowitz, H. Capel,

G. Contopoulos, B. Dorizzi, A. Fokas, J. Hietarinta, R. Hirota, M. Kruskal, F. Nijhoff, V. Papageorgiou, J. Satsuma, H. Segur, J.-M. Strelcyn, K.M. Tamizhmani and P. Winternitz. To all of the above (and to those whom we have unintentionally left out of the above list) the authors wish to express their sincere gratitude.

References

1. We shall not present here references that are of strict historical interest. They will be amply documented in these proceedings and the interested reader can surely track them down.
2. This time-scale (several decades) is characteristic of the fact that the interest in integrable equations had somehow been lost at the beginning of this century, only to reappear in the '70s.
3. This is probably the explanation of the time lag. The important piece that was lacking at the beginning of the century was the knowledge of IST techniques that would allow the complete solution of the problems.
4. There exist, of course, other, simpler types of integrability where one can transform a nonlinear PDE into a linear purely differential one or even reduce it to quadratures: F. Calogero, in: P. A. Clarkson (ed.), *Applications of Analytical and Geometric Methods to Nonlinear Differential Equations*, NATO ASI C413, Kluwer (1993), p. 109. However, integrability through IST is the generic situation.
5. Hirota, R.: *Phys. Rev. Lett.* **27** (1971), 1192.
   Hirota, R.: in: R. K. Bullough and P. J. Caudrey (Eds), *Soliton*, Springer (1980), p. 157.
6. Hietarinta, J.: Hirota's bilinear method and partial integrability, NATO ASI C310 (1990), p. 459.
7. This is an oversimplified statement that does not accurately reproduce Painlevé's argument. We refer the reader to Kruskal's review of the Painlevé approach and its extensions: Kruskal, M. D.: *Flexibility in Applying the Painlevé Test*, in: D. Levi and P. Winternitz (ed.), *Painlevé Transcendents*, NATO ASI series B278, Plenum (1992) p. 175. Still, as far as the present review is concerned, this presentation of the Painlevé property will suffice.
8. Before proceeding further, let us make certain points clear. The term integrability will be used loosely, in this paper, to indicate integrability through IST techniques. It will, however, cover the cases where the nonlinear equation can be linearized into a purely differential one through a Cole–Hopf transformation. What it does *not* certainly cover are the cases where the system is reduced to quadratures. In this case there does not seem to exist any relation between the Painlevé property and integrability. This question was examined in detail in our course at les Houches, '89: Kruskal, M. D., Grammaticos, B., and Ramani, A.: *Singularity Analysis and its Relation to Complete, Partial and Nonintegrability*, NATO ASI C310 (1990) p. 321. While the whole Painlevé business can be formulated as a conjecture, we do not claim that a large-applicability theorem can be obtained. There even exist counterarguments to that: Ramani, A., Grammaticos, B., and Yoshida, H.: *Acta Appl. Math.* **8** (1987), 75. However, our experience over more than a decade working on integrable systems has shown that once a system passes the Painlevé test it is most probably integrable. Counterexamples do exist but in every case one can point to some pathological feature of the system as an explanation. In this paper we shall not worry about these (marginal) cases but will focus on the use of the Painlevé criterion as an integrability predictor.
9. The 'Lax pairs' for the Painlevé equations have been known for quite some time: Garnier, R.: *Ann. Sci. Ec. Norm. Sup.* **29** (1912), 1. What is really astonishing is that nobody at that time had the idea that they could be used for the integration of the Painlevé equations.
10. Ablowitz, M. J. and Segur, H.: *Phys. Rev. Lett.* **38** (1977), 13.
11. This may seem, *a posteriori*, to be a trivial or evident remark. However, at the time, the whole Painlevé approach was simply forgotten and, in a sense, Ablowitz and Segur were reinventing it.
12. Painlevé, P.: *Acta Math.* **25** (1902), 1 and Gambier, B.: *Acta Math.* **33** (1910), 1. See also, Ince, E. L.: *Ordinary Differential Equations*, Dover, London (1956). Although the latter is the

standard reference on Painlevé equations, it contains many misprints and the analysis of most equations is given in an elliptic way.
13. Ablowitz, M. J., Ramani, A., and Segur, H.: *Lett. Nuov. Cim.* **23** (1978), 333.
14. The interested reader can find a detailed account in our review article: Ramani, A., Grammaticos, B., and Bountis, A.: *Phys. Rep.* **180** (1989), 159. Although this work is more than five years old its reference list can still be used profitably.
15. Weiss, J., Tabor, M., and Carnevale, J.: *J. Math. Phys.* **24** (1983), 522.
16. Ward, R. S.: *Phys. Lett. A* **102** (1984), 279.
17. Complemented by Kruskal's simplifying ansatz (Jimbo, M., Kruskal, M. D., and Miwa, T.: *Phys. Lett. A* **92** (1982), 59) it has made singularity analysis of PDE really accessible. This has, unfortunately, given birth to a useless production line where authors lacking inspiration choose some PDE (most of the time for no apparent reason), apply the ARS–Weiss algorithm, and write a paper on the results.
18. Weiss himself has extensively studied integrable PDEs from this point of view: *J. Math. Phys.* **24** (1983), 1405; **25** (1984), 13, 2226; **26** (1985), 258, 2174; **27** (1986), 1293, 2467; **28** (1987), 2025 and *The Singular Manifold Method*, in: D. Levi and P. Winternitz (ed.), *Painlevé Transcendents*, NATO ASI series B278, Plenum (1992) p. 175. A reformulation of this approach, introducing a more convenient expansion can be found in: R. Conte, *Unification of PDE and ODE versions of Painlevé analysis into a single invariant version*, in the same proceedings of the workshop on Painlevé transcendents.
19. Hietarinta, J.: *J. Math. Phys.* **28** (1987), 1732, 2094, 2586; **29** (1988), 628.
20. Grammaticos, B., Ramani, A., and Hietarinta, J.: *J. Math. Phys.* **31** (1990), 2572.
21. Ablowitz, M. J. and Ladik, F. J.: *Stud. Appl. Math.* **55** (1976), 213; **57** (1978), 1.
22. Hirota, R.: *J. Phys. Soc. Japan* **43** (1977), 1424, 2079; **50** (1981), 3785.
23. Grammaticos, B., Ramani, A., and Papageorgiou, V. G.: *Phys. Rev. Lett.* **67** (1991), 1825.
24. We shall not here go into all the details concerning the singularity confinement criterion, its precise definition, the investigation of its applicability, etc. The interested reader can find the details in our review: *Singularity confinement*, to appear in the proccedings of the 1994 Esterel workshop on discrete equations.
25. Ramani, A., Grammaticos, B., and Hietarinta, J.: *Phys. Rev. Lett.* **67** (1991), 1829. For a more detailed presentation of the discrete Painlevé equations we refer to our review: Grammaticos, B. and Ramani, A.: *Discrete Painlevé Equations: Derivation and Properties*, NATO ASI C413 (1993), 299.
26. Ramani, A., Grammaticos, B., and Satsuma, J.: *Phys. Lett.* **A169** (1992), 323.
27. Satsuma, J., Ramani, A., and Grammaticos, B.: *Phys. Lett.* **A174** (1993), 387.
28. Grammaticos, B., Ramani, A., and Hietarinta, J.: *Phys. Lett.* **A190** (1994), 65.
29. Ablowitz, M. J. and Segur, H.: *Solitons and the Inverse Scattering Transform*, SIAM, Philadelphia 1981.
30. Ablowitz, M. J. and Ladik, J.: *J. Math. Phys.* **16** (1975), 598.
31. Kac, M. and Van Moerbeke, P.: *Adv. Math.* **16** (1975), 160.
32. Ramani, A., Grammaticos, B., and Tamizhmani, K. M.: *J. Phys.* **A26** (1993), L53.
33. Grammaticos, B., Ramani, A., and Moreira, I. C.: *Physica* **A196** (1993), 574.
34. Papageorgiou, V., Grammaticos, B., and Ramani, A.: *Phys. Lett.* **A179** (1993), 111.
35. Brezinski, C.: *Numer. Math.* **35** (1980), 175.
36. Capel, H. W., Nijhoff, F. W., and Papageorgiou, V. G.: *Phys. Lett.* **A155** (1991), 337.
37. The authors are convinced that a bicentennial KdV conference will take place by 2095. Their major regret is that they will be unable to participate in this important event.

Acta Applicandae Mathematicae **39:** 349–360, 1995.

Generalized Self-Dual Yang–Mills Flows, Explicit Solutions and Reductions*

CHAOHAO GU
Institute of Mathematics, Fudan University, Shanghai, China

(Received: 12 September 1994)

Abstract. An evolution equation is added to the generalized self-dual Yang–Mills equations. The evolution equation contains terms of negative powers of the spectral parameter as well as terms of positive powers. The Darboux matrix method is used to obtain explicit solutions, especially single and multiple solitons. All integrable soliton equations in the framework of AKNS system (in $\mathbf{R}^{n+1}$ or $\mathbf{R}^{1+1}$) can be derived from the generalized Yang–Mills flows by reduction.

Mathematics Subject Classifications (1991): 58F07, 35Q51.

Key words: evolution equations, generalized Yang–Mills flows, Darboux transformations.

1. Introduction

The self-dual Yang–Mills equations (SDYM) [1] are among the most important differential equations in mathematics as well as in physics. In the aspect of mathematics, SDYM play essential roles in topology and geometry [2, 3]. Moreover, SDYM are integrable systems in $\mathbf{R}^4$ [4–6]. Recently, the relationship between SDYM and soliton equations becomes a central problem in the theory of integrable systems [7]. Many famous soliton equations, such as KdV, NLS, KP, and DS equations [7] and harmonic maps from $\mathbf{R}^2$ and $\mathbf{R}^{1+1}$ to $\mathrm{U}(N)$ [8] can be deduced from SDYM. It is expected that SDYM contain 'all' integrable soliton systems [9].

On the other hand, the integrable and soliton systems in higher-dimensional space-time are attracting more attentions than before, e.g., SDYM have been generalized to $4n$-dimensional space by Takasaki [10], a kind of generalization of the nonlinear wave equation and sine-Gordon equation were studied in [11, 12], and the generalized AKNS systems in $\mathbf{R}^{n+1}$ were systematically studied in [13, 14].

In a previous paper [15], a class of integrable systems of equations based on generalized self-dual Yang–Mills equation was studied. The systems considered may be considered as evolution systems in the space-time $\mathbf{R}^{2n+1}$.

It is found that the Darboux matrix method is very useful for obtaining explicit solutions to these systems. The first and successive Darboux transformations of

* Supported by Chinese research project 'Nonlinear Science' and K.C. Wong Education Foundation, Hong Kong.

the trivial solution are called single solitons and multi-solitons, respectively, since they are almost localized and behaves asymptotically as a travelling wave, and as $t \to \pm\infty$, a double soliton (resp., k-soliton) almost splits up into 2 (resp., k) solitons and the interaction of solitons is almost elastic.

In the present paper we consider more general flows for which the Lax equation contains negative powers of the spectral parameter as well as the positive powers. It is shown that all the results of the previous paper hold true. We also indicate a natural way of reduction.

2. Generalized Self-Dual Yang–Mills Flows

Let $(x,p) = (x_1,\dots,x_n;\ p_1,\dots,p_n)$ be coordinates of points in $\mathbf{R}^{2n}$ $(n \geqslant 2)$. (x,p) may be real coordinates or conjugate complex coordinates with $p_i = \bar{x}_i$ $(i = 1,2,\dots,n)$ as in the original self-dual Yang–Mills theory in $\mathbf{R}^4$ [1].

We start with the Lax equations [10]

$$L_i\Psi \equiv \left(\frac{\partial}{\partial p_i} - \lambda\frac{\partial}{\partial x_i}\right)\Psi = -A_i\Psi. \tag{2.1}$$

Here Ψ and A_is are $N \times N$ matrices and λ is a spectral parameter. Moreover, the A_is are independent of λ. The integrability condition of (2.1) is

$$\frac{\partial A_i}{\partial p_j} - \frac{\partial A_j}{\partial p_i} - [A_i, A_j] = 0 \tag{2.2}$$

and

$$\frac{\partial A_i}{\partial x_j} - \frac{\partial A_j}{\partial x_i} = 0. \tag{2.3}$$

This is the self-dual Yang–Mills equations in $\mathbf{R}$-gauge for $n = 2$ [1].

We consider (2.1)–(2.3) in $\mathbf{R}^{2n}$ (or some simply connected region of $\mathbf{R}^{2n}$). It should be noted that the self-duality makes sense for even n. However, the results obtained below are valid for odd n, too. From (2.2) there exist matrix-valued functions J such that

$$A_i = -\frac{\partial J}{\partial p_i}J^{-1}. \tag{2.4}$$

Hence (2.2) and (2.3) are equivalent to

$$\frac{\partial}{\partial x_i}\left(\frac{\partial J}{\partial p_j}J^{-1}\right) - \frac{\partial}{\partial x_j}\left(\frac{\partial J}{\partial p_i}J^{-1}\right) = 0. \tag{2.5}$$

We assume the gauge group to be $\mathrm{GL}(N,\mathbf{C})$ for complex (x,p), and $\mathrm{GL}(N,\mathbf{C})$, $\mathrm{GL}(N,\mathbf{R})$ or $\mathrm{U}(N)$ for real (x,p).

Assuming that Ψ, A_is depend on another parameter, denoted by t, we introduce an evolution equation

$$\frac{\partial \Psi}{\partial t} = V\Psi = \sum_{a=0}^{m+q} V_a \lambda^{m-a} \Psi \quad (q \geqslant 0). \tag{2.6}$$

Here V_as are independent of the spectral parameter λ and can be determined in the following way.

Considering the condition of integrability for (2.1) and (2.7) we obtain a system of equations for V_as

$$\begin{aligned} \frac{\partial V_0}{\partial x_i} &= 0, \\ \frac{\partial V_a}{\partial x_i} &= \frac{\partial V_{a-1}}{\partial p_i} + [A_i, V_{a-1}] \quad (a = 1, 2, \dots, m) \end{aligned} \tag{2.7}$$

and

$$\begin{aligned} \frac{\partial V_{m+q}}{\partial p_i} &= 0, \\ \frac{\partial V_{a-1}}{\partial p_i} &= \frac{\partial V_a}{\partial x_i} + [V_{a-1}, A_i] \quad (a = m+q, \dots, m+2) \end{aligned} \tag{2.8}$$

and a system of evolution equations for A_is

$$\frac{\partial A_i}{\partial t} + \frac{\partial V_m}{\partial p_i} - \frac{\partial V_{m+1}}{\partial x_i} + [A_i, V_m] = 0. \tag{2.9}$$

THEOREM 2.1. *If A_is satisfy the generalized self-dual Yang–Mills equations* (2.2), (2.3), *then* (2.7) *and* (2.8) *are completely integrable and* V_a $(a = 0, 1, \dots, m, m+1, \dots, m+q)$ *can be expressed as integro-differential expressions of A_is.*

Proof. From the first equation of (2.8) we have

$$V_0 = V_0(p, t).$$

Assume that $V_1, V_2, \dots, V_{a-1}$ $(a < m)$ are constructed already. By differentiating (2.7) with respect to x_j, we have

$$\frac{\partial^2 V_a}{\partial x_i \partial x_j} = \frac{\partial^2 V_a}{\partial x_j \partial x_i}$$

if (2.2) and (2.3) hold. Hence $V_a(x, p)$ can be obtained by integrating the right-hand sides of (2.7) along the paths on which $p_i = \text{const}$, i.e.

$$V_a = \sum_i \int_{(0,p)}^{(x,p)} \left(\frac{\partial V_{a-1}}{\partial p_i} + [A_i, V_{a-1}] \right) \mathrm{d}x_i + V_a^0(p, t). \tag{2.10}$$

Similarly, for the system (2.8) we have

$$V_{m+q} = V_{m+q}(x,t). \tag{2.11}$$

If (2.2) and (2.3) hold, we have

$$\frac{\partial^2 V_{a-1}}{\partial p_i \partial p_j} = \frac{\partial^2 V_{a-1}}{\partial p_j \partial p_i} \quad (a > m+1).$$

Hence (2.8) is completely integrable. More precisely, let

$$V_a = J W_a J^{-1}, \quad a = m+q, \ldots, m+1, \tag{2.12}$$

(2.8) becomes

$$\frac{\partial W_{a-1}}{\partial p_i} = \frac{\partial W_a}{\partial x_i} + \left[J^{-1}\frac{\partial J}{\partial x_i}, W_a\right], \quad a = m+q, \ldots, m+2. \tag{2.13}$$

Hence

$$W_{a-1} = \sum_i \int_{(x,0)}^{(x,p)} \left(\frac{\partial W_a}{\partial x_i} + \left[J^{-1}\frac{\partial J}{\partial x_i}, W_a\right]\right) \mathrm{d}p_i + W^0_{a-1}(x,t). \tag{2.14}$$

Here, along the paths of integration, $x_i = \text{const}$, $V_0(p,t)$, $V^0_a(p,t)$ $(0 < a \leqslant m)$, $V_{m+q}(x,t)$, $W^0_a(x,t)$ $(m < a < m+q)$ are arbitrary functions of integration.

The theorem is proved. □

Thus we have a system of Equations (2.2), (2.3) and (2.9), and (2.9) is a system of integro-differentiable evolution equations for A_is.

3. Darboux Transformations

The Darboux transformation is actually a special gauge transformation

$$\widetilde{\Psi} = S\Psi \tag{3.1}$$

of the solutions of Lax equations. We take the gauge matrix

$$S = \lambda I + \alpha \tag{3.2}$$

and call it a Darboux matrix. Here I is the unit matrix and α is to be determined. It is required that $\widetilde{\Psi}$ satisfies (2.1) and (2.7) with some $\tilde{A}_i$ and $\widetilde{V}_a$, i.e.

$$L_i\widetilde{\Psi} = \left(\frac{\partial}{\partial p_i} - \lambda\frac{\partial}{\partial x_i}\right)\widetilde{\Psi} = -\tilde{A}_i\widetilde{\Psi}, \tag{3.3}$$

$$\frac{\partial \widetilde{\Psi}}{\partial t} = \sum \widetilde{V}_a \lambda^{m-a}\widetilde{\Psi}. \tag{3.4}$$

Substituting (3.2) into (3.3), we obtain

$$L_i\widetilde{\Psi} = \left(\left(\frac{\partial}{\partial p_i} - \lambda\frac{\partial}{\partial x_i}\right)\alpha - (\lambda I + \alpha)A_i\right),$$

$$\Psi = -\tilde{A}_i(\lambda I + \alpha)\Psi. \tag{3.5}$$

Hence

$$\tilde{A}_i = A_i + \frac{\partial \alpha}{\partial x_i} \tag{3.6}$$

and

$$\frac{\partial \alpha}{\partial p_i} = \alpha A_i - \tilde{A}_i\alpha = [\alpha, A_i] - \frac{\partial \alpha}{\partial x_i}\alpha. \tag{3.7}$$

Similarly, from (3.2) and (3.4) we obtain

$$\begin{aligned}
&\widetilde{V}_0 = V_0,\\
&\widetilde{V}_a = V_a + \alpha V_{a-1} - \widetilde{V}_{a-1}\alpha \quad (a = 1, 2, \ldots, m),\\
&\widetilde{V}_{m+q} = \alpha V_{m+q}\alpha^{-1}, \qquad\qquad (3.8)\\
&\widetilde{V}_{m+k} = \alpha V_{m+k}\alpha^{-1} + [V_{m+k+1} - \widetilde{V}_{m+k+1}]\alpha^{-1} \quad (k = q-1, \ldots, 1),
\end{aligned}$$

and

$$\frac{\partial \alpha}{\partial t} + \alpha V_m - \widetilde{V}_m\alpha + V_{m+1} - \widetilde{V}_{m+1} = 0. \tag{3.9}$$

Since the generalized self-dual Yang–Mills equations (2.2), (2.3) and the evolution Equation (2.9) are the consequence of the Lax equations (2.1) and (2.6) (provided that det $\Psi \not\equiv 0$). $\tilde{A}_i$s are solutions to the same system as well. We have

THEOREM 3.1. *Let A_i be a solution to the generalized self-dual Yang–Mills equations* (2.2), (2.3) *and the evolution Equations* (2.9), *and Ψ be a fundamental solution to the Lax equations* (2.1), (2.6). *If α is a solution to* (3.7) *and* (3.9), *then*

$$\tilde{A}_i = A_i + \frac{\partial \alpha}{\partial x_i} \tag{3.10}$$

satisfy (2.2), (2.3) *and* (2.9), *in which V_a are replaced by $\widetilde{V}_a$ in* (3.8), *and*

$$\widetilde{\Psi} = (\lambda I + \alpha)\Psi,$$

$\tilde{A}_i$ satisfy (2.1) *and* (2.6).

The transformation $(\Psi, A_i) \to (\widetilde{\Psi}, \tilde{A}_i)$ defined by (3.1), (3.2) and (3.10) is called a Darboux transformation.

System (3.7) and (3.9) is nonlinear and very complicated if m or q are not very small. However, we are able to construct its explicit solutions as follows.

Let λ_α $(\alpha = 1, 2, \ldots, N)$ be N numbers (at least two of them are not equal, $\lambda_\alpha \neq 0$) and $h_\alpha(\lambda_\alpha, x, p)$ be N columns, satisfying (2.1) and (2.7) for $\lambda = \lambda_\alpha$ $(\alpha = 1, 2, \ldots, N)$, i.e.

$$\frac{\partial h_\alpha}{\partial p_i} = \lambda_\alpha \frac{\partial h_\alpha}{\partial x_i} - A_i h_\alpha, \qquad \frac{\partial h_\alpha}{\partial t} = \sum_a V_a \lambda_\alpha^{m-a} h_\alpha. \tag{3.11}$$

Define

$$H = [h_1, h_1, \ldots, h_N]. \tag{3.12}$$

Here we require that $\det H \neq 0$.

THEOREM 3.2. *The matrix* α *defined by*

$$\alpha = -H\Lambda H^{-1} \quad \textit{with} \quad \Lambda = \begin{pmatrix} \lambda_1 & & & \\ & \lambda_2 & & \\ & & \ddots & \\ & & & \lambda_N \end{pmatrix} \tag{3.13}$$

satisfies Equations (3.7) *and* (3.9).

Proof. From (3.11) it is easily seen that

$$\frac{\partial H}{\partial p_i} = \frac{\partial H}{\partial x_i}\Lambda - A_i H. \tag{3.14}$$

Hence

$$\begin{aligned} \frac{\partial \alpha}{\partial p_i} &= -\frac{\partial H}{\partial x_i}\Lambda^2 H^{-1} + H\Lambda H^{-1}\frac{\partial H}{\partial x_i}\Lambda H^{-1} - A_i\alpha + \alpha A_i, \\ \frac{\partial \alpha}{\partial x_i} &= -\frac{\partial H}{\partial x_i}\Lambda H^{-1} + H\Lambda H^{-1}\frac{\partial H}{\partial x_i} H^{-1}. \end{aligned} \tag{3.15}$$

By using (3.13), the definition of α, it is easily seen that (3.7) is satisfied.

Similarly we have

$$\frac{\partial H}{\partial t} = \sum_{a=0}^{m+q} V_a H \Lambda^{m-a}. \tag{3.16}$$

Hence

$$\begin{aligned} \frac{\partial \alpha}{\partial t} &= -\sum_{a=0}^{m+q} V_a H \Lambda^{m-a+1} H^{-1} + \sum_{a=0}^{m+q} H\Lambda H^{-1} V_a H \Lambda^{m-a} H^{-1} \\ &= \sum_{a=0}^{m+q} V_a(-\alpha)^{m-a}\alpha - \alpha\sum_{a=0}^{m+q} V_a(-\alpha)^{m-a}. \end{aligned} \tag{3.17}$$

From (3.8) it is seen that the right-hand side of (3.17) is just $\widetilde{V}_m\alpha - \alpha V_m + V_{m+1} - \widetilde{V}_{m+1}$. Thus (3.9) is satisfied.

Remark 3.1. Just as many other cases, we can use the Darboux transformation successively and an infinite sequence of solutions can be obtained by using the purely algebraic algorithm, indicated in Theorem 3.2, provided that a seed solution $(\Psi(\Lambda), A_i)$ is known.

Remark 3.2. The column solutions h_α in (3.11) can be obtained from $\Psi(\lambda)$ by setting $h_\alpha = \Psi(\lambda)l_\alpha$, where l_α satisfies

$$L_i l_\alpha = \frac{\partial l_a}{\partial p_i} - \lambda \frac{\partial l_\alpha}{\partial x_i} = 0,$$

i.e. $l_\alpha = l_\alpha(\lambda p + x)$.

Remark 3.3. The theorem of permutability [13] holds as well.

For the case of $U(N)$, we consider (x, p) as real coordinates of $\mathbf{R}^{2n}$. The potential A_is should be valued in the Lie algebra $u(N)$, i.e.

$$A_i^* + A_i = 0. \tag{3.18}$$

Besides, we can choose the functions of integration $V_0(p)$, $V_a^0(p)$ $(a = 1, 2, \ldots, m)$ such that $V_0, V_1, \ldots, V_m, V_{m+1}, \ldots, V_{m+q}$ are valued in $u(N)$, too. In this case we can apply a Darboux transformation as before. However, we should consider the additional requirement

$$\tilde{A}_i^* + \tilde{A}_i = 0 \tag{3.19}$$

which is equivalent to

$$\frac{\partial \alpha^*}{\partial x_i} + \frac{\partial \alpha}{\partial x_i} = 0 \tag{3.20}$$

since (3.18) and (3.6) hold. For this purpose we choose parameters λ_αs and l_αs as follows. Let μ be a complex number $(\mu \neq \bar{\mu})$ and

$$\lambda_\alpha = \mu \quad \text{or} \quad \bar{\mu}. \tag{3.21}$$

We choose l_α such that

$$h_\alpha^* h_\beta = 0 \quad \text{if} \quad \lambda_\alpha \neq \lambda_\beta. \tag{3.22}$$

In fact, for $\lambda_\alpha \neq \lambda_\beta$, we have $\bar{\lambda}_\alpha = \lambda_\beta$, and

$$\begin{aligned} L_i(h_\alpha^* h_\beta) &= -h_\alpha^* A_i^* h_\beta - h_\alpha^* A_i h_\beta = 0, \\ \frac{\partial}{\partial t}(h_\alpha^* h_\beta) &= h_\alpha^*(V^*(\lambda_\beta) + V(\lambda_\beta))h_\beta = 0. \end{aligned} \tag{3.23}$$

(3.22) holds everywhere if it holds on $p_i = 0$, $t = 0$. Hence we can choose l_α such that (3.22) holds on $p_i = 0$, $t = 0$ and then we set $l_\alpha = l_\alpha(\lambda_\alpha p + x)$ by continuing the function $l_\alpha(x)$ to complex regions. Besides, we choose the sets $\{l_\alpha \mid \lambda_\alpha = \mu\}$ and $\{l_\alpha \mid \lambda_\alpha = \bar{\mu}\}$ such that they consist of linearly independent vectors.

From (3.13), the definition of α, we have

$$\alpha h_\beta = -\lambda_\beta h_\beta, \qquad h_\beta^* \alpha^* = -\lambda_\beta^* h_\beta^*. \tag{3.24}$$

Hence

$$h_\beta^*(\alpha^* + \alpha)h_\gamma = -h_\beta^*(\mu^* + \mu)h_\gamma \tag{3.25}$$

holds for all $\beta, \gamma = 1, 2, \ldots, N$, if (3.22) is also considered. Since the set of columns $\{h_\alpha\}$ are linearly independent, we have

$$\alpha^* + \alpha = -(\mu + \bar{\mu})I \tag{3.26}$$

and (3.20) holds. Thus we obtain

THEOREM 3.2. *If $A_i \in u(N)$, $V_a \in u(N)$ and parameters $\lambda_1, \ldots, \lambda_N; l_1, \ldots, l_N$ for constructing the Darboux matrix satisfy* (3.21) *and* (3.22) *and the condition of linear independence, then $\tilde{A}_i \in u(N)$.*

4. Examples and Solutions

We start with the trivial solution $A_i = 0$. In order to obtain the solution $\Psi_0(\lambda)$ to the Lax equations, we have to determine $V_0, V_1, \ldots, V_{m+q}$. For $A_i = 0$, (2.8), (2.9) become

$$V_0 = V_0(p),$$

$$\frac{\partial V_a}{\partial x_i} = \frac{\partial V_{a-1}}{\partial p_i} \quad (a = 1, 2, \ldots, m+q), \tag{4.1}$$

$$\frac{\partial V_{m+q}}{\partial p_i} = 0.$$

We assume that $V_0, V_a^0, W_{m+k}^0, W_{m+q}$ are independent of t. It is easily seen that

$$V(\lambda) = U_1(\lambda p + x) + U_2\left(p + \frac{x}{\lambda}\right), \tag{4.2}$$

where U_1, U_2 are matrix-valued polynomials of degrees m and q. Thus

$$\Psi_0 = \exp\left[V(\lambda)t\right]. \tag{4.3}$$

Just as in [15], we can get the explicit expressions for H and α as follows. Here we assume $N = 2$ for simplifying the expressions and let

$$V(\lambda) = \begin{pmatrix} v(\lambda) & 0 \\ 0 & -v(\lambda) \end{pmatrix}$$

with

$$v(\lambda) = u_1(\lambda p + x) + u_2\left(p + \frac{x}{\lambda}\right).$$

Here u_1, u_2 are polynomials of degrees m and q.

The matrix H which consists of two column solutions for $\lambda = \mu$ and $\bar{\mu}$ of the Lax equations, respectively, can be written as

$$H = \begin{pmatrix} \mathrm{e}^{iv(\mu)t} & -g^*(\bar{\mu})\mathrm{e}^{iv(\bar{\mu})t} \\ g(\mu)\mathrm{e}^{-iv(\mu)t} & \mathrm{e}^{-iv(\bar{\mu})t} \end{pmatrix}. \tag{4.4}$$

Here we use the notation

$$\begin{aligned} v(\mu) &= U_0(\lambda p + x) + U_1\left(p + \frac{x}{\lambda}\right), \\ g(\mu) &= g(\mu p + x), \qquad g^*(\bar{\mu}) = g^*(\bar{\mu}p + x) = \overline{g(\mu)}. \end{aligned} \tag{4.5}$$

The condition $h_2^* h_1 = 0$ holds. By calculations we have

$$\Delta = \det H = \mathrm{e}^{wt} + |g(\mu)|^2 \mathrm{e}^{-wt}, \tag{4.6}$$

where

$$w = i(v(\mu) - v(\bar{\mu})) = \text{a real function of } p \text{ and } x,$$

and

$$H^{-1} = \frac{1}{\Delta}\begin{pmatrix} \mathrm{e}^{-iv(\bar{\mu})t} & g^*(\bar{\mu})\mathrm{e}^{iv(\bar{\mu})t} \\ -g(\mu)\mathrm{e}^{-iv(\mu)t} & \mathrm{e}^{iv(\mu)t} \end{pmatrix}. \tag{4.7}$$

Hence

$$\alpha = -\frac{1}{\Delta}\begin{pmatrix} \mu\mathrm{e}^{wt} + \bar{\mu}|g|^2\mathrm{e}^{-wt} & (\mu - \bar{\mu})g^*(\bar{\mu})\mathrm{e}^{iut} \\ (\mu - \bar{\mu})g(\mu)\mathrm{e}^{-iut} & \mu|g|^2\mathrm{e}^{-wt} + \bar{\mu}\mathrm{e}^{wt} \end{pmatrix}, \tag{4.8}$$

where

$$u = v(\mu) + v(\bar{\mu}) = \text{a real function of } p \text{ and } x. \tag{4.9}$$

Let

$$\mu = \sigma + i\tau, \qquad g(\mu) = \mathrm{e}^{\rho(\mu) + i\theta(\mu)}, \tag{4.10}$$

where $\sigma, \tau, \rho, \theta$ are real and $\tau \neq 0$. We can write

$$\begin{aligned}\alpha &= \begin{pmatrix} \alpha_{11} & \alpha_{12} \\ \alpha_{21} & \alpha_{22} \end{pmatrix} \\ &= -\begin{pmatrix} \sigma - i\tau\,\mathrm{th}(\rho - wt) & i\tau\,\mathrm{sech}(\rho - wt)\mathrm{e}^{i(-\theta+ut)} \\ i\tau\,\mathrm{sech}(\rho - wt)\mathrm{e}^{i(\theta-ut)} & \sigma + i\tau\,\mathrm{th}(\rho - wt) \end{pmatrix}. \end{aligned} \tag{4.11}$$

The nontrivial solution obtained is

$$A_i = \frac{\partial \alpha}{\partial x_i} = \begin{pmatrix} a_{11} & a_{12} \\ -\bar{a}_{12} & -a_{11} \end{pmatrix}$$

with

$$a_{11} = -i\tau\,\mathrm{sech}^2(\rho - wt)\left(\frac{\partial \rho}{\partial x_i} - \frac{\partial w}{\partial x_i} t\right)$$

$$\begin{aligned} a_{12} &= i\tau\,\mathrm{sech}(\rho - wt) \times \\ &\quad \times \left[\mathrm{th}(\rho - wt)\left(\frac{\partial \rho}{\partial x_i} - \frac{\partial w}{\partial x_i} t\right) + i\left(\frac{\partial \theta}{\partial x_i} - \frac{\partial u}{\partial \xi_i} t\right)\right]\mathrm{e}^{i(ut-\theta)}. \end{aligned} \tag{4.12}$$

To understand the behavior of α or A_i, the key point is to understand the behavior of

$$|\alpha_{12}| = |\tau|\,\mathrm{sech}(\rho - wt) \tag{4.13}$$

since we have $|\alpha_{11}|^2 + |\alpha_{12}|^2 = \sigma^2 + \tau^2$ and $|\alpha_{21}| = |\alpha_{12}|$, $|\alpha_{22}| = |\alpha_{11}|$.

For general consideration, there are many choices of the function g. For convenience we choose g to be the exponential of a polynomial of $\mu p + x$ of arbitrary degree k. We denote the set of obtained solutions $(\Psi(\lambda), A_i, \alpha)$ by $\Sigma(\mu, g, V)$.

For $t = \text{const}$, $|\alpha_{12}| \to 0$ quickly as $|p| + |x| \to \infty$ for almost all directions. Hence the solution is almost localized. To consider the asymptotic behavior of $|\alpha_{12}|$ as $t \to \pm\infty$, we set

$$p = \pi^1 t + \pi^0, \qquad x = \xi^1 t + \xi^0 \tag{4.14}$$

($\pi^1, \pi^0, \xi^1, \xi^0$ are constant n-vectors, $(\pi^1, \xi^1) \neq (0,0)$). Then

$$|\alpha_{12}| = |\tau|\,\mathrm{sech}[B_r t^r + B_{r-1} t^{r-1} + \cdots + B_1 t + B_0], \tag{4.15}$$

where r is $\max\{\deg g, (\deg u_1) + 1, (\deg u_2) + 1\}$.

If along the ray (4.14) at least one of $B_r, B_{r-1}, \ldots, B_1 \neq 0$, then $|\alpha_{12}| \to 0$. If along the ray (4.14) $B_r = B_{r-1} = \cdots = B_1 = 0$, then $|\alpha_{12}|$ behaves as $|\alpha_{12}(\xi^0, \pi^0, 0)|$ asymptotically. Hence the solution has asymptotic behavior with fixed wave shape. Thus we have

THEOREM 4.1. *All solutions in $\Sigma(\mu, g, V)$ have the properties*

(i) *They are almost localized for fixed t;*
(ii) *$|\alpha_{12}|$ approaches zero or behaves like $|\alpha_{12}(x,p,0)|$ as $t \to \pm\infty$.*

We call the solutions in the class $\Sigma(\mu, g, V)$ single solitons.

For a solution $S_1 \in \Sigma(\mu, g, V)$ we call the ray l defined by (5.18) generic if at least one of $B_r, B_{r-1}, \dots, B_1$ does not equal zero, and special if $B_r = B_{r-1} = \dots = B_1 = 0$. We have

THEOREM 4.2. *Along a generic ray $|\alpha_{12}| \to 0$ as $t \to \pm\infty$, and along a special ray $|\alpha_{12}|$ behaves as $|\alpha_{12}(x,p,0)|$ as $t \to \pm\infty$.*

We will denote the set of special rays for the solution S_1 by $T(S_1)$.

We turn to the case of double solitons. Let $S_1 = (\Psi_1(\lambda), A_i^1, \alpha^1)$, $S_2 = (\Psi_1'(\lambda), A_i^{1'}, \alpha^{1'})$ be solutions in the class $\Sigma(\mu_1, g_1, V)$ and $\Sigma(\mu_2, g_2, V)$, respectively ($\mu_2 \neq \mu_1$). We apply Darboux transformation to $(\Psi_1(\lambda), A_i^1, \alpha^1)$ by using the parameters (μ_2, g_2) and obtain a solution $(\Psi_2(\lambda), A_i^2, \alpha^2)$. Consider the asymptotic behaviors of this solution. We only need to examine

$$\alpha^2 = \begin{pmatrix} \alpha_{11}^{(2)} & \alpha_{12}^{(2)} \\ -\overline{\alpha_{12}^{(2)}} & -\alpha_{11}^{(2)} \end{pmatrix}.$$

We have the following theorem.

THEOREM 4.3 (theorem of splitting). *A double soliton is splitting up into two single solitons along the ray $l \notin T(S_1) \cap T(S_2)$, i.e. as $t \to \pm\infty$,*

(i) *$|\alpha_{12}^{(2)}| \to 0$ if $l \notin T(S_1) \cup T(S_2)$;*
(ii) *$|\alpha_{12}^{(2)}|$ behaves as $|\alpha_{12}^{(1)'}|$ with phase shift $h^{\pm}$ if $l \in T(S_2)$ and $l \notin T(S_1)$;*
(iii) *$|\alpha_{12}^{(2)}|$ behaves as $|\alpha_{12}^{(1)}|$ with phase shift $h^{\pm'}$ if $l \in T(S_1)$ and $l \notin T(S_2)$.*

This theorem shows that the interaction is almost elastic, since the set $T(S_1) \cap T(S_2)$ is of lower dimensions in the space of rays. For the proof see [15].

We can define the kth solitons as the kth Darboux transformation of the trivial solution and prove the splitting theorem by using mathematical induction.

5. Reductions

Let J_i be n constant diagonal matrices ($n = 1, 2, \dots, n$). Define

$$\begin{aligned} \Phi(p,t) &= \mathrm{e}^{-\Sigma x_i J_i}\Psi, \\ P_i(p,t) &= -\mathrm{e}^{-\Sigma x_i J_i} A_i \mathrm{e}^{\Sigma x_l J_l}, \\ U_a(p,t) &= \mathrm{e}^{-\Sigma x_i J_i} V_a \mathrm{e}^{\Sigma x_l J_l} \quad (a = 0, 1, \dots, n+q), \end{aligned} \tag{5.1}$$

and assume that Φ, P_i, U_a are independent of x_i. Substituting in (2.1) and (2.6), we obtain

$$\frac{\partial\Phi}{\partial p_i} = (\lambda J_i + P_i)\Phi, \qquad \frac{\partial\Phi}{\partial t} = \sum_{a=0}^{m+q} U_a \lambda^{m-a}\Phi. \tag{5.2}$$

If $q = 0$, this is just the generalized AKNS system studied in [13, 14]. Here the negative powers of λ may appear. In particular, the generalized nonlinear wave equations and sine-Gordon equations in [11] can be obtained from (2.1) and (2.6).

If Φ depends on p_1 and t only, then we can derive quite a lot of integrable soliton equations in $\mathbf{R}^{1+1}$, such as KdV hierarchy, MKdV hierarchy, N-wave equation, NLS hierarchy, sine-Gordon equation etc. Most of these reductions have been done before (e.g., see [7]). However, our approach suits the reductions to all equations which belong to the framework of AKNS system in $\mathbf{R}^{n+1}$ and contain negative powers of the spectral parameters.

References

1. Yang, C. N.: Condition of self-duality for SU_2 gauge fields on Euclidean 4-dimensional space, *Phys. Rev. Lett.* **38** (1977), 377.
2. Atiyah, M. F.: *Geometry of Yang–Mills Fields*, Lezioni Fermiane, Accademia Nazionale dei Lincei Scuola Normaie Superione, Pisa, 1979.
3. Freed, D. S. and Uhlenbech, K. K.: *Instantons and Four-Manifolds*, Springer-Verlag, Math. Sciences Res. Inst. Publ. I, 1984.
4. Corrigan, E. F., Fairlie, D. B., Yates, R. G., and Goddard, P.: The construction of self-dual solutions to $SU(2)$ gauge theory, *Commun. Math. Phys.* **58** (1978), 223–240.
5. Chau, L. L., Prasard, M. K., and Sinha, A.: Some aspects of the linear systems for self-dual Yang–Mills fields, *Phys. Rev.* **D24** (1981), 1574–1580.
6. Takasaki, K.: A new approach to the self-dual Yank–Mills equations, *Commun. Math. Phys.* **94** (1984), 35–59.
7. Ablowitz, M. J., Chakrayarty, S., and Takhtajan, L. A.: A self-dual Yang–Mills hierarchy and its reduction to integrable systems in $1+1$ and $2+1$ dimensions, *Commun. Math. Phys.* **159** (1993), 289–314.
8. Uhlenbeck, K.: On the connection between harmonic maps and the self-dual Yang–Mills and the sine-Gordon equation, *J. Geom. Phys.* **8** (1992), 283–316.
9. Ward, R. S.: Intebrable and solvable systems and relations among them, *Phil. Trans. Roy. Soc. London* **A315** (1985), 451–457.
10. Takasaki, K.: Integrable systems in gauge theory, Kähler geometry and super KP hierarchy – symmetries and algebraic points of view, in: *Proc. International Congr. Math.*, Kyoto 1990, Vol. II, pp. 1205–1214.
11. Beals, R. and Tenenblat, K.: An intrinsic generalization for the wave and sine-Gordon equation, in: *Differential Geometry*, Pitman Monograph 52, Longman, Harlow, pp. 25–46, 1991.
12. Gu, C. H. and Hu, H. S.: Explicit solutions to the intrinsic generalization for the wave and sine-Gordon equations, *Lett. Math. Phys.* **29** (1993), 1–11.
13. Gu, C. H.: On the interaction of solitons for a class of integrable systems in the space-time $\mathbf{R}^{n+1}$, *Lett. Math. Phys.* **26** (1992), 199–209.
14. Gu, C. H. and Zhou, Z. X.: On Darboux transformations for soliton equations in high-dimensional space-time, *Lett. Math. Phys.* **32** (1994), 1–10.
15. Gu, C. H.: Integrable evolution systems based on generalized self-dual Yang–Mills equations and their soliton-like solutions, Preprint IDMF 9402.

Acta Applicandae Mathematicae **39**: 361–378, 1995.

Symbolic Software for Soliton Theory

W. HEREMAN* and W. ZHUANG
Department of Mathematical and Computer Sciences, Colorado School of Mines, Golden, CO 80401, U.S.A.

(Received: 15 September 1994)

Abstract. Four symbolic programs, in Macsyma or Mathematica language, are presented. The first program tests forthe existence of solitons for nonlinear PDEs. It explicitly constructs solitons using Hirota's bilinear method. In the second program, the Painlevé integrability test for ODEs and PDEs is implemented. The third program provides an algorithm to compute conserved densities for nonlinear evolution equations. The fourth software package aids in the computation of Lie symmetries of systems of differential and difference-differential equations. Several examples illustrate the capabilities of the software.

Mathematics Subject Classifications (1991): Primary 35Q51, 35Q53, 35C05, 68Q40; Secondary 34-04, 35-04.

Key words: soliton theory, symbolic programs, Hirota method, Painlevé test, Lie symmetries, conserved densities.

1. Introduction

The search for special properties (such as integrability), exact solutions, Lie symmetries and conserved densities of nonlinear partial differential equations (PDEs) becomes increasingly attractive due to the availability of symbolic manipulation programs such as MACSYMA, MAPLE, REDUCE, MATHEMATICA, AXIOM, MuPAD and DERIVE, which allow one to perform the tedious algebra and routine computations.

Four computer programs for symbolic computations relevant to soliton theory are presented. The programs are written in MACSYMA and/or MATHEMATICA syntax. They are entirely symbolic and produce exact analytic output.

The first program tests for the existence of solitary wave and soliton solutions of nonlinear PDEs in bilinear form. It also explicitly constructs multi-soliton solutions via Hirota's method. The second program automatically carries out the Painlevé integrability test for ODEs and PDEs. The third program offers a simple algorithm to compute conserved densities for nonlinear evolution equations. The fourth software package aids in the computation of various types of Lie symmetries of ODEs, PDEs and difference-differential equations.

* Research supported in part by NSF under Grant CCR-9300978.

The purpose, methods, algorithms of the four programs are outlined. Implementation issues are briefly discussed and several examples illustrate the capabilities of the software.

2. Hirota's Method

2.1. PURPOSE

Hirota's method [11, 12] allows one to construct exact soliton solutions of nonlinear evolution and wave equations, provided the equations can be brought in bilinear form.

Our MACSYMA program, called HIROTA_SINGLE.MAX, symbolically computes the one-, two- and three-soliton solutions of well-known nonlinear PDEs. Thus far, the bilinear representation of the PDE must be of Korteweg–de Vries (KdV) type [9]. The program also automatically verifies whether three- and four-soliton solutions will exist for such equations. The development and implementation of a comprehensive algorithm that could handle other types of bilinear representations [9] is in progress.

2.2. ALGORITHM AND IMPLEMENTATION

Details of the method can be found in almost any book on soliton theory [1, 2, 4]: here we outline the procedure.

Hirota's method requires: (i) a clever change of dependent variable; (ii) the introduction of a novel differential operator; (iii) a perturbation expansion to solve the resulting bilinear equation.

Our leading example is the Korteweg–de Vries equation,

$$u_t + 6uu_x + u_{3x} = 0. \tag{1}$$

Substitution of

$$u(x,t) = 2\,\frac{\partial^2 \ln f(x,t)}{\partial x^2} = 2\Big(\frac{f f_{xx} - f_x^2}{f^2}\Big) \tag{2}$$

into (1) and one integration with respect to x yields,

$$f f_{xt} - f_x f_t + f f_{4x} - 4 f_x f_{3x} + 3 f_{2x}^2 = 0. \tag{3}$$

This quadratic equation in f can taken be written in *bilinear form*,

$$B(f \cdot f) \stackrel{\text{def}}{=} P(D_x, D_t)(f \cdot f) \stackrel{\text{def}}{=} (D_x D_t + D_x^4)(f \cdot f) = 0, \tag{4}$$

where the new operator is given by

$$D_x^m D_t^n (f \cdot g) = (\partial x - \partial x')^m (\partial t - \partial t')^n f(x,t) g(x',t')|_{x'=x,\ t'=t}. \tag{5}$$

In (4), P should be interpreted as a polynomial in its arguments, B abbreviates the bilinear operator for the KdV equation.

Introducing a bookkeeping parameter ε, we seek a solution

$$f = 1 + \sum_{n=1}^{\infty} \varepsilon^n f_n. \tag{6}$$

Substituting (6) into (4) and equating the powers of ε to zero, yields

$$O(\varepsilon^0) \ : \ B(1 \cdot 1) = 0, \tag{7}$$
$$O(\varepsilon^1) \ : \ B(1 \cdot f_1 + f_1 \cdot 1) = 0, \tag{8}$$
$$O(\varepsilon^2) \ : \ B(1 \cdot f_2 + f_1 \cdot f_1 + f_2 \cdot 1) = 0, \tag{9}$$
$$O(\varepsilon^3) \ : \ B(1 \cdot f_3 + f_1 \cdot f_2 + f_2 \cdot f_1 + f_3 \cdot 1) = 0, \tag{10}$$
$$O(\varepsilon^4) \ : \ B(1 \cdot f_4 + f_1 \cdot f_3 + f_2 \cdot f_2 + f_3 \cdot f_1 + f_4 \cdot 1) = 0, \tag{11}$$
$$O(\varepsilon^n) \ : \ B\Big(\sum_{j=0}^{n} f_j \cdot f_{n-j}\Big) = 0, \quad \text{with } f_0 = 1. \tag{12}$$

If the original PDE admits a N-soliton solution then (6) will truncate at level $n = N$ provided f_1 is the sum of N exponential terms. For simplicity, consider the case of a three soliton solution ($N = 3$), where

$$f_1 = \sum_{i=1}^{3} \exp(\theta_i) = \sum_{i=1}^{3} \exp(k_i x - \omega_i t + \delta_i), \tag{13}$$

with k_i, ω_i and δ_i constant. Of course, (7) is trivially, satisfied, whereas (8) determines the dispersion law, $\omega_i = k_i^3$, $i = 1, 2, 3$.

The terms generated by $B(f_1, f_1)$ in (9) justify the choice

$$\begin{aligned} f_2 &= a_{12} \exp(\theta_1 + \theta_2) + a_{13} \exp(\theta_1 + \theta_3) + a_{23} \exp(\theta_2 + \theta_3) \\ &= a_{12} \exp\left[(k_1 + k_2)x - (\omega_1 + \omega_2)t + \delta_1 + \delta_2\right] + \\ &\quad + a_{13} \exp\left[(k_1 + k_3)x - (\omega_1 + \omega_3)t + \delta_1 + \delta_3\right] + \\ &\quad + a_{23} \exp\left[(k_2 + k_3)x - (\omega_2 + \omega_3)t + \delta_2 + \delta_3\right], \end{aligned} \tag{14}$$

and (9) allows one to calculate the coupling constants

$$a_{ij} = \frac{(k_i - k_j)^2}{(k_i + k_j)^2}, \quad i, j = 1, 2, 3. \tag{15}$$

Then, $B(f_1 \cdot f_2 + f_2 \cdot f_1)$ in (10) motivates the particular solution

$$\begin{aligned} f_3 &= b_{123} \exp(\theta_1 + \theta_2 + \theta_3) \\ &= b_{123} \exp\left[(k_1 + k_2 + k_3)x - (\omega_1 + \omega_2 + \omega_3)t + \delta_1 + \delta_2 + \delta_3\right], \end{aligned} \tag{16}$$

and one computes

$$b_{123} = a_{12}a_{13}a_{23} = \frac{(k_1 - k_2)^2(k_1 - k_3)^2(k_2 - k_3)^2}{(k_1 + k_2)^2(k_1 + k_3)^2(k_2 + k_3)^2}. \tag{17}$$

Subsequently, (11) allows one to verify that indeed $f_4 = 0$. In the sixth equation of the scheme $B(f_2 \cdot f_3 + f_3 \cdot f_2)$ should equal zero in order to assure that $f_5 = 0$. If so, it will be possible to take $f_i = 0$ for $i > 6$. Finally, setting $\varepsilon = 1$ in (6), we obtain

$$\begin{aligned} f &= 1 + \exp\theta_1 + \exp\theta_2 + \exp\theta_3 + a_{12}\exp(\theta_1 + \theta_2) + \\ &\quad + a_{13}\exp(\theta_1 + \theta_3) + a_{23}\exp(\theta_2 + \theta_3) + b_{123}\exp(\theta_1 + \theta_2 + \theta_3), \end{aligned} \tag{18}$$

which upon substitution in (2) generates the well-known three-soliton solution of (1).

For the N-soliton solution, Hirota [12] starts with a generalization of (18):

$$f = \sum_{\mu=0,1} \exp\left[\sum_{i<j}^{(N)} A_{ij}\mu_i\mu_j + \sum_{i=1}^{N} \mu_i\theta_i\right], \tag{19}$$

where the θ_i are as in (13) for $i = 1, 2, \ldots, N$, and $a_{ij} = \exp A_{ij}$ and $b_{123} = a_{12}a_{13}a_{23}$. The summation $\sum_{\mu=0,1}$ is over all combinations of $\mu_1 = 0, 1$, $\mu_2 = 0, 1, \ldots$, $\mu_N = 0, 1$. The summation $\sum_{i<j}^{(N)}$ stands for all possible combinations under the condition $0 < i < j < N$.

Concerning the computer implementation, Hirota's bilinear operators (5) are redefined as

$$\begin{aligned} &Dxt[m,n](f,g) \\ &= \sum_{j=0}^{m}\sum_{i=0}^{n} \frac{(-1)^{(m+n-j-i)}m!}{j!(m-j)!} \frac{n!}{i!(n-i)!} \frac{\partial^{i+j} f}{\partial t^i \partial x^j} \frac{\partial^{n+m-i-j} g}{\partial t^{n-i}\partial x^{m-j}}, \end{aligned} \tag{20}$$

with similar expression for $Dt[n](f,g)$, $Dx[n](f,g)$ and $Dy[n](f,g)$. The observant reader will recognize the Leibniz formula for the derivatives of products, up to an alteration in signs.

In practice, the dispersion law and the coupling coefficients (15) and (17) can be computed as follows. First determine $P(k, -\omega)$ from

$$B(\mathrm{e}^{kx-\omega t+\delta} \cdot 1) = P(k, -\omega)\mathrm{e}^{kx-\omega t\delta}, \tag{21}$$

and solve $P(k, -\omega) = 0$ to get the dispersion law. In our example, $P(k, -\omega) = k^4 - \omega k = 0$, yields $\omega = k^3$.

It is easy to show [17] that, once P is known,

$$a_{ij} = -\frac{P(k_i - k_j, -\omega_i + \omega_j)}{P(k_i + k_j, -\omega_i - \omega_j)}, \qquad b_{123} = a_{12}a_{13}a_{23}. \tag{22}$$

Note that the knowledge of the forms of P and f in (19) allow one to circumvent the explicit solution of the bilinear scheme (7)–(12).

With the polynomial P at hand, the conditions for the existence of a one- and two-soliton solution are $P(D_x, D_t) = P(-D_x, -D_t)$, and $P(0,0) = 0$.

To have a N-soliton solution, Hirota [12] and Matsuno [14] showed that the condition

$$\begin{aligned} S[P,n] &= \sum_{\sigma=\pm 1} P\left(\sum_{i=1}^{n} \sigma_i k_i, -\sum_{i=1}^{n} \sigma_i \omega_i\right) \times \\ &\quad \times \prod_{i<j}^{(n)} P(\sigma_i k_i - \sigma_j k_j, -\sigma_i \omega_i + \sigma_j \omega_j)\sigma_i \sigma_j = 0 \end{aligned} \tag{23}$$

must be satisfied for $n = 2, 3, \ldots, N$. Recall that $\sum_{\sigma=\pm 1}$ indicates the summation over all possible combinations of $\sigma_1 = \pm 1$, $\sigma_2 = \pm 1, \ldots,$ $\sigma_n = \pm 1$ and that $\prod_{i<j}^{(n)}$ refers to the product over all possible combinations of n elements under the condition $i < j$, and all ω_i eliminated via the dispersion law.

The verification that $S[P,n] = 0$ for $n = 3$ and $n = 4$ is nontrivial. The user can choose to test the conditions with constrained random numbers (fast) or symbolically (much slower).

2.3. Examples

• For the Boussinesq equation (BE) [4]

$$u_{2t} - u_{2x} - 3(u^2)_{2x} - u_{4x} = 0, \tag{24}$$

the bilinear operator is

$$B(f,g) = Dxt[0,2](f,g) - Dx[2](f,g) - Dx[4](f,g). \tag{25}$$

The BE admits three- and four-soliton solutions, confirmed during testing. With $\theta_i = k_i x - \omega_i t + \delta_i$, the dispersion relation is $\omega = -k\sqrt{1+k^2}$, and the coupling coefficients of the three-soliton solution are

$$a_{ij} = \frac{\sqrt{1+k_i^2}\sqrt{1+k_j^2} - 2k_i^2 + 3k_i k_j - 2k_j^2 - 1}{\sqrt{1+k_i^2}\sqrt{1+k_j^2} - 2k_i^2 - 3k_i k_j - 2k_j^2 - 1}, \tag{26}$$

$$i, j = 1, 2, 3 \text{ and } i < j,$$

and $b_{123} = a_{12}a_{13}a_{23}$.

• In this seconds example, we illustrate the use of the program HIROTA_SINGLE.MAX in the search for integrable equations. Consider the bilinear equation [10]

$$(kD_t^2 + D_x^3 D_t + D_x^6)(f \cdot f) = 0, \tag{27}$$

with an arbitrary parameter k. It is known that this equation passes the Painlevé test for $k = -1/5$. With our program we calculate the condition for the existence of a three-soliton solution:

$$\frac{5k+1}{k^3}\left(2k^2 + 2k + \sqrt{1-4k} - 1\right)\times \times (k_1 k_2 k_3)^4 \left[(k_2^2 - k_1^2)(k_3^2 - k_1^2)(k_3^2 - k_2^2)\right]^2 = 0, \tag{28}$$

fueling the conjecture that the existence of a three-soliton solution implies complete integrability [18]. With $k = -1/5$, we continue the calculation of the two- and three-soliton solutions of (27). We obtained completely automatically

$$a_{ij} = \frac{(k_i - k_j)^2(\sqrt{5}k_i^2 + 3k_i^2 + \sqrt{5}k_i k_j + k_i k_j + \sqrt{5}k_j^2 + 3k_j^2)}{(k_i + k_j)^2(\sqrt{5}k_i^2 + 3k_i^2 + \sqrt{5}k_i k_j + k_i k_j + \sqrt{5}k_j^2 + 3k_j^2)}, \tag{29}$$

$$b_{123} = a_{12} a_{13} a_{23}. \tag{30}$$

Then, with the program, we verified that the equation also admits a four-soliton solution, suggesting that the equation is indeed completely integrable.

3. The Paintlevé Test

3.1. Purpose

The MACSYMA and PAINLEVE_SINGLE.MAX allows one to determine whether or not a given single nonlinear ODE or PDE with (real) polynomial terms fulfills the necessary conditions for having the Painlevé property.

A PDE is said to possess the Painlevé property [1] if its solutions in the complex plane are single-valued in the neighborhood of noncharacteristic, movable singular manifolds. For ODEs, this means that solutions should have no worse singularities than movable poles. Such equations are prime candidates for being completely integrable.

3.2. Algorithm and implementation

We illustrate the algorithm for a single PDE. The solution u, say in only two independent variables (x, t), expressed as a Laurent series,

$$u(x,t) = g^{\alpha} \sum_{k=0}^{\infty} u_k(x,t) g^k(x,t) \tag{31}$$

should only have movable poles. In (31), $u_0(t,x) \neq 0$, α a negative integer, and $u_k(x,t)$ are analytic functions in a neighborhood of the singular, noncharacteristic manifold $g(x,t) = 0$, with $g_x(x,t) \neq 0$. For an ODE $g(x) = x - x_0$, where x_0 is the initial value for x.

The Painlevé test is carried out in three steps:

Step 1. Determine the negative integer α and u_0 from the leading order 'ansatz'. This is done by balancing the minimal power terms after substituting of $u \propto u_0 g^\alpha$ into the given PDE.

Step 2. Calculate the non-negative integer powers r, called the *resonances*, at which arbitrary functions u_r enter the expansion. This is done by requiring that u_r is arbitrary after substitution of $u \propto u_0 g^\alpha + u_r g^{\alpha+r}$ into the equation, only retaining its most singular terms.

Step 3. Verify that the correct number of arbitrary functions u_r indeed exists by substituting the truncated expansion

$$u(x,t) = g^\alpha \sum_{k=0}^{r_{\max}} u_k(x,t) g^k(x,t)$$

into the given equation ($r_{\max}$ is the largest resonance). At nonresonance levels, determine all u_k unambiguously. At resonance levels, u_r should be arbitrary, and since we are dealing with a nonlinear equation, a *compatibility condition* must be satisfied.

An equation (or system) for which the above steps can be carried out consistently, and for which the compatibility conditions at all resonances are satisfied, is said to have the Painlevé property and is conjectured to be completely integrable.

The reader should be warned that the above algorithm does not detect essential singularities. In other words, for an equation to be integrable it is *necessary* but *not* yet *sufficient* that it passes the Painlevé test. There are indeed integrable equations, such as the Dym–Kruskal equation, $u_t = u^3 u_{xxx}$, that only have the Painlevé property after a suitable chance of variables.

3.3. Examples

Instead of just testing some equations for the Painlevé property (see, e.g. [8]), we show how to use the symbolic program PAINLEVE_SINGLE.MAX in the investigation of integrability and the search for exact solutions.

• Consider a 2D KdV equation [20] with time-dependent coefficients $a(t)$ and $b(t)$,

$$(u_t + 6uu_x + u_{xxx})_x + a(t)u_x + b(t)u_{yy} = 0, \tag{32}$$

and determine the coefficients for which the equation will pass the Painlevé test.

In this case we seek a solution of the form

$$u(x,y,t) = g(x,y,t)^\alpha \sum_{k=0}^{\infty} u_k(x,y,t) g(x,y,t)^k, \tag{33}$$

with singularity $g(x,y,t)$. The program PAILEVE_SINGLE.MAX determines $\alpha = -2$, $u_0 = -2(g_x)^2$, and computed the resonances $r = 4, 5$ and 6. The compatibility conditions at levels 4 and 5 are identically satisfied. At resonance level $r = 6$ the compatibility condition

$$[a_t + 2a^2]g_x^3 + [b_t + 4ab](g_x^2 g_{yy} + g_{xx}g_y^2 - 2g_x g_{xy} g_y) = 0 \tag{34}$$

must hold irrespective the form of the singularity manifold $g(x,y,t)$. Thus, (32) will have the Painlevé property if

$$a_t + 2a^2 = 0, \qquad b_t + 4ab = 0. \tag{35}$$

Two cases are possible. The first case with $a(t) = 0$, $b(t) = c$, for any constant c, leads to the KdV equation itself ($c = 0$), or to the Kadomtsev–Petviashvili (KP) equation ($c = \pm 1$).

The more interesting second case, with

$$a(t) = \frac{1}{2(t-t_0)}, \qquad b(t) = \frac{k}{(t-t_0)^2}, \tag{36}$$

where k and t_0 are integration constants, leads for $k = 0$ and upon integration with respect to x, to the cylindrical KdV equation [1],

$$u_t + 6uu_x + u_{xxx} + \frac{u}{2t} = 0, \tag{37}$$

which describes solitary waves in a channel with slowly varying depth. For $k \neq 0$, upon appropriate scaling of the variable y, Equation (32) reduces to the well-known cylindrical KP equation [1],

$$(u_t + 6uu_x + u_{xxx})_x + \frac{1}{2t}u_x + \frac{3\sigma^2}{t^2}y_{yy} = 0, \quad \sigma^2 = \pm 1, \tag{38}$$

which describes surface waves in a fluid which are characterized by small deviation from axial symmetry. The KdV and KP equations and their cylindrical generalizations (37) and (38) are all known to be completely integrable [1].

• In this second example, we seek an exact solution of the Fitzhugh–Nagumo (FHN) equation with convection term [16],

$$u_t + kuu_x - u_{xx} - u(1-u)(a-u) = 0. \tag{39}$$

For (39), $\alpha = -1$ and u_0 must satisfy $u_0^2 - ku_0 f_x - 2f_x^2 = 0$. The resonances are $r = -1$ and $r = 4 + k(u_0/f_x) = 2 + (u_0/f_x)^2$. In order to have integer resonances, we must have $u_0(x,t) = \sqrt{m} f_x$ and $k = (m-2)/\sqrt{m}$, for any positive integer m.

To find a traveling wave solution, we substitute the Laurent series, truncated at the constant level term,

$$u(x,t) = \sqrt{m}\,\frac{\partial \ln f(x,t)}{\partial x} + u_1(x,t) = \sqrt{m}\,\frac{f_x(x,t)}{f(x,t)} + u_1(x,t) \tag{40}$$

into (39) with $k = (m-2)/\sqrt{m}$, and, we set the different power terms in f separately equal to zero. This results in an overdetermined system in $f(x,t)$ and $u_1(x,t)$:

$$0 = f_t - (1+m)f_{xx} - \sqrt{m}\left[\frac{2}{m}(1+m)u_1 - 1 - a\right] f_x, \tag{41}$$

$$\begin{aligned} 0 = {} & f_{xt} - f_{xxx} + \frac{1}{\sqrt{m}}(m-2)u_1 f_{xx} + \\ & + \left[3u_1^2 - 2(1+a)u_1 + a + \frac{1}{\sqrt{m}}(m-2)(u_1)_x\right] f_x, \end{aligned} \tag{42}$$

$$0 = (u_1)_t + \frac{1}{\sqrt{m}}(m-2)u_1(u_1)_x - (u_1)_{xx} + u_1(1-u_1)(a-u_1). \tag{43}$$

The above system is linear if u_1 is any of the constant solutions of the FHN equation (43), i.e. $u_1 = 0, 1$, or a. For $u_1 = 0$ and

$$f = c + \sum_{i=1}^{N} f_i = c + \sum_{i=1}^{N} \exp(\theta_i) = c + \sum_{i=1}^{N} \exp(k_i x - \omega_i t + \delta_i), \tag{44}$$

we find that $N = 2$ and $k_1 = 1/\sqrt{m}$, $\omega_1 = (am-1)/m$, $k_2 = a/\sqrt{m}$, and $\omega_2 = a(m-a)/m$. Hence,

$$\begin{aligned} f = {} & c + \exp\left[\frac{1}{\sqrt{m}}x + \frac{a(a-m)}{m}t + \delta_1\right] + \\ & + \exp\left[\frac{a}{\sqrt{m}}x + \frac{(1-am)}{m}t + \delta_2\right]. \end{aligned} \tag{45}$$

Returning to $u(x,t)$ via (40) yields

$$\begin{aligned} & u(x,t) \\ & = \frac{\exp\left[\frac{1}{\sqrt{m}}x + \frac{(1-am)}{m}t + \delta_1\right] + a\exp\left[\frac{a}{\sqrt{m}}x + \frac{a(a-m)}{m}t + \delta_2\right]}{c + \exp\left[\frac{1}{\sqrt{m}}x + \frac{(1-am)}{m}t + \delta_1\right] + \exp\left[\frac{a}{\sqrt{m}}x + \frac{a(a-m)}{m}t + \delta_2\right]}. \end{aligned} \tag{46}$$

This solution reduces for $m = 2$ to the known solution of the FHN equation without convection term.

4. Conserved Densities

4.1. Purpose

Our MATHEMATICA program DENSITY_SINGLE.M allows one to compute polynomial-type conservation laws of single PDEs. A similar program, called DENSITY_SYSTEM.M applies to systems of PDEs.

Recall that for a density ρ, such that there exists an associated flux J, obeying the conservation law

$$\rho_t + J_x = 0, \tag{47}$$

one finds that

$$P = \int_{-\infty}^{+\infty} \rho \, \mathrm{d}x = \text{constant}, \tag{48}$$

provided J vanishes at infinity. The existence of a sufficiently large (in principal infinite) number of conservation laws of type (47) assures completely integrable of an evolution equation. Our MATHEMATICA programs automatically construct the pairs ρ and J.

4.2. Algorithm and Implementation

We explain the algorithm, based on ideas of Kruskal and coworkers [13, 15] and Verheest and Hereman [21], for the KdV equation

$$u_t + uu_x + u_{3x} = 0. \tag{49}$$

Information about the building blocks of $\rho(u, u_x, u_{2x}, \ldots, u_{nx})$ can be obtained from the scaling or symmetry properties of the equation. The scaling of (49) is such that

$$u \sim \frac{\partial^2}{\partial x^2}, \qquad \frac{\partial}{\partial t} \sim \frac{\partial^3}{\partial x^3}. \tag{50}$$

Ignoring the first two trivial conserved densities, corresponding to $\rho_1 = u$, and $\rho_2 = u^2$, we start with building block u^3, which has three factors u.

Keeping (50) in mind, this is equivalent to two factors u and two derivatives (to be distributed over these two factors as uu_{2x} or as u_x^2), or one factor u and four derivatives (obviously u_{4x}). But, uu_{2x} can be integrated by parts to yield $(uu_x)_x - u_x^2$, and $u_{4x} = (u_{3x})_x$. Removing any density (or part thereof) that is a total x-derivative, only u_x^2 must be kept. Therefore, ρ_3 is a linear combination of the building blocks u^3 and u_x^2. It is then straightforward to find the appropriate numerical coefficient, resulting in $\rho_3 = u^3 - 3u_x^2$.

In the computer implementation, the building blocks of $\rho(u, u_x, u_{2x}, \ldots, u_{nx})$ can easily be computed. To compute the building blocks of, say, ρ_6 one proceeds as follows:

(i) Start with u^6, divide it by u and, taking into account the scaling (50), compute $(u^5)_{2x}$. This produces the following list of terms $[u^3u_x^2, u^4u_{2x}] \to [u^3u_x^2]$, where in the second list we have removed the terms that can be written either as a total derivative with respect to x, or as a total derivative up to terms appearing earlier in the list.

Next, divide u^6 by u^2, and compute $(u^4)_{4x}$, the corresponding lists are

$$[u_x^4, uu_x^2u_{2x}, u^2u_{2x}^2, u^2u_xu_{3x}, u^3u_{4x}] \longrightarrow [u_x^4, u^2u_{2x}^2].$$

In a similar fashion, proceed with $(u^6/u^3)_{6x} = (u^3)_{6x}$, $(u^6/u^4)_{8x} = (u^2)_{8x}$, and $(u^6/u^5)_{10x} = (u)_{10x}$, to obtain the lists:

$$[u_{2x}^3, u_xu_{2x}u_{3x}, uu_{3x}^2, u_x^2u_{4x}, uu_{2x}u_{4x}, uu_xu_{5x}, u^2u_{6x}] \longrightarrow [u_{2x}^3, uu_{3x}^2],$$

$$[u_{4x}^2, u_{3x}u_{5x}, u_{2x}u_{6x}, u_xu_{7x}, uu_{8x}] \longrightarrow [u_{4x}^2], \quad \text{and} \quad [u_{10x}] \rightarrow [\,],$$

where [ ] refers to the empty list.
Gathering the terms in the simplified lists yields

$$\rho_6 = u^6 + c_1u^3u_x^2 + c_2u_x^4 + c_3u^2u_{2x}^2 + c_4u_{2x}^3 + c_5uu_{3x}^2 + c_6u_{4x}^2, \tag{51}$$

where the constants c_i have to be determined.

(ii) Now compute $(\partial/\partial t)\rho_6$, and replace $u_t, u_{xt}, \ldots, u_{nx,t}$ by $-(uu_x + u_{xxx})$ and its differential consequences.

(iii) Integrate the result with respect to x, carry out all integrations by parts, and require that the part that can no longer be integrated vanishes. Setting equal to zero all the coefficients of the independent combinations involving powers of u and its derivatives with respect to x, leads to a linear system in the unknowns $c_1, c_2, \ldots, c_6$. After solving that system, we found

$$\begin{aligned}\rho_6 &= u^6 - 60u^3u_x^2 - 30u_x^4 + 108u^2u_{2x}^2 + \\ &\quad + \frac{720}{7}u_{2x}^3 - \frac{648}{7}uu_{3x}^2 + \frac{216}{7}u_{4x}^2.\end{aligned} \tag{52}$$

(iv) Once the constants c_i are determined, the form of the flux J_6 can be computed by substituting these constants into the integrable part of ρ_6, and reversing the sign.

For simplicity, let us illustrate steps (ii)–(iv) for ρ_3. Applying step (i) leads to $\rho_3 = u^3 + c[1]u_x^2$. After replacement of u_t and u_{xt} in $\partial\rho_3/\partial t$ by $-(uu_x + u_{xxx})$ and $-(uu_x + u_{xxx})_x$, and integration with respect to x, we obtain

$$\begin{aligned}\frac{\partial\rho_3}{\partial t} &= -\left[\frac{3}{4}u^4 + (c_1 - 3)uu_x^2 + 3u^2u_{xx} - c_1u_{xx}^2 + 2c_1u_xu_{xxx}\right]_x - \\ &\quad -(c_1 + 3)u_x^3.\end{aligned} \tag{53}$$

The last term in (53) must vanish. Hence, $c_1 = -3$, and the expression $[\ldots]$ gives

$$J_3 = \frac{3}{4}u^4 - 6uu_x^2 + 3u^2u_{xx} + 3u_{xx}^2 - 6u_xu_{xxx}, \tag{54}$$

corresponding to $\rho_3 = u^3 - 3u_x^2$.

4.3. Examples

• Let us study the conservation laws, and consequently the integrability of a generalized Schamel (gS) equation [21]

$$n^2 u_t + (n+1)(n+2)u^{2/n}u_x + u_{xxx} = 0, \tag{55}$$

with n a positive integer. We our MATHEMATICA program DENSITY_SINGLE.M we computed $\rho_1 = u$, $\rho_2 = u^2$, $\rho_3 = (1/2)u_x^2 - (n^2/2)u^{2+(2/n)}$. It has be shown [21] that these are the only conserved densities. Hence, establishing that for arbitrary values of n the gS equation is not completely integrable.

• Our MATHEMATICA program DENSITY_SYSTEM.M, was used to compute the first five nontrivial conserved densities of a modified vector derivative nonlinear Schrödinger equation (MVDNLS) [22],

$$\frac{\partial \mathbf{B}_\perp}{\partial t} + \frac{\partial}{\partial x}(B_\perp^2 \mathbf{B}_\perp) + \alpha \mathbf{B}_{\perp 0}\mathbf{B}_{\perp 0} \cdot \frac{\partial \mathbf{B}_\perp}{\partial x} + \mathbf{e}_x \times \frac{\partial^2 \mathbf{B}_\perp}{\partial x^2} = \mathbf{0}. \tag{56}$$

The parameter α characterizes the extra term which distinguishes the MVDNLS from the derivative NLS. Equation (56), which comes from plasma physics, can be replaced by

$$u_t + \left(u(u^2+v^2) + \beta u - v_x\right)_x = 0, \qquad v_t + \left(v(u^2+v^2) - u_x\right)_x = 0, \tag{57}$$

where u and v denote the components of $\mathbf{B}_\perp$ parallel and perpendicular to $\mathbf{B}_{\perp 0}$, and $\beta = \alpha B_{\perp 0}^2$.

The first five conserved densities were computed. The first three are

$$\rho_1 = u^2 + v^2, \tag{58}$$

$$\rho_2 = \frac{1}{2}(u^2+v^2)^2 - uv_x + u_x v + \beta u^2, \tag{59}$$

$$\rho_3 = \frac{1}{4}(u^2+v^2)^3 + \frac{1}{2}(u_x^2 + v_x^2) - u^3 v_x + v^3 u_x + \frac{\beta}{4}(u^4 - v^4), \tag{60}$$

the remaining two can be found in [22].

5. Lie-Point Symmetries

5.1. Purpose

The program SYMMGRP.MAX [3, 5–7] automatically computes the determining equations for the coefficients in the vector field that realizes the Lie algebra of point symmetries. With a feedback mechanism, these determining equations can then be solved explicitly.

The program SYMMGRP.MAX has been adapted in two ways. It now allows one to compute nonclassical symmetries of systems of differential equations, and Lie-point symmetries (classical and nonclassical) of system of difference-differential equations. Examples of the latter include the 1D Toda lattice,

$$u_{tt}(n) - \mathrm{e}^{u(n-1)-u(n)} + \mathrm{e}^{u(n)-u(n+1)} = 0,$$

the 2D Toda lattice where $u_{xt}(n)$ replaces $u_{tt}(n)$, and the discrete version of the KdV equation,

$$u_t(n) + u(n)[u(n-1) - u(n+1)] = 0.$$

5.2. Algorithm and Implementation

The method used in the algorithm or our program SYMMGRP.MAX [3], is based on prolonged vector fields described in [19], from which we also adopted the notations and terminology.

For notational simplicity, let us consider the case of Lie-point symmetries. We start with a system of m differential equations,

$$\Delta^i(x, u^{(k)}) = 0, \quad i = 1, 2, \ldots, m, \tag{61}$$

of order k, with p independent and q dependent (real) variables, denoted by

$$x = (x_1, x_2, \ldots, x_p) \in \mathbb{R}^p, \qquad u = (u^1, u^2, \ldots, u^q) \in \mathbb{R}^q.$$

We stress that m, k, p and q are *arbitrary* positive integers. The partial derivatives of u^l are represented using a multi-index notation, for $J = (j_1, j_2, \ldots, j_p) \in \mathbb{N}^p$, we denote

$$u_J^l \equiv \frac{\partial^{|J|} u^l}{\partial x_1^{j_1} \partial x_2^{j_2} \ldots \partial x_p^{j_p}},$$

where $|J| = j_1 + j_2 + \cdots + j_p$. Finally, let $u^{(k)}$ denote a vector whose components are all the partial derivatives of order 0 up to k of all the u^l.

The group transformations have the form $\tilde{x} = \Lambda_g(x, u)$, $\tilde{u} = \Omega_g(x, u)$, where the functions Λ_g and $\Omega_g$are to be determined. Note that the subscript g refers to the group parameters. Instead of looking for a Lie group G, we look for its Lie algebra $\mathcal{L}$, realized by vector fields of the form

$$\alpha = \sum_{i=1}^{p} \eta^i(x, u) \frac{\partial}{\partial x_i} + \sum_{l=1}^{q} \varphi_l(x, u) \frac{\partial}{\partial u^l}. \tag{62}$$

The problem is now reduced to finding the coefficients $\eta^i(x, u)$ and $\varphi_l(x, u)$. In essence, the computer constructs the kth prolongation $\mathrm{pr}^{(k)}\alpha$ of the vector field α, applies it to the system (61), and requests that the resulting expression vanishes on the solution set of (61). Although that sounds straightforward, the method involves tedious calculations because the length and complexity of the expressions increase rapidly as p, q, m, and especially k, increase. Here are the details and the steps to be performed:

1. Construct the kth prolongation of the vector field α in (62) by means of the formula

$$\mathrm{pr}^{(k)}\alpha = \alpha + \sum_{l=1}^{q}\sum_{J}\psi_l^J(x, u^{(k)})\frac{\partial}{\partial u_J^l}, \quad 1 \leqslant |J| \leqslant k, \tag{63}$$

where the coefficients ψ_l^J are defined as follows. The coefficients of the first prolongation are

$$\psi_l^{J_i} = D_i\varphi_l(x, u) - \sum_{j=1}^{p} u_{J_j}^l D_i\eta^i(x, u), \tag{64}$$

where J_i is a p-tuple with 1 on the ith position and zeros elsewhere, and D_i is the total derivative operator

$$D_i = \frac{\partial}{\partial x_i} + \sum_{l=1}^{q}\sum_{J} u_{J+J_i}^l \frac{\partial}{\partial u_J^l}, \quad 0 \leqslant |J| \leqslant k. \tag{65}$$

The higher-order prolongations are defined recursively as

$$\psi_l^{J+J_i} = D_i\psi_l^J - \sum_{j=1}^{p} u_{J+J_j}^l D_i\eta^j(x, u), \quad |J| \geqslant 1. \tag{66}$$

2. Apply the prolonged operator $\mathrm{pr}^{(k)}\alpha$ to each equation $\Delta^i(x, u^{(k)})$ and require that

$$\mathrm{pr}^{(k)}\alpha\Delta^i|_{\Delta^j=0} = 0, \quad i, j = 1, \ldots, m. \tag{67}$$

The condition (67) expresses that $\mathrm{pr}^{(k)}\alpha$ vanishes on the solution set of the originally given system (61). Precisely, this condition assures that α is an infinitesimal symmetry generator of the group transformation; $\tilde{x} = \Lambda_g(x, u)$, $\tilde{u} = \Omega_g(x, u)$, i.e., that $u(x)$ is a solution of (61) whenever $\tilde{u}(\tilde{x})$ is one.
3. Choose, if possible, m components of the vector $u^{(k)}$, say $v^1, \ldots, v^m$, such that:
   (a) Each v^i is equal to a derivative of a u^l $(l = 1, \ldots, q)$ with respect to at least one variable $x_i (i = 1, \ldots, p)$.
   (b) None of the v^i is the derivative of another one in the set.
   (c) The system (61) can be solved algebraically for the v^i in terms of the remaining components of $u^{(k)}$, which we denote by w. Hence, $v^i = S^i(x, w)$, $i = 1, \ldots, m$.
   (d) The derivatives of v^i, $v_J^i = D_J S^i(x, w)$, where $D_J \equiv D_1^{j_1} D_2^{j_2} \ldots D_p^{j_p}$, can all be expressed in terms of the components of w and their derivatives, without ever reintroducing the v^i or their derivatives.

The requirements in Step 3 put some restrictions on the system (61), but for many systems the choice of the appropriate v^i is quite obvious. For example, for a system of evolution equations

$$\frac{\partial u^i}{\partial t}(x_1, \ldots, x_{p-1}, t) = F^i(x_1, \ldots, x_{p-1}, t, u^{(k)}), \quad i = 1, \ldots, m, \tag{68}$$

where $u^{(k)}$ involves derivatives with respect to the variables x_i but not t, an appropriate choice is $v^i = \partial u^i/\partial t$.

4. Use $v^i = S^i(x, w)$ to eliminate all v^i and their derivatives from the expression (67), so that all the remaining variables are now independent of each other. It is tacitly assumed that the resulting expression is now a polynomial in the u^l_J.
5. Obtain the determining equations for $\eta^i(x, u)$ and $\varphi_l(x, u)$ by equating to zero the coefficients of all functionally independent expressions (monomials) in the remaining derivatives u^l_J.

In the above algorithm the variables x_i, u^l, and u^l_J are treated as independent; the dependent ones are η^i and φ_l.

In summary, the result of implementing (63) is a system of linear homogeneous PDEs for η^i and φ_l, in which x and u are independent variables. These are the so-called determining or defining equations for the symmetries of the system. Solving these by hand, interactively or automatically with a symbolic package, will give the explicit forms of the $\eta^i(x, u)$ and $\varphi_l(x, u)$.

The procedure, which is thoroughly discussed in [19], consists of two major steps: *deriving* the determining equations, and *solving* them explicitly. Details on how this can be done by computer can be found in [5–7]. In these papers we also presented a survey of available codes, including a discussion of their strengths and weaknesses.

5.3. Examples

• Consider the Dym–Kruskal equation [1]

$$u_t - u^3 u_{xxx} = 0. \tag{69}$$

The program SYMMGRP.MAX automatically computes the determining equations for the coefficients eta[1] $= \eta^x$, eta[2] $= \eta^t$ and phi[1] $= \varphi^u$ of the vector field

$$\alpha = \eta^x \frac{\partial}{\partial x} + \eta^t \frac{\partial}{\partial t} + \varphi^u \frac{\partial}{\partial u}. \tag{70}$$

There are only eight determining equations,

$$\frac{\partial \text{eta2}}{\partial u[1]} = 0, \qquad \frac{\partial \text{eta2}}{\partial x[1]} = 0, \qquad \frac{\partial \text{eta1}}{\partial u[1]} = 0, \qquad \frac{\partial^2 \text{phi1}}{\partial u[1]^2} = 0, \tag{71}$$

$$\frac{\partial^2 \text{phi1}}{\partial u[1] \partial x[1]} - \frac{\partial^2 \text{eta1}}{\partial x[1]^2} = 0, \qquad \frac{\partial \text{phi1}}{\partial x[2]} - u[1]^3 \frac{\partial^3 \text{phi1}}{\partial x[1]^3} = 0, \tag{72}$$

$$3u[1]^3 \frac{\partial^3 \text{phi1}}{\partial u[1] \partial x[1]^2} + \frac{\partial \text{eta1}}{\partial x[2]} - u[1]^3 \frac{\partial^3 \text{eta1}}{\partial x[1]^3} = 0, \tag{73}$$

$$u[1]\frac{\partial \text{eta2}}{\partial x[2]} - 3u[1]\frac{\partial \text{eta1}}{\partial x[1]} + 3\text{phi1} = 0.$$

These determining equations are easily solved explicitly with the freedback mechanism within SYMMGRP.MAX. The general solution, rewritten in the original variables, is

$$\eta^x = k_1 + k_3 x + k_5 x^2, \tag{74}$$
$$\eta^t = k_2 - 3k_4 t, \tag{75}$$
$$\varphi^u = (k_3 + k_4 + 2k_5 x)u, \tag{76}$$

where $k_1, \ldots, k_5$ are arbitrary constants. The five infinitesimal generators then are

$$\begin{aligned} G_1 &= \partial_x, & G_2 &= \partial_t, \\ G_3 &= x\partial_x + u\partial_u, & G_4 &= -3t\partial_t + u\partial_u, \\ G_5 &= x^2\partial_x + 2xu\partial_u. \end{aligned} \tag{77}$$

Clearly, (69) is invariant under translations (G_1 and G_2) and scaling (G_3 and G_4). Computation of the flow corresponding to G_5 requires integration of the system

$$\begin{aligned} \frac{d\tilde{x}}{d\varepsilon} &= \tilde{x}^2, & \tilde{x}(0) &= x, \\ \frac{d\tilde{t}}{d\varepsilon} &= 0, & \tilde{t}(0) &= t, \\ \frac{d\tilde{u}}{d\varepsilon} &= 2\tilde{x}\tilde{u}, & \tilde{u}(0) &= u, \end{aligned} \tag{78}$$

where ε is the parameter of the transformation group.

One readily obtains $\tilde{x}(\varepsilon) = x/(1-\varepsilon x)$, $\tilde{t}(\varepsilon) = t$, and $\tilde{u}(\varepsilon) = u/(1-\varepsilon x)^2$. We therefore conclude that for any solution $u = f(x,t)$ of Equation (69) the transformed solution

$$\tilde{u}(\tilde{x},\tilde{t}) = (1+\varepsilon\tilde{x})^2 f\left(\frac{\tilde{x}}{1+\varepsilon\tilde{x}}, \tilde{t}\right) \tag{79}$$

will solve $\tilde{u}_{\tilde{t}} - \tilde{u}^3\tilde{u}_{\tilde{x}\tilde{x}\tilde{x}} = 0$.

• As a second example, we take the nonlinear Schrödinger (NLS) equation [1],

$$iu_t + u_{xx} + u|u|^2 = 0. \tag{80}$$

The complex equation can be replaced by a coupled system,

$$v_t + w_{xx} + w(v^2 + w^2) = 0,$$

$$w_t - v_{xx} - v(v^2 + w^2) = 0, \tag{81}$$

for the real and imaginary parts v, w of the complex variable u.

Now, SYMMGRP.MAX quickly generates the twenty determining equation for the coefficients of the vector field

$$\alpha = \eta^x \frac{\partial}{\partial x} + \eta^t \frac{\partial}{\partial t} + \varphi^v \frac{\partial}{\partial v} + \varphi^w \frac{\partial}{\partial w}. \tag{82}$$

The first eleven single-term determining equations are similar to (71), and provide information about the dependencies of the ηs and the ϕs on x, t, v and w, and their linearity in the latter two independent variables. The remaining nine determining equations are more complicated, but the entire system is readily solved.

In the original variables, the solution reads

$$\eta^x = k_1 + 2k_4 t + k_5 x, \tag{83}$$
$$\eta^t = k_2 + 2k_5 t, \tag{84}$$
$$\varphi^v = k_3 w - k_4 x w - k_5 v, \tag{85}$$
$$\varphi^w = -k_3 v + k_4 x v - k_5 w, \tag{86}$$

where $k_1, \ldots, k_5$ are arbitrary constants. As in the previous examples, the complete symmetry algebra is spanned by five vector fields (generators):

$$\begin{aligned} G_1 &= \partial_x, & G_2 &= \partial_t, \\ G_3 &= w\partial_v - v\partial_w, & G_4 &= 2t\partial_x - x(w\partial_v - v\partial_w), \\ G_5 &= x\partial_x + 2t\partial_v - v\partial_v - w\partial_w. \end{aligned} \tag{87}$$

Clearly, (80) is invariant under translations in space and time (G_1 and G_2). Generator G_3 corresponds to adding an arbitrary constant to the phase of u. The Galilean boost is generated by G_4. Finally, G_5 indicates invariance of the equation under scaling (or dilation).

References

1. Ablowitz, M. J. and Clarkson, P. A.: *Solitons, Nonlinear Evolution Equations and Inverse Scattering*, London Mathematical Society Lecture Notes on Mathematics, Vol. 149, Cambridge University Press, Cambridge, UK, 1991.
2. Ablowitz, M. J. and Segur, H.: Solitons and the Inverse Scattering, *SIAM Studies in Appl. Math.* Vol. 4, SIAM, Philadelphia, 1981.
3. Champagne, B., Hereman, W., and Winternitz, P.: The computer calculation of Lie point symmetries of large systems of differential equations, *Comp. Phys. Comm.* **66** (1991), 319–340.
4. Drazin, P. G. and Johnson, R. S.: *Solitons: An Introduction*, Cambridge University Press, Cambridge, 1989.
5. Hereman, W.: SYMMGRP.MAX and other symbolic programs for symmetry analysis of partial differential equations, in: E. Allgower, K. Georg and R. Miranda (Eds), *Exploiting Symmetry in Applied and Numerical Analysis*, Lectures in Applied Mathematics, Vol. 29, American Mathematical Society, Providence, Rhode Island, pp. 241–257, 1993.
6. Hereman, W.: Review of symbolic software for the computation of Lie symmetries of differential equations, *Euromath Bulletin* **1**(2) (1994), 45–82.

7. Hereman, W.: Symbolic software for Lie symmetry analysis, in: N. H. Ibragimov (Ed.), *CRC Handbook of Lie Group Analysis of Differential Equations*, Vol. 3, CRC Press, Boca Raton, Florida, Ch. 13, in press (1995).
8. Hereman, W. and Angenent, S.: The Painlevé test for nonlinear ordinary and partial differential equations, *MACSYMA Newsletter* **6** (1989), 11–18.
9. Hereman, W. and Zhuang, W.: Symbolic computation of solitons with MACSYMA, in: W. F. Ames and P. J. van der Houwen (Eds), *Computational and Applied Mathematics II: Differential Equations*, North-Holland, Amsterdam, The Netherlands, pp. 287–296, 1992.
10. Hietarinta, J.: A search for bilinear equations passing Hirota's three-soliton condition, Part I. KdV-type bilinear equations, *J. Math. Phys.* **28** (1987), 1732–1742.
11. Hirota, R.: Direct method of finding exact solutions of nonlinear evolution equation, in: R. M. Miura (Ed.), *Bäcklund Transformations, the Inverse Scattering Method, Solitons, and Their Applications*, Lecture Notes in Mathematics, Vol. 515, Springer-Verlag, Berlin, pp. 40–68, 1976.
12. Hirota, R.: Direct methods in soliton theory, in: R. K. Bullough and P. J. Caudrey (Eds), *Solitons, Topics in Physics*, Vol. 17, Springer-Verlag, Berlin, pp. 157–176, 1980.
13. Kruskal, M. D., Miura, R. M., Gardner, C. S., and Zabusky, N. J.: Korteweg–de Vries equations and generalizations. V. Uniqueness and nonexistence of polynomial conservation laws, *J. Math. Phys.* **11** (1970), 952–960.
14. Matsuno, Y.: *Bilinear Transformation Method*, Academic Press, Orlando, 1984.
15. Miura, R. M., Gardner, C. S., and Kruskal, M. D.: Korteweg–de Vries equations and generalizations. II. Existence of conservation laws and constants of motion, *J. Math. Phys.* **9** (1968), 1204–1209.
16. Murray, J. D.: *Mathematical Biology, Biomathematics Texts*, Vol. 19, Springer-Verlag, Berlin, 1989.
17. Newell, A. C.: *Solitons in Mathematics and Physics*, SIAM, Philadelphia, 1985.
18. Newell, A. C. and Zeng, Y. B.: The Hirota conditions, *J. Math. Phys.* **29** (1987), 2016–2021.
19. Olver, P. J.: *Applications of Lie Groups to Differential Equations*, Graduate Texts in Mathematics, Vol. 107, Springer-Verlag, New York (1986 and 1993 2nd edition).
20. Steeb, W.-H. and Euler, N.: *Nonlinear Evolution Equations and Painlevé Test*, World Scientific, Singapore, 1988.
21. Verheest, F. and Hereman, W.: Conservation laws and solitary wave solutions for generalized Schamel equations, *Physica Scripta* **50** (1994), 611–614.
22. Willox, R., Hereman, W., and Verheest, F.: Complete integrability of a modified vector derivative nonlinear Schrödinger equation, *Physica Scripta* **51** (1995) in press.

Acta Applicandae Mathematicae **39:** 379–387, 1995.

Solitons of Curvature

B. G. KONOPELCHENKO
Dipartimento di Fisica, Università di Lecce e Sezione INFN, 73100 Lecce, Italy

(Received: 10 August 1994)

Abstract. An intristic geometry of surfaces is discussed. In geodesic coordinates the Gauss equation is reduced to the Schrödinger equation where the Gaussian curvature plays the role of a potential. The use of this fact provides an infinite set of explicit expressions for the curvature and metric of a surface. A special case is governed by the KdV equation for the Gaussian curvature. We consider the integrable dynamics of curvature via the KdV equation, higher KdV equations and $(2+1)$-dimensional integrable equations with breaking solitons.

Mathematics Subject Classification (1991): 58F07.

Key words: Gaussian curvature, inverse spectral transform.

The dynamics of interfaces, surfaces, and fronts forms an important ingredient of numerous nonlinear phenomena in physics, such as surface waves, the growth of crystals, the propagation of flame fronts, the deformation of membranes, the dynamics of vortex sheets, and many problems of hydrodynamics associated with the motion of boundaries between regions of differing densities and viscosities (see, e.g., [1–3]). In some cases such dynamics can be modelled by the nonlinear partial differential equations (PDEs) which describe the evolution of surfaces in time. Solvable cases for which the corresponding PDEs can be integrated analytically are, of course, of great interest (see [1–3] and recent papers [4–6]).

In mathematics the differential geometry of surfaces was completed, in essence, at the end of the 19th and the beginning of the 20th centuries (see, e.g., [7–10]). Basic differential equations which describe surfaces in three-dimensional space have been studied in detail from various points of view. A study of the interrelation between the special type of surfaces and nonlinear PDEs has been one of the classical problem of differential geometry [7–10]. The Liouville and sine-Gordon equations which describe minimal and pseudospherical surfaces, respectively, are the best known examples.

A new tool to solve nonlinear PDEs was discovered in 1967 [11]. This method (inverse spectral transform (IST) method) allows one effectively to solve a number of nonlinear PDEs which appear in various fields of physics and mathematics (see, e.g., [12–17]). A key element of the IST method is the representation of the nonlinear PDE as the compatibility condition of certain system of linear equations for the so-called eigenfunctions. Nonlinear PDEs which are integrable by the IST method possess a number of remarkable properties [12–17]. In particular, it was shown that the Liouville and sine-Gordon equations are solvable by this method.

The sine-Gordon and Liouville equations were the first nonlinear PDEs which revealed a deep connection between soliton equations and the differential geometry of surfaces. After that, such a connection has been discussed many times from the various points of view (see, e.g., [18–30]). A common feature of all these different approaches was that they treated surfaces extrinsically, i.e. as embedded into the three-dimensional Euclidean space. The PDEs which describe this imbedding (Gauss–Codazzi equations) have then been treated and used in one manner or another.

In the present paper, in contrast, we will consider the intristic geometry of surfaces which is formulated without any reference to the enveloping space. Intrisically, a surface is characterized by the metric and the Gaussian curvature, which are connected to each other by the Gauss equation. The fact that this Gauss equation is reduced in geodesic coordinates to the one-dimensional stationary Schrödinger equation, in which the Gaussian curvature plays the role of a potential, is the starting point of our approach. Using this observation, we construct the Gaussian curvature and metric explicitly in terms of multi-Bargmann potentials or Riemann θ-functions parametrised by arbitrary functions of a single variable. Then we fix the dependence on this variable (y) by the requirement that an additional linear equation for metric be satisfied. Consequently, the curvature obeys the Korteweg–de Vries (KdV) equation. As a result, we construct surfaces whose Gaussian curvatures are nothing but solitons of the KdV equation. We also present the integrable dynamics of surfaces of revolution and generic surfaces via the KdV and higher KdV equations. The solitonic and more general evolutions of the Gaussian curvature are given. An important type of integrable evolutions is given by the $(2+1)$-dimensional integrable equations with breaking solitons. The exact solutions of such equations provide the explicit evolutions of the curvature of the shock wave type.

Solitonic curvatures and their integrable dynamics inherit all remarkable properties of the KdV equation (see [12–17]), among them being an infinite set of integrals of motion. In our case they are the global characteristics of surfaces which are preserved by integrable dynamics.

Thus we will consider two-dimensional surfaces with the linear element (first fundamental form)

$$\mathrm{d}s^2 = E\,\mathrm{d}x^2 + 2F\,\mathrm{d}x\,\mathrm{d}y + G\,\mathrm{d}y^2 \tag{1}$$

where E, F and G are functions on the local coordinates x and y of a surface (see [7–10]). The basic invariant characteristic of the intristic geometry of a surface is the Gaussian curvature $K = R_{1212}/H$, where R_{1212} is the nonzero component of the Riemann tensor $R_{\alpha\beta\gamma\delta}$ and $H = \sqrt{EG - F^2}$ [7–10]. The Gaussian curvature

$K(x, y)$ and the metric of a surface are related by the Gauss equation [7–10]

$$K = \frac{1}{2H}\left[\left(\frac{F}{HE}E_y - \frac{1}{H}G_x\right)_x + \left(\frac{2}{H}F_x - \frac{1}{H}E_y - \frac{F}{EH}E_x\right)_y\right]. \tag{2}$$

In a special coordinate system on a surface the Gauss equations (2) may have a very simple form. The Liouville and sine-Gordon equations are the classical examples [7–10]. If one chooses $E = G = 0$, then the coordinate curves are the minimal lines and the Gauss equation (2) is reduced to $(\ln F)_{xy} + KF = 0$ or to the equation $\varphi_{xy} + K \exp \varphi = 0$ where $F = \exp \varphi$. For the constant Gaussian curvature K this is the Liouville equation. Further for the choice $F = 0,\ E = \cos^2 \theta,\ G = \sin^2 \theta$ one has the lines of curvature as the coordinate curves. Then, for constant negative curvature, $K = -2a^2$ the Gauss equation (2) is nothing but the sine-Gordon equation $\theta_{xx} - \theta_{yy} = a^2 \sin \theta$ [7–10].

Thus, for different choices of metric (or systems of coordinates) on a surface, the Gauss equation (2) is reduced to different simple PDEs. Note that these equations arise within the analysis of the pure intristic geometry of surfaces. We emphasize also that the only well-studied such nonlinear PDEs (Liouville and sine-Gordon equations) correspond to the case of constant Gaussian curvature.

In the present paper we describe surfaces in a system of coordinates which is also well-known in differential geometry but, to the best of our knowledge, has not been used for the study of nonlinear PDEs associated with the surfaces.

This is the geodesic system of coordinates for which $E = I,\ F = 0$ and the linear element (1) is of the form $\mathrm{d}s^2 = \mathrm{d}x^2 + G\,\mathrm{d}y^2$ (see, e.g., [7–10]). The name is due to the fact that the curve $y = \text{const}$ is a geodesic. Under the conditions $G(x = 0, y) = 0,\ G_x(x = 0) = 1$ one has the so-called geodesic polar coordinates [7–10]. Note that the metric $\mathrm{d}s^2 = \mathrm{d}x^2 + G\,\mathrm{d}y^2$ is the metric of a genetic surface, but referred to the geodesic coordinates. In such a metric the Gauss Equation (2) is of the form $K = -G^{-1/2}(G^{1/2})_{xx}$. In terms of the variable $H = \sqrt{G}$ the linear element of the surface and the Gauss equation look like

$$\mathrm{d}s^2 = \mathrm{d}x^2 + H^2\,\mathrm{d}y^2 \tag{3}$$

and

$$H_{xx} + K(x, y)H = 0. \tag{4}$$

Equation (4) is linear and seems too simple and not so attractive as the Liouville and sine-Gordon equations discussed above. However, just its linearity and nonrigidity (the variable y is involved as a parameter) provide us a freedom and flexibility in constructing the surfaces explicitly and in formulating the associated nonlinear PDEs.

In the particular case $K = \text{const}$ the corresponding surfaces are well studied (see [7–10]). They are the spherical or pseudospherical surfaces of constant curvature. Here we will consider the case of variable $K(x, y)$.

First we note that Equation (4) is closely connected with the one-dimensional stationary Schrödinger equation

$$-\psi_{xx} - K(x,y)\psi = \lambda^2\psi, \tag{5}$$

where the Gaussian curvature $K(x,y)$ plays the role of a potential and $H(x,y) = \psi(x,y,\lambda = 0)$. Thus all exactly solvable cases for the Schrödinger Equation (5) provide us the surfaces (referred to their geodesic coordinates) with the explicit expressions for the Gaussian curvature $K(x,y)$ and the metric $(H(x,y))$. Note that the variable y *is* contained in these formulae as a parameter. There are a number of solvable cases for the Schrödinger equation (see, e.g., [31, 12–14]). A broad class of exact solutions of Equation (5) is given by the multi-Bargmann potentials (see, e.g., [31, 12–14])

$$K(x,y) = 2(\ln \det A)_{xx} \tag{6}$$

where A is an $N \times N$ matrix with the elements

$$\begin{aligned} A_{nk} &= \delta_{nk} + \frac{\beta_n(y)}{\alpha_n(y)+\alpha_k(y)} \times \\ &\quad \times \exp\left[-(\alpha_n(y)+\alpha_k(y))_x\right], \quad n,k = 1,\ldots,N, \end{aligned} \tag{7}$$

and $\alpha_n(y)$, $\beta_n(y)$ $(n = 1,\ldots,N)$ are arbitrary distinct positive functions. In our case the known formula for $\psi(x,y,\lambda)$ [31, 12–17] also gives

$$H(x,y) = 1 + \sum_{n=1}^{N} \alpha_n^{-1} e^{-\alpha_n x} \frac{\det A^{(n)}}{\det A}, \tag{8}$$

where the matrix elements of the matrix $A^{(n)}$ are given by (7) with the substitution of the last column by the column $\beta_n \exp(-\alpha_n x)$ $(n = 1,\ldots,N)$. In the simplest case $N = 1$ one has

$$K(x,y) = \frac{2\alpha^2(y)}{\mathrm{ch}^2(\alpha(y)x - \gamma(y))}, \qquad H(x,y) = \mathrm{th}(\alpha(y)x - \gamma(y)), \tag{9}$$

where

$$\gamma(y) = \frac{1}{2} \ln \frac{\beta(y)}{2\alpha(y)}$$

and $\alpha(y)$, $\beta(y)$ are arbitrary functions.

More general, periodic on x, the so-called, finite-gap solutions of (5) are of the form (see, e.g., [12, 14, 15])

$$K(x,y) = 2\left[\theta(U_{(y)}x + V_{(y)})\right]_{xx} + \mathrm{const} \tag{10}$$

and

$$H(x,y) = \frac{\theta(A(\gamma) + Ux + V)\theta(V)}{\theta(A(\gamma) + V)\theta(Ux + V)} \exp\left(x \int \omega\right), \tag{11}$$

where θ is the Riemann θ-function, U, V, A are matrices of periods for the hyperelliptic curve γ. Note that the dependence of all parameters in (6)–(11) on the variable y is not fixed. In particular, in the formulae (10), (11), if one chooses the parameters with periodic dependence on y, one gets the compact surfaces referred to their geodesics.

In general one can prescribe any dependence of $K(x,y)$ and $H(x,y)$ on the variable y. The only requirement is that such dependence has to preserve Equation (5).

One way to fix the dependence on y is to require that the function H obeys the additional linear equation $H_y = A(K, K_x, \partial_x)H$, where A is some differential operator. So, having in mind that $H = \psi(x, y, \lambda = 0)$, where ψ obeys Equation (5), we require that in addition to Equation (5) the function ψ obeys an equation of the type

$$\psi_y = A(K, K_x, \ldots, \partial_x)\psi, \tag{12}$$

where A is the linear differential operator in $\partial = \partial/\partial x$. The compatibility of (12) and (5) guarantees the preservation of (5) in time. On the other hand, the compatibility condition of (5) and (12) is equivalent to the nonlinear PDE for K. Choosing $A = c\partial_x$ where c is a constant, one gets $K = K\ (x + cy)$. A nonlinear equation arises with the choice of A as a third-order operator, namely, if

$$\psi_y + 4\psi_{xxx} + 6K\psi_x + 3K_x\psi = 0. \tag{13}$$

The compatibility condition of (5) and (13) is equivalent to the equation

$$K_y + K_{xxx} + 6KK_x = 0. \tag{14}$$

This is nothing other than the famous Korteweg–de Vries (KdV) equation. This equation in integrable by the inverse spectral transform (IST) method [11–17]. The basic idea of this method is to generate solvable nonlinear PDE via the compatible system of linear PDEs with variable coefficients [11–17]. Our approach to surfaces is just motivated by this main idea of the IST method.

Thus, the choice (13) means that the Gaussian curvature of a surface obeys the KdV Equation (14) while the coefficient $H(x,y)$ of the metric (3) is given by the KdV eigenfunction ψ evaluated at $\lambda = 0$. The KdV Equation (14) has remarkable, exact solutions, called solitons. Consequently, one has surfaces with solitonic curvature and corresponding solitonic metric. The simplest soliton surface has

$$\begin{aligned} K(x,y) &= \frac{2\alpha^2}{\mathrm{ch}^2[\alpha(x - 4\alpha^2 y - x_0)]}, \\ H(x,y) &= \mathrm{th}[\alpha(x - 4\alpha^2 y - x_0)] \end{aligned} \tag{15}$$

where α and x_0 are arbitrary constants. The multi-soliton solutions of the KdV Equation (14) and, thus, the multi-soliton surfaces are given by the formulas

(6)–(8) with $\alpha_n = \text{const}$ and $\beta_n = \beta_{n0} \exp 8\alpha_n^3 y$, where β_{n0} are constants. Note that the Gaussian curvature $K(x, y)$ described by these formulas is bounded on the whole surface. Then $K(x, y) \to 0$ as $x^2 + y^2 \to \infty$ except the lines $x = 4\alpha_n^2 y$ $(n = 1, \ldots, N)$ along which $K \to \text{const}$. More general finite-gap solutions of the KdV equation periodic in x and the corresponding surfaces are given by the formulas (10), (11) with $U = \text{const}$, $V = V_0 y + W_0$, where V_0 and W_0 are constants. The n-gap elliptic potentials of Lame and Ince (see, e.g., [12]) $K(x, y) = n(n+1)/2\ P(x - \alpha(y))$, where P is the Weierstrass function is the particular case of the above solutions.

The KdV equation has a number of remarkable properties: infinite sets of integrals of motion, Darboux and Backlund transformations and corresponding nonlinear superposition formula, Lagrangian, Hamiltonian and bi-Hamiltonian structure and so on [12–17]. The solitonic surfaces inherit all these specific properties. For instance, using the Darboux transformation, one finds the following Wronskian representation (see, e.g., [32]) of soliton surfaces

$$\begin{aligned} K(x,y) &= K_0(x,y) + (\ln W(\psi_1, \ldots, \psi_n)_{xx}), \\ H(x,y) &= \frac{W(\psi_1, \ldots, \psi_n, \psi)}{W(\psi_1, \ldots, \psi_n)}, \end{aligned} \tag{16}$$

where $\psi_1, \ldots, \psi_n, \psi$ are different solutions of Equation (45) with the potential $K_0(x, y)$.

There is an infinite family of nonlinear PDEs associated with the problem (5) the so-called KdV hierarchy. Higher KdV equations are equivalent to the compatibility condition of (5) and Equation (12) where A is the operator of $(2n+1)$th order [12–17]. All these higher KdV equations can be used to fix dependence of K and H on y. The corresponding explicit K and H are given by the formulas (6)–(11) with $\alpha_n = \text{const}$ and $\beta_n = \beta_{n0} \exp((2\alpha_n)^{2n+1})$. The whole KdV hierarchy is characterised by a very important quantity, the so-called τ-function (see, e.g., [14–17]). One has $K(x, y) = 2\ (\ln \tau)_{xx}$.

Now we will describe the integrable evolutions of surfaces referred to the geodesic coordinates. We will consider three different cases.

Surfaces of revolution. By definition a surface of revolution is the surface generated by a plane curve when the plane of the curve is made to rotate about a line in the plane (see [7–10]). For such surfaces H and, consequently, K depend only on x. To construct integrable evolutions of such surfaces in time t we assume that $K = (x, t)$, $H = H(x, t)$ and look for evolutions which preserve (5) in time.

These evolutions, obviously, are given by the KdV hierarchy for the curvature K and corresponding eigenfunction equations for H, i.e. by Equations (12), (14) with the substitution $y \to t$. The evolution given by the KdV equation $K_t + K_{xxx} + 6KK_x = 0$ is a very specific one. First, one has the wide class of explicit evolutions. They are given by the formulas (15), (6)–(8) with $y \to t$.

Then, using the well-known results for the KdV equation [11–17], one can solve the initial value problem $(H(x,t=0),\ K(x,t=0) \to H(x,t), K(x,t)$ for the evolutions of these type of surfaces. Further the integrals of motion for the KdV equation (and higher KdV equations) provide us the infinite set of the characteristics of surface which are preserved in time. The simplest such invariants are

$$C_1 = \int \mathrm{d}x\, K(x,t), \quad C_2 = \int \mathrm{d}x\, K^2(x,t), \quad C_3 = \int \mathrm{d}x\, (K_x^2 + 2K^3).$$

The integrable evolutions of surfaces of revolution are closely connected with the integrable evolutions of plane curves which have been discussed in [33–36]. Indeed, let us consider, following to [33–36], a plane curve and its integrable evolution in time. The simplest nonlinear integrable evolution is described by the mKdV equation $v_t + v_{sss} - 12v^2 v_s = 0$ where $v(x,t)$ is the curvature of the curve and s is arclength. Then, at each moment t we rotate this curve about the fixed line in the plane. As a result at each t we get the surface of revolution and, consequently, the time-evolution of such surface. Our generating curve is now lying on the surface and $v(y,t) = K_g^{(x)}$, where $K_g^{(x)}$ is the geodesic curvature of the curve. For the metric of the form (3) the Gaussian curvature K and the geodesic curvature of the curve $x = \text{const}$ are connected by the Liouville formula $K = -(K_g^{(x)})_x - K_g^{(x)2}$ (see, e.g., [9, 10]). But it is nothing but the well-known Miura transformation which relates the solutions of the MKdV and KdV equations [11]. Note that the above consideration demonstrates the geometric meaning of the Miura transformation.

Generic surfaces. Other way of introducing the dynamics of a surface is to fix the y-dependence of $K(x,y)$ and $H(x,y)$ by one of the equation from the KdV hierarchy (say by the KdV Equation (14) itself) and then fix the evolution in t by the other equation from this hierarchy. The common solution $K(x,y,t)$ of the KdV Equation (14) and such higher KdV equation $(K_t + \partial_x^{2n+1} K \ldots = 0)$ gives us the evolution of the generic surface. Explicit evolutions of this type are of the form (6)–(9) with $\alpha_n = \text{const}$ and $\beta_n = \beta_{n0} \exp(8\alpha_n^3 y + (2\alpha_n)^{2n+1} t)$.

Another possibility for the fixing the dependence of K and H on y and simultaneously for the determination of the evolution in time t consists in the consideration of the compatible system of Equation (5) and an equation of the form

$$\psi_t + f(\partial_x^2 + K)\psi_y + \sum_{n=1}^{M} U_n(x,y,t)\partial_x^n \psi = 0, \tag{17}$$

where $f(\alpha)$ in an arbitrary polynom and U_n are functions. The compatibility condition of (5) and (17) is equivalent to the $(2+1)$-dimensional integrable equation for K. Such integrable equations have been proposed in [37] and then have been discussed by several authors (see, e.g., [38, 39]). These integrable

PDEs possess the so-called breaking solitons (see [39]). One of the simplest equations of this type looks like [39]

$$K_t + K_{xxy} + 4KK_y + 2K_x(\partial_x^{-1}K_y) = 0. \tag{18}$$

For Equation (18) [39]

$$\psi_t + 2(\partial_x^2 + K)\psi_y + 2(\partial_x^{-1}K_y)\psi_x + 3K_y\psi = 0. \tag{19}$$

Equation (18) has an infinite set of integrals of notion $C_n = \iint \mathrm{d}x\,\mathrm{d}y P_n(x,y)$, where the densities P_n coincide with those for the KdV Equation [39]. So the first such integrals of motion are

$$C_1 = \iint \mathrm{d}x\,\mathrm{d}y\,K(x,y), \quad C_2 = \iint \mathrm{d}x\,\mathrm{d}y\,K^2(x,y),$$
$$C_3 = \iint \mathrm{d}x\,\mathrm{d}y\,(K_x^2 + 2K^3), \ldots .$$

Thus, there exists an infinite set of global characteristics of surface which are invariant under the evolution of the type (18).

The multi-soliton like solutions of Equation (18) and the corresponding $H(x,y)$ are given by the formulas (6), (8), where the functions $\alpha_n(y,t)$ and $\beta_n(y,t)$ obey the equations $\alpha_{nt} + 4\alpha_n^2\alpha_{ny} = 0$, $\beta_{nt} + 4\alpha_n^2\beta_{ny} = 0$ [39]. The one-soliton solution is of the form (9) where α and γ obey the equations $\alpha_t + 4\alpha^2\alpha_y = 0$, $\gamma_t + 4\alpha^2\gamma_y = 0$ [39]. The characteristic feature of these equations is that their soliton solutions are breaking solitons. Consequently the solitons of Equation (18) are breaking solitons too [39]. The multi-valuedness of such solutions may be reinterpretated as the appearance of shock waves (see, e.g., [39]). Thus the surfaces and their integrable dynamics described by Equation (18) admit the formation of shock waves of curvature.

The nonlinear soliton equations for the Gaussian curvature presented above manifest the existence of the integrable intristic geometries of surfaces and their integrable dynamics. An important problem now is to find the physical phenomena or models which may be described by such solitonic curvatures.

References

1. Langer, J. S.: *Rev. Mod. Phys.* **52** (1980), 1.
2. Bishop, A. R., Campbell, L. J., and Channel, P. J. (Eds): *Fronts, Interfaces and Patterns,* North-Holland, 1984.
3. Pelce, P. (Ed.): *Dynamics of Curved Fronts,* Academic Press (1986).
4. Brower, R. C., Kessler, D. A., Koplik, J., and Levine, H.: *Phys. Rev.* **A29** (1984), 1335.
5. Nakayama, K. and Wadati, M.: *Phys. Soc. Japan* **62** (1993), 1895.
6. Konopelchenko, B. G.: *Phys. Lett.* **183A** (1993), 153.
7. Darboux, G.: *Lecons sur la théorie des surfaces et les applications geométriques du calcul infinitesimal* t. 1–4, Paris, Gauthier-Villars (1877–1896).
8. Bianchi, L.: *Lezioni di geometria differentiale* (2nd edn), Piza, Spoerri, 1902.
9. Eisenhart, L. P.: *A Treatise on the Differential Geometry of Curves and Surfaces,* Dover Publ., New York, 1909.

10. Forsyth, A. R.: *Lectures on the Differential Geometry of Curves and Surfaces*, Cambridge Univ. Press, 1920.
11. Gardner, G. S., Greene, J. M., Kruskal, M. D. and Miura, R. M.: *Phys. Rev. Lett.* **19** (1967), 1015.
12. Zakharov, V. E., Manakov, S. V., Novikov, S. P. and Pitaevsky, L. P.: *Theory of Solitons*, Nauka, 1980; Consultant Bureau, 1984.
13. Lamb, G. L.: *Elements of Soliton Theory*, Wiley, New York, 1980.
14. Ablowitz, M. J. and Segur, H.: *Solitons and Inverse Scattering Transform*, SIAM, Philadelphia, 1985.
15. Newell, A. C.: *Solitons in Mathematics and Physics*, SIAM, Philadelphia, 1985.
16. Ablowitz, M. J. and Clarkson, P. A.: *Solitons, Nonlinear Evolution Equations and Inverse Scattering*, Cambridge Univ. Press, 1991.
17. Konopelchenko, B. G.: *Introduction to Multidimensional Integrable Equations*, Plenum Press, New York, 1992.
18. Lund, F.: 'Solitons and geometry', in A. O. Barut (Ed.), *Nonlinear Equations in Physics and Mathematics*, Reidel, Dordrecht, 1978.
19. Sasaki, R.: *Nucl. Phys.* **B154** (1979), 434.
20. Konopelchenko, B. G.: *Phys. Lett.* **72A** (1979), 101.
21. Chern, S. S. and Tenenblat, K.: *Stud. Appl. Math.* **74** (1986), 55.
22. Tenenblat, K. and Terng, C.-L.: *Ann. Math.* **111** (1980), 477; **111** (1980), 491.
23. Ablowitz, M., Beals, R., and Tenenblat, K.: *Stud. Appl. Math.* **74** (1986), 177.
24. Saveliev, M. V.: *Teor. Mat. Fyz.* **69** (1986), 411.
25. Sym, A.: *Soliton Surfaces and Their Applications,* in Lecture Notes in Physics, Vol. 239, Springer-Verlag, 1985, p. 154.
26. Bobenko, A. I.: *Math. Ann.* **290** (1990), 207.
27. Dryuma, V. S.: *Mathematical Studies* **124** (1992), 56.
28. Konopelchenko, B. G. and Schief, W. K.: Lamé and Zakharov–Manakov systems: Combescure, Darboux and Backlund transformations, Preprint AM 93/9, Sydney, 1993.
29. Konopelchenko, B. G.: 'Induced surfaces and their integrable dynamics', Preprint 93-114, Novosibirsk, 1993.
30. Schrief, W. K. and Rogers, C.: *Inverse Problems*, 1994.
31. Chadan, K. and Sabatier, P. C.: *Inverse Problems in Quantum Scattering Theory*, Springer, New York, 1977.
32. Matveev, V. B. and Salle, M. A.: *Darboux Transformation and Solitons*, Springer, Berlin, 1991.
33. Lamb, G. L.: *J. Math. Phys.* **18** (1977), 1654.
34. Goldstein, R. E. and Petrich, D. M.: *Phys. Rev. Lett.* **67** (1991), 3203.
35. Langer, J. and Perline, R.: *J. Nonlinear Sci.* **1** (1991), 71.
36. Nakayama, K., Segur, H., and Wadati, M.: *Phys. Rev. Lett.* **60** (1992), 2603.
37. Calogero, F.: *Lett. Nuovo Cim.* **14** (1975), 443.
38. Zakharov, V. E. and Manakov, S. V.: *Teor. Mat. Fyz.* **27** (1976), 283.
39. Bogoyavlensky, O. I.: *Breaking Solitons,* Nauka, Moscow, 1991.

Acta Applicandae Mathematicae **39**: 389–403, 1995.

The Reductive Perturbation Method and the Korteweg–de Vries Hierarchy

R. A. KRAENKEL, J. G. PEREIRA
Instituto de Física Teórica, Universidade Estadual Paulista, Rua Pamplona 145, 01405-900 São Paulo, SP–Brazil

and

M. A. MANNA
Physique Mathématique et Théorique, Université de Montpellier II, 34095 Montpellier, Cedex 05, France

(Received: 2 March 1995)

Abstract. By using the reductive perturbation method of Taniuti with the introduction of an infinite sequence of slow time variables τ_1, τ_3, τ_5, $\ldots$, we study the propagation of long surface-waves in a shallow inviscid fluid. The Korteweg–de Vries (KdV) equation appears as the lowest order amplitude equation in slow variables. In this context, we show that, if the lowest order wave amplitude ζ_0 satisfies the KdV equation in the time τ_3, it must satisfy the $(2n+1)$th order equation of the KdV hierarchy in the time τ_{2n+1}, with $n = 2, 3, 4, \ldots$. As a consequence of this fact, we show with an explicit example that the secularities of the evolution equations for the higher-order terms $(\zeta_1, \zeta_2, \ldots)$ of the amplitude can be eliminated when ζ_0 is a solitonic solution to the KdV equation. By reversing this argument, we can say that the requirement of a secular-free perturbation theory implies that the amplitude ζ_0 satisfies the $(2n+1)$th order equation of the KdV hierarchy in the time τ_{2n+1}. This essentially means that the equations of the KdV hierarchy do play a role in perturbation theory. Thereafter, by considering a solitary-wave solution, we show, again with an explicit, example that the elimination of secularities through the use of the higher order KdV hierarchy equations corresponds, in the laboratory coordinates, to a renormalization of the solitary-wave velocity. Then, we conclude that this procedure of eliminating secularities is closely related to the renormalization technique developed by Kodama and Taniuti.

Mathematics Subject Classification (1991): 58F07.

Key words: reductive perturbation method, multiple time formalism, higher-order evolution equations.

1. Introduction

As is well known, the Korteweg–de Vries (KdV) equation [1] describes nonlinear long-wave dynamics in general weak dispersive systems. It is then natural to ask whether the higher order equations of the KdV hierarchy [2] could also play some similar role in the description of such systems. In this paper, we will show that the equations of the KdV hierarchy do appear in the description of those systems. In order to do that, we will consider the classical problem of long surface waves in shallow water, but introducing an infinite sequence of time-scales, defined from the basic time variable t by $\tau_1 = \varepsilon^{1/2}t$, $\tau_3 = \varepsilon^{3/2}t$, $\tau_5 = \varepsilon^{5/2}t, \ldots$, with ε a small parameter. The primary reason for introducing them is that they may be used as a tool to eliminate the soliton related secularities that appear in the solutions for the higher order terms of the amplitude in perturbation theory. However, when a function of a single time variable $F(t)$ is extended to a function of several time variables $F(\tau_1, \tau_3, \tau_5, ..)$, the problem of knowing the allowed evolution in each one of the different time-scales is posed [3]. In fact, as we are going to see, these evolutions are not arbitrary, but they are determined by a scale invariance requirement which must hold to ensure the ordering of the expansions, and also by the compatibility condition

$$\frac{\partial^2 F}{\partial\tau_3 \partial\tau_{2n+1}} = \frac{\partial^2 F}{\partial\tau_{2n+1} \partial\tau_3},$$

which arises naturally in a multiple time formalism. If, for a physical system, the evolution of a certain amplitude is governed by the KdV equation in the time τ_3, we will show that the above compatibility condition constraints the evolution of that amplitude on higher order times to be governed by the higher order equations of the KdV hierarchy, leaving one free parameter at each order, which is related to the possibility of redefining the time τ_{2n+1} by a multiplicative factor α_{2n+1}. By choosing specific values for the free parameters, the task of secularity elimination can then be accomplished. As a consequence, our construction of a *bona fide* perturbative expansion will provide a link between the equations of the KdV hierarchy and the evolution of a physical quantity. The physical system studied here, that of long surface water waves, is a kind of classical system where our ideas are well illustrated. However, the results obtained are, to a certain extend, model independent, and in this sense the study made here can be considered as representative of a general physical situation.

The paper is organized as follows. In Section 2, we obtain the basic equations describing surface water-waves in the multiple time formalism. In Section 3, the first few evolution equations are obtained. In Section 4, by using the multiple time formalism, we examine how the symmetry of the time derivatives determines the evolution of the wave amplitude in any time scale. The use of the KdV hierarchy equations to eliminate the soliton related secularities in the evolution of the higher order terms of the wave amplitude is discussed in Section 5. In Section 6, we show with an explicit example that the method of eliminating the soliton related secularities by using these equations is connected to the renormalization technique of Kodama and Taniuti [4]. Finally, in Section 7, we summarize and discuss the results obtained.

2. The Multiple Time Formalism for Water Waves

We consider a two-dimensional inviscid incompressible fluid in a constant gravitational field. The space coordinates are denoted by (x, z) and the corresponding components of the velocity $\mathbf{v}$ by (u, w). The gravitational acceleration $\mathbf{g}$ is in the negative z direction. If we assume the fluid to be irrotational, the equations describing it are [5]

$$\nabla^2 \phi = 0, \tag{1}$$

and

$$\frac{p - p_0}{\rho} = -\frac{\partial}{\partial t}\phi - \frac{1}{2}(\nabla\phi)^2 - gz, \tag{2}$$

where ρ is the fluid density, p is the presure, ϕ is the velocity potential and p_0 is an integration constant.

We consider the case of a fluid of height h, limited above by a passive gas exerting a constant pressure p_0 on it, and we describe the upper surface by

$$z = \zeta(x, t). \tag{3}$$

The kinematic boundary condition at this surface is then written in the form

$$\frac{D\zeta}{Dt} \equiv \zeta_t + \phi_x \zeta_x = \phi_z, \tag{4}$$

with the subscripts denoting partial derivatives. However, there is also a dynamic boundary condition obtained from Equation (2), which reads

$$\phi_t + \tfrac{1}{2}\left[(\phi_x)^2 + (\phi_z)^2\right] + g\rho = 0, \tag{5}$$

on $z = \zeta(x, t)$. Finally, the lower boundary is supposed to be a rigid horizontal flat botton, localized at $z = -h$. In this case, the corresponding boundary condition implies that the normal velocity of the fluid must vanish:

$$\phi_z = 0. \tag{6}$$

We now wish to consider the long-wave in shallow-water approximation to the above equations. This may be done, for instance, by using the the reductive perturbation method of Taniuti [6], which introduces slow space and time variables. The slow space variable is defined by:

$$\xi' = \varepsilon^{1/2} x, \tag{7}$$

with ε a small parameter. Next, we introduce an infinite sequence of slow time variables [7]:

$$\tau_1 = \varepsilon^{1/2} t, \qquad \tau_3 = \varepsilon^{3/2} t, \qquad \tau_5 = \varepsilon^{5/2} t, \ldots . \tag{8}$$

In addition, we expand ζ and ϕ in a suitable power series in the parameter ε:

$$\zeta = \varepsilon\hat{\zeta} \equiv \varepsilon\left(\zeta_0 + \varepsilon\zeta_1 + \varepsilon^2\zeta_2 + \ldots\right) \tag{9}$$

$$\phi = \varepsilon^{1/2}\hat{\phi} \equiv \varepsilon^{1/2}\left(\phi_0 + \varepsilon\phi_1 + \varepsilon^2\phi_2 + \cdots\right). \tag{10}$$

Introducing these expansions and the slow variables into the water wave equations, we obtain:

$$\varepsilon\hat{\phi}_{\xi'\xi'} + \hat{\phi}_{zz} = 0, \quad -h < z < \varepsilon\hat{\zeta} \tag{11}$$

$$\hat{\phi}_z = 0, \qquad z = -h \tag{12}$$

$$\hat{\phi}_z = \varepsilon\hat{\zeta}_{\tau_1} + \varepsilon^2\hat{\zeta}_{\tau_3} + \varepsilon^3\hat{\zeta}_{\tau_5} + \cdots + \varepsilon^2\hat{\phi}_{\xi'}\hat{\zeta}_{\xi'}, \qquad z = \varepsilon\hat{\zeta} \tag{13}$$

$$2g\hat{\zeta} + 2\hat{\phi}_{\tau_1} + 2\varepsilon\hat{\phi}_{\tau_3} + 2\varepsilon^2\hat{\phi}_{\tau_5} + \cdots + \varepsilon\hat{\phi}_{\xi'}\hat{\phi}_{\xi'} + \hat{\phi}_z\hat{\phi}_z = 0, \quad z = \varepsilon\hat{\zeta}. \tag{14}$$

These are the basic equations we are going to use to study the evolution of surface-waves in this system.

3. The First Evolution Equations

Equation (11) for the velocity potential can be solved for the boundary condition (12), independently of Equations (13) and (14). At order ε^0, the solution is

$$\phi_0 = \mathcal{F}, \tag{15}$$

with $\mathcal{F} = \mathcal{F}(\xi'; \tau_1, \tau_3, \tau_5, \ldots)$ an arbitrary function. At order ε^1, it is given by

$$\phi_1 = -\left(\frac{z^2}{2} + hz\right)\mathcal{F}_{\xi'\xi'} + \mathcal{G}, \tag{16}$$

with $\mathcal{G} = \mathcal{G}(\xi'; \tau_1, \tau_3, \tau_5, \ldots)$ another arbitrary function. At order ε^2, we get

$$\phi_2 = -\tfrac{1}{24}(z^4 + 4hz^3 - 8h^3z)\mathcal{F}_{(4\xi')} - \tfrac{1}{2}(z^2 + 2hz)\mathcal{G}_{\xi'\xi'} + \mathcal{H}, \tag{17}$$

with $\mathcal{H} = \mathcal{H}(\xi'; \tau_1, \tau_3, \tau_5, \ldots)$ another arbitrary function. At order ε^3, the solution is

$$\begin{aligned}\phi_3 = {} & -\tfrac{1}{720}(z^6 + 6hz^5 - 40h^3z^3 + 16h^6)\mathcal{F}_{(6\xi')} \\ & + \tfrac{1}{24}(z^4 + 4hz^3 - 8h^3z)\mathcal{G}_{(4\xi')} - \tfrac{1}{2}(z^2 + 2hz)\mathcal{H}_{\xi'\xi'} + \mathcal{I},\end{aligned} \tag{18}$$

with $\mathcal{I} = \mathcal{I}(\xi'; \tau_1, \tau_3, \tau_5, \ldots)$ again an arbitrary function. We could proceed further and calculate $\phi_4, \phi_5, \ldots$. But, we stop here since this is all we are going to need.

We pass now to Equations (13) and (14), and solve the first few orders in our perturbative scheme. At order ε^0, Equation (13) gives nothing, while Equation (14) reads

$$g\zeta_0 + \phi_{0\tau_1} = 0. \tag{19}$$

Substituting ϕ_0, we get

$$\zeta_0 = -\frac{1}{g}\mathcal{F}_{\tau_1}. \tag{20}$$

At order ε^1, Equation (13) is

$$\mathcal{F}_{\tau_1\tau_1} - gh\mathcal{F}_{\xi'\xi'} = 0. \tag{21}$$

Defining

$$gh \equiv c^2, \tag{22}$$

with c a velocity, its solution can be written in the form

$$\mathcal{F} = F(\xi' - c\tau_1) + G(\xi' + c\tau_1),$$

where the sign $-(+)$ refers to a wave moving to the right (left) with a velocity c. For definiteness, let us choose the wave moving to the right, and let us define a new coordinate system by

$$\xi = \xi' - c\tau_1. \tag{23}$$

Therefore,

$$\mathcal{F}_{\tau_1} = -c\mathcal{F}_{\xi}, \tag{24}$$

and from Equation (20) we get

$$\zeta_0 = \frac{h}{c}\mathcal{F}_{\xi}. \tag{25}$$

Consequently,

$$\zeta_{0\tau_1} + c\zeta_{0\xi} = 0, \tag{26}$$

which is the usual linear wave equation.

Now, using Equations (7) and (8), we can rewrite Equation (23) in the form

$$\xi = \varepsilon^{1/2}(x - ct), \tag{27}$$

from which we see that ξ is a coordinate system moving with the velocity c of the linear waves, in relation to the laboratory coordinate x. Therefore, from now on, all dependent variables will be assumed to depend on (ξ', τ_1) through ξ only, which automatically implies that they satisfy Equation (24).

At order ε^1, Equation (14) gives

$$2g\zeta_1 - 2c\mathcal{G}_{\xi} + 2\mathcal{F}_{\tau_3} + \mathcal{F}_{\xi}\mathcal{F}_{\xi} = 0, \tag{28}$$

where we have already used Equation (24) for $\mathcal{G}$. Derivating in relation to ξ, and using Equation (25), we obtain

$$c\zeta_{1\xi} - h\mathcal{G}_{\xi\xi} = -\zeta_{0\tau_3} - \frac{c}{h}\zeta_0\zeta_{0\xi}. \tag{29}$$

Now we go to the order ε^2. From Equation (13), we have

$$c\zeta_{1\xi} - h\mathcal{G}_{\xi\xi} = \zeta_{0\tau_3} + 2\frac{c}{h}\zeta_0\zeta_{0\xi} + \frac{ch^2}{3}\zeta_{0\xi\xi\xi}. \tag{30}$$

Substituting into Equation (29), we get

$$\zeta_{0\tau_3} + \frac{3c}{2h}\zeta_0\zeta_{0\xi} + \frac{ch^2}{6}\zeta_{0\xi\xi\xi} = 0, \tag{31}$$

which is the KdV equation. It is the first nontrivial equation in the KdV hierarchy. At this point we can see the importance of introducing multiple times: while the linear wave Equation (26) describes the evolution of ζ_0 in the time τ_1, the KdV equation describes the evolution of ζ_0 in the time τ_3. In other words, the phenomena described by each one of these equations occur in a different time scale. Now, from Equation (14) at order ε^2, we obtain

$$2g\zeta_2 + 2c^2\zeta_0\zeta_{0\xi\xi} - 2c\mathcal{H}_\xi + 2\mathcal{G}_{\tau_3} + 2\mathcal{F}_{\tau_5} + 2\mathcal{F}_\xi\mathcal{G}_\xi + c^2\zeta_{0\xi}\zeta_{0\xi} = 0. \tag{32}$$

Derivating in relation to ξ, and using Equation (25), it becomes

$$\begin{aligned} &c\zeta_{2\xi} - h\mathcal{H}_{\xi\xi} + \frac{h}{c}\mathcal{G}_{\xi\tau_3} + (\zeta_{0\xi}\mathcal{G}_\xi + \zeta_0\mathcal{G}_{\xi\xi}) \\ &\quad = -\zeta_{0\tau_5} - 2ch\zeta_{0\xi}\zeta_{0\xi\xi} - ch\zeta_0\zeta_{0\xi\xi\xi}. \end{aligned} \tag{33}$$

We pass now to the order ε^3. From Equation (13), we have

$$\begin{aligned} &c\zeta_{2\xi} - h\mathcal{H}_\xi - (\zeta_0\mathcal{G}_{\xi\xi} + \zeta_{0\xi}\mathcal{G}_\xi) \\ &\quad - \frac{c}{h}(\zeta_{0\xi}\zeta_1 + \zeta_0\zeta_{1\xi}) - \frac{h^3}{3}\mathcal{G}_{(4\xi)} = \zeta_{0\tau_5} + \frac{2}{15}ch^4\zeta_{0(5\xi)}. \end{aligned} \tag{34}$$

Equations (33) and (34) can be combined to yield an evolution equation involving $\zeta_{1\tau_3}$, $\zeta_{0\tau_5}$ and $\mathcal{G}_{\xi\tau_3}$:

$$\begin{aligned} &2\zeta_{0\tau_5} + \tfrac{2}{15}ch^4\zeta_{0(5\xi)} + 2ch\zeta_{0\xi}\zeta_{0\xi\xi} + ch\zeta_0\zeta_{0\xi\xi\xi} + \zeta_{1\tau_3} + \\ &\quad + \frac{c}{h}(\zeta_{0\xi}\zeta_1 + \zeta_0\zeta_{1\xi}) + \frac{h}{c}\mathcal{G}_{\xi\tau_3} + 2(\zeta_0\mathcal{G}_{\xi\xi} + \zeta_{0\xi}\mathcal{G}_\xi) + \frac{h^3}{3}\mathcal{G}_{(4\xi)} = 0. \end{aligned} \tag{35}$$

Now, making use of Equation (31) to describe $\mathcal{F}_{\tau_3}$, Equation (28) can be put in the form

$$\zeta_1 - \frac{h}{c}\mathcal{G}_\xi = \frac{1}{4h}\zeta_0{}^2 + \frac{h^2}{6}\zeta_{0\xi\xi}. \tag{36}$$

Then, we use this equation to eliminate $\mathcal{G}_\xi$ from Equation (35). The result is

$$\zeta_{1\tau_3} + \frac{3c}{2h}(\zeta_0\zeta_1)_\xi + \frac{ch^2}{6}\zeta_{1\xi\xi\xi} = S_1(\zeta_0), \tag{37}$$

where

$$S_1(\zeta_0) = -\zeta_{0\tau_5} - \tfrac{19}{360}ch^4\zeta_{0(5\xi)} - \tfrac{5}{12}ch\zeta_0\zeta_{0\xi\xi\xi} - \tfrac{23}{24}ch\zeta_{0\xi}\zeta_{0\xi\xi} + \\ + \tfrac{3c}{8h^2}{\zeta_0}^2\zeta_{0\xi}. \tag{38}$$

Equation (37), as it stands, can not be viewed as an evolution equation for ζ_1 in the time τ_3 because the nonhomogeneous term involves $\zeta_{0\tau_5}$, and the evolution of ζ_0 in τ_5 is not known up to this point. Moreover, the term proportional to $\zeta_{0(5\xi)}$ in $S_1(\zeta_0)$ is a resonant term when ζ_0 is a solitary wave solution to the KdV equation. In other words, it is a secular producing term to the solution ζ_1 [4]. We will show in the next sections how the evolution of ζ_0 in τ_5 may be determined, and how it may be used to cancel the secular terms.

4. The Symmetry of Time Derivatives and the Korteweg–de Vries Hierarchy

The infinite sequence of equations of the KdV hierarchy can be expressed in terms of the integrals of motion I_n [8, 9] of the KdV equation according to [10, 11]

$$\zeta_t = \frac{\partial}{\partial x}\left(\frac{\delta I_n}{\delta \zeta}\right). \tag{39}$$

For $n = 1$, the KdV equation is obtained, which is the first nontrivial member of the KdV hierarchy. By using the explicit form of I_n [9], it is easy to show that each one of Equations (39) is invariant, up to a Galilean term, under the transformations

$$\xi = \varepsilon^{1/2}(x - ct), \tag{40}$$

$$\tau_{2n+1} = \varepsilon^{n+(1/2)}t; \quad n = 1, 2, 3, \ldots, \tag{41}$$

$$\zeta = \varepsilon\zeta_0. \tag{42}$$

In these new coordinates, called slow variables, Equation (39) is written in the form

$$\zeta_{0\tau_{2n+1}} = \frac{\partial}{\partial \xi}\left(\frac{\delta I_n}{\delta \zeta_0}\right). \tag{43}$$

Notice that an infinite sequence of slow time variables τ_3, τ_5, τ_7, ... was introduced in the above transformations. As a consequence, any dependent variable $F(x; t)$ will be a function of these multiple times:

$$F(x; t) = F(x; \tau_1, \tau_3, \tau_5, \ldots). \tag{44}$$

Therefore, the time derivative of $F(x;t)$ turns out to be

$$\frac{\partial F}{\partial t} = \left(\varepsilon^{3/2}\frac{\partial}{\partial \tau_3} + \varepsilon^{5/2}\frac{\partial}{\partial \tau_5} + \varepsilon^{7/2}\frac{\partial}{\partial \tau_7} + \cdots\right) F. \tag{45}$$

Let us now return to our physical system. As we have already seen, the evolution of ζ_0 in the time τ_3 was found to be the KdV equation

$$\zeta_{0\tau_3} = \alpha_1 \zeta_{0\xi\xi\xi} + \beta_1 \zeta_0 \zeta_{0\xi}, \tag{46}$$

with $\alpha_1 = -ch^2/6$ and $\beta_1 = -3c/2h$. Let us then examine how the evolution of ζ_0 in times $\tau_5, \tau_7, \ldots$, can be obtained. The first crucial point of this analysis is to note that, for the evolution of ζ_0 in the time τ_3, the two terms in the r.h.s. of Equation (46) exhaust all possible terms such that, when transforming back from the slow to the laboratory variables, the resulting equation does not depend on ε.

In the very same way, the evolution equation of ζ_0 in the time τ_5 must only involve terms presenting the same property as the terms in the KdV equation. In other words, this evolution equation must be formally the same in the slow (ξ, τ_5, ζ_0) as well as in the laboratory coordinates (x, t, ζ), establishing thus a scale invariance requirement. If this were not so, the ordering of the perturbative series would not be ensured. A simple analysis shows that the only possible terms are:

$$\zeta_{0\tau_5} = \alpha_2 \zeta_{0(5\xi)} + \beta_2 \zeta_0 \zeta_{0\xi\xi\xi} + (\beta_2 + \gamma_2)\, \zeta_{0\xi} \zeta_{0\xi\xi} + \delta_2 {\zeta_0}^2 \zeta_{0\xi}, \tag{47}$$

where $\alpha_2, \beta_2, \gamma_2, \delta_2$ are constants. Now comes the second crucial point: the coefficients $\alpha_2, \beta_2, \gamma_2, \delta_2$ are not completely arbitrary if ζ_0 is to satisfy also the KdV equation in the time τ_3. Instead, they are constrainted by the relations arising when the natural (in the multiple time formalism) compatibility condition

$$\left(\zeta_{0\tau_3}\right)_{\tau_5} = \left(\zeta_{0\tau_5}\right)_{\tau_3} \tag{48}$$

is imposed. This condition only makes sense in the multiple time formalism, since otherwise it would be redundant and would lead to a trivial identity. A straightforward calculation shows that the relations arising from this condition are:

$$\frac{\beta_2}{\alpha_2} = \frac{5}{3}\frac{\beta_1}{\alpha_1}, \qquad \frac{\gamma_2}{\alpha_2} = \frac{5}{3}\frac{\beta_1}{\alpha_1}, \qquad \frac{\delta_2}{\alpha_2} = \frac{5}{6}\left(\frac{\beta_1}{\alpha_1}\right)^2. \tag{49}$$

Substituting into Equation (47) leads to

$$\begin{aligned} \zeta_{0\tau_5} &= \alpha_2 \left[\zeta_{0(5\xi)} + \frac{5}{3}\left(\frac{\beta_1}{\alpha_1}\right)\zeta_0\zeta_{0\xi\xi\xi} + \right. \\ &\quad \left. + \frac{10}{3}\left(\frac{\beta_1}{\alpha_1}\right)\zeta_{0\xi}\zeta_{0\xi\xi} + \frac{5}{6}\left(\frac{\beta_1}{\alpha_1}\right)^2 \zeta_0^2\zeta_{0\xi}\right], \end{aligned} \tag{50}$$

which is the fifth order equation of the KdV hierarchy. Different choices of the free parameter α_2 correspond to different normalizations of the time τ_5. In particular, for $\alpha_2 = 6$ it acquires the canonical form [12]

$$\zeta_{0\tau_5} = 6\zeta_{0(5\xi)} + 10\alpha\zeta_0\zeta_{0\xi\xi\xi} + 20\alpha\zeta_{0\xi}\zeta_{0\xi\xi} + 5\alpha^2\zeta_0^2\zeta_{0\xi}, \tag{51}$$

with $\alpha = \beta_1/\alpha_1$.

This procedure can be extended to any higher order. In other words, if ζ_0 satisfies the KdV equation in the time τ_3, by making use of the scale invariance described above, we can first select all possible terms appearing in the evolution of ζ_0 in the time τ_{2n+1}. Then, by imposing the general time symmetry condition

$$\frac{\partial^2\zeta_0}{\partial\tau_3\partial\tau_{2n+1}} = \frac{\partial^2\zeta_0}{\partial\tau_{2n+1}\partial\tau_3}, \tag{52}$$

the parameters $\beta_n, \gamma_n, \ldots$ appearing in that equation can be uniquely determined in terms of α_n, α_1 and β_1, and the resulting equation is found to be the $(2n+1)$th order equation of the KdV hierarchy. As before, α_n is left as a free parameter responsible for the arbitrariness in the scale of the time τ_{2n+1}.

Before proceeding further, let us remark that one could *a priori* envisage, instead of Equation (47) a new one including also a dependence on ζ_1, and still keep the agreement with the scale invariance requirement. However, when using the physical Equation (37), the compatibility condition (48) makes all the coefficients of these supplementary terms to vanish.

5. Higher Order Evolution Equations: Elimination of the Secular Producing Terms

Let us now return to the evolution equation for ζ_1 in the time τ_3, which is given by Equations (37), (38). As we have seen, there were two problems related to it: the evolution of ζ_0 in τ_5, and the secular producing term. The last section gave the clue to the first problem. As we have seen, the evolution of ζ_0 in τ_{2n+1} can not be chosen arbitrarily, but it is restricted to the KdV hierarchy equations. We show now how the secular producing term related to the soliton part can be eliminated. A comparison between Equation (38) for $S_1(\zeta_0)$ and the fifth order KdV Equation (50), immediately shows that the choice

$$\alpha_2 = -\tfrac{19}{360}ch^4 \tag{53}$$

in Equation (65) has the property of eliminating the terms

$$\zeta_{0\tau_5} - \tfrac{19}{360}ch^4\zeta_{0(5\xi)} \tag{54}$$

from the nonhomogeneous term $S_1(\zeta_0)$. After we do that, $S_1(\zeta_0)$ acquires the form

$$S_1(\zeta_0) = \tfrac{9}{24}ch\zeta_0\zeta_{0\xi\xi\xi} + \tfrac{5}{8}ch\zeta_{0\xi}\zeta_{0\xi\xi} + \tfrac{63}{16}\frac{c}{h^2}\zeta_0{}^2\zeta_{0\xi}, \tag{55}$$

and we see that it does not present a secular producing term anymore. The secular-free solution ζ_1 can then be found by solving the linear Equation (37) with $S_1(\zeta_0)$ given by Equation (55).

In the next orders [13], the elimination of the soliton related secular terms becomes more and more involved, but the general procedure is basically the same. For example, in the next order we obtain for ζ_2 in the time τ_3 a linear evolution equation of the form:

$$\zeta_{2\tau_3} + \frac{3c}{2h}(\zeta_0\zeta_2)_\xi + \frac{ch^2}{6}\zeta_{2\xi\xi\xi} = S_2(\zeta_0, \zeta_1), \tag{56}$$

where

$$S_2(\zeta_0, \zeta_1) = -\zeta_{0\tau_7} - \zeta_{1\tau_5} - s_2^{(1)}\zeta_{0(7\xi)} - s_2^{(2)}\zeta_{1(5\xi)} - \cdots, \tag{57}$$

with $s_2^{(1)}$ and $s_2^{(2)}$ known constants. All the remaining terms of $S_2(\zeta_0, \zeta_1)$ are nonlinear terms involving ζ_0 and ζ_1. At this order, there appears two terms that will give rise to secularities, namely $\zeta_{0(7\xi)}$ and $\zeta_{1(5\xi)}$. These secularities come from the solitonic nature assumed for ζ_0, and from the solution of the nonhomogeneous linear equation for ζ_1 in the time τ_3, which can be found in [4]. For the elimination of these secularities, the first step is to determine the evolution of ζ_1 in the time τ_5. By making use of the already discussed scale invariance between slow and laboratory variables, we are able to write all the terms to possibly appear in this equation. A simple analysis shows that they are

$$\begin{aligned}\zeta_{1\tau_5} - a_2\Big[\zeta_{1(5\xi)} + b_2\zeta_1\zeta_{1\xi} + c_2\left(\zeta_0\zeta_{1\xi}\right)_\xi + \\ + d_2\left(\zeta_0\zeta_{1\xi\xi\xi} + \zeta_1\zeta_{0\xi\xi\xi}\right) + e_2\left({\zeta_0}^2\zeta_1\right)_\xi\Big] = T_2(\zeta_0),\end{aligned} \tag{58}$$

where

$$\begin{aligned}T_2(\zeta_0) \;=\; & f_2\zeta_{0(7\xi)} + g_2\zeta_{0\xi\xi}\zeta_{0\xi\xi\xi} + h_2\zeta_{0\xi}\zeta_{0(4\xi)} + i_2\zeta_0\zeta_{0(5\xi)} + \\ & + j_2{\zeta_0}^2\zeta_{0\xi\xi\xi} + k_2\zeta_0\zeta_{0\xi\xi} + l_2(\zeta_{0\xi})^3 + m_2{\zeta_0}^3\zeta_{0\xi}\end{aligned} \tag{59}$$

with $a_2, b_2, \ldots, m_2$ constants to be determined. Then, since we already know the equation for $\zeta_{1\tau_3}$, we can use the symmetry condition

$$\left(\zeta_{1\tau_3}\right)_{\tau_5} = \left(\zeta_{1\tau_5}\right)_{\tau_3} \tag{60}$$

to obtain all the constants appearing in the evolution equation for ζ_1 in the time τ_5. This is a laborious step, and the result is

$$\begin{aligned} a_2 &= -\tfrac{19}{360}\,ch^4; \qquad b_2 = 0; \\ c_2 &= \tfrac{10}{3}\alpha; \quad d_2 = \tfrac{5}{3}\alpha; \qquad e_2 = \tfrac{5}{6}\alpha^2, \end{aligned} \tag{61}$$

where $\alpha = \beta_1/\alpha_1$, which shows that the l.h.s. of Equation (58) is the linearized fifth order equation of the KdV hierarchy. The remaining constants are given by

$$g_2 = \frac{247}{288} ch^3 + \frac{105}{h^3} f_2;$$

$$h_2 = \frac{19}{36} ch^3 + \frac{63}{h^3} f_2; \qquad i_2 = \frac{19}{96} ch^3 + \frac{21}{h^3} f_2;$$

$$j_2 = \frac{57}{16} c + \frac{315}{2h^6} f_2; \qquad k_2 = \frac{437}{32} c + \frac{630}{h^6} f_2;$$

$$l_2 = \frac{209}{64} c + \frac{315}{2h^6} f_2; \qquad m_2 = \frac{285}{16} \frac{c}{h^3} + \frac{945}{h^9} f_2;$$

with f_2 a free parameter. Notice that the nature of f_2 is completely different from that of the free parameter α_2 which arose from the symmetry condition (48). While the latter is related to different normalizations of τ_5, the former expresses the relevance of the term $\zeta_{0(7\xi)}$ in the equation for $\zeta_{1(\tau_5)}$. The importance of this fact can be seen in the following way. Considering ζ_0 as a pure one-soliton solution, we see that $\zeta_{0(7\xi)}$ will be a secular term for the evolution of ζ_1 in the time τ_5. Therefore, if we want to make sense of Equations (58), (59) as a physical equation, we may choose $f_2 = 0$ so to eliminate that secularity. Then, substituting this equation into Equation (57), $S_2(\zeta_0, \zeta_1)$ acquires the form

$$S_2(\zeta_0, \zeta_1) = -\zeta_{0\tau_7} - s_2^{(1)} \zeta_{0(7\xi)} - \tilde{s}_2^{(2)} \zeta_{1(5\xi)} - \cdots . \tag{62}$$

Note that the coefficient of $\zeta_{1(5\xi)}$ has been modified due to the contribution coming from the substitution of $\zeta_{1\tau_5}$. Now, for $\zeta_0 \sim \operatorname{sech}^2\eta$, both terms $\zeta_{0(7\xi)}$ and $\zeta_{1(5\xi)}$ will contribute with secular terms of the same kind to $S_2(\zeta_0, \zeta_1)$. The sum of these secular contributions can be written in the form [4] $s_2^{(3)} \operatorname{sech}^2\eta \operatorname{tg} h\eta$, with $s_2^{(3)}$ a known constant. The next step is to remember that, according to the discussion presented in Section 4, the evolution of ζ_0 in the time τ_7 is given by the seventh order equation of the KdV hierarchy, which is of the form

$$\zeta_{0\tau_7} = \alpha_3 \left[\zeta_{0(7\xi)} + \cdots \right], \tag{63}$$

with α_3 a free parameter. Since $\zeta_{0(7\xi)}$ contains a term proportional to $\operatorname{sech}^2\eta \operatorname{tg} h\eta$, a properly chosen α_3 turns possible the use of this equation to eliminate all the secular terms from $S_2(\zeta_0, \zeta_1)$. In other words, the seventh order equation of the KdV hierarchy is able to eliminate all the secular terms at this order. As the resonant terms at any order of the perturbative scheme are always the linear terms of S_n [4], the elimination of the secular producing terms by choosing appropriate values for the free parameters in the higher order KdV hierarchy equations remains possible at any higher order.

6. Renormalization of the Soliton Velocity

We have already assumed, in the context of the multiple time formalism, that

$$\zeta_0 = \zeta_0(\xi; \tau_3, \tau_5, \tau_7, \ldots), \tag{64}$$

where

$$\xi = \xi' - c\tau_1, \tag{65}$$

so that ζ_0 automatically satisfies the linear wave equation

$$\zeta_{0\tau_1} + c\zeta_{0\xi} = 0. \tag{66}$$

Our concern now will be the solutions to the higher order equations of the KdV hierarchy. Hereafter, for simplicity, we will assume that c and h can be set equal to unity, so that the KdV equation reads

$$\zeta_{0\tau_3} = \alpha_1 \zeta_{0\xi\xi\xi} + \beta_1 \zeta_0 \zeta_{0\xi}, \tag{67}$$

with $\alpha_1 = -1/6$ and $\beta_1 = -3/2$. We now look for a traveling wave solution ζ_0 that satisfies this equation in the time τ_3, but that satifies also the higher order equations of the KdV hierarchy in the times $\tau_5, \tau_7, \ldots$. This *multi-solution* can be written in the form

$$\zeta_0 = \frac{k^2}{3} \operatorname{sech}^2(k\Lambda) \tag{68}$$

with the argument Λ given by

$$\Lambda = \xi - \frac{k^2}{A_1}\tau_3 - \frac{k^4}{A_2}\tau_5 - \frac{k^6}{A_3}\tau_7 - \cdots, \tag{69}$$

and with $A_1, A_2, A_3, \ldots$ parameters depending respectively on the constants α_1, α_2, α_3, ... This argument is to be interpreted in the following way: when the evolution equation under consideration is concerned to the time τ_{2n+1}, the relevant argument is

$$\Lambda = \xi - \sum_{i=1}^{n} \frac{k^{2i}}{A_i}\tau_{2i+1} - \theta_n,$$

with θ_n a phase involving all the remaining terms of the sum. Then, for the KdV Equation (67), $A_1 = (1/\alpha_1) = 6$, and the solution is

$$\zeta_0 = \frac{k^2}{3} \operatorname{sech}^2 \left[k \left(\xi - \frac{k^2}{6}\tau_3 - \theta_3(k;\ \tau_5, \ldots) \right) \right], \tag{70}$$

where θ_3 is a phase. On the other hand, the traveling wave solution satisfying simultaneously the KdV Equation (67) in the time τ_3, and the fifth order KdV Equation (50) in the time τ_5, is given by

$$\zeta_0 = \frac{k^2}{3}\operatorname{sech}^2\left[k\left(\xi - \frac{k^2}{6}\tau_3 - \frac{k^4}{A_2}\tau_5 - \theta_5(k;\tau_7,\ldots)\right)\right], \tag{71}$$

and so on, to any higher order.

Once we have found the solutions to any member of the KdV hierarchy, let us return to the equation for ζ_1 in the time τ_3, which is now $(c = h = 1)$ written in the form

$$\zeta_{1\tau_3} + \tfrac{3}{2}(\zeta_0\zeta_1)_\xi + \tfrac{1}{6}\zeta_{1\xi\xi\xi} = S_1(\zeta_0), \tag{72}$$

with

$$S_1(\zeta_0) = \tfrac{9}{24}\zeta_0\zeta_{0\xi\xi\xi} + \tfrac{5}{8}\zeta_{0\xi}\zeta_{0\xi\xi} + \tfrac{63}{16}\zeta_0{}^2\zeta_{0\xi}. \tag{73}$$

It is important to remember that the linear term of $S_1(\zeta_0)$, which is the secular producing term in the equation for ζ_1, was eliminated through the use of the fifth order KdV Equation (50), with α_2 given by Equation (53). Therefore, the ζ_0 appearing in Equations (72), (73) is that given by Equation (71), since it must be a solution of KdV as well as of the KdV hierarchy fifth order Equation (50).

Finally, let us take the solutions ζ_0, and let us make the transformation back from the slow (ξ, τ, ζ_0) to the laboratory coordinates (x, t, ζ). For the case of solution (70) to the KdV equation, we get

$$\zeta = \frac{k^2}{3}\varepsilon\operatorname{sech}^2\left[k\varepsilon^{1/2}(x - V_3 t)\right], \tag{74}$$

where

$$V_3 = c + \varepsilon\frac{k^2}{6} \tag{75}$$

is the solitary-wave velocity in the laboratory coordinates. For the case of solution (71) to the fifth order KdV equation, we get

$$\zeta = \frac{k^2}{3}\varepsilon\operatorname{sech}^2\left[k\varepsilon^{1/2}(x - V_5 t)\right], \tag{76}$$

where now

$$V_5 = V_3 + \varepsilon^2\frac{k^4}{A_2}. \tag{77}$$

Since A_2 depends on the parameter α_2, to choose α_2 means to define the velocity renormalization itself. In this sense we can say that there is a unique velocity renormalization leading to a secular-free perturbation scheme for ζ_1. From these properties, we can see now that this method is equivalent to the renormalization technique developed by Kodama and Taniuti [4], in which the secular-free higher order effects were also given by the renormalization of the KdV soliton velocity. Moreover, in the same way as in the method of Kodama and Taniuti, if higher order scales are introduced, it is possible to continue the secular-free perturbation to higher orders by using the higher order equations of the KdV hierarchy. And in general, depending on the order considered in the perturbative scheme, the renormalized solitary-wave velocity in the laboratory coordinates is given by

$$V_{2n+1} = c + \sum_{i=1}^{n} \varepsilon^i \frac{k^{2i}}{A_i}. \tag{78}$$

7. Final Remarks

By using the reductive perturbation method of Taniuti [6], with the introduction of an infinite sequence of slow time variables $\tau_1, \tau_3, \tau_5, \ldots$, we studied the propagation of long surface waves in a shallow inviscid fluid. The three main ingredients of our analysis were: (i) the scale-invariance argument, which restricts the possible forms of the evolution equations, and which is necessary for the coherence of the perturbative expansion; (ii) the compatibility condition (48), which appears when extending a function of a single time-variable to a multiple time-variable; (iii) the secularity elimination procedure, without which the perturbative expansion would be meaningless. By using (i) and (ii) we have shown that, if the amplitude ζ_0 satisfies the KdV equation, it also satisfies the higher-order equations of the KdV hierarchy. Then, by using the fifth order equation of the hierarchy with a properly chosen α_2, we have shown that the soliton related secular producing terms of the equation for ζ_1 in the time τ_3 could be eliminated. As higher orders are reached, the number of terms contributing to the secularity increases. Despite of this, a similar procedure will be able to remove all these secularities up to any higher order. As the secularity elimination is mandatory for a physical theory, being essencially a finiteness requirement, the results obtained in this paper allowed us to describe the role played by the KdV hierarchy equations. Thereafter, by considering a solitary wave solution, we have shown that the elimination of secularities through the use of higher order KdV equations corresponds, in the laboratory coordinates, to a renormalization of the soliton velocities, as obtained previously by Kodama and Taniuti [4].

Acknowledgements

The authors would like to thank J. Léon and J. C. Montero for useful discussions. Two of the authors (RAK and JGP), would also like to thank CNPq, Brazil, and CAPES, Brazil, for financial support.

References

1. Korteweg, D. J. and de Vries, G.: *Phil. Mag.* **39** (1895), 422.
2. Lax, P. D.: *Comm. Pure Appl. Math.* **21** (1968), 467.
3. Sandri, G.: *Il Nuovo Cim.* **36** (1965), 67.
4. Kodama, Y. and Taniuti, T.: *J. Phys. Soc. Japan* **45** (1978), 298.
5. Whitham, G. B.: *Linear and Nonlinear Waves*, Wiley, New York, 1974.
6. Taniuti, T.: *Suppl. Prog. Theor. Phys.* **55** (1974), 1.
7. Jeffrey, A. and Kawahara, T.: *Asymptotic Methods in Nonlinear Wave Theory*, Pitman, London, 1982.
8. Miura, R. M.: *J. Math. Phys.* **9** (1968), 1202.
9. Miura, R. M., Gardner, C. S., and Kruskal, M. D.: *J. Math. Phys.* **9** (1968), 1204.
10. Gardner, C. S.: *J. Math. Phys.* **12** (1971), 1548.
11. Novikov, S., Manakov, S. V., Pitaevskii, L. P., and Zakharov, V. E.: *Theory of Solitons*, Plenum, New York, 1984.
12. Dodd, R. K., Eilbeck, J. C., Gibbon, J. D., and Morris, H. C.: *Solitons and Nonlinear Waves*, Academic, London, 1982.
13. We thank A. Degasperis and P. M. Santini for useful comments on this point.

Acta Applicandae Mathematicae **39:** 405–433, 1995.

Darboux Transformations for Higher-Rank Kadomtsev–Petviashvili and Krichever–Novikov Equations

GEOFF A. LATHAM
Burdett, Buckeridge and Young Ltd., Level 45, Rialto South Tower, 525 Collins St., Melbourne Vic 3000, Australia

and

EMMA PREVIATO
Department of Mathematics, Boston University, Boston, MA 02215, U.S.A.

(Received: 8 January 1995)

Abstract. It is shown that the action of a special 'rank 2' or 'rank 3' Darboux transformation, called *transference*, upon a pair of commuting ordinary differential operators of orders 4 and 6 implements the Abelian sum on their elliptic joint spectrum. A consequence of this is that, under the deformation of these commuting operators by the KP flow, every 'rank 2' KP solution corresponds to a solution of the Krichever–Novikov (KN) equation, and vice versa, with the transference process providing the correspondence between $(2+1)$ and $(1+1)$ dimensions. For a singular joint spectrum, it is further shown that transference at the singular point produces a correspondence between solutions of the singular KN equation and those of the KdV equation. These correspondences are illustrated by considering examples of a nondecaying rational KdV or Boussinesq solutions and the corresponding rational, singular-KN and rational KP solutions which the transference process generates.

Mathematics Subject Classification (1991): 58F07.

Key words: Darboux transformations, Kadomtsev–Petviashvili equations, Krichever–Novikov equations.

1. Introduction

The *algebro–geometric* solutions of the Kadomtsev–Petviashvili (KP) equation arise from the deformation, under the KP flow, of the *algebraic data*: namely, $\mathcal{S}$, a genus g algebraic curve; $P_\infty \in \mathcal{S}$, a smooth point; $z^{-1}(P_\infty)$, a local parameter near P_∞; $\mathcal{E}$, a rank r degree rg generic vector bundle over $\mathcal{S}$; and ϕ, a local trivialization of $\mathcal{E}$ at P_∞ [18, 19, 25, 27, 29]. $\mathcal{E}$ consists of the common eigenspace of commuting ordinary differential operators (ODOs), of Lax type, associated with the KP flow. $\mathcal{S}$ is the Burchnall–Chaundy [2–4] (joint) *spectral curve* for a (rank r) algebra of commuting ODOs and remains fixed under the KP flow.

When $r = 1$, the algebro-geometric solutions are well-understood and can be written down explicitly as having τ-functions given by the ϑ-functions for the compact genus g Riemann surface $\mathcal{S}$. Hence, in rank 1, a well-known dictionary between the algebraic data and KP solutions exists (cf. [25, 29]). For $r > 1$ it is more difficult to characterize *explicitly* the algebro-geometric solutions, however, some success has been forthcoming in ranks 2 [18, 19, 12, 14] and 3 [20, 24] but only for $g = 1$. Part of the underlying difficulty in these higher rank cases is that posed by the requirement to determine explicitly the coefficients of the commuting ODOs whose joint eigenfunctions give the vector bundle $\mathcal{E}$. Although this part of the problem can be formulated as a Riemann–Hilbert problem on $\mathcal{S}$ [17], this approach fails to provide the required explicit formulae and so other methods must be used.

For $r = 2$ and $g = 1$ (the elliptic spectrum) Krichever and Novikov in [18, 19] were able to obtain the necessary solution of the rank 2 commuting ODO problem and thereby formulate an explicit description, by what is now called the Krichever–Novikov (KN) equation, of the deformation of the algebraic data by the KP flow. For $r = 3$, $g = 1$, a similar evolutionary system is found [24]. The aim of this paper is to explore the role, within the Krichever–Novikov framework, of a beautiful geometric transformation (upon the underlying commuting ODOs) which Burchnall and Chaundy called *transference*. We now describe the situation more concretely in these special cases.

Given a formal first-order pseudo-differential operator

$$\mathcal{L} = D + \tfrac{1}{4} U D^{-1} + W D^{-2} + \cdots, \tag{1}$$

where $D = \partial_x$ and whose coefficients depend on the variables $x_1 = x, x_2, x_3, \ldots$, the KP hierarchy is defined by the flows

$$\partial_{x_s} \mathcal{L} = [(\mathcal{L}^s)_+, \mathcal{L}], \quad s = 2, 3, \ldots, \tag{2}$$

where $(\cdot)_+$ denotes the operation of taking the differential operator part. The compatibility of the flows (2) for $s = 2$ and $s = 3$ leads to the usual KP equation, namely,

$$\tfrac{3}{4} U_{x_2 x_2} = \left(U_{x_3} - \tfrac{3}{4} U U_x - \tfrac{1}{4} U_{xxx} \right)_x. \tag{3}$$

Suppose that $L = \mathcal{L}^4$ ($\mathcal{L}^6$, resp.) is an ODO which satisfies

$$[\mathcal{L}^4, (\mathcal{L}^6)_+] = 0, \quad ([\mathcal{L}^6, (\mathcal{L}^9)_+] = 0, \text{ resp.}). \tag{4}$$

We say that the *commuting pair* (L, B), $B = (\mathcal{L}^6)_+$, $((\mathcal{L}^9)_+$, resp.) has rank 2 (3, resp.) if the greatest common divisor (gcd) of the orders of all differential operators which commute with L is 2 (3, resp.) (r = rank $\mathcal{E}$ = 2, 3) [32]. According to the main result of Burchnall–Chaundy theory, any two commuting ODOs satisfy identically an algebraic relation. This paper is concerned only with the genus one case of this result in which L and B, satisfying (4), define an elliptic spectral curve $\mathcal{S}$:

$$B^2 = L^3 + \tfrac{1}{4} g_2 L + \tfrac{1}{4} g_3. \tag{5}$$

Given a point $P = (\lambda, \mu)$, an application of the Euclidean algorithm [16, Section 5.4] produces the factorizations

$$L - \lambda = Q_1 \cdot G,$$

$$B - \mu = Q_2 \cdot G, \tag{6}$$

where G is the greatest common divisor of $L - \lambda$ and $B - \mu$. When (4) holds and P is a generic point on the spectrum $\mathcal{S}$, G has order 2 (3, resp.), while for all other P, $G \equiv 1$. All of the geometric structure of the vector bundle $\mathcal{E}$ of common eigenfunctions of L and B is encapsulated by G, see [28]. For the commuting pair (L, B), the process of *transference at* $P = (\lambda, \mu)$ on $\mathcal{S}$ is defined through

$$\tilde{L} - \lambda = G \cdot Q_1,$$

$$\widetilde{B} - \mu = G \cdot Q_2, \tag{7}$$

which is the equivalent of successively performing two (three, resp.) Darboux transformations (conjugations) by the first-order factors of G. The transference process creates, from the commuting pair (L, B) and the point (λ, μ) on the spectrum, a new commuting pair $(\tilde{L}, \widetilde{B})$ with the same spectrum. What changes is the bundle, $\widetilde{\mathcal{E}} = \mathcal{E} \otimes \mathcal{O}_{\mathcal{S}}(\infty - P)$ [11].

With regard to KP flows, it was shown in [18, 19], by combining asymptotic arguments on the spectral curve with explicit formulae for commuting differential operators, that any KP solution U, whose $\mathcal{L}$ in addition satisfies the commutativity condition (4) in the $r = 2$ case, must be given by

$$U = 2\partial_x^2 \ln\left[\tau^2/\partial_x \wp(c)\right] + \{\wp(c);\ x\} - 1/(2c_x^2), \tag{8}$$

where $\tau = \wp(x_2) - \wp(c)$,

$$\{f; x\} = f_{xxx}/f_x - \tfrac{3}{2}\, f_{xx}^2/f_x^2$$

denotes the (partial) Schwarzian derivative, $\wp$ is the Weierstrass function associated with the elliptic curve (5) and $c(x, x_3)$ is a solution of the 'Krichever–Novikov (KN) equation'

$$\frac{c_{x_3}}{c_x} = \frac{1}{4}\{c; x\} + \frac{3}{8c_x^2} - \frac{3\wp(2c)c_x^2}{2}, \tag{9}$$

in which c_x is not identically zero. Conversely, when U is given by (8) with c satisfying (9), $\mathcal{L}^4$ deforms according to KP flows, thus fulfilling the commutativity condition (4).

The aim of this paper is to explore the correspondence between the KP and KN equations in terms of transference at certain fixed points on the spectral curve. This is done by first showing how transference acts on the commuting pair in (4), and second, by interpreting this action as the reduction process from (3) to (9). It turns out (Theorem 3) that, in this broad sense, the transference process is the x_2 KP flow (from preliminary remarks, one could argue that it is the x_3 flow in rank 3, but we cannot produce an explicit formula yet). If S is singular, then a further application of the transference at the singular point is shown to reduce the singular KN equation (i.e. (9) with $\wp(2c) = 1/(4c^2)$) to the KdV equation. In this case, at the level of the commutative algebra of ODOs, resolving the singularity of S simply transforms the ring $\mathbb{C}[L, B]$ into $\mathbb{C}[K^2, K^3]$ where $K = \partial^r +$ (lower-order-terms) is still a KP solution, but now of simpler type (KdV if $r = 2$, Boussinesq if $r = 3$)! Hence, in the singular case, a mechanism exists for generating from any KdV solution, first a singular KN solution, and then a KP solution; namely, two successive steps of transference, first at the singular point and then at a nonsingular point. The details of the process can be carried out quite explicitly for the vanishing rational solutions of the KdV equation [23] and gives an explicit alternative implementation of the work in [31].

The explicit formulae and parametrizing data for the coefficients of commuting differential operators of orders four and six are recalled in Subsection 2.1 (see also [22]). From these, explicit expressions for G and Q_1 in (6) are derived. The effect of performing transference, i.e. passing from (6) to (7), is derived in Subsection 2.2. These results represent an extension to rank 2 of those known by Burchnall and Chaundy in rank 1. Subsection 2.3 translates the results for ODOs to the context of the KP and KN equations and shows transference to be the key which links these two equations. In Subsection 2.4, we show that transference at the singular point of the spectrum reduces the singular KN equation to the KdV equation and illustrate this process with a true rank 2 example. In Subsections 3.1–3.3, we extend the analysis of the singular case to rank 3, providing examples of both a nonvanishing rational KP solution and a family of vanishing rational solutions computed recursively by transference of (rank 1) Boussinesq operators (Subsection 3.3.3), which are patterned after the recursion relations for *general* (vanishing) rational KdV solutions [1]. As with KdV, the main tool of the recursive computation is the first-order Darboux transformation of the Boussinesq operators.

2. Transference on Commuting Operators of Orders 4 and 6

The fundamental problem underlying the consideration of rank 2 KP solutions is that of studying the deformation, by the KP flow, of a commuting pair of ODOs of orders four and six. An explicit study of this problem requires a corresponding explicit knowledge of the coefficients of such operators. The question of determining the coefficients of commuting operators is a formidable problem in itself and has been studied extensively in the past [2–4], with considerable recent progress being made under the impetus of related problems in integrable systems [10, 17, 25, 29, 32].

2.1. Explicit formulae in rank 2

The problem of determining explicitly the coefficients of commuting ODOs of orders 4 and 6 has been solved independently in three different ways [6, 13, 18]. The forms of the results differ due to the presence of an arbitrary function in the resulting formulae, with different parametrizations of the arbitrary function leading to different (but equivalent) forms of the final answer. The formulae given here are those of [12, 18, 19] since they relate more directly to the geometry of the underlying elliptic Burchnall–Chaundy curve, or spectrum, and allow the use of $\wp$-function identities to simplify the calculations in various proofs.

After, if necessary, a Liouville transformation and change of independent variable, the most general (regular) ODO of order 4 can be written in the *normalized* form (with leading coefficient 1 and second coefficient 0)

$$L_4 = \left(D^2 + \tfrac{1}{2}u\right)^2 + a_1 D + D a_1 + a_0, \tag{10}$$

where the coefficients u, a_1 and a_0 are locally (at $x = 0$, say) smooth functions of x. L_4 will be called (formally) *self-adjoint* if $a_1 \equiv 0$ and *nonself-adjoint* otherwise. It was shown in [13] and [18] that any ODO L_6, of order 6, which commutes with L_4 and defines the elliptic curve (5), must be given by $L_6 = (L_4^{3/2})_+$. Hence, the condition $[L_4, L_6] = 0$ becomes a (complicated) set of nonlinear ordinary differential equations in the coefficients of L_4. The common eigenvalues (λ, μ) of such a commuting pair (L_4, L_6) then lie on the *spectral curve* $\mathcal{S}$ (cf. (5)):

$$\mu^2 = \lambda^3 + \tfrac{1}{4}g_2\lambda + \tfrac{1}{4}g_3.$$

That the differential equations, equivalent to the commutativity condition, are explicitly integrable, determines the coefficients of L_4.

THEOREM 1 ([6, 13, 18]). *A commuting pair (L_4, L_6) which corresponds to the elliptic curve $\mathcal{S}$ is given by $L_6 = (L_4^{3/2})_+$ with*

$$\begin{aligned} u &= 2\partial_x^2 \ln\left[(\wp(\gamma_0) - \wp(c))^2/\partial_x\wp(c)\right] + \{\wp(c); x\} - 1/(2c'^2), \\ a_1 &= c'\left(\wp(\gamma_0 - c) - \wp(\gamma_0 + c)\right), \\ a_0 &= -\wp(\gamma_0 + c) - \wp(\gamma_0 - c), \end{aligned} \tag{11}$$

where γ_0 is a constant, $c(x)$ is an arbitrary function of x and $\wp$ is the Weierstrass $\wp$-function for $\mathcal{S}$.

The commuting operators in Theorem 1 are parametrized by the data γ_0 and c. These two quantities define a divisor on $\mathcal{S}$ with Abelian coordinates $\gamma_1 := \gamma_0 + c$ and $\gamma_2 := \gamma_0 - c$.

Remark 1. Note that the commuting pair in the theorem is parametrized by the three constants γ_0, g_2 and g_3 and the arbitrary function $c(x)$. The correspondence of this parametrization with that in [13] is established via $g(x) = \wp'(\gamma_0)/(\wp(\gamma_0)-\wp(c))$, $K_{10} = -2\wp'(\gamma_0)$, $K_{11} = \wp''(\gamma_0)/\wp'(\gamma_0)$ and $K_{12} = -2\wp(\gamma_0)$.

2.2. The Greatest Common Divisor and Transference

Given a point $P = (\lambda, \mu)$ on S, the gcd of the commuting pair $(L_4-\lambda, L_6-\mu)$ of Theorem 1, is the second-order differential operator G obtained by the Euclidean algorithm [16, Subsection 5.4];

$$L_4 - \lambda = Q_1 \cdot G,$$

$$L_6 - \mu = Q_2 \cdot G. \tag{12}$$

Since the coefficients of L_4 and L_6 are given by (11), corresponding explicit expressions for the coefficients of G and the Q_i, $i = 1, 2$, can be computed.

LEMMA 1 ([12, 22]). *For $P = (\lambda, \mu) = (\wp(\rho), \frac{1}{2}\wp'(\rho))$ on S and L_4 and L_6 as given in Theorem 1, the factorization (12) is given by*

$$G = D^2 + \Gamma_1 D + \Gamma_0,$$

$$Q_1 = D^2 - \Gamma_1 D + u - \Gamma_0 - 2\Gamma_1' + \Gamma_1^2,$$

where

$$\Gamma_1 = \partial_x \ln\left(\frac{\sigma(\gamma_0 + c)\sigma(\gamma_0 - c)}{\sigma(\gamma_0 + \rho + c)\sigma(\gamma_0 + \rho - c)}\right),$$

$$\begin{aligned}\Gamma_0 &= F' - F^2 + F\Gamma_1 - 1/(4c'^2) - \\ &\quad -1/2(\zeta(\gamma_2 + \rho) - \zeta(\gamma_2) + \zeta(\gamma_1 + \rho) - \zeta(\gamma_1) - 2\zeta(\rho)),\end{aligned}$$

and

$$F = \partial_x \ln\left(\frac{\sigma(\gamma_0 + c)\sigma(\gamma_0 - c)}{\sqrt{c'}\sigma(2c)}\right). \tag{13}$$

Proof. An explicit application of the Euclidean algorithm to the Grünbaum parametrization followed by the use of the dictionary in Remark 1 produces the above formulae which agree with those obtained by a more geometric argument in [12]. □

Remark 2. The exact correspondence with the notation in [12] is established by identifying the Grinevich local curve coordinate 'P' with $-\rho$ and by noting that two of the Tyurin parameters are $\alpha_1 = 1/(2c') + F$ and $\alpha_2 = -1/(2c') + F$.

We use, as convenient, either a prime or ∂_x to denote the derivative of a function with respect to x but reserve D for differentiation as an operator.

DEFINITION 1. Given a point $P = (\lambda, \mu)$ on $\mathcal{S}$, the process of *transference at* P creates from a commuting pair (L_4, L_6) a new commuting pair $(\tilde{L}_4, \tilde{L}_6)$ (with the same spectrum $\mathcal{S}$) defined by

$$\begin{aligned} \tilde{L}_4 - \lambda &= G \cdot Q_1, \\ \tilde{L}_6 - \mu &= G \cdot Q_2, \end{aligned} \tag{14}$$

where G, Q_1, and Q_2 are as given in (12) and Lemma 1. An analogous definition applies for the commuting pair (L_6, L_9).

Since $\tilde{L}_4$ and $\tilde{L}_6$ commute and have spectrum $\mathcal{S}$, they will be given by formulae similar to (11). The precise effect of transference on the parametrization of the commuting differential operators defined by Theorem 1 is quite simple and yet displays some deep and beautiful geometry.

THEOREM 2. *Let $P = (\lambda, \mu) = (\wp(\rho), \frac{1}{2}\wp'(\rho))$ be a point on the spectrum $\mathcal{S}$. Then transference at P produces the commuting pair $(\tilde{L}_4, \tilde{L}_6)$, with spectrum $\mathcal{S}$, whose algebraic data is $\tilde{\gamma}_0 = \gamma_0 + \rho$, $\tilde{c} = c$, or, more explicitly, whose coefficients are given by*

$$\begin{aligned} \tilde{u} &= 2\partial_x^2 \ln[(\wp(\gamma_0 + \rho) - \wp(c))^2/\partial_x\wp(c)] + \{\wp(c); x\} - 1/(2c'^2), \\ \tilde{a}_1 &= c'(\wp(\gamma_0 + \rho + c) - \wp(\gamma_0 + \rho - c)), \\ \tilde{a}_0 &= -\wp(\gamma_0 + \rho + c) - \wp(\gamma_0 + \rho - c). \end{aligned} \tag{15}$$

Proof. This involves some short calculations which utilize the addition formulae for the Weierstrass functions. For example, from Lemma 1 and (14), we easily obtain $\tilde{u} = u - 4\Gamma_1'$ from which the expression for $\tilde{u}$ easily follows upon using the identity $\wp(\gamma_0) - \wp(c) = -\sigma(\gamma_1)\sigma(\gamma_2)/\sigma^2(\gamma_0)\sigma^2(c)$. □

Remark 3. We often denote the transformation under transference described in the theorem by $L_4 \xrightarrow{P} \tilde{L}_4$ with the understanding that L_6 also transforms accordingly. This is sometimes called *transference at P applied to L_4* (or (L_4, L_6)).

Interpreted geometrically, Theorem 2 states that transference implements the Abelian sum on the elliptic curve $\mathcal{S}$, that is, the local curve coordinates for the points

$$P_0 = \left(\wp(\gamma_0),\ \tfrac{1}{2}\wp'(\gamma_0)\right) \quad \text{and} \quad P = \left(\wp(\rho),\ \tfrac{1}{2}\wp'(\rho)\right)$$

are added to give the new parametrizing point $\tilde{P}_0 = P_0 + P$, while the arbitrary function $c(x)$ remains unchanged. By again using $\wp$-function identities, it can be shown that, in Grünbaum's parametrization (Remark 1), the arbitrary function g undergoes a Möbius transformation, $\tilde{g} = K_0 + K_2/(g + K_1)$ where the constants K_i depend on P and P_0 but are independent of x.

The limiting case $\rho = -\gamma_0$ of Theorem 2 deserves special attention since the calculation produces the self-adjoint case (see [22]).

COROLLARY 1. *If $P = (\wp(\gamma_0), -\frac{1}{2}\wp(\gamma_0)) = -P_0$, then transference at $-P_0$ applied to (L_4, L_6) produces the self-adjoint commuting pair $(\tilde{L}_4, \tilde{L}_6)$ for which*

$$\begin{aligned}\tilde{u} &= -2\partial_x^2 \ln\left[\partial_x \wp(c)\right] + \{\wp(c);\ x\} - 1/(2c'^2),\\ \tilde{a}_1 &= 0,\\ \tilde{a}_0 &= -2\wp(c).\end{aligned} \tag{16}$$

Because of its ubiquity, we will denote the fourth-order self-adjoint operator with coefficients given by (16) by just L. Note that the coefficients in (16) can be interpreted as the limit $\gamma_0 \to 0$ of those in Theorem 1. Also, the substitution $\wp(c) = -x$ in (16) produces the well-known example in [7] of a commuting pair of orders four and six with polynomial coefficients.

COROLLARY 2. *If L is formally self-adjoint with coefficients given by (16), then the general nonself-adjoint case (11) can be obtained as the operator $\tilde{L}$ generated by transference at $P_0 = (\wp(\gamma_0), \frac{1}{2}\wp'(\gamma_0))$ applied to L. Conversely, if L_4 is nonself-adjoint with coefficients given by (11), then the self-adjoint case (16) can be obtained as the operator $\tilde{L}_4 = L$ generated by transference at $-P_0 = (\wp(\gamma_0), -\frac{1}{2}\wp'(\gamma_0))$ applied to L_4.*

Remark 4. The geometric significance of self-adjointness is the following: the bundle corresponding to the adjoint operator $L_4^\dagger$ is (cf. [5]) $\mathcal{E}^\dagger \otimes \mathcal{K}$, where $\mathcal{E}^\dagger$ is the dual bundle of $\mathcal{E}$ and $\mathcal{K}$ is the canonical bundle, in this case $\mathcal{K} = \mathcal{O}_S$. Thus, $L_4^\dagger = L_4$ if and only if $\mathcal{E}$ is self-dual, a condition that is always possible to achieve by an appropriate Abelian sum, and translated into KN data means $\gamma_0 = 0$.

Remark 5. That transference takes the self-adjoint case to the general nonself-adjoint case can be easily verified explicitly by using the factorization in Lemma 1 with $\gamma_0 = 0$. In fact, one finds that $\tilde{a}_1 = \partial_x(2\mu/(\lambda + a_0/2))$, where $a_0 = -2\wp(c)$, which is exactly the expression for a_1 in (11).

Clearly, the process of transference can be iterated for different points on the spectral curve. The result of n steps of such an iteration, beginning with the self-adjoint case, can be found in [22]. It is of some interest for later applications to the KP equation to consider transference at the branchpoints of the spectral curve. Let S factor as

$$(2\mu)^2 = 4(\lambda - e_1)(\lambda - e_2)(\lambda - e_3),$$

where $e_1 + e_2 + e_3 = 0$, $e_i = \wp(\omega_i)$, $2\omega_1$ and $2\omega_3$ are the basic periods of $\wp$ and $\omega_2 = -\omega_1 - \omega_3$. Consider a self-adjoint L whose coefficients are given by (16). Since transference amounts to addition, the points of order 2 preserve self-adjointness (cf. Remark 5).

LEMMA 2. *The application of transference at* $(e_i, 0)$, $i = 1, 2, 3$, *to the self-adjoint commuting pair* $(L, L_+^{3/2})$ *produces the self-adjoint commuting pair* $(\tilde{L}, \tilde{L}_+^{3/2})$ *for which*

$$\begin{aligned}
\tilde{u} &= u + 4\partial_x^2 \ln[e_i - \wp(c)],\\
\tilde{a}_1 &= 0,\\
\tilde{a}_0 &= -2e_i + \frac{1}{2}\frac{h'(e_i)}{e_i - \wp(c)},
\end{aligned}$$

where $h(\lambda) = 4\lambda^3 + g_2\lambda + g_3$.

Proof. From Theorem 2, we see that $\tilde{a}_1 = c'(\wp(\omega_i - c) - \wp(\omega_i + c))$, but, by periodicity, $\wp(\omega_i - c) = \wp(\omega_i + c)$ and so $\tilde{a}_1 = 0$. Again from Theorem 2 and periodicity, $\tilde{a}_0 = -\wp(\omega_i - c) - \wp(\omega_i + c) = -2\wp(\omega_i + c)$. The addition theorem for the $\wp$-function now gives

$$\wp(\omega_i + c) = \wp(\omega_i) + \frac{1}{4}\frac{h'(e_i)}{\wp(c) - \wp(\omega_i)},$$

which in turn implies the required expression for $\tilde{a}_0$. □

The fundamental relationship of the Burchnall–Chaundy theory to hierarchies of integrable equations such as KP hinges upon the fact that commuting ODOs, such as those above, can be deformed with integrable flows without destroying the commutativity condition. In the other direction, commuting differential operators arise from the expression of integrable systems as the compatibility for flows on Lax operators together with the requirement that their solutions be stationary under linear combinations of evolutionary variables.

2.3. FROM KP TO KN BY TRANSFERENCE

Assume now that the coefficients of L_4 and L_6 depend on added parameters x_2, $x_3, \ldots$, as well as $x_1 = x$. Further, let $\mathcal{L}$ in (1) satisfy $\mathcal{L}^4 = L_4$ where L_4 is given by (10). The coefficient $u(x)$ in (10) is now renamed $U(x, x_2, x_3)$ to highlight its dependence upon added variables. As mentioned in the Introduction, we are interested in the case where U evolves according to the KP equation

$$\tfrac{3}{4}U_{x_2x_2} = \left(U_{x_3} - \tfrac{3}{4}UU_x - \tfrac{1}{4}U_{xxx}\right)_x. \tag{17}$$

It was shown in [14, 18, 19] that, under the identification $\gamma_0 \equiv x_2$, and by allowing $c(x, x_3)$ to depend on x_3 (and higher KP variables), then when $[\mathcal{L}^4, \mathcal{L}^6_+] = 0$ holds, U, now given by (11), solves the KP equation if and only if c evolves according to the KN equation

$$\frac{c_{x_3}}{c_x} = \frac{1}{4}\{c;x\} + \frac{3}{8c_x^2} - \frac{3\wp(2c)c_x^2}{2}, \tag{18}$$

provided c_x is nonzero. It is often convenient to remove the $\wp$-function by the substitution $v = \wp(c)$ [31] which transforms (18) into

$$\frac{v_{x_3}}{v_x} = \frac{1}{4}\{v;x\} + \frac{3}{8}\frac{h(v)}{v_x^2}, \quad \text{where} \quad h(v) = 4v^3 + g_2v + g_3. \tag{19}$$

The Lax operator for the KN equation has been determined in [30] and assumes an immediate relevance in the light of Corollary 1.

LEMMA 3. *The KN equation is equivalent to*

$$\partial_{x_3}L = [(L^{3/4})_+, L], \tag{20}$$

where $L = (D^2 + \frac{1}{2}\varphi)^2 + a_0$ *with*

$$\begin{aligned} \varphi &= -2\partial_x^2 \ln[\partial_x\wp(c)] + \{\wp(c);x\} - 1/(2c_x^2), \\ &= -2\partial_x^2 \ln[v_x] + \{v;x\} - \frac{1}{2}\frac{h(v)}{v_x^2}, \\ a_0 &= -2\wp(c) = -2v. \end{aligned} \tag{21}$$

Thus, from Corollary 1, the self-adjoint case of commuting pairs of orders 4 and 6 defines the Lax operator for the KN equation. In view of this, Theorem 2 gives the immediate connection between 'rank 2' KP solutions and KN solutions. We will refer equivalently to any of c, v or $L^{1/4}$ as a KN solution. We also say that U is a KP solution for brevity, although it is only a solution of the KP equation (17), for the operator $\mathcal{L} = L^{1/4}$ cannot be recovered from U alone.

THEOREM 3. *Any solution of the KP equation, for which* $[\mathcal{L}^4, \mathcal{L}^6_+] = 0$, *determines a solution* c *(or* v *or* $\tilde{\mathcal{L}} = L^{1/4}$*) of the KN equation, by transference at* $-P_0 = (\wp(x_2), -\frac{1}{2}\wp'(x_2))$ *applied to* $(\mathcal{L}^4, \mathcal{L}^6_+)$, *for which* $\tilde{\mathcal{L}}^4 = L$ *is self-adjoint. Conversely, any solution* c *(or* v *or* $\tilde{\mathcal{L}} = L^{1/4}$*) of the KN equation determines a solution of the KP equation, by transference at* $P_0 = (\wp(x_2), \frac{1}{2}\wp'(x_2))$ *applied to* $(L, L^{3/2}_+)$, *for which* $\mathcal{L}^4 = \tilde{L}$ *is a differential operator and* $[\mathcal{L}^4, \mathcal{L}^6_+] = 0$.

The theorem illustrates that transference at P_0 implements the x_2-flow by introducing x_2-dependence into the differential Lax operator (L), for the KN equation, which hitherto carried only x_1 and x_3 dependence. Similarly, the reduction of the KP equation, under the 'rank 2' ansatz (4), to the KN equation is implemented by transference at $-P_0$. The operator $\tilde{\mathcal{L}}^4 = L$, so produced, is exactly the Lax operator for the KN equation [30] with coefficients given by (21).

2.4. From KN to KdV: The Singular Spectrum

The case of transference at zeros in the singular case ($h'(e_i) = 0$ above) is of considerable interest. To explore the consequences of such transference requires the following useful lemma.

LEMMA 4. *Assume that v solves the KN Equation* (19) *and let*

$$V = 4\partial_x^2 \ln[v - e_i] - 2\partial_x^2 \ln[v_x] + \{v; x\} - \frac{1}{2}\frac{h(v)}{v_x^2}. \tag{22}$$

Then V satisfies

$$V_{x_3} - \frac{3}{4}VV_x - \frac{1}{4}V_{xxx} = \frac{3}{2}\frac{h'(e_i)v_x}{(v - e_i)^2}. \tag{23}$$

Proof. This is proven by substituting the expression (22) into the left-hand side of (23) and using (19) to eliminate v_{x_3}. □

This lemma now makes the connection with the KdV equation rather obvious.

COROLLARY 3. *Assume that v solves the KN equation* (19), *L is the corresponding Lax operator and $\mathcal{S}$ is singular, i.e. $\mu^2 = (\lambda - e_1)^2(\lambda + 2e_1)$. Then transference at the singular point $(e_1, 0)$ applied to the self-adjoint commuting pair $(L, L_+^{3/2})$ produces the (translate of the) square of the KdV Lax operator; namely,*

$$\tilde{L} = (D^2 + \tfrac{1}{2}V)^2 - 2e_1,$$

where V, given by (22) *with $i = 1$ and $h(v) = 4(v - e_1)^2(v + 2e_1)$, is a solution of the KdV equation*

$$V_{x_3} - 3VV_x/4 - V_{xxx}/4 = 0. \tag{24}$$

Schematically, when $\mathcal{S}$ is singular, we now have the following picture for the connections among the rank 2 KP, singular-KN and KdV solutions:

$$L_4, \quad U \underset{P_0}{\overset{-P_0}{\longleftrightarrow}} L, \quad v \underset{(e_1,0)}{\overset{(e_1,0)}{\longleftrightarrow}} K^2 - 2e_1, \quad V, \tag{25}$$

where $K = D^2 + V/2$. Since the entire process is reversible, it is possible to begin with a KdV solution V and by successive transferences at $(e_1, 0)$ and P_0, generate a KP solution U. This is indicated by the left pointing arrows in (25). The major practical difficulty in explicitly implementing this procedure is that, given a KdV solution V, one must then solve the nonlinear Equation (22) for v. One case where this procedure can be carried through explicitly (apart from the example below) is for the vanishing rational solutions of KdV. Then, the Adler–Moser [1] rational solutions $V_n = 4\partial_x^2 \ln \vartheta_n$ generate a corresponding infinite sequence v_n of solutions of the singular KN equation and, after transference at $P_0 = (x_2^{-2}, -x_2^{-3})$, a corresponding sequence U_n of vanishing rational KP solutions. The details of this construction are presented in [23]. For this sequence of examples however, the associated commuting pairs from (25) are 'fake rank 2' by which we mean that they are rank 1 because of the presence of an odd order operator which commutes with them.

As an illustration of the construction in (25) we consider Grünbaum's 'example (c)' [14, 15]. In particular, this example explicitly demonstrates the *existence* of rank 2 KP solutions (there is no odd order differential operator commuting with the corresponding L_4) – albeit for a singular (cuspidal) spectral curve.

EXAMPLE. By searching directly for solutions of the singular KN equation, Grünbaum in [14] found the following candidate rank 2 KP solution:

$$U = -\frac{2x}{k + \frac{3}{2}x_3} - \frac{4(k + \frac{3}{2}x_3)^2}{((k + \frac{3}{2}x_3)x - x_2)^2}, \tag{26}$$

where k is a constant. This is the KP solution computed from (8) using $\wp(x_2) = x_2^{-2}$, and

$$c(x, x_3) = \sqrt{(k + \tfrac{3}{2}x_3)x} \quad \text{or} \quad v = 1/\Big(k + \tfrac{3}{2}x_3\Big)x$$

and therefore corresponds to the cuspidal spectral curve $\mu^2 = \lambda^3$. Performing transference at $-P_0 = (x_2^{-2}, x_2^{-3})$ takes the operator L_4 with coefficient (26) into a self-adjoint operator $\tilde{L}_4 = L$ whose D^2 coefficient is

$$\widetilde{U} = -\frac{2x}{k + \frac{3}{2}x_3} - \frac{4}{x^2}. \tag{27}$$

Clearly, $\widetilde{U}$ is computed from (16) with $v = 1/(k + \frac{3}{2}x_3)x$. As in Remark 4, $\widetilde{U}$ can also be obtained as the limit of U as $x_2 \to 0$ and is not a KdV solution. According to (22) with $e_i = 0$ and $h(v) = 4v^3$, V is computed as

$$V = -\frac{2x}{k + \frac{3}{2}x_3}, \tag{28}$$

which solves the KdV equation (24). Thus, a further transference at $(0, 0)$ transforms L into K^2, where $K = D^2 + \frac{1}{2}V$ with V given by (28) (cf. (25)). The commuting pair (K^2, K^3) is rank 2 and since, as was observed in [22], transference does not change the rank, the commuting pairs $(L, L_+^{3/2})$ corresponding to (27) and (L_4, L_6) corresponding to (27) are of rank 2, and hence, (26) is a (true) rank 2 solution of the KP equation.

3. Transference on Commuting Operators of Orders 6 and 9

In principle, similar results to those presented above for rank 2 apply in still higher rank. In particular, since the corresponding problems of finding commuting ODOs and an analog of the KN equation (the M-system) have been solved in rank 3 for nonsingular spectrum $\mathcal{S}$ [24], it should be possible to construct similar explicit correspondences in that case. Moreover, when the spectrum is singular [20], the end result of transference would be the Boussinesq equation. The following subsections treat this case where much can be done explicitly. Because of the singular spectrum, $\wp$-functions do not appear and many of the formulae have a different character to their counterparts in rank 2.

3.1. Transference in Rank 3: The Singular Spectrum

We now describe the action of transference on the underlying commuting operators and KP solutions for the singular spectrum $\mu^2 = \lambda^3$ in rank 3. The results show a complete parallel to the picture in rank 2 [23] except that the singular Krichever–Novikov (s-KN) equation is more complicated and the end result of two transferences is the square of a Boussinesq (Bq) Lax operator.

First, we recall some of the relevant results from [20, 21] pertaining to commuting ODOs of orders 6 and 9 and the deformation of such operators by the KP flows. As in the Introduction, let $\mathcal{L}$ be a first-order pseudo such that $\mathcal{L}^6 = L_6$ is the differential operator given by

$$L_6 = \left(D^3 + \tfrac{3}{8}\, uD + \tfrac{3}{8}\, Du + a_3\right)^2 + a_2 D^2 + D^2 a_2 + a_1 D + D a_1 + a_0,$$

where u and the a_i are locally smooth functions.

PROPOSITION 1 ([21]). *If a_1 and h are arbitrary smooth functions of x, with h' not identically zero, and*

$$\begin{aligned}
\tfrac{3}{8}\, u &= -\frac{a_2''}{2a_2} + \frac{3{a_2'}^2}{8a_2^2} - \frac{3a_1^2}{2a_2^2} + \frac{K_1 + \frac{3}{4}\, K_2 h - \frac{1}{2}\, h^2}{a_2},\\
a_3 &= -\frac{a_1''}{a_2} + \frac{a_1' a_2'}{2a_2^2} + \frac{a_1}{2a_2}\left(\frac{a_2''}{a_2} - \frac{{a_2'}^2}{a_2^2}\right) + \frac{2a_1^3}{a_2^3} + \frac{a_1(h^2 - \frac{3}{2}\, K_2 h - 2K_1)}{a_2^2},\\
a_2 &= h',\\
a_0 &= h'u + \tfrac{5}{6}\, h''' - \tfrac{2}{3}\, h^2 + K_2 h - \tfrac{3}{8}\, K_2^2,
\end{aligned} \tag{29}$$

where K_1 and K_2 are independent of x, then $[L_6, (L_6^{3/2})_+] = 0$. The Burchnall–Chaundy spectrum is singular and is given by

$$\mu^2 = (\lambda - \tfrac{1}{48}\beta)^2(\lambda + \tfrac{1}{24}\beta), \tag{30}$$

where $\beta = 9K_2^2 + 32K_1$.

The commuting pair (L_6, L_9), $L_9 = (L_6^{3/2})_+$, of Proposition 1 can be evolved according to the KP flow in the following manner. The coefficient $u(x)$ in (29) is again renamed $U(x, x_2, x_3)$ to highlight its dependence on added variables.

THEOREM 4 ([20]). *Let*

$$\begin{aligned} h &= \partial_{x_3} \ln [p(x_3) - q(x, x_2)], \\ a_2 &= h_x, \\ a_1 &= h_{x_2}/2, \\ K_1 &= -\tfrac{1}{4}\,\partial_{x_3}^2 \ln p'(x_3) \quad \text{and} \quad K_2 = \tfrac{2}{3}\,\partial_{x_3} \ln p'(x_3). \end{aligned} \tag{31}$$

Then

$$\tfrac{3}{8}U = -\frac{1}{2}\frac{h_{xxx}}{h_x} + \frac{3}{8}\frac{h_{xx}^2 - h_{x_2}^2}{h_x^2} - \frac{h^2 - (p''/p')h + \frac{1}{2}(p''/p')'}{2h_x}, \tag{32}$$

solves the KP equation:

$$\tfrac{3}{4}U_{x_2x_2} = \left(U_{x_3} - \tfrac{3}{4}UU_x - \tfrac{1}{4}U_{xxx}\right)_x, \tag{33}$$

if and only if

$$\partial_{x_3}\{p; x_3\} = 0, \tag{34}$$

and q is a solution of the s-KN equation

$$\left(\frac{q_{x_2}}{q_x}\right)_{x_2} + \frac{1}{2}\left(\frac{q_{x_2}^2}{q_x^2}\right)_x + \frac{1}{3}\{q; x\}_x + \left(\frac{H(q)}{q_x}\right)_x = 0, \tag{35}$$

where

$$H(q) = \begin{cases} \frac{4}{3\Delta}(cq - a)^2, & \text{if } \{p; x_3\} = 0; \\ -\frac{4K_3}{3\Delta}(cq - a)(dq - b), & \text{if } \{p; x_3\} = -\frac{1}{2}K_3^2. \end{cases} \tag{36}$$

Further, the spectrum (30) is given by $\beta = -8\{p; x_3\} = 4K_3^2$. Here $\binom{a\,b}{c\,d}$ characterizes the invariance of the Schwarzian in (34) and $\Delta = ad - bc \neq 0$.

Remark 6. See [8] for a similar evolution equation to (35), derived by a generalized Miura transformation.

For the remainder of this section, we will avoid carrying unnecessary cluttering constants by specializing to the particular case $p(x_3) = x_3$ ($a = d = 1$, $b = c = 0$), call $x_2 = y$, and treat the corresponding case of (35) for which $H(q) = 4/3$;

$$\left(\frac{q_y}{q_x}\right)_y + \frac{1}{2}\left(\frac{q_y^2}{q_x^2}\right)_x + \frac{1}{3}\{q; x\}_x + \frac{4}{3}\left(\frac{1}{q_x}\right)_x = 0. \tag{37}$$

For this p, we have $K_1 = K_2 = 0$ and so (30) reduces to the cuspidal $\mathcal{S}$:

$$\mu^2 = \lambda^3. \tag{38}$$

Specializing (29), (31) and (32) to these choices yields

$$\begin{aligned}
\tfrac{3}{8} U &= \tfrac{3}{2} \partial_x^2 \ln[x_3 - q] + F(q), \\
a_3 &= \tfrac{3}{2} \partial_x \partial_y \ln[x_3 - q] + G(q), \\
a_2 &= \partial_x \left(\frac{1}{x_3 - q}\right), \\
a_1 &= \frac{1}{2} \partial_y \left(\frac{1}{x_3 - q}\right), \\
a_0 &= -\frac{1}{2}\frac{q_{xxx}}{(x_3 - q)^2} + \frac{q_{xx}^2}{q_x (x_3 - q)^2} + \frac{q_x q_{xx}}{(x_3 - q)^3} + \frac{q_x^3}{(x_3 - q)^4} - \\
&\quad - \frac{q_y^2}{q_x (x_3 - q)^2} - \frac{2}{(x_3 - q)^2},
\end{aligned} \tag{39}$$

where

$$\begin{aligned}
F(q) &= -\frac{1}{2}\frac{q_{xxx}}{q_x} + \frac{3}{8}\frac{q_{xx}^2}{q_x^2} - \frac{3}{8}\frac{q_y^2}{q_x^2} - \frac{1}{2q_x}, \\
G(q) &= -\frac{1}{2}\frac{q_{xxy}}{q_x} + \frac{1}{4}\frac{q_y q_{xxx}}{q_x^2} + \frac{1}{4}\frac{q_{xy} q_{xx}}{q_x^2} - \frac{1}{4}\frac{q_y q_{xx}^2}{q_x^3} + \frac{1}{4}\frac{q_y^3}{q_x^3} + \frac{1}{2}\frac{q_y}{q_x^2}.
\end{aligned} \tag{40}$$

Remark 7. Notice that these particular commuting pairs $(L_6, (L_6^{3/2})_+)$ are parametrized by a point $P_0 = (x_3^{-2}, -x_3^{-3})$ on the singular spectrum and the function q, whereas what is predicted by the general construction [27] are 3 points and 2 arbitrary functions.

For $L_9 = (L_6^{3/2})_+$ and $(\lambda, \mu) \neq (0, 0)$ on (38), consider the factorization which defines the gcd of $L_6 - \lambda$ and $L_9 - \mu$:

$$\begin{aligned}
L_6 - \lambda &= Q_1 \cdot G, \\
L_9 - \mu &= Q_2 \cdot G.
\end{aligned} \tag{41}$$

The operation of transference at (λ, μ) implies relations among the coefficients of all the operators involved.

LEMMA 5. *Let $\tilde{L}_6$, with coefficients $\tilde{u}$ and $\tilde{a}_i$, be related to L_6, with coefficients u and a_i, by $\tilde{L}_6 G = G L_6$, where $G = D^3 + c_2 D^2 + c_1 D + c_0$, then*

$$\begin{aligned} u - \tilde{u} &= 4c_2', \\ a_3 - \tilde{a}_3 &= \tfrac{3}{2} c_2'' - 3c_2 c_2' + 3c_1' - \tfrac{9}{4} u', \\ a_2 - \tilde{a}_2 &= c_2''' - 3(c_2 c_2')' + (c_2^3)' + 3c_1'' - 3(c_1 c_2)' + \\ &\quad + 3c_0' + \tfrac{3}{4}(c_2 u)' - \tfrac{27}{8} u'' - 3a_3'. \end{aligned} \tag{41a}$$

These relations can now be used to prove the following rank 3 analog of Lemma 1 and Theorem 2.

LEMMA 6. *If $(\lambda, \mu) = (\rho^{-2}, -\rho^{-3})$, the coefficients of $G = D^3 + c_2 D^2 + c_1 D + c_0$ for the commuting operators L_6, L_9 defined by (39) are*

$$\begin{aligned} c_2 &= \partial_x \ln\left[\frac{x_3 - q}{x_3 + \rho - q}\right], \\ c_1 &= \frac{1}{2}\left(c_2^2 - \partial_x c_2 + \partial_y \ln\left[\frac{x_3 - q}{x_3 + \rho - q}\right] + \frac{3}{2} U\right), \\ c_0 &= -\tfrac{1}{3}\partial_x^2 c_2 - \tfrac{1}{3} c_2^3 + c_2 \partial_x c_2 - \partial_x c_1 + c_1 c_2 + \tfrac{9}{8} U_x + \\ &\quad + a_3 - \tfrac{1}{4} c_2 U + \tfrac{1}{3}\left(\frac{1}{x_3 - q} - \frac{1}{x_3 + \rho - q}\right), \end{aligned} \tag{41b}$$

so that, for $\rho \neq 0$, transference at $(\rho^{-2}, -\rho^{-3})$ implements the Abelian sum on (38).

Hence, as in rank 2, by choosing the transference point (λ, μ) to be $-P_0$ (cf. Remark 1) we arrive at a rank 3 analog of the self-adjoint case in Corollary 1.

PROPOSITION 2. *Let L_6 have coefficients given by (39) and let $(\lambda, \mu) = (x_3^{-2}, x_3^{-3})$. Then the action of transference at (λ, μ) on L_6 is to produce the differential operator $L := \tilde{L}_6$ with coefficients*

$$\tfrac{3}{8}\tilde{U} = \tfrac{3}{2}\partial_x^2 \ln q + F(q),$$

$$\tilde{a}_3 = \tfrac{3}{2}\partial_x \partial_y \ln q + G(q),$$

$$\tilde{a}_2 = -\partial_x\left(\frac{1}{q}\right),$$

$$\tilde{a}_1 = -\frac{1}{2}\partial_y\left(\frac{1}{q}\right),$$

$$\tilde{a}_0 = -\frac{1}{2}\frac{q_{xxx}}{q^2} + \frac{q_{xx}^2}{q_x q^2} - \frac{q_x q_{xx}}{q^3} + \frac{q_x^3}{q^4} - \frac{q_y^2}{q_x q^2} - \frac{2}{q^2}. \tag{42}$$

The significance of L, which is $\lim_{x_3\to 0} L_6$, in relation to (37) is now fairly evident.

THEOREM 5. *The s-KN Equation* (37) *is completely integrable and has the equivalent Lax representation*

$$\partial_y L = [L_+^{1/3}, L], \tag{43}$$

where L is the operator whose coefficients are given by (42). *In particular, higher s-KN flows can be defined by*

$$\partial_{x_j} L = [L_+^{j/6}, L], \quad j \not\equiv 0 \ (\mathrm{mod}\, 3). \tag{44}$$

A further transference at the singular point of $\mathcal{S}$ now takes L to a Bq Lax operator.

PROPOSITION 3. *Let L be given by* (42). *Then transference at* $(0,0)$ *transforms L into the differential operator*

$$\tilde{L} = \left(D^3 + \tfrac{3}{8}\tilde{\tilde{U}}D + \tfrac{3}{8}D\tilde{\tilde{U}} + \tilde{\tilde{a}}_3\right)^2, \tag{45}$$

where

$$\tfrac{3}{8}\tilde{\tilde{U}} = F(q),$$

$$\tilde{\tilde{a}}_3 = G(q). \tag{46}$$

Moreover, if q solves (37) (*i.e.* (43) *holds*), *then* $\tilde{\tilde{U}}$ *is a solution of the Boussinesq equation:*

$$3U_{yy} = -(3UU_x + U_{xxx})_x, \tag{47}$$

i.e. $\tilde{L}$ *is the square of the Lax operator for the Bq equation.*

Remark 8. The above relationships between the coefficients of L and $\tilde{L}$ can be summarized succinctly by

$$\begin{aligned}
\tilde{U} &= \tilde{\tilde{U}} + 4\partial_x^2 \ln q, \\
\tilde{a}_3 &= \tilde{\tilde{a}}_3 + \tfrac{3}{2}\,\partial_x\partial_y \ln q, \\
\tilde{a}_2 &= \tilde{\tilde{a}}_2 - \partial_x\left(\frac{1}{q}\right), \\
\tilde{a}_1 &= \tilde{\tilde{a}}_1 - \frac{1}{2}\partial_y\left(\frac{1}{q}\right),
\end{aligned}$$

where (46) holds and, of course, $\tilde{\tilde{a}}_2 = \tilde{\tilde{a}}_1 = 0$.

To see that solutions of the s-KN equation (37) and associated commuting pairs abound, we now give a few examples of rational solutions of KP which illustrate the above results. However, a more complete treatment of fake rank 3 vanishing rational solutions generated by transference from Boussinesq rational solutions is postponed until Subsection 3.3.

EXAMPLE (a). *Nonvanishing rational solution.* This example comes from the Example 4.2 in [20] and corresponds to the s-KN solution $q = 4y^2/(9x)$ and $p(t) = t$ ($t \equiv x_3$).

$$U = 4\partial_x^2 \ln\left[t - 4y^2/(9x)\right] - \left(\frac{x^2}{y^2} + \frac{4}{x^2}\right) \quad \text{(KP solution)},$$

$$\tilde{U} = -\frac{x^2}{y^2} \quad (q = 4y^2/(9x) \text{ is the s-KN solution}),$$

$$\tilde{\tilde{U}} = -\left(\frac{x^2}{y^2} + \frac{4}{x^2}\right) \quad \text{(Bq solution)},$$

$$\tilde{\tilde{a}}_3 = \frac{1}{4}\frac{x^3}{y^3}. \tag{48}$$

In view of the results in Lemma 5, Proposition 2 and Proposition 3, there are third-order operators G_L and G_R such that $G_L L_6 = \tilde{L} G_L$ and $L_6 G_R = G_R \tilde{L}$, where $\tilde{L}$ and L_6 are given by (45) and (39), respectively, with the choices which lead to (48). We compute for G_L,

$$c_2 = \frac{4y^2}{x(9xt - 4y^2)},$$

$$c_1 = -\frac{3}{4}\frac{x^2}{y^2} + \frac{4xy + 9t}{x(4y^2 - 9xt)} - \frac{162t^2}{(4y^2 - 9xt)^2}, \tag{49L}$$

$$c_0 = \frac{1}{4}\frac{x^3}{y^3} - \frac{3}{4}\frac{x}{y^2} - \frac{4y + x^2}{x(4y^2 - 9xt)} + \frac{9t(4xy + 9t)}{x(4y^2 - 9xt)^2} - \frac{1458t^3}{(4y^2 - 9xt)^3},$$

and for G_R,

$$c_2 = \frac{4y^2}{x(4y^2 - 9xt)},$$

$$c_1 = -\frac{3}{4}\frac{x^2}{y^2} - \frac{2}{x^2} - \frac{4xy - 9t}{x(4y^2 - 9xt)},$$

$$c_0 = \frac{1}{4}\frac{x^3}{y^3} - \frac{3}{4}\frac{x}{y^2} + \frac{2}{x^3} - \frac{4xy - x^3 + 9t}{x^2(4y^2 - 9xt)}. \tag{49R}$$

The coefficients of G_L and G_R are related by reversing the roles of L_6 and $\tilde{L}$ in the transference process. This example is actually a simplification of the case in [20] where the more general function $q = 4/(9x\wp(y))$, where $g_2 = 0$, is taken as the solution of (37).

CONJECTURE. This example is true rank 3 and is also rank 3 bi-spectral [10]. Moreover, setting $\tilde{L} = K^2$, the centralizer of K is $C(K) = \mathbb{C}[K]$.

EXAMPLE (b). *Fake rank* 3 (*rank* 1) *vanishing rational solutions.* We follow the τ-function naming convention in [22] which is reiterated for convenience in Subsection 3.3.1 below:

(i) $U = 4\partial_x^2 \ln \tau_{(2,1)}$, $q = x^3/3$; $\tilde{\tilde{U}} = 0$.
(ii) $U = 4\partial_x^2 \ln \tau_{(2,2)}$, $q = x^3/12 + y^2/x$; $\tilde{\tilde{U}} = 4\partial_x^2 \ln \tau_{(1)}$.
(iii) $U = 4\partial_x^2 \ln \tau_{(3,2)}$, $q = \tau_{(1,1)}^{-1}(x^5/24 - xx_4 + x^3y/6 + xy^2/2)$;
$\tilde{\tilde{U}} = 4\partial_x^2 \ln \tau_{(1,1)}$.
(iv) $U = 4\partial_x^2 \ln \tau_{(3,3,1)}$, $q = \tau_{(3,1)}^{-1}(x^7/240 + \cdots + x^2x_5/2 + yx_5)$;
$\tilde{\tilde{U}} = 4\partial_x^2 \ln \tau_{(3,1)}$.
(v) $U = 4\partial_x^2 \ln \tau_{(3,3,2)}$, $q = \tau_{(3,1,1)}^{-1}(x^8/960 + \cdots + x_4^2)$; $\tilde{\tilde{U}} = 4\partial_x^2 \ln \tau_{(3,1,1)}$.
(vi) $U = 4\partial_x^2 \ln \tau_{(4,1)}$, $q = -\tau_{(1,1)}^{-1}(x^5/30 + x^3y/3 - x_5)$; $\tilde{\tilde{U}} = 4\partial_x^2 \ln \tau_{(1,1)}$.

This double pattern will be explained in Corollaries 6–8 below. The rational solutions (i)–(v) all have the form $\tau_{\text{KP}} = -\tau_{\text{Bq}}x_3 + \Phi(x, y, x_4, \ldots)$ resulting in the rational solution $q = \Phi/\tau_{\text{Bq}}$ of the s-KN equation and, since $F(\Phi/\tau_{\text{Bq}}) = 3/2\partial_x^2 \ln \tau_{\text{Bq}}$, the Bq solution $\frac{3}{8}\tilde{\tilde{U}} = F(\Phi/\tau_{\text{Bq}})$ with (x_3-independent) KP τ-function τ_{Bq}. The generation of these expressions for q, starting with the Boussinesq operator, is the subject of the next section.

3.2. The Formula for q

We have seen that two successive steps of transference applied to L_6 at the points $-P_0 = (x_3^{-2}, x_3^{-3})$ and $(0, 0)$ on the singular spectrum has produced the square of a Boussinesq Lax operator. The *explicit* implementation of the reverse of this process, namely generating KP solutions from Bq solutions, rests upon being able to obtain the correct factorization (flag of zero eigenfunctions) of the (square of the) Bq Lax operator. Similarly to the singular rank 2 case [9, 23], the solution of the s-KN Equation (37) can be constructed as a suitable Wronskian of the Boussinesq zero eigenfunctions.

Let ϕ_0, ϕ_1 be (independent) zero eigenfunctions of the Bq operator

$$K = D^3 + \tfrac{3}{8}\tilde{\tilde{U}}D + \tfrac{3}{8}D\tilde{\tilde{U}} + \tilde{\tilde{a}}_3, \quad \text{where } \tfrac{3}{8}\tilde{\tilde{U}}_y = (\tilde{\tilde{a}}_3)_x$$

and factor K as

$$K = (D - v_0 - v_1)(D + v_1)(D + v_0), \tag{50}$$

where $v_0 = \partial_x \ln(1/\phi_0)$ and $v_1 = \partial_x \ln[\phi_0/W(\phi_0, \phi_1)]$ (see [32, Proposition 4.10).

Let ϕ_2 be the third (independent) zero eigenfunction of K chosen so that $W(\phi_0, \phi_1, \phi_2) = \pm 1$. Assume that $\partial_y \phi_i = K_+^{2/3} \phi_i = (D^2 + \tilde{\tilde{U}}/2)\phi_i$ for $i = 0, 1$. Let ψ_0, ψ_1, ψ_2, be the remaining zero eigenfunctions of K^2 which satisfy $K\psi_i = \phi_i$ for $i = 0, 1, 2$. Consider the *flag* $\mathcal{F}_3$: $\{0\} \subset \{\psi_1\} \subset \{\psi_1, \phi_0\} \subset \{\psi_1, \phi_0, \phi_2\} \subset \{\psi_1, \phi_0, \phi_2, \psi_0\} \subset \{\psi_1, \phi_0, \phi_2, \psi_0, \phi_1\} \subset \{\psi_1, \phi_0, \phi_2, \psi_0, \phi_1, \psi_2\}$, and the associated factorization of K^2:

$$K^2 = (D + \nu_5)(D + \nu_4)(D + \nu_3)(D + \nu_2)(D + \nu_1)(D + \nu_0), \tag{51}$$

where $\nu_i = \partial_x \ln[W_i/W_{i+1}]$, $i = 0, \ldots, 5$, with W_i denoting the Wronskian determinant of the first i eigenfunctions of the flag $\mathcal{F}_3$. Note that W_6 is constant because of the normalization of K^2 and that, trivially, $\sum_i \nu_i = 0$ (cf. (50)). The claim now is that the factorization (51) gives the gcd G and quotient Q_1, both at the cusp, for the commuting operators $(L, L_+^{3/2})$ of the s-KN equation (Theorem 5).

PROPOSITION 4. *The factorization* (51), *by the flag* $\mathcal{F}_3$, *of the square of the Bq Lax operator* K *is given by* $K^2 = G \cdot Q_1$ *where* $L = Q_1 \cdot G$ *is the Lax operator for the s-KN Equation* (37). *Moreover,*

$$q = W(\psi_1, \phi_0, \phi_2), \tag{52}$$

gives the explicit correspondence between the coefficients of K *and* L, *so that, necessarily,* $\frac{3}{8}\tilde{\tilde{U}} = F(q)$ *and* $\tilde{\tilde{a}}_3 = G(q)$.

Thus, the rank of a KP solution produced by transference on the singular spectrum is determined by the underlying Boussinesq operator as follows.

COROLLARY 4. *Any Bq operator* K *for which the centralizer* $\mathcal{C}(K)$ *is rank* 3 *generates a (true) rank* 3 *KP solution of the form* (39). *This (rank* 3*) KP solution is obtained from* K *by transference of* K^2, *first at* $(0, 0)$ *and then at* $P_0 = (x_3^{-2}, -x_3^{-3})$, *on the singular spectrum* (38) *with* (52) *giving the explicit formula for* $q(x, y)$.

Remark 10. The ordering of the eigenfunctions ϕ_i used here is that which is produced by first-order Darboux transformations of the Boussinesq solutions as in Subsection 3.3.3 below.

We now show explicitly how some of Examples (b)(i)–(vi) fit into this framework.

EXAMPLE (b) (continued).

(i) Consider the Bq operator $K = D^3$ for which $\tilde{\tilde{U}} = \tilde{a}_3 = 0$. Take for the eigenfunctions ϕ_0 and ϕ_1, 1 and x, respectively, so that the flag $\mathcal{F}_3$ is given by (top subspace) $\{x^4/4!, 1, x^2/2! + y, x, x^3/3!, x^5/5! + yx^3/3!\}$. We then get that $q = W(\psi_1, \phi_0, \phi_2) = x^3/3$. The corresponding factorization (51) is

$$\begin{aligned} K^2 &= D^6 = (D + 4/x)(D - 1/x)D \cdot D(D + 1/x)(D - 4/x) \\ &= \left(D^3 + \frac{3}{x}D^2 - \frac{3}{x^2}D\right) \cdot \left(D^3 - \frac{3}{x}D^2 + \frac{3}{x^2}D\right) \\ &= x^{-3}\delta(\delta - 2)(\delta + 2) \cdot x^{-3}\delta(\delta - 2)(\delta - 4) \\ &= G \cdot Q_1, \end{aligned} \tag{53}$$

where $\delta = xD$. Forming the product $L = Q_1 \cdot G$ we get

$$\begin{aligned} L &= D^6 - 18x^{-2}D^4 + 72x^{-3}D^3 - 144x^{-4}D^2 + 144x^{-5}D \\ &= x^{-6}\delta(\delta - 2)(\delta - 3)(\delta - 5)(\delta - 7)(\delta + 2), \end{aligned} \tag{54}$$

which is the (Euler) Lax operator for the s-KN solution $q(x, y) = x^3/3$. The coefficients of L in (54) are just those in (42) computed for this particular q. Transference at (x_3^{-2}, x_3^{-3}) applied to L now gives the operator L_6 with KP solution $U = 4\partial_x^2 \ln \tau_{(2,1)}$ which is just Example (b)(i).

(ii) Consider the Bq operator with $\tilde{\tilde{U}} = 4\partial_x^2 \ln \tau_{(1)}$, i.e.

$$K = D^3 - \frac{3}{x^2}D + \frac{3}{x^3} = x^{-3}(\delta - 1)(\delta + 1)(\delta - 3), \tag{55}$$

where $\delta = xD$. We take for the basis ϕ_i the zero eigenfunctions $\{3/(8x), x - 2y/x, x^3/3! + 2xy/3\}$ which have unit Wronskian. The corresponding zero eigenfunctions ψ_i of K^2 which satisfy $K\psi_i = \phi_i$ are then

$$\{-x^2/8, x^4/15 + 2yx^2/3, x^6/378 + 2yx^4/45\}.$$

One then computes that $q = W(\psi_1, \phi_0, \phi_2) = x^3/12 + y^2/x$, which is Example (b)(ii) above. The factorization analagous to (53) can also be computed for this example and transference then results in the KP solution $U = 4\partial_x^2 \ln \tau_{(2,2)}$.

(iii) Consider the Bq operator K with $\tilde{\tilde{U}} = 4\partial_x^2 \ln \tau_{(1,1)}$, i.e.

$$\begin{aligned} K &= D^3 + 3(Q^{-1} - x^2Q^{-2})D + 3(x^3Q^{-3} - xQ^{-2}) \\ &= (D - xQ^{-1} + 1/x)(D - 1/x)(D + xQ^{-1}), \end{aligned} \tag{56}$$

where $Q = \tau_{(1,1)} = x^2/2 - y$. This example is the first one with nontrivial y-dependence in K. For the basis ϕ_i of zero eigenfunctions we take $\{Q^{-1}, Q +$

$2(x_4 - y^2)Q^{-1}, 1/2(x^5/20 - xy^2)Q^{-1}\}$, which have unit Wronskian, and as the corresponding basis ψ_i:

$$\left\{\left(\frac{x^3}{6} - xy\right)Q^{-1}, \frac{1}{2}\left(\frac{x^7}{84} - \frac{2x^5y}{15} + \frac{y^2x^3}{3}\right)Q^{-1} + \right.$$
$$\left. +2(x_4 - y^2)\left(\frac{x^3}{6} - xy\right)Q^{-1}, \frac{1}{2}\left(\frac{x^8}{1920} + \frac{x^6y}{80} - \frac{x^4y^2}{24}\right)Q^{-1}\right\}.$$

One then computes that $q = W(\psi_1, \phi_0, \phi_2) = (x^5/24 + x^3y/6 + xy^2/2 - xx_4)Q^{-1}$ which is precisely Example (b)(iii) above.

(vi) This case can be constructed with the same eigenfunctions as in (b)(iii) since K is again given by (56). For the basis ϕ_i we take $\{Q^{-1}, 1/2(x^5/20 - y^2x)Q^{-1} + x_5Q^{-1}/2, Q\}$, which have Wronskian -1 and for the corresponding basis ψ_i:

$$\left\{\left(\frac{x^3}{6} - xy\right)Q^{-1}, \frac{1}{2}\left(\frac{x^8}{1920} + \frac{x^6y}{80} - \frac{x^4y^2}{24}\right)Q^{-1} + \right.$$
$$\left. +\frac{x_5}{2}\left(\frac{x^3/6 - xy)Q^{-1}, 1/2(x^7}{84} - \frac{2x^5y}{15} + \frac{y^2x^3}{3}\right)Q^{-1}\right\},$$

from which we compute $q = W(\psi_1, \phi_0, \phi_2) = (x^5/30 + x^3y/3 - x_5)Q^{-1}$. This is the q from Example (b)(vi) above.

All the Schur–Sato KP solutions of Example (b) can be generated in this manner and are clearly rank 1 (fake rank 3) – they are all Darboux conjugates of D^3 in the spirit of Adler–Moser (see Subsection 3.3 below).

3.3. First-order Rational Darboux Transformations for Boussinesq and Fake Rank 3 KP Solutions

We now briefly describe the generation of the vanishing rational solutions of the Bq equation (47) by using first-order Darboux transformations, beginning with the operator $K = D^3$. These rational solutions are described by x_3-independent KP τ-functions and are parametrized by certain partitions $(n_1, n_2, \ldots, n_k)$ of certain integers $\sum_i n_i$, with corresponding Bq solution $\tilde{\tilde{U}} = 4\partial_x^2 \ln \tau$, $\tau \equiv \tau_{(n_1, n_2, \ldots, n_k)}$. The problem considered here is to precisely determine the Bq partitions (those KP partitions independent of x_{3i}, $i \geqslant 1$) and to draw the directional road map between these solutions as determined by first-order Darboux transformations (see Figure 1 below). In the process, we generate corresponding KP partitions, whose τ-functions are *linear* in x_3, and hence, are the result of two third order Darboux transformations (transferences) of the underlying Bq solutions. Coincidentally, this process also generates rational solutions of the s-KN Equation (37) (see Corollaries 6 and 7 below).

Before describing the Bq and KP partitions, we need some simple facts about τ-functions and their manipulation.

3.3.1. *Preliminaries on Tau-Functions*

Recall the standard notation for τ-functions: given the partition $(n_1, n_2, \ldots, n_k)$, $n_i > 0$, $n_1 \geqslant n_2 \geqslant \cdots \geqslant n_k$, and the Schur polynomials p_j with generating function

$$\exp\left(\sum_{i=1}^{\infty} x_i k^i\right) = \sum_{j=0}^{\infty} p_j k^j,$$

we denote by $\tau_{(n_1,n_2,\ldots,n_k)}$ the τ-function given by the Wronskian determinant $\det(p_{n_i-i+j})_{1\leqslant i,j,\leqslant k}$. Invariance of the set of Schur polynomials under differentiation leads to some useful facts about τ-functions.

PROPOSITION 5. (i) *The Schur polynomials satisfy* $\partial_{x_\ell} p_j = p_{j-\ell}$, *where we interpret* $p_i \equiv 0$ *if* $i < 0$.

(ii) *For the partition* $(n_1, n_2, \ldots, n_k)$, $n_i > 0$, $n_1 \geqslant n_2 \geqslant \cdots \geqslant n_k$, *of* $\sum_i n_i$, *then*

$$\partial_{x_\ell} \tau_{(n_1,n_2,\ldots,n_k)} = \sum_{j=1}^{k} \tau_{(n_1,\ldots,n_j-\ell,\ldots,n_k)}. \tag{57}$$

(iii) *The* τ*-function* $\tau_{(n_1,n_2,\ldots,n_k)}$ *is independent of* x_ℓ *if for each* j, $1 \leqslant j \leqslant k$, *satisfying* $n_j + k - j \geqslant \ell$, *there exists a* $j' > j$ *such that*

$$n_j - n_{j'} + j' - j - \ell = 0. \tag{58}$$

Remark 10. The x_ℓ-independence in (iii) is easy to check explicitly for any given partition. Namely, for each row of the (Wronskian) determinant that is the τ-function for the partition, subtracting ℓ from the index of every Schur polynomial in that row should produce a row identical to one below it, and in this way every term in the sum of (57) vanishes. Hence, (iii) is a consequence of (i) and (ii).

3.3.2. *A Sequence of KP and Bq Partitions*

Using Proposition 5, it is easy enough to verify the following assertions which describe, by virtue of (39), a sequence of corresponding (fake rank 3) vanishing rational solutions to the Bq, s-KN, and KP equations.

COROLLARY 5. *The KP partitions* $(n, n-1, n-1, \ldots, 2, 2, 1, 1)$ *of the integer* n^2 *and* $(n, n, n-1, n-1, \ldots, 2, 2, 1, 1)$ *of the integer* $n^2 + n$ *have* τ*-functions independent of* x_{3j} *for* $j \geqslant 1$ *and are therefore Bq partitions.*

Sitting above each of these Bq partitions (denoted $Bq(n)$) are two KP partitions (denoted $KP^{\pm}(n)$) whose τ-functions are linear in x_3.

COROLLARY 6. *The (pairwise) corresponding KP partitions, denoted by* $KP^{-}(n)$, $(n+1,n+1,n-1,\ldots,2,2,1,1)$ *of* n^2+3 *and* $(n+2,n+1,n-1,n-1,\ldots,2,2,1,1,)$ *of* n^2+n+3, *and the corresponding KP partitions, denoted by* $KP^{+}(n)$, $(n+3,n-1,n-1,\ldots,2,2,1,1)$ *of* n^2+3 *and* $(n+3,n,n-1,n-1,\ldots,2,2,1,1)$ *of* n^2+n+3 *have* τ*-functions which satisfy*

$$\tau_{KP^{\pm}(n)} = \pm x_3 \tau_{Bq(n)} + \Phi_n^{\pm}, \tag{59}$$

where the polynomials $\Phi_n^{\pm}$ *are independent of* x_{3j} *for* $j \geqslant 1$.

COROLLARY 7. *With the notation of Corollary 6, the rational functions* $q_n^{\pm} = \mp\Phi_n^{\pm}/\tau_{Bq(n)}$ *are solutions of the s-KN Equation* (37). *Moreover, there exists a correspondence through singular (fake) rank* 3 *transference between the solutions* $U_n^{\pm} = 4\partial_x^2 \ln \tau_{KP^{\pm}(n)}$ *of KP,* $q_n^{\pm}$ *of s-KN, and* $\tilde{\tilde{U}}_n = 4\partial_x^2 \ln \tau_{Bq(n)}$ *of Bq, i.e. there exists the correspondence*

$$U_n^{\pm} \longleftrightarrow q_n^{\pm} \longleftrightarrow \tilde{\tilde{U}}_n,$$

where the arrows represent third order transference/Darboux.

Remark 11. Since the partitions $Bq(n)$ in Corollary 5 are related, for different n, by first-order Darboux at the Bq level (see the right of Figure 1), these partitions give an infinite sequence of Bq, s-KN, and KP solutions all connected by Darboux. This picture is reminiscent of rank 2, the only complication being the added dimension at the Bq level (see Subsections 3.3.3 and 3.3.4).

3.3.3. *First-Order Darboux for Bq*

Consider the process of generating all the vanishing rational solutions of Bq by first-order Darboux, starting from the trivial solution with Lax operator $K = D^3$ and zero eigenfunctions $\phi_0 = 1$, $\phi_1 = x$ and $\phi_2 = x^2/2 + y$. It will be helpful to visualize this process by examining Figure 1. For notational convenience, we temporarily drop the double tilde used above for Bq solutions.

Suppose at some stage of this Darboux process we are at $K = D^3 + \frac{3}{4}UD + \frac{3}{8}U' + a_3$ with $U = 4\partial_x^2 \ln \tau$, $a_3 = 3/2\partial_x\partial_y \ln \tau$, and zero eigenfunctions ϕ_i, $i = 0, 1, 2$, satisfying $\partial_y \phi_i = K_+^{2/3}\phi_i = (D^2 + U/2)\phi_i$. These (rational) eigenfunctions all have the form $\phi_i = \tau_i/\tau$ for certain Bq τ-functions τ_i. Fix i, and consider the first order Darboux transformation using ϕ_i. We pass from $K = (D^2 + \phi_i'/\phi_i D + b)(D - \phi_i'/\phi_i)$ to $\tilde{K} = (D - \phi_i'/\phi_i)(D^2 + \phi_i'/\phi_i D + b)$ with $\tilde{U} = 4\partial_x^2 \ln \tau_i$ and $\tilde{a}_3 = 3/2\partial_x\partial_y \ln \tau_i$ (these formulae follow since $\tilde{U} =$

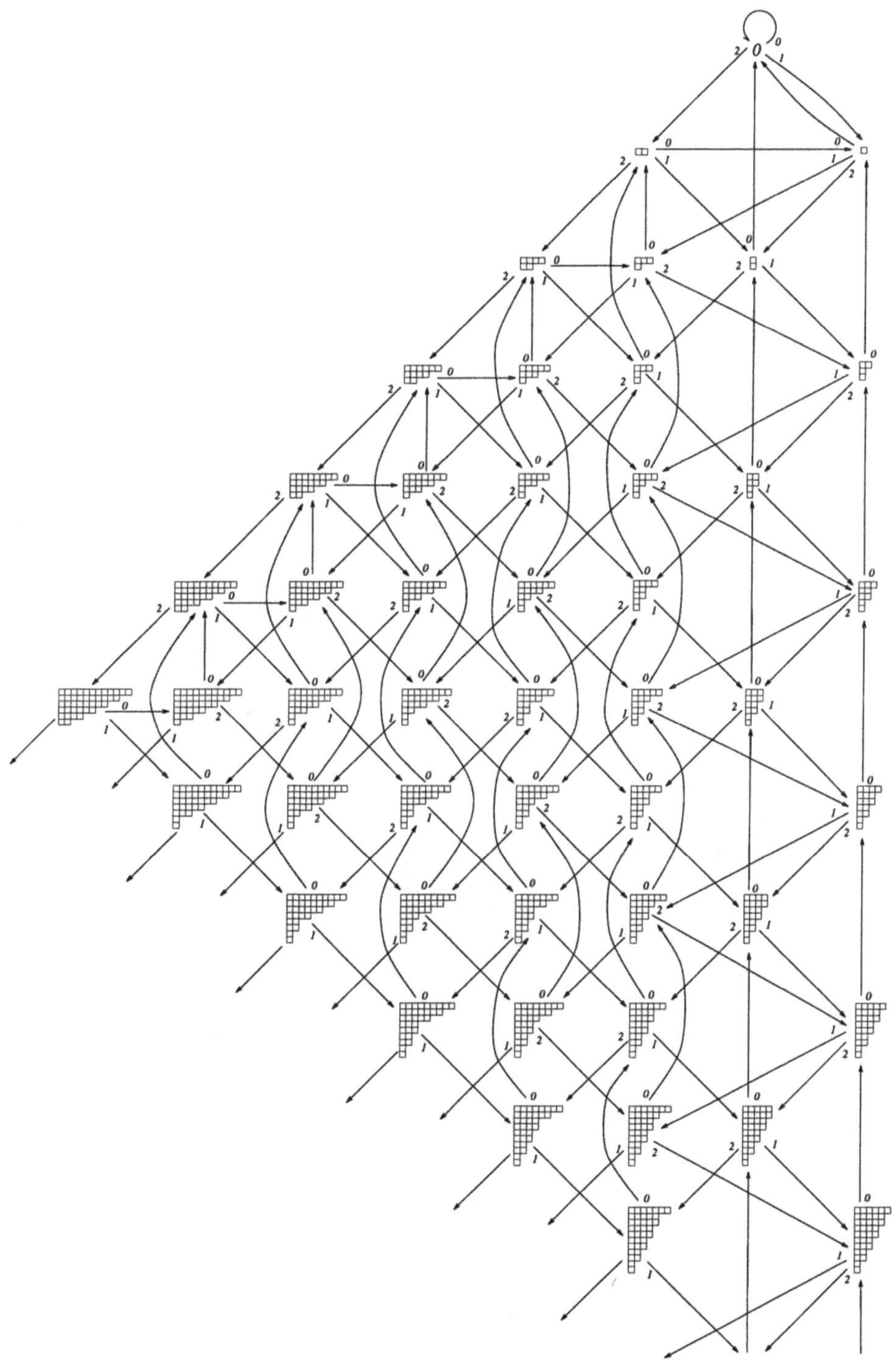

Fig. 1. Vanishing rational Boussinesq solutions generated by first-order Darboux transformations.

$4\partial_x^2 \ln(\tau\phi_i)$ and, provided $\partial_y\phi_i = K_+^{2/3}\phi_i$, $\tilde{a}_3 = 3/2\partial_x\partial_y \ln(\tau\phi_i))$. Two of the zero eigenfunctions $\tilde{\phi}_j$ of $\tilde{K}$, for $j \neq i$, are given directly by $\tilde{\phi}_j = (D - \phi_i'/\phi_i)\phi_j$. Let $\tilde{\phi}_j = \tilde{\tau}_j/\tau_i$, then we obtain the relation

$$\tilde{\tau}_j\tau = \tau_i\tau_j' - \tau_i'\tau_j, \tag{60}$$

which is an analog of the Adler–Moser differential recursion for KdV. Hence, (60) can be used to generate Bq τ-functions starting from the case $\tau = 1$, $\tau_0 = 1$, $\tau_1 = x$ and $\tau_2 = x^2/2 + y$ (the zero eigenfunctions of D^3).

The zero eigenfunction $\tilde{\phi}_i$ (and hence $\tilde{\tau}_i = \tau_i\tilde{\phi}_i$), can be obtained by resorting to the factorization (50) of $\tilde{K}$ given by the already known zero eigenfunctions $\tilde{\phi}_j$, $j \neq i$. This gives

$$\tilde{\phi}_i = \tilde{\phi}_j \int W/\tilde{\phi}_j^2 \int \tilde{\phi}_j/W^2 + c_0\tilde{\phi}_j + c_1\tilde{\phi}_k, \quad i \neq j \neq k, \tag{61}$$

where $W = \tilde{\phi}_j\tilde{\phi}_k' - \tilde{\phi}_j'\tilde{\phi}_k$ is the Wronskian of the already known eigenfunctions and the c_r are constants with respect to x.

Remark 12. To obtain for $\tilde{\tau}_i = \tau_i\tilde{\phi}_i$ the correct Bq partition involving all the higher KP variables, it is necessary to take the c_r as nonzero functions of x_r for $r \geqslant 2$ and $r \not\equiv 0 \pmod 3$. In other words, knowing two eigenfunctions in (61) will not give the correct 'flow outs' in the higher KP variables as computed from the partition unless the c_r are suitably chosen to depend on these higher variables. This fact is just a mechanistic computational detail which can be ignored without loss of continuity.

With this general description in place, we can now draw the road map of Darboux transformations among the partitions which characterize the vanishing rational solutions of the Bq equation. This is done in Figure 1. The tail of each arrow in Figure 1 carries a label i indicating which eigenfunction (ϕ_i) is used to Darboux from the source partition to the destination partition connected by (resp.) the tail and head of the arrow. In particular, for any partition in the figure, the eigenfunction ϕ_i, for the Bq operator corresponding to that partition, can be identified as the ratio of the τ-functions for the partitions at the head and tail of the arrow with label i. Looking at such a Darboux road map, the following encompassing result of Corollary 5 is clear.

LEMMA 7. *Modulo the substitution $y \to -y$, the Bq partitions are enumerated as the partitions $(n+2k, \ldots, n+2, n, n-1, n-1, \ldots, 1, 1)$ of the integer $n^2 + k(n+k+1)$ and $(n+2k, \ldots, n+2, n, n, n-1, n-1, \ldots, 1, 1)$ of the integer $n^2 + (n+k)(k+1)$, both for $n \geqslant 1$ and $k \geqslant 0$.*

Remark 13. The integers $n^2 + k(n+k+1)$ and $n^2 + (n+k)(k+1)$, for $n \geqslant 1$ and $k \geqslant 0$, are the analogs of the KdV 'triangular numbers' for Bq. The cases $k = 0$ are those of Corollary 5 (cf. also the Remark 4.4.2 in [26]).

Remark 14. The general pattern of Darboux transformations (Figure 1) is that starting from the partitions of Corollary 5, a new destination partition is created by adding a new first row with two more plaquettes than the first row of the source partition. The τ-functions appearing in (60) are just those connected by arrows in the figure.

3.3.4. *Fake Rank 3 KP Partitions Generated From All Bq Partitions*

As in Item 3.3.2, Lemma 7 implies an analog of Corollary 6 which describes a pair of KP and s-KN solutions sitting above every Bq solution given by the partitions of Lemma 7 (denoted by $Bq(n,k)$).

LEMMA 8. *Corresponding to the partitions $Bq(n,k)$ of Lemma 7, the KP partitions, denoted by $KP^-(n,k)$, $(n+2k,\dots,n+2,n+1,n+1,n-1,\dots,1,1)$ of the integer $n^2+k(n+k+1)+3$ and $(n+2k,\dots,n+2,n+2,n+1,n-1,n-1,\dots,1,1)$ of the integer $n^2+(n+k)(k+1)+3$, and the KP partitions, denoted $KP^+(n,k)$, $(n+2k+3,\dots,n+2,n,n-1,n-1,\dots,1,1)$ of the integer $n^2+k(n+k+1)+3$ and $(n+2k+3,\dots,n+2,n,n,n-1,n-1,\dots,1,1)$ of the integer $n^2+(n+k)(k+1)+3$, all for $n\geqslant 1$ and $k\geqslant 0$, have τ-functions which satisfy*

$$\tau_{KP^\pm(n,k)} = \pm x_3\tau_{Bq(n,k)} + \Phi^\pm_{n,k}, \tag{62}$$

where the polynomials $\Phi^\pm_{n,k}$ are independent of x_{3j} for $j\geqslant 1$.

The analog of Corollary 7 above also applies.

LEMMA 9. *With the notation of Lemma 8, the rational functions*

$$q^\pm_{n,k} = \mp\Phi^\pm_{n,k}/\tau_{Bq(n,k)}$$

are solutions of the s-KN Equation (37). Moreover, there exists a correspondence through singular (fake) rank 3 transference between the solutions $U^\pm_{n,k} = 4\partial_x^2\ln\tau_{KP^\pm(n,k)}$ of KP, $q^\pm_{n,k}$ of s-KN, and $\tilde{\tilde{U}}_{n,k} = 4\partial_x^2\ln\tau_{Bq(n,k)}$ of Bq, i.e. there exists the correspondence

$$U^\pm_{n,k} \longleftrightarrow q^\pm_{n,k} \longleftrightarrow \tilde{\tilde{U}}_{n,k},$$

where the arrows represent third-order transference/Darboux.

Again $k=0$ reproduces the cases of Corollaries 7 and 8. Note that the rational solutions of Lemma 9 satisfy the identity $F(q^\pm_{n,k}) = 3/2\partial_x^2\ln\tau_{Bq(n,k)}$, where $F(q)$ is as defined in (40).

Acknowledgement

The second author is grateful for the support of NSF Grant DMS-9404087, while both authors would like to thank Shane Latham whose XFIG skills created the diagram in Figure 1.

References

1. Adler, M. and Moser, J.: On a class of polynomials connected with the Korteweg–de Vries equations, *Comm. Math. Phys.* **61** (1978), 1–30.
2. Burchnall, J. L. and Chaundy, T. W.: Commutative ordinary differential operators, *Proc. London Math. Soc.* **211** (1923), 420–440.
3. Burchnall, J. L. and Chaundy, T. W.: Commutative ordinary differential operators, *Proc. Royal Soc. London (A)* **118** (1928), 557–583.
4. Burchnall, J. L. and Chaundy, T. W.: Commutative ordinary differential operators, II. The identity $P^n = Q^m$, *Proc. Royal Soc. London (A)* **134** (1932), 471–485.
5. Cherednik, I. V.: Differential equations for the Baker–Akhiezer functions of algebraic curves, *Funct. Anal. Appl.* **12** (1978), 195–203.
6. Dehornoy, P.: Opérateurs différentiels et courbes elliptiques, *Compositio Math.* **43** (1981), 71–99.
7. Dixmier, J.: Sur les Algebras de Weyl, *Bull. Soc. Math. France* **96** (1968), 209–242.
8. Dorfman, I. Ya. and Nijhoff, F. W.: On a (2+1)-dimensional version of the Krichever–Novikov equation, Clarkson University, Preprint INS, No. 167.
9. Drinfeld, V. G. and Sokolov, V. V.: On equations related to the Korteweg–de Vries equation, *Soviet Math. Dokl.* **32** (1985), 361–365.
10. Duistermaat, J. J. and Grünbaum, F. A.: Differential equations in the spectral parameter, *Comm. Math. Phys.* **103** (1986), 177–240.
11. Ehlers, F. and Knörrer, H.: An algebro-geometric interpretation of the Bäcklund transformation for the Korteweg–de Vries equation, *Comment. Math. Helvetici* **57** (1982), 1–10.
12. Grinevich, G. P. G.: Rational solutions for the equation of commutation of differential operators, *Funct. Anal. Appl.* **16** (1982), 19–24.
13. Grünbaum, F. A.: Commuting pairs of linear ordinary differential operators of orders four and six, *Physica* **31D** (1988), 424–433.
14. Grünbaum, F. A.: The Kadomtsev–Petviashvili equation: an elementary approach to the 'rank 2' solutions of Krichever and Novikov, *Phys. Lett. A* **139** (1989), 146–150.
15. Grünbaum, F. A.: Darboux's method and some 'rank two' explicit solutions of the KP equation, in A. Degasperis, A. P. Fordy and M. Lakshmanan (eds), *Nonlinear Evolution Equations: Integrability and Spectral Methods*, Manchester University Press, Manchester, 1990, pp. 271–277.
16. Ince, E. L.: *Ordinary Differential Equations*, Dover Publications, New York, 1956.
17. Krichever, I. M.: Commutative rings of ordinary differential operators, *Funct. Anal. Appl.* **12** (1978), 175–185.
18. Krichever, I. M. and Novikov, S. P.: Holomorphic fiberings and nonlinear equations, finite zone solutions of rank 2, *Sov. Math. Dokl.* **20** (1979), 650–654.
19. Krichever, I. M. and Novikov, S. P.: Holomorphic bundles over algebraic curves and nonlinear equations, *Soviet Math. Surveys* **35** (1980), 53–79.
20. Latham, G. A.: Solutions of the KP equation associated to rank 3 commuting differential operators over a singular elliptic curve, *Physica* **D41** (1990), 55–66.
21. Latham, G. A.: The computational approach to commuting ordinary differential operators of orders six and nine, *J. Austral. Math. Soc., Series B* **35** (1994), 399–419.
22. Latham, G. A. and Previato, E.: Higher rank Darboux transformations, in N. M. Ercolani, I. R. Gabitov, C. D. Levermore and D. Serre (eds), *Singular Limits of Dispersive Waves*, NATO ASI Series, Series B: Physics, Vol. 320, Plenum, New York, 1994, pp. 117–134.

23. Latham, G. A. and Previato, E.: KP solutions generated from KdV by 'rank 2' transference, submitted to *Physica D*.
24. Mokhov, I. O.: Commuting differential operators of rank 3 and nonlinear equations, *Math. USSR Izvestiya* **35** (1990), 629–655.
25. Mumford, D.: An algebro-geometric construction of commuting operators and of solutions to the Toda lattice equation, KdV equation and related nonlinear equations, in M. Nagata (ed.), *Proc. Int. Symp. Alg. Geom.* Kinokuniya, Tokyo, 1978, pp. 115–153.
26. Previato, E.: 70 years of spectral curves: 1923-1993, in *Proc. CIME 1993*, Lecture Notes in Physics, Springer-Verlag (to appear).
27. Previato, E. and Wilson, G.: Vector bundles over curves and solutions of the KP equations, in *Proc. Sympos. Pure Math.* **49** (1989), 553–569.
28. Previato, E. and Wilson, G.: Differential operators and rank two bundles over elliptic curves, *Compositio Math.* **81** (1992), 107–119.
29. Segal, G. B. and Wilson, G.: Loop groups and equations of KdV type, *Publ. Math. I.H.E.S.* **61** (1985), 5–65.
30. Sokolov, V. V.: On the Hamiltonian property of the Krichever–Novikov equation, *Dokl. Akad. Nauk SSSR* **277** (1984), 48–50; English translation in *Sov. Math. Dokl.* **30** (1984), 44–46.
31. Svinolupov, S. I., Sokolov, V., and Yamilov, R.: On Bäcklund transformations for integrable evolution equations, *Soviet Math. Dokl.* **28** (1983), 165–168.
32. Wilson, W. G.: Algebraic curves and soliton equations, in E. Arbarello, C. Procesi and E. Strickland (eds), *Geometry Today*, Birkhäuser, Boston, 1985, pp. 303–329.

Acta Applicandae Mathematicae **39:** 435–443, 1995.

Moment Problem of Hamburger, Hierarchies of Integrable Systems, and the Positivity of Tau-Functions

YOSHIMASA NAKAMURA
Applied Mathematics Laboratory, Doshisha University, Tanabe, Kyoto 610-03, Japan

and

YUJI KODAMA
Department of Mathematics, The Ohio State University, Columbus, OH 43210, U.S.A.

(Received: 15 September 1994)

Abstract. A moment problem of Hamburger is studied to find a parametric Stieltjes measure from given moments. It is shown that if a deformation, or a dynamics, of moments is governed by a hierarchy of a Kac–van Moerbeke system, then the Stieltjes measure can be constructed explicitly by integrating a hierarchy of Moser's nonlinear dynamical system. The positivity of tau-functions is related to the existence of the Stieltjes measure at a deep level.

Mathematics Subject Classifications (1991): Primary 58F07; Secondary 34A05.

Key words: Hamburger moment problem, Moser's system, Kac–van Moerbeke system, tau functions.

1. Introduction

A hierarchy of the finite nonperiodic Toda equation has recently been presented [11] which describes a motion of zeros of certain rational functions. The dynamics of residues at eigenvalues of a Jacobi matrix, which appears in a Lax representation of the Toda equation, is governed by a dynamical system introduced by Moser [10]. The Markov parameters of rational functions are shown [12] to satisfy a dynamical system of Kac and van Moerbeke [7]. By linearizing the Kac–van Moerbeke system a tau-function of the Toda equation is found to be a symmetric Wronskian matrix [12].

On the other hand it is well-known [3, 13] that the inverse spectral problem of the Jacobi operator is closely related to the theory of orthogonal polynomials. There is a theorem by Markov, see [13], that a rational function of degree n defined by orthogonal polynomials converges to a Stieltjes integral as $n \to \infty$, where its Stieltjes measure determines the orthogonal polynomials. It should be noted that the moments associated with the measure are essentially equivalent to the Markov parameters of rational functions. On the other hand Kato and

Aomoto [8] showed that a linear evolution equation of spectral density in a Stieltjes integral induces a Lax type equation for a Jacobi operator.

It seems to be worthwhile to shed new light on these subjects from the viewpoints of integrable systems and tau-functions. In this paper we consider a restricted version of the moment problem of Hamburger [1]. The purpose is to design an algorithm to find a parametric Stieltjes measure from given moments. To this end new hierarchies of Moser's nonlinear dynamical system and the Kac–van Moerbeke system will be introduced. Moser's system and the Kac–van Moerbeke system describe the dynamics of Stieltjes measure and moments, respectively. A necessary and sufficient condition for the existence of solutions of the moment problem was already given by Toeplitz, see [1]. It will be shown in this paper that the condition is essentially equivalent to positivity of tau-functions of the finite and infinite Toda equations. The discussions here are based on the classical theorems of Markov, Toeplitz and Szegö and some recent developments in integrable systems.

2. Preliminaries and Problem

Let $\alpha(\lambda)$ be a fixed nondecreasing function with infinitely many points of increase in the infinite interval $(-\infty, \infty)$. Let $\{p_n(\lambda)\}$ be a set of orthogonal polynomials determined by $\alpha(\lambda)$, namely,

$$\int_{-\infty}^{\infty} p_n(\lambda) p_m(\lambda)\, \mathrm{d}\alpha(\lambda) = \delta_{n,m}, \tag{2.1}$$

where n is the degree of $p_n(\lambda)$ and coefficient of λ^n of $p_n(\lambda)$ is positive. We suppose that the moments

$$\langle \lambda^k \rangle = \int_{-\infty}^{\infty} \lambda^k\, \mathrm{d}\alpha(\lambda), \quad k = 0, 1, \ldots \tag{2.2}$$

exist. The moment problem of Hamburger is an 'inverse' problem to find a non-decreasing function $\alpha(\lambda)$ from a given set of moments $\{\langle \lambda^k \rangle\}$.

Next let us consider a rational function

$$f_n(\lambda) = \frac{p_{n-1}(\lambda)}{p_n(\lambda)} \tag{2.3}$$

for some n. Here we briefly sketch the facts from the book [13] by Szegö. The polynomial $p_n(\lambda)$ has n real zeros, denoted by $\{\lambda_j\}$, which are mutually distinct and located in $(-\infty, \infty)$. We set $\lambda_1 < \lambda_2 < \cdots < \lambda_n$. Moreover, each interval $[\lambda_j, \lambda_{j+1}]$ contains exactly one zero of $p_{n-1}(\lambda)$ for $j = 1, \ldots, n-1$. The rational function $f_n(\lambda)$ of degree n always admits a partial fraction expansion

$$f_n(\lambda) = \sum_{j=1}^{n} \frac{\ell_j}{\lambda - \lambda_j}, \quad \ell_j > 0, \tag{2.4}$$

where the residues $\{\ell_j\}$ are called Christoffel numbers. Thus there exists a step function $\alpha_n(\lambda)$ satisfying

$$f_n(\lambda) = \int_{-\infty}^{\infty} \frac{d\alpha_n(\mu)}{\lambda - \mu}. \tag{2.5}$$

The nondecreasing function $\alpha_n(\lambda)$ can be expressed by $d\alpha_n(\lambda) = \sum_{j=1}^{n} \ell_j \delta(\lambda - \lambda_j) d\lambda$. Hence the associated moments $\langle\lambda^k\rangle_n$ defined by $\langle\lambda^k\rangle_n = \int_{-\infty}^{\infty} \lambda^k \, d\alpha_n(\lambda)$ take the following form

$$\langle\lambda^k\rangle_n = \sum_{j=1}^{n} \lambda_j^k \ell_j, \quad k = 0, 1, \ldots. \tag{2.6}$$

It should be noted that $\langle\lambda^k\rangle_n$ are equivalent to Markov parameters [13] of rational functions derived in [12], namely, $f_n(\lambda) = \sum_{k=0}^{\infty} \langle\lambda^k\rangle_n / \lambda^{k+1}$. Markov's limit theorem [13] says that $\alpha_n(\lambda) \to \alpha(\lambda)$ as $n \to \infty$ for an arbitrary point λ in the complex plane cut along $(-\infty, \infty)$. This presents us with the Padé approximation of a Stieltjes integral

$$\int_{-\infty}^{\infty} \frac{d\alpha(\mu)}{\lambda - \mu} \tag{2.7}$$

in terms of the rational function $f_n(\lambda)$.

Let us introduce constants r_j^2 by

$$\sum_{j=1}^{n} \frac{r_j^2}{\lambda - \lambda_j} = \frac{1}{\sum_{j=1}^{n} \ell_j} f_n(\lambda), \quad r_j > 0. \tag{2.8}$$

Obviously, r_j are normalized as $\sum_{j=1}^{n} r_j^2 = 1$.

Markov (see [13]) also proved a theorem concerning the dependence of zeros of the orthogonal polynomials $\{p_n(\lambda)\}$ on a deformation parameter $t \in \mathbb{R}$ which appears in the Stieltjes measure $\alpha(\lambda) = \alpha(\lambda; t)$. Various applications of this theorem are discussed in [13]. In the subsequent sections we consider the following restricted version of moment problem of Hamburger.

PROBLEM. Let t be a set of parameters

$$t = (t_1, t_2, \ldots) \in \mathbb{R}^{\infty} \tag{2.9}$$

which describes a deformation of moments $\langle\lambda^k\rangle$ and $\langle\lambda^k\rangle_n$ as well as Stieltjes measures $\alpha(\lambda)$ and $\alpha_n(\lambda)$. Let $\alpha(\lambda) = \alpha(\lambda; 0)$ and $\alpha_n(\lambda) = \alpha_n(\lambda; 0)$ be given as initial values, which determine $\langle\lambda^k\rangle = \langle\lambda^k\rangle(0)$ and $\langle\lambda^k\rangle_n = \langle\lambda^k\rangle_n(0)$, respectively. The problem is to show how to construct parametric Stieltjes measures $\alpha(\lambda; t)$ and $\alpha_n(\lambda; t)$, where the deformations (or, time evolutions) from $\langle\lambda^k\rangle(0)$ and $\langle\lambda^k\rangle_n(0)$ to $\langle\lambda^k\rangle(t)$ and $\langle\lambda^k\rangle_n(t)$, respectively, are governed by certain integrable dynamical systems.

3. Construction of the Stieltjes Measure $\alpha_n(\lambda;t)$

Let $\langle\lambda^k\rangle_n(0)$ and $\alpha_n(\lambda;0)$ be given data, and then $\lambda_j(0)$, ℓ_j and $r_j(0)$ be known. We consider a hierarchy of dynamical system of Volterra type

$$\begin{cases} \partial r_j^2/\partial t_m &= \left(\lambda_j^m - \sum_{k=1}^n \lambda_k^m r_k^2\right) r_j^2, \\ \partial\lambda_j/\partial t_m &= 0, \quad m = 1,2,\dots, \; j = 1,2,\dots,n. \end{cases} \tag{3.1}$$

When $m = 1$, this was first introduced in Moser [10] to linearize the finite nonperiodic Toda equation. The hierarchy (3.1) is related to the so-called Jacobi–Toda hierarchy discussed in [11] and describes a motion of zeros of the polynomial $p_{n-1}(\lambda)$ and leaves invariant those of $p_n(\lambda)$ in (2.3). Here we investigate a detail of the motion along a line similar to [10]. Note that

$$\sum_{j=1}^n r_j^2(t) = 1 \tag{3.2}$$

for any t. The first term of (3.1) is a projection of gradient vector field to S^{n-1} for the potential $V_m(r) = \frac{1}{4}\langle\lambda^m\rangle_n/\sum_{j=1}^n r_j^2$. This is because we can express (3.1) as

$$\frac{\partial r_j}{\partial t_m} = \mathrm{grad}_{r_j} V_m(r)|_{S^{n-1}}. \tag{3.3}$$

It follows from (3.3) that $\partial V_m/\partial t_m = |\mathrm{grad} V_m(r)|^2 \geqslant 0$. Since $\lambda_1 < \cdots < \lambda_n$, we can show

$$\sum_{j=1}^n \frac{r_j^2}{\lambda-\lambda_j} \longrightarrow \frac{1}{\lambda-\lambda_n} \quad \text{as} \quad t_m \to \infty. \tag{3.4}$$

Let ν_j be zeros of $p_{n-1}(\lambda)$, where $\nu_1 < \cdots < \nu_{n-1}$. It is easy to see from (3.4) that $\nu_j \to \lambda_j$ as $t_m \to \infty$ for $j = 1,\dots,n-1$.

Now we proceed to 'time evolution' of the moments $\langle\lambda^k\rangle_n$. Multiplying Moser's dynamical system (3.1) by λ_j^k and taking a summation we obtain

$$\frac{\partial\langle\lambda^k\rangle_n}{\partial t_m} = \langle\lambda^{k+m}\rangle_n - \langle\lambda^k\rangle_n\langle\lambda^m\rangle_n, \quad k,m = 1,2,\dots. \tag{3.5}$$

When $m = 1$, (3.5) is a reduced form of the Kac–van Moerbeke system [7, 12] under a boundary condition. Immediately, $\partial\langle\lambda^k\rangle_n/\partial t_m = \partial\langle\lambda^m\rangle_n/\partial t_k$ from (3.5). We have

LEMMA 3.1. *There exists a positive function $g = g(t)$ satisfying*

$$\langle\lambda^k\rangle_n(t) = \frac{\partial \log g(t)}{\partial t_k}, \quad k = 1,2,\dots. \tag{3.6}$$

Substituting (3.6) is to (3.5) we derive

$$\frac{\partial^2 \log g}{\partial t_k \partial t_m} = \frac{\partial \log g}{\partial t_{k+m}} - \frac{\partial \log g}{\partial t_k}\frac{\partial \log g}{\partial t_m}. \tag{3.7}$$

The system (3.7) implies

$$\frac{\partial^2 g}{\partial t_k \partial t_m} = \frac{\partial g}{\partial t_{k+m}}. \tag{3.8}$$

Hence we have

LEMMA 3.2. *For the positive function $g = g(t)$ the following linear system holds*

$$\frac{\partial g}{\partial t_m} = \frac{\partial^m g}{\partial t_1^m}, \quad m = 1, 2, \dots . \tag{3.9}$$

Integrating (3.9) we obtain

$$g(t) = \sum_{j=1}^{n} \exp\left(\sum_{m=1}^{\infty} t_m \lambda_j^m + \omega_j\right), \tag{3.10}$$

where ω_j are constants satisfying

$$\langle \lambda^k \rangle_n(0) = \frac{\sum_{j=1}^n \lambda_j^k \exp \omega_j}{\sum_{i=1}^n \exp \omega_i}. \tag{3.11}$$

From (3.1) and Lemma 3.1 with (3.10) we have a linear system

$$\frac{\partial r_j^2}{\partial t_m} = \left(\lambda_j^m - \frac{\partial \log g(t)}{\partial t_m}\right) r_j^2, \quad r_j^2(0) = \frac{\ell_j}{\sum_{i=1}^n \ell_i}. \tag{3.12}$$

Now we state the main result of this section.

THEOREM 3.1. *For a set of moments $\langle \lambda^k \rangle_n$ and fixed spectrum $\{\lambda_j\}$ a parametric Stieltjes measure is given by*

$$\mathrm{d}\alpha_n(\lambda; t) = \sum_{k=1}^{n} \ell_k \sum_{j=1}^{n} r_j^2(t) \delta(\lambda - \lambda_j) \mathrm{d}\lambda. \tag{3.13}$$

In the remainder of this section we discuss some aspects of dynamical systems introduced here as integrable systems. First, it is noted that the hierarchy of Moser's system (3.1) itself admits a Lax representation of double bracket form

$$\frac{\partial L}{\partial t_m} = [[C^m, L], L], \tag{3.14}$$

where L and C are $n \times n$ matrices defined by

$$L = \begin{pmatrix} r_1^2 & r_1 r_2 & \dots & r_1 r_n \\ r_2 r_1 & r_2^2 & \dots & r_2 r_n \\ \vdots & \vdots & & \vdots \\ r_n r_1 & r_n r_2 & \dots & r_n^2 \end{pmatrix}, \qquad C = \begin{pmatrix} \lambda_1 & & & 0 \\ & \lambda_2 & & \\ & & \ddots & \\ 0 & & & \lambda_n \end{pmatrix}. \tag{3.15}$$

Recently, it has been shown that a class of integrable gradient systems takes the double bracket form (cf. [2]).

Secondly, we remark that the functions $u_k(t)$ defined by $u_k = \langle \lambda^k \rangle_n$ satisfy the so-called Burgers–Hopf hierarchy [5]. For example,

$$\begin{aligned} \frac{\partial u_1}{\partial t_2} &= \frac{\partial^2 u_1}{\partial t_1^2} + 2u_1 \frac{\partial u_1}{\partial t_1}, \\ \frac{\partial u_1}{\partial t_3} &= \frac{\partial^3 u_1}{\partial t_1^3} + 3u_1 \frac{\partial^2 u_1}{\partial t_1^2} + 3\left(\frac{\partial u_1}{\partial t_1}\right)^2 + 3u_1^2 \frac{\partial u_1}{\partial t_1}. \end{aligned} \tag{3.16}$$

Next let us consider a tau-function for the Jacobi–Toda hierarchy [11]. Here the notion of tau-functions has not only been a key to constructing solutions of soliton equations but a useful tool in infinite dimensional analysis of, for example, the conformal field theory. A tau-function of a finite dimensional integrable system and its meanings in information spaces is discussed in [12]. A tau-function of the Jacobi–Toda hierarchy is introduced as follows. Let us define a sequence of functions $\{g_m(t)\}$ by

$$g_m = \frac{\partial g}{\partial t_m}, \quad m = 1, 2, \dots . \tag{3.17}$$

Using this we set a sequence of $k \times k$ Hankel determinants

$$\tau_{n,k}(t) = \begin{vmatrix} g & g_1 & \cdots & g_{k-1} \\ g_1 & g_2 & \cdots & g_k \\ \vdots & \vdots & & \vdots \\ g_{k-1} & g_k & \cdots & g_{2k-2} \end{vmatrix}, \quad k = 1, 2, \dots . \tag{3.18}$$

From (3.9) $\tau_{n,k}$ are determinants of symmetric Wronskian matrices. If $\tau_{n,k}$ depends only on t_1 and t_2, then $\tau_{n,k}(t_1, t_2)$, $k = 1, \dots,$ is the same as the known tau-function of the two-dimensional finite Toda equation (or the affine Toda equation) [6]. Here we can immediately check that $\tau_{n,k}(t_1)$ satisfy the finite nonperiodic Toda equation in bilinear form

$$\tau_{n,k} \frac{\partial^2 \tau_{n,k}}{\partial t_1^2} - \left(\frac{\partial \tau_{n,k}}{\partial t_1}\right)^2 - \tau_{n,k-1}\tau_{n,k+1} = 0, \quad \tau_{n,0} = 1. \tag{3.19}$$

Finally in this section we discuss a relationship between the tau-function $\tau_{n,k}$ and the Hankel determinant $D_{n,k}$ of moments which plays a fundamental role in moment problems [1], inverse spectral problem [3] and orthogonal polynomials [13]. We can show

LEMMA 3.3.

$$\begin{aligned} D_{n,k}(t) &= \begin{vmatrix} 1 & \langle\lambda\rangle & \dots & \langle\lambda^{k-1}\rangle \\ \langle\lambda\rangle & \langle\lambda^2\rangle & \dots & \langle\lambda^k\rangle \\ \vdots & \vdots & & \vdots \\ \langle\lambda^{k-1}\rangle & \langle\lambda^k\rangle & \dots & \langle\lambda^{2k-2}\rangle \end{vmatrix} \\ &= g^{-k}(t)\tau_{n,k}(t) \\ &= \sum_{1\leqslant j_1<\cdots<j_k\leqslant k} r_{j_1}^2\cdots r_{j_k}^2 \begin{vmatrix} 1 & 1 & \dots & 1 \\ \lambda_{j_1} & \lambda_{j_2} & \dots & \lambda_{j_k} \\ \vdots & \vdots & & \vdots \\ \lambda_{j_1}^{k-1} & \lambda_{j_2}^{k-1} & \dots & \lambda_{j_k}^{k-1} \end{vmatrix}^2 \\ &> 0. \end{aligned} \tag{3.20}$$

The positivity of $D_{n,k}(t)$ for any t, proved in Lemma 3.3, guarantees the existence of Stieltjes measure $\alpha_n(\lambda;t)$ through the Toeplitz theorem [1]. As an application we re-prove (cf. [12]) a positivity of the tau-function, $\tau_{n,k}(t) > 0$, for $k = 1, 2, \dots, n$ and $\tau_{n,k}(t) = 0$, for $k = n+1, \dots$ for any t. Hence, $\tau_{n,k}$ is the determinant of totally positive symmetric Wronskian matrix of rank n.

4. Construction of the Stieltjes Measure $\alpha(\lambda;t)$

In this section we generalize the hierarchy of Moser's dynamical system (3.1) so that we can formulate a time evolution of the moments $\langle\lambda^k\rangle$ by a generalized Kac–van Moerbeke system. Let us begin with an analogue of the hierarchy of Moser's system

$$\begin{cases} \partial \mathrm{d}\alpha(\lambda)/\partial t_m &= (\lambda^m - \langle\lambda^m\rangle)\,\mathrm{d}\alpha(\lambda), \\ \partial\lambda/\partial t_m &= 0, \quad m = 1, 2, \dots, \end{cases} \tag{4.1}$$

where $\alpha(\lambda) = \alpha(\lambda;t)$. We suppose that $\int_{-\infty}^{\infty} \mathrm{d}\alpha(\lambda;0) = 1$. Then

$$\int_{-\infty}^{\infty} \mathrm{d}\alpha(\lambda;t) = 1 \tag{4.2}$$

for any t. Multiplying (4.1) by λ^k and integrating it to derive

$$\frac{\partial\langle\lambda^k\rangle}{\partial t_m} = \langle\lambda^{k+m}\rangle - \langle\lambda^k\rangle\langle\lambda^m\rangle, \quad k, m = 1, 2, \dots. \tag{4.3}$$

Note that the higher flows all commute, namely, $\partial\langle\lambda^k\rangle/\partial t_m = \partial\langle\lambda^m\rangle/\partial t_k$. By a discussion similar to that in the previous section we see

LEMMA 4.1. *There exists a positive function $g = g(t)$ such that*

$$\langle\lambda^k\rangle(t) = \frac{\partial \log g(t)}{\partial t_k}, \quad k = 1, 2, \dots. \tag{4.4}$$

LEMMA 4.2. *The positive function $g = g(t)$ satisfies*

$$\frac{\partial g}{\partial t_m} = \frac{\partial^m g}{\partial t_1^m}, \quad m = 1, 2, \ldots . \tag{4.5}$$

The Burgers–Hopf hierarchy (3.16) holds again for the moments $u_k(t) = \langle \lambda^k \rangle(t)$. Let us introduce a sequence of Hankel determinants defined by

$$\tau_k(t) = \begin{vmatrix} g & g_1 & \cdots & g_{k-1} \\ g_1 & g_2 & \cdots & g_k \\ \vdots & \vdots & & \vdots \\ g_{k-1} & g_k & \cdots & g_{2k-2} \end{vmatrix}, \quad g_k = \frac{\partial g}{\partial t_k}, \quad k = 1, 2, \ldots . \tag{4.6}$$

From Lemma 4.2 we see that τ_k is a tau-function of the semi-infinite Toda equation. It should be noted that D_k defined by $D_k(t) = g^{-k}(t)\tau_k(t)$ are determinants of moments $\langle \lambda^j \rangle$,

$$D_k(t) = \begin{vmatrix} 1 & \langle \lambda \rangle & \ldots & \langle \lambda^{k-1} \rangle \\ \langle \lambda \rangle & \langle \lambda^2 \rangle & \ldots & \langle \lambda^k \rangle \\ \vdots & \vdots & & \vdots \\ \langle \lambda^{k-1} \rangle & \langle \lambda^k \rangle & \ldots & \langle \lambda^{2k-2} \rangle \end{vmatrix}. \tag{4.7}$$

Markov's limit theorem, namely, $\alpha_n(\lambda) \to \alpha(\lambda)$ as $n \to \infty$, implies that $\tau_{n,k} \to \tau_k$ as $n \to \infty$, where $\tau_{n,k}$ is the tau-function (3.18) of the finite Toda equation. Indeed, there is an integral formula by Szegö [13]

$$\begin{aligned} D_k(t) &= \frac{1}{k!} \underbrace{\int_{-\infty}^{\infty} \cdots \int_{-\infty}^{\infty}}_{k \text{ times}} \prod_{1 \leqslant \ell < m \leqslant k} (\mu_\ell - \mu_m)^2 \mathrm{d}\alpha(\mu_1; t) \cdots \mathrm{d}\alpha(\mu_k; t) \\ &> 0, \end{aligned} \tag{4.8}$$

which guarantees the existence of the Stieltjes measure $\alpha(\lambda; t)$ for any t in terms of the Toeplitz theorem. As a direct consequence of (4.8) we prove the positivity of tau-function $\tau_k(t) > 0$ for $k = 1, 2, \ldots$. The solution of (4.5) can be expressed as

$$g(t) = \int_{-\infty}^{\infty} \exp\left(\sum_{m=1}^{\infty} t_m \lambda^m + \omega(\lambda) \right) \mathrm{d}\alpha(\lambda; 0), \tag{4.9}$$

where $\omega(\lambda)$ is a constant in t given by the initial value $\langle \lambda^k \rangle(0)$. Inserting (4.9) into (4.1) through (4.4) we derive a linear evolution equation

$$\frac{\partial \mathrm{d}\alpha(\lambda)}{\partial t_m} = \left(\lambda^m - \frac{\partial \log g(t)}{\partial t_m} \right) \mathrm{d}\alpha(\lambda). \tag{4.10}$$

This brings us the parametric Stieltjes measure $\alpha(\lambda; t)$, a solution of a moment problem of Hamburger, from the given set of moments $\{\langle \lambda^k \rangle\}$. We have proved

THEOREM 4.1. *For a set of moments* $\{\langle\lambda^k\rangle\}$ *a parametric Stieltjes measure* $\alpha(\lambda;t)$ *is given by the quadrature of the nonlinear integrable system* (4.1).

It would be interesting to compare (4.1) with the linear differential equation [8] for spectral densities being equivalent to a Lax type equation for a Jacobi operator. An asymptotic formula of τ_k/τ_{k-1} as $k \to \infty$ can be derived from that of D_k/D_{k-1} given in [13] providing that the Stieltjes measure $\alpha(\lambda)$ belongs a suitable class of functions. An asymptotic formula of D_k as $k \to \infty$ also was found in [3] as an application of scattering theory. However, no closed form of D_∞ has been known. In the same limit τ_k approaches to a solution of a dispersionless (or continuous) limit of the Toda equation [9]. The corresponding parametric moment problem remains to be pursued.

Acknowledgement

The authors are grateful to K. Kajiwara of Doshisha University for stimulating discussions. Y. N. was supported in part by Grant-in-Aid for Scientific Research No. 06221111 from the Japan Ministry of Education, Science, and Culture and Aid of Doshisha University's Research Promotion Fund. The work of Y. K. was partially supported by NSF Grant DMS-9109041.

References

1. Ahiezer, N. I. and Kreĭn, M.: *Some Questions in the Theory of Moments*, Transl. Math. Monographs, Vol. 2, Amer. Math. Soc., Providence, 1962.
2. Bloch, A. M., Brockett, R. W., and Ratiu, T.: Completely integrable gradient flows, *Commun. Math. Phys.* **147** (1992), 57–74.
3. Case, K. M.: Orthogonal polynomials from the viewpoint of scattering theory, *J. Math. Phys.* **15** (1974), 2166–2174.
4. Gantmacher, F. R.: *The Theory of Matrices*, Vol. 2, Chelsea, New York, 1959.
5. Harada, H.: New subhierarchies of the KP hierarchy in the Sato theory. I. Analysis of the Burgers–Hopf hierarchy by the Sato theory, *J. Phys. Soc. Japan* **12** (1985), 4507–4512.
6. Hirota, R., Ohta, Y., and Satsuma, J.: Solutions of the Kadomtsev–Petviashvili equation and the two-dimensional Toda equations, *J. Phys. Soc. Japan* **57** (1988), 1901–1904.
7. Kac, M. and van Moerbeke, P.: On an explicitly soluble system of nonlinear differential equations related to certain Toda lattice, *Adv. Math.* **16** (1975), 160–169.
8. Kato, Y. and Aomoto, K.: Jacobi–Perron algorithms, bi-orthogonal polynomials and inverse scattering problems, *Publ. RIMS, Kyoto Univ.* **20** (1984), 635–658.
9. Kodama, Y.: Solutions of the dispersionless Toda equation, *Phys. Lett.* **A 147** (1990), 477–482.
10. Moser, J.: Finitely many points on the line under the influence of an exponential potential – An integrable system, in: J. Moser (Ed.), *Dynamical Systems, Theory and Applications*, Lec. Notes Phys., Vol. 38, Springer-Verlag, Berlin, New York, 1975, pp. 467–497.
11. Nakamura, Y.: Geometry of rational functions and nonlinear integrable systems, *SIAM J. Math. Anal.* **22** (1991), 1744–1754. The level manifolds of a generalized Toda equation hierarchy, *Trans. Amer. Math. Soc.* **333** (1992), 83–94.
12. Nakamura, Y.: A tau-function for the finite Toda molecule, and information spaces, in: Y. Maeda, H. Omori and A. Weinstein (Eds), *Symplectic Geometry and Quantization*, Contemp. Math. Vol. 179, Amer. Math. Soc., Providence, 1994.
13. Szegö, G.: *Orthogonal Polynomials*, 4th edn, Colloq. Publ., Vol. 23, Amer. Math. Soc., Providence, 1975.

Acta Applicandae Mathematicae **39**: 445–455, 1995.

New Features of Soliton Dynamics in $2+1$ Dimensions*

F. PEMPINELLI
Dipartimento di Fisica dell'Università, Lecce, Italy
I.N.F.N., Sezione di Lecce, Italy

(Received: 12 September 1994)

Abstract. Exponentially localized soliton solutions have been found recently for all the equations of the hierarchy related to the Zakharov–Shabat hyperbolic spectral problem in the plane. In particular the N^2-soliton solution of the Davey–Stewartson I equation is considered. It is shown that the boundaries fix the kinematics of solitons, while the dynamics of their mutual interaction is determined by the chosen initial condition. The interacting solitons can have, asymptotically, zero mass and can simulate quantum effects as inelastic scattering, fusion and fission, creation and annihilation.

Mathematics Subject Classifications (1991): 35Q51, 35Q55, 58F07.

Key words: bidimensional solitons, Davey–Stewartson equations.

It has recently been shown that localized solitons exist in two spatial and one temporal dimensions $(2+1)$. More precisely, in 1988 Boiti, Léon, Martina and Pempinelli [1] discovered that all the equations in the hierarchy related to the Zakharov–Shabat (ZS) hyperbolic spectral problem in the plane have exponentially localized soliton solutions. The most representative equation in the hierarchy is the Davey–Stewartson I (DSI) equation, which provides a two dimensional generalization of the non-linear Schrödinger equation. Moreover, as a relevant application of the spectral method proposed by Sabatier [2] and further developed by Boiti, Pempinelli and Sabatier [3], it has been stated that an additional non-linear evolution equation, called that Davey–Stewartson III equation, also admits localized soliton solutions with properties similar to those of the DSI equation [4]. Localized soliton solutions for the DSI equation were first found by using gauge Bäcklund transformations (BT) [1, 5]. Later these solutions were also obtained by means of the inverse spectral transform [6–8] and direct methods [9, 10].

These coherent structures display a richer phenomenology than the one dimensional solitons. Different effects have been reported successively, due to the fact that the soliton solution is structurally unstable with respect to special choices of

* Work supported in part by M.U.R.S.T.

the parameters [7]. Some quantum-like effects, such as the nonconservation of the number of solitons, have also been discovered. In particular there exist solitons with zero mass at large time (virtual solitons), which interact with the other visible solitons with nonzero mass (real solitons). This special effect is possible because, although the total mass of solitons is conserved, the mass of the single soliton, in general, is not preserved by the interaction. Solitons can simulate inelastic scattering processes of quantum particles as creation and annihilation, fusion and fission, and interaction with virtual particles.

An extended review of multidimensional localized solitons and a large reference list are given in [11].

The solitons we are considered are solutions of the so-called Davey–Stewartson I (DSI) equation, which can be written in characteristic coordinates $u = x + y$ and $v = x - y$ as follows

$$iQ_t + \sigma_3(Q_{uu} + Q_{vv}) + [A, Q] = 0 \tag{1}$$

where Q is a $2+2$ off diagonal matrix field

$$Q = \begin{pmatrix} 0 & q(u,v,t) \\ r(u,v,t) & 0 \end{pmatrix} \tag{2}$$

and the diagonal 2×2 matrix field A, called the auxiliary field, is chosen to have arbitrary boundary values a_1 and a_2 according to the formula

$$A = \begin{pmatrix} -\frac{1}{2}\int_{-\infty}^{u} \mathrm{d}u'(Q^2)_v + a_1(v,t) & 0 \\ 0 & \frac{1}{2}\int_{-\infty}^{v} \mathrm{d}v'(Q^2)_u + a_2(u,t) \end{pmatrix}. \tag{3}$$

Of interest is the reduced case $r = \varepsilon\bar{q}$ (the overbar means complex conjugation and $\varepsilon^2 = 1$), describing physical situations as in hydrodynamics or in plasma physics. Originally it was proposed as a model of the evolution of shallow water waves that are weakly nonlinear and nearly monochromatic, when the effects of the surface tension are important. The field q is the complex envelope of the free surface wave and the auxiliary field $\sigma_3 A$ is the real velocity for the mean motion generated by the surface wave. However, more generally, the DSI equation is obtained as a universal multiscale limit of nonlinear dispersive wave equations.

The DSI equation can be obtained [3, 4] as the compatibility condition between two Lax operators T_1 and T_2 which commute in the 'weak' sense [12]

$$T_1\psi = 0, \qquad [T_1, T_2]\psi = 0. \tag{4}$$

T_1 is the ZS hyperbolic operator in the plane

$$T_1\psi \equiv \left\{2\begin{pmatrix} \partial_u & 0 \\ 0 & \partial_v \end{pmatrix} + Q\right\}\psi = 0 \tag{5}$$

and T_2 has the form

$$2T_2\psi \equiv \left\{ i\partial_t + \partial_v^2 - \partial_u^2 + A + \begin{pmatrix} 0 & -q_u \\ r_v & 0 \end{pmatrix} \right\} \psi = -k^2 \psi \sigma_3. \tag{6}$$

Let us remember that the boundary conditions of the auxiliary field A can be arbitrary chosen. The DSI equation in his standard version, i.e. with the boundary written as in (3), admits localized soliton solutions [1]. The one-soliton solution has the form

$$q = -\frac{2}{D}\lambda_I \eta e^{i\theta}, \qquad r = -\frac{2}{D}\mu_I \rho e^{-i\theta}, \tag{7}$$

where:

$$D = 2\gamma(\cosh \xi_1 + \cosh \xi_2) + e^{\xi_2} \tag{8}$$

$$\xi_1 = -\mu_I u - \lambda_I v + 2(\lambda_R \lambda_I + \mu_R \mu_I)t \tag{9}$$

$$\xi_2 = \mu_I u - \lambda_I v + 2(\lambda_R \lambda_I - \mu_R \mu_I)t \tag{10}$$

$$\theta = \mu_R u + \lambda_R v + (\lambda_I^2 - \lambda_R^2 + \mu_I^2 - \mu_R^2)t \tag{11}$$

$$\gamma = \frac{1}{4}\eta\rho. \tag{12}$$

The complex parameters $\lambda = \lambda_R + i\lambda_I$, $\mu = \mu_R + i\mu_I$ are the discrete eigenvalues of the associated ZS spectral problem and ρ, η are arbitrary complex constants satisfying the conditions $\gamma \in \mathbb{R}$ and $\gamma(1+\gamma) > 0$.

The elastic scattering of two solitons is shown in Figure 1, where $|q|$ of the first soliton, in its reference frame, is represented during the interaction with the second (outside) soliton. The only memory of the interaction is a bidimensional shift of the position.

These results were first obtained by using Bäcklund gauge transformations, but can be reobtained by defining a new spectral transform (ST). Our choice, in analogy with the $1+1$ dimensional case, is that it must satisfy the following requirements:

1. the discrete part of the spectrum corresponds to solitons and the continuous part to the radiation;
2. the time evolution can be explicitly integrated.

Alternative STs with different requirements are allowed in multidimensional case.

The boundaries a_1 and a_2 can be considered as potentials in two different time-dependent Schrödinger equations

$$(i\partial_t + \partial_v^2 + a_1)\phi_1 = 0 \tag{13}$$

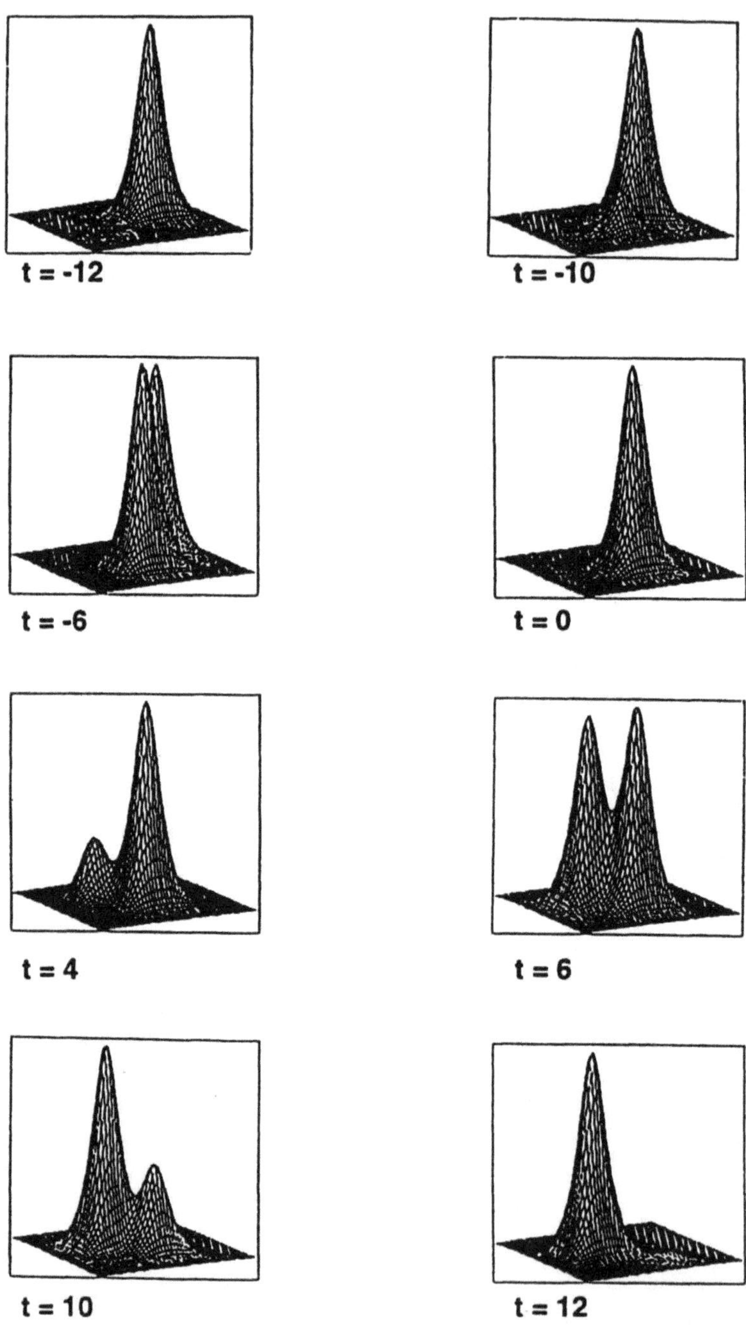

Fig. 1. The time sequence of $|q_1|$ of one soliton in its reference frame describes the interaction with another outside soliton. The soliton, during the interaction, for $t < 0$ generates a new bump, which grows and the old bump decreases. At $t = 0$ there is only the new bump. For $t > 0$ again a new bump is generated and it grows and the previous bump decreases. At the end of the interaction there is only one bump of the same form of the original bump, but shifted in its position.

$$(i\partial_t - \partial_u^2 + a_2)\phi_2 = 0. \tag{14}$$

The complex spectral parameter k is introduced by requiring that the eigenfunctions ψ and ϕ_i satisfy, as $k \to \infty$, the asymptotic properties

$$\psi \exp\left[-ik(\sigma_3 x - y)\right] = 1 + \mathcal{O}\Big(\frac{1}{k}\Big) \tag{15}$$

$$\phi_1 \exp\left[-ikv + ik^2 t\right] = 1 + \mathcal{O}\Big(\frac{1}{k}\Big) \tag{16}$$

$$\phi_2 \exp\left[iku - ik^2 t\right] = 1 + \mathcal{O}\Big(\frac{1}{k}\Big). \tag{17}$$

The ST is defined as the measure of the departure from analyticity of the eigenfunctions

$$\frac{\partial\psi}{\partial\bar{k}} = \iint \mathrm{d}\ell \wedge \mathrm{d}\bar{\ell}\psi(\ell)R(k,\ell) \tag{18}$$

$$\frac{\partial\phi_i}{\partial\bar{k}} = \iint \mathrm{d}\ell \wedge \mathrm{d}\bar{\ell}\phi_i(\ell)r_i(k,\ell), \quad i = 1,2. \tag{19}$$

The integral equation for ψ is of Volterra type and therefore the singularities of ψ in the complex plane are those of the homogeneous term and of the sectionally holomorphic Green function defining the integral equation. Consequently, ψ has simple poles only if the homogeneous term, which is related to the eigenfunctions of the time dependent Schrödinger equations, has simple poles. We deduce that solutions Q of the DSI equation have a discrete ST only for boundaries a_i which are wave-solitons of the Kadomtsev–Petviashvili I equation.

We therefore need to consider the most general discrete spectral transforms of a_1 and a_2

$$r_1(k,\ell) = \sum_{n,m} r_{nm}^{(1)} \exp\left[-i(k^2 - \ell^2)t\right]\delta(\ell - \bar{\lambda}_m)\delta(k - \lambda_n) \tag{20}$$

$$r_2(k,\ell) = \sum_{n,m} r_{nm}^{(2)} \exp\left[i(k^2 - \ell^2)t\right]\delta(\ell - \bar{\mu}_m)\delta(k - \mu_n). \tag{21}$$

They are furnishing real a_1 and a_2 if the constant matrices $r^{(i)}$ are Hermitian. It is convenient to parametrize these matrices as follows

$$r_{nm}^{(1)} = 2\pi i \exp\left[-i\bar{\lambda}_m v_{om} + i\lambda_n v_{on}\right] D_{nm}\tilde{\lambda}_{mm} \tag{22}$$

$$r_{nm}^{(2)} = 2\pi i \exp\left[i\bar{\mu}_m u_{om} - i\mu_n u_{on}\right] C_{nm}\tilde{\mu}_{mm}. \tag{23}$$

The complex parameters λ_n and μ_n are the discrete eigenvalues, the real parameters v_{on} and u_{on} fix the position of the nth wave soliton in the corresponding plane, $\tilde{\lambda}_{mn} = \bar{\lambda}_m - \lambda_n$, $\tilde{\mu}_{mn} = \bar{\mu}_m - \mu_n$ and the two matrices D_{nm}, C_{nm} satisfy

$$D_{nn}^2 = C_{nn}^2 = 1, \qquad D_{nm}\tilde{\lambda}_{mm} = \overline{D}_{mn}\tilde{\lambda}_{nn}, \qquad C_{nm}\tilde{\mu}_{mm} = \overline{C}_{mn}\tilde{\mu}_{nn}. \tag{24}$$

For definiteness we choose $D_{nn} = C_{nn} = 1$, $\lambda_{nI} < 0$, $\mu_{nI} > 0$ and the same number N of λs and μs. By solving the $\bar{\partial}$-equation one gets the ϕ_i and, consequently, the following boundaries

$$a_1 = 2\partial_v^2 \ln \det \mathcal{B}, \qquad a_2 = -2\partial_u^2 \ln \det \mathcal{A} \tag{25}$$

where

$$\mathcal{B} = \mathbf{1} + D\beta, \qquad \beta_{nm} = \frac{\tilde{\lambda}_{nn}}{\tilde{\lambda}_{nm}} \exp\left[\bar{\beta}_n + \beta_m\right] \tag{26}$$

$$\mathcal{A} = \mathbf{1} + C\alpha, \qquad \alpha_{nm} = \frac{\tilde{\mu}_{nn}}{\tilde{\mu}_{nm}} \exp[\bar{\alpha}_n + \alpha_m] \tag{27}$$

with

$$\beta_n = -i\lambda_n(v - v_{on}) + i\lambda_n^2 t, \qquad \alpha_n = i\mu_n(u - u_{on}) - i\mu_n^2 t. \tag{28}$$

By solving the $\bar{\partial}$-equation (18) one gets the corresponding eigenfunction ψ and, then, by expanding the obtained ψ in powers of $1/k$ in the two spectral equations (5), (6), one derives the following explicit algebraic formulae for Q and A

$$Q = 2i\sigma_3 \sum_n \begin{pmatrix} 0 & \Omega_n^{(2)} \exp[\alpha_n] \\ \Omega_n^{(1)} \exp[\beta_n] & 0 \end{pmatrix} \tag{29}$$

$$Q^2 = -4\partial_u\partial_v \ln \Delta \tag{30}$$

$$A = 2\begin{pmatrix} \partial_v^2 & 0 \\ 0 & -\partial_u^2 \end{pmatrix} \ln \Delta \tag{31}$$

where

$$\left[\mathbf{1} + \tfrac{1}{4}\eta\mathcal{B}\rho\mathcal{A}\right]\Omega^{(2)} = -\tfrac{1}{2}\eta\delta \tag{32}$$

$$\left[\mathbf{1} + \tfrac{1}{4}\rho\mathcal{A}\eta\mathcal{B}\right]\Omega^{(1)} = \tfrac{1}{2}\rho\gamma \tag{33}$$

$$\Delta = \det\left[\mathbf{1} + \tfrac{1}{4}\eta\mathcal{B}\rho\mathcal{A}\right] = \det\left[\mathbf{1} + \tfrac{1}{4}\rho\mathcal{A}\eta\mathcal{B}\right] \tag{34}$$

with

$$\gamma_n = \sum_m C_{nm}\tilde{\mu}_{mm} \exp\bar{\alpha}_m, \qquad \delta_n = \sum_m D_{nm}\tilde{\lambda}_{mm} \exp\bar{\beta}_m \tag{35}$$

and ρ and η are arbitrary complex constant matrices. One can prove that the reduced case

$$r = \varepsilon \bar{q}, \qquad \varepsilon^2 = 1, \tag{36}$$

the one we consider in the following, is obtained if and only if

$$\sum_m \rho_{nm} C_{ms} \tilde{\mu}_{ss} = \varepsilon \sum_m D_{nm} \tilde{\lambda}_{mm} \bar{\eta}_{sm}. \tag{37}$$

In the so-called focusing case $\varepsilon = -1$ one can prove that if the Hermitian matrices $iC_{nm}\tilde{\mu}_{mm}$ and $-iD_{nm}\tilde{\lambda}_{mm}$ have positive eigenvalues then the N^2-soliton solution and the auxiliary field A are regular.

One can show that the solution q describes N^2 localized coherent structures interacting in a complicated way at finite times, but moving in the far past and in the far future with constant velocities $\mathbf{V}_{in} = (2\lambda_{iR}, 2\mu_{nR})$ $(n, i = 1, 2, \ldots, N)$. It is therefore natural to call this solution the N^2-soliton solution. One can also show that this N^2-soliton solution is structurally unstable at $t = -\infty$ and at $t = +\infty$ when any couple of discrete eigenvalues λ_n, λ_m or μ_n, μ_m have the same real part or are equal [8].

If both coefficients ρ_{in} and η_{ni} are equal to zero the corresponding soliton moving with velocity $\mathbf{V}_{in} = (2\lambda_{iR}, 2\mu_{nR})$ at large times disappears. In particular, if the matrices ρ and η are diagonal and $C = D = \mathbf{1}$ one recovers the N-soliton solution described in [7], where the explicit expressions of q and A are given.

Let us examine more in detail the N-soliton solution in the cases when is structurally unstable at $t = -\infty$ and at $t = +\infty$.

In the first case $(\lambda_{iR} = \lambda_{jR}$ or $\mu_{iR} = \mu_{jR})$ the solitons after the interaction:

1. have a bidimensional shift;
2. change their form but do not change their mass.

Figure 2 shows the evolution of $|q|$ in the two soliton case. Note the form of these new localized solitons which have two bumps.

In the second case $(\lambda_i = \lambda_j$ or $\mu_i = \mu_j)$ the solitons after the interaction:

1. have a bidimensional shift;
2. change their form;
3. change their mass but the total mass is conserved.

Figure 3 shows the evolution of $|q|$ in the two soliton case.

Moreover, in both cases, the form and the mass of solitons depend on the initial position u_{0n}, v_{0n} of every soliton.

A relevant information in the global dynamical behaviour of the solitons is furnished by the mass (energy, charge or number of particles according to the physical context) of the solution q

$$M = \iint |q|^2 \, du \, dv \tag{38}$$

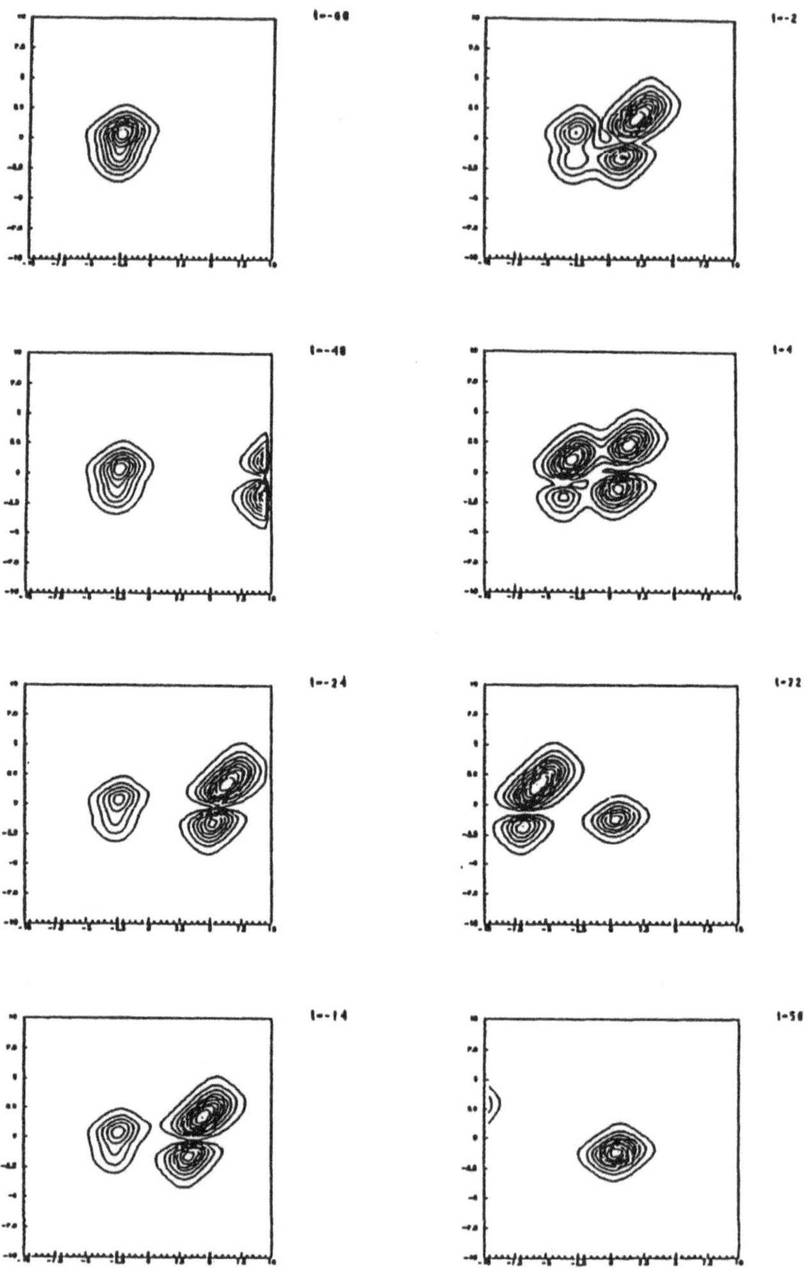

Fig. 2. The case $\mu_{1R} = \mu_{2R}$ ($\lambda_1 = -0.1 - i$, $\mu_1 = i$, $\rho_1 = 2.004$, $\lambda_2 = -1.01i$, $\mu_2 = 1.3i$, $\rho_2 = 2.108$). The time sequence of the level contours of $|q|$ is described for two soliton solution in the frame of reference of the soliton 1.

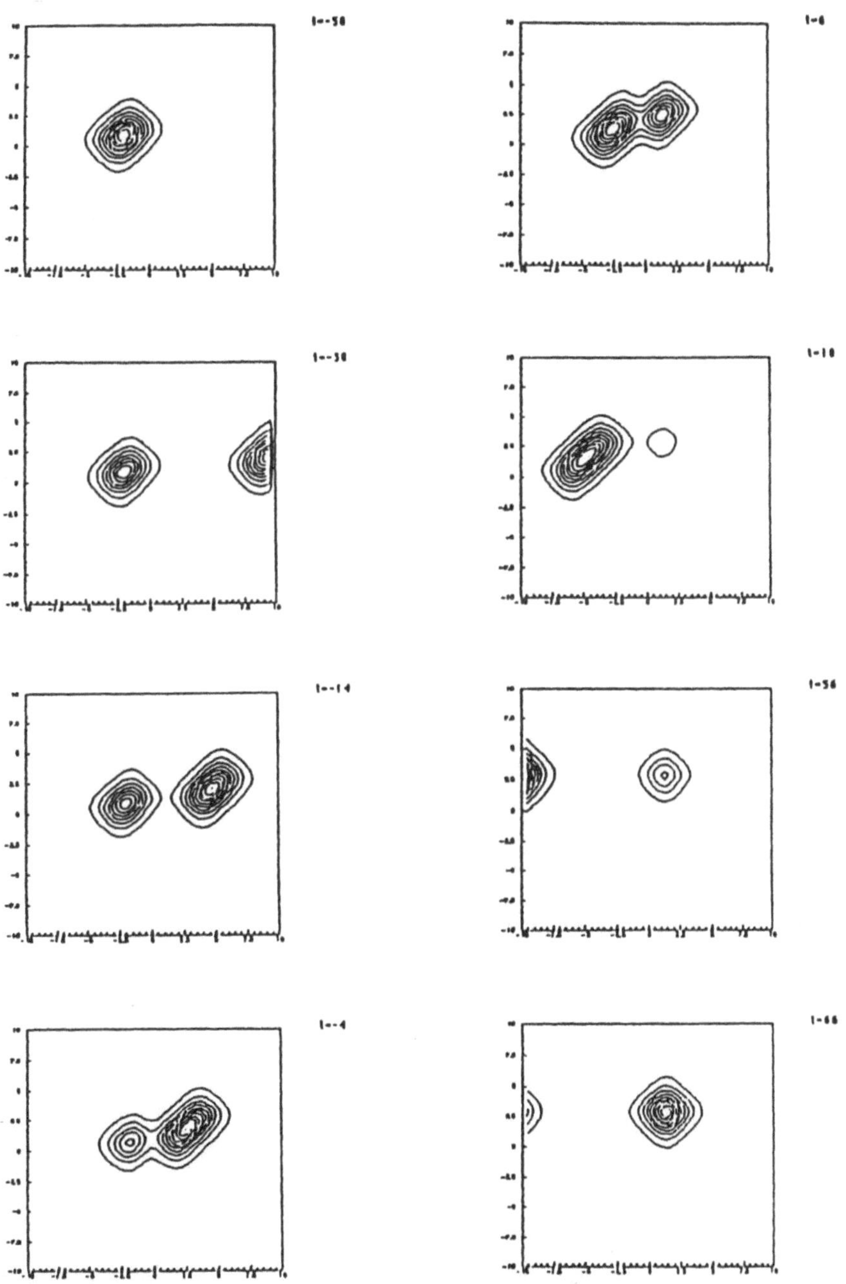

Fig. 3. The time sequence in the case $\mu_1 = \mu_2$ ($\lambda_1 = -0.1 - i$, $\mu_1 = \mu_2 = i$, $\rho_1 = 2.004$, $\lambda_2 = -1.01i$, $\rho_2 = 2.108$).

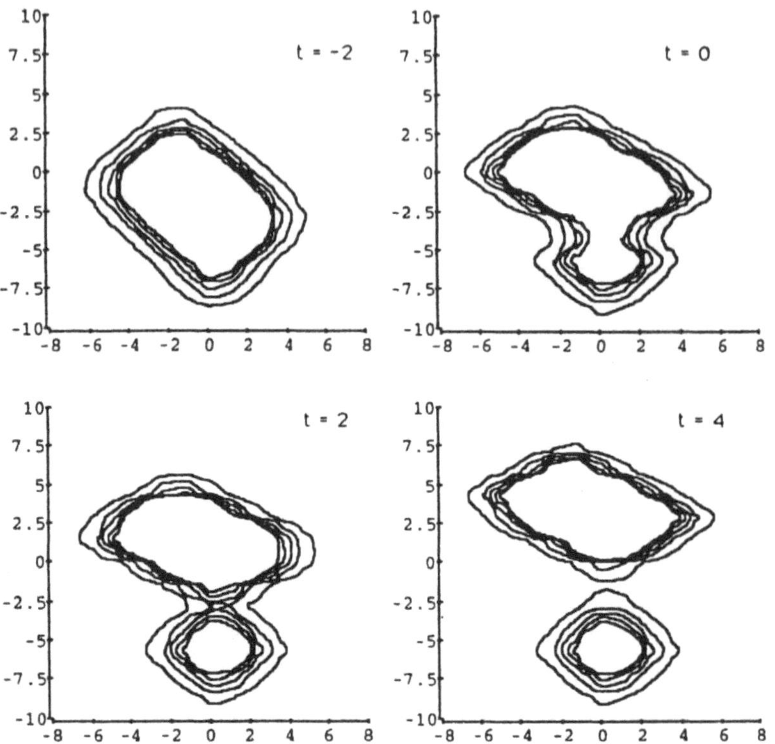

Fig. 4. The fission of a soliton flashed at four different times. The contour levels are drawn for $|q|$ ranging between 0 and 0.1. The soliton parameters are $\lambda_1 = -1$, $\lambda_2 = 0.5 - 1.5i$, $\mu_1 = \mu_2 = i$, $C = D = 1$, $u_{01} = u_{02} = v_{01} = v_{02} = 0$ and $\rho = \begin{pmatrix} 0.1 & 0.1 \\ 0.2 & 0.3 \end{pmatrix}$, $q = \begin{pmatrix} 0.1 & 4/30 \\ 0.1 & 0.2 \end{pmatrix}$. If the time is inverted, the sequence shows the fusion of two solitons.

and by the masses of the solitons at $t = \pm\infty$

$$M_{in}^{(\pm)} = \iint |q_{in}^{(\pm)}|^2 \, du \, dv \quad (i = 1, 2, \ldots, N) \tag{39}$$

where $q_{in}^{(\pm)}$ is the asymptotic behaviour of q at $t = \pm\infty$ computed in the rest reference frame of the (i, n) soliton.

In general, it results that

$$M = \sum_{in} M_{in}^{(-)} = \sum_{in} M_{in}^{(+)}. \tag{40}$$

However, the mass of the single soliton is not conserved and in particular it can be zero at $t = +\infty$ or at $t = -\infty$. For a special choice of the parameters the mass of the (i, n) soliton can be zero at $t = \pm\infty$ also when the coefficients ρ_{in} and η_{ni} are not both equal to zero. We call these solitons with zero mass virtual solitons and they generate peculiar effects.

For different choices of the parameters the four soliton solution can describe the creation or the annihilation of a soliton, the creation or the annihilation of a pair of solitons, two interacting solitons and inelastic scattering of solitons. All these dynamical behaviours can be obtained by choosing the same boundaries. Boundaries do not give any information on the dynamics of solitons, but fix only their asymptotic kinematics, i.e. their possible locations in the plane and their velocities at large times.

Of special interest is the case in which two spectral data are equal, say $\mu_1 = \mu_2$. In general, this solution describes the interaction of two solitons and the masses $M_{ii}^{(\pm)}$ $(i = 1, 2)$ in this degenerate case depend also on the initial position of the solitons. With a special choice of the parameters one can get a solution describing the fission of a soliton and the fusion of two solitons (see Figure 4). If in addition one chooses $\det(\rho\eta) = 0$ the solution describes a single soliton that by the interaction with a virtual soliton is forced to change velocity. These two dynamical processes can be obtained by choosing the same boundaries and different matrices ρ and η.

References

1. Boiti, M., Léon, J., Martina, L., and Pempinelli, F.: *Phys. Lett.* **A 132** (1988), 432.
2. Sabatier, P. C.: *Inverse Problems* **8** (1992), 263;
   Sabatier, P. C.: *Phys. Lett.* **A 161** (1992), 345.
3. Boiti, M., Pempinelli, F., and Sabatier, P. C.: *Inverse Problems* **9** (1993), 1.
4. Pempinelli, F.: Localized soliton solutions for Davey–Stewartson I and Davey–Stewartson III equations, in: Clarkson, P. A. (Ed.) *Application of Analytic and Geometric Methods to Nonlinear Differential Equations*, NATO ASI Series C 413, Kluwer Acad. Publ., Dordrecht, 1993, pp. 207–215.
5. Boiti, M., Léon, J., and Pempinelli, F.: *J. Math. Phys.* **31** (1990), 2612.
6. Fokas, A. S. and Santini, P. M.: *Phys. Rev. Lett.* **63** (1989), 1329;
   Fokas, A. S. and Santini, P. M.: *Physica* **D 44** (1990), 99.
7. Boiti, M., Léon, J., and Pempinelli, F.: *Inverse Problems* **6** (1990), 715.
8. Boiti, M., Martina, L., Pashaev, O. K., and Pempinelli, F.: *Phys. Lett.* **A 160** (1991), 55.
9. Hietarinta, J. and Hirota, R.: *Phys. Lett.* **A 145** (1990), 237.
10. Hernandez Heredero R., Martinez Alonso L., and Medina Reus E.: *Phys. Lett.* **A 152** (1991), 37.
11. Boiti, M., Martina, M., and Pempinelli, F.: Multidimensional localized solitons, *Chaos, Solitons and Fractals* **4** (1994).
12. Boiti, M., Léon, J., Manna, M., and Pempinelli, F.: *Inverse Problems* **2** (1986), 271;
   Boiti, M., Léon, J., Manna, M., and Pempinelli, F.: *Inverse Problems* **3** (1987), 25.

Acta Applicandae Mathematicae **39**: 457–475, 1995.

Cnoidal Wave Trains and Solitary Waves in a Dissipation-Modified Korteweg–de Vries Equation

A. YE. REDNIKOV[1], M. G. VELARDE[1], YU. S. RYAZANTSEV[1,2], A. A. NEPOMNYASHCHY[3] and V. N. KURDYUMOV[1,2]

[1] *Instituto Pluridisciplinar, Universidad Complutense, Paseo Juan XXIII 1, Madrid 28040, Spain*

[2] *E.T.S.I. Aeronàuticos, Universidad Politecnica de Madrid, Ciudad Universitaria, Madrid 28040, Spain*

[3] *Department of Mathematics, Technion, Haifa 32000, Israel*

(Received: 27 February 1995)

Abstract. A generalization of the Korteweg–de Vries equation incorporating an energy input–output balance, hence a dissipation-modified KdV equation is considered. The equation is relevant to describe, for instance, nonlinear Marangoni–Bénard oscillatory instability in a liquid layer heated from above. Cnoidal waves and solitary waves of this equation are obtained both asymptotically and numerically.

Mathematics Subject Classification (1991): 58F07.

Key words: cnoidal waves, solitary waves, dissipation-modified KdV equation, stationary waves.

1. Introduction

Recently, dissipation-modified Korteweg–de Vries (KdV) equations have been derived [1–4]. Such equations have been associated with long wave oscillatory instability in problems where an input–output energy balance exists besides the usual KdV balance between nonlinearity and dispersion. Hence they are natural extensions to the treatment of real fluids of the original dissipation-free, inviscid problem considered by Korteweg and de Vries [5]. One particular case is the Marangoni–Bénard oscillatory convection in a horizontal layer heated from above [2–4]. Some of the solutions of these equations have been studied, such as solitary waves [3, 6, 7], periodic wave trains [8] as well as the (head-on and oblique) interaction of waves [3, 9]. Experimental results support the relevance of the theoretical approach [10–13]. A particular dissipation-modified KdV equation is

$$h_t + (h^2)_y + h_{yyy} + \delta\left[h_{yy} + h_{yyyy} + D(h^2)_{yy} + \alpha h\right] = 0, \tag{1a}$$

where $h(y,t)$ is a scaled elevation of the surface in the study of one-directional waves in Marangoni–Bénard layers. We assume that the mean value of h along the layer is zero (mass conservation). The coefficient D can be either positive or negative, while α is always nonnegative. We consider the dissipation as a perturbation upon the ideal, viscous-free case, hence $\delta \ll 1$. The objective of the following analysis is to study both asymptotically and numerically the stationary waves solutions of Equation (1). Note that besides the addition of dissipative terms in Marangoni–Bénard problem [2, 3] the particular case of Equation (1a) also contains a generalization of the surface tension effects already correctly introduced by Korteweg and de Vries [5].

2. Stationary Waves

2.1. Cnoidal Waves

For stationary waves propagating with (phase) velocity c Equation (1a) reduces to

$$-ch + h^2 + h'' + \delta\left[h' + h''' + D(h^2)' + \alpha \int^x h\,\mathrm{d}x\right] = \text{const}, \tag{1b}$$

where the prime denotes derivative with respect to $x = y - ct$, in the moving frame.

We look for a solution in the form

$$h = h_0 + \delta h_1 + \cdots, \qquad c = c_0 + \delta c_1 + \cdots.$$

In the zeroth approximation we get the KdV equation. The solution is [5]

$$h_0 = A\left[-\frac{E(s)}{K(s)} + \mathrm{dn}^2\left(\left(\frac{A}{6}\right)^{1/2} x; s\right)\right], \tag{2}$$

where $K(s)$ and $E(s)$ are the complete elliptic integrals of the first and second kind, respectively; dn is a Jacobian elliptic function; s is the modulus ($0 \leqslant s \leqslant 1$); A is some non-negative parameter. The wave number $k = 2\pi/\lambda$ (λ is the wavelength) as well as the (phase) velocity c_0 may be expressed in terms of A and s:

$$k = \pi(A/6)^{1/2}/K(s), \tag{3}$$

$$c_0 = 2A\left[\frac{2 - s^2}{3} - \frac{E(s)}{K(s)}\right]. \tag{4}$$

The equation for h_1 is

$$-c_0 h_1 + h_1'' + 2h_0 h_1 = f, \tag{5}$$

where

$$f = -h_0' - h_0''' - D(h_0^2)' - \alpha \int^x h_0\,\mathrm{d}x + c_1 h_0 + \text{const}.$$

As h_0' is a solution of the homogeneous Equation (5), we look for a solution in the form

$$h_1 = g h_0',$$

where g is some auxiliary function. We get

$$g' = (h_0')^{-2} \int_0^x f h_0' \, \mathrm{d}x. \tag{6}$$

Using properties of the elliptic functions [14], one can show that h_1 is bounded and periodic only when

$$\int_0^{K(s)} \Big\{ (2/3)A^3 s^4 \mathrm{sn}^2\mathrm{cn}^2\mathrm{dn}^2 \big[-1 + (2/3)A(s^2\mathrm{cn}^2 - s^2\mathrm{sn}^2 + \mathrm{dn}^2) + \\ +2D\,A\,E(s)/K(s) - 2D\,A\,\mathrm{dn}^2\big] + \alpha A^2 \Big[-\frac{E(s)}{K(s)} + \mathrm{dn}^2 \Big]^2 \Big\} \mathrm{d}x = 0, \tag{7}$$

where the elliptic functions sn, cn and dn have, for simplicity, the arguments $(x; s)$. After integration, (7) reduces to

$$b_2 A^4 + b_1 A^3 + b_0 A^2 = 0, \tag{8}$$

where

$$\begin{aligned} b_2 = {} & \frac{2}{315K(s)} [15s^4 K(s)E(s) + 5s^6 K^2(s) - 10s^6 K(s)E(s) + 5s^4 K^2(s) - \\ & -78s^4 D\, K(s)E(s) + 42s^4 D\, E^2(s) + 30s^4 D\, K^2(s) - 20s^2 K^2(s) + \\ & +15s^2 K(s)E(s) - 10K(s)E(s) + 10K^2(s) - 36s^2 D\, K^2(s) + \\ & +90s^2 D\, K(s)E(s) - 42s^2 D\, E^2(s) - 60D\, K(s)E(s) + 42D\, E^2(s) - \\ & -12s^6 D\, K^2(s) + 24s^6 D\, K(s)E(s) + 18D\, K^2(s)], \end{aligned} \tag{9}$$

$$b_1 = [s^4 K(s) - 2s^4 E(s) - 3s^2 K(s) + 2s^2 E(s) + 2K(s) - 2E(s)]/15, \tag{10}$$

$$b_0 = \frac{\alpha}{2K(s)} [s^2 K^2(s) - K^2(s) + 4K(s)E(s) - 2s^2 K(s)E(s) - 3E^2(s)]. \tag{11}$$

Along with the trivial solutions, (8) has two other solutions:

$$A = \left(-b_1 \pm \sqrt{b_1^2 - 4b_0 b_2} \right) / (2b_2). \tag{12}$$

Equation (12) provides an additional relation between A and s. It shows that due to the dissipative terms only a countable number of amplitudes is possible for a given wavelength, while in the original KdV equation a continuum of amplitudes is possible.

Let us introduce the amplitude $Amp \equiv s^2 A$, which is the level difference between the maximum and minimum values of h_0. The results will be presented in terms of the parametric plots of the allowed amplitudes versus wavenumber, $Amp(k)$, s being a parameter, where $k(s)$ is defined by (3) with $A(s)$ given by (12). First, let us consider the limit cases $s \to 0$ (small amplitude sinusoidal waves limit) and $s \to 1$ (solitary wave limit).

In the limit $s \to 0$, (3) and (12) give the wavenumbers of the excitable modes according to linear theory

$$k_L^2 = [1 \pm (1 - 4\alpha)^{1/2}]/2.$$

In addition we get that $\mathrm{d}\,Amp/\mathrm{d}k = O(1/s^2)$ as $s \to 0$, hence the slope of $Amp(k)$ at the points $Amp = 0$, $k = k_L$, diverges to infinity. The sign of the leading coefficient in the series defines the concavity/convexity of the curve at those points.

Consider now the limit case $s \to 1$. Then from (9)–(12) we have

$$Amp = \frac{\frac{2}{15} \pm \left[\frac{4}{225} - \frac{16}{21}\alpha\left(\frac{1}{3} - \frac{4}{5}D\right)\right]^{1/2}}{\frac{8}{21}\left(\frac{1}{3} - \frac{4}{5}D\right)} \tag{13}$$

which shows that as the only physically relevant case is when Amp is real and non-negative, no solitary waves exist if $D < D_*$, where $D_* = (100\alpha - 7)/(240\alpha) < 5/12$ for all positive α. If $D_* < D < 5/12$ there exist two different solitary waves and if $D \geqslant 5/12$ there is only one solitary wave.

On the other hand the asymptotics of $\mathrm{d}Amp/\mathrm{d}k$ at $s \to 1$ yields that the slope is, in general, finite. When the pair of solitary waves appears at $D = D_*$, $\mathrm{d}Amp/\mathrm{d}k = +\infty$ at the point with higher amplitude, and $\mathrm{d}Amp/\mathrm{d}k = -\infty$ at the point with smaller amplitude if $0 < \alpha < 13/100$, and vice versa if $\alpha > 13/100$. This implies two different mechanisms of appearance of solitary waves as we shall see later on.

These results as well as the results of computer calculations permit us to schematically outline how $Amp(k)$ behaves in various regions of the parameter space $\{\alpha, D\}$. Figure 1 (a–d) corresponds to $0 < \alpha < 13/100$. Small enough D brings to the case (a) (solid line). As D is increased the dotted line appears. When D passes above D_*, the curve $Amp(k)$ crosses the ordinate axis and solitary wave solutions appear (b). At $D = 5/12$ one of the solitary wave solutions disappears at infinity and for $D > 5/12$ we have (c). Finally, for larger D the diagram looks like (d).

Figure 2 (a, b) corresponds to $13/100 < \alpha < 0.236$. For $D < D_*$ we have (a). As D is increased over D_*, a new branch with solitary waves appears (b). The solid line is transformed into dotted one. Unlike Figure 1, here the transition has a different topological character. Other possibilities analogous to those of Figure 2 are also possible as D is increased. In particular, the two branches merge to yield a diagram qualitatively similar to Figure 1 (b).

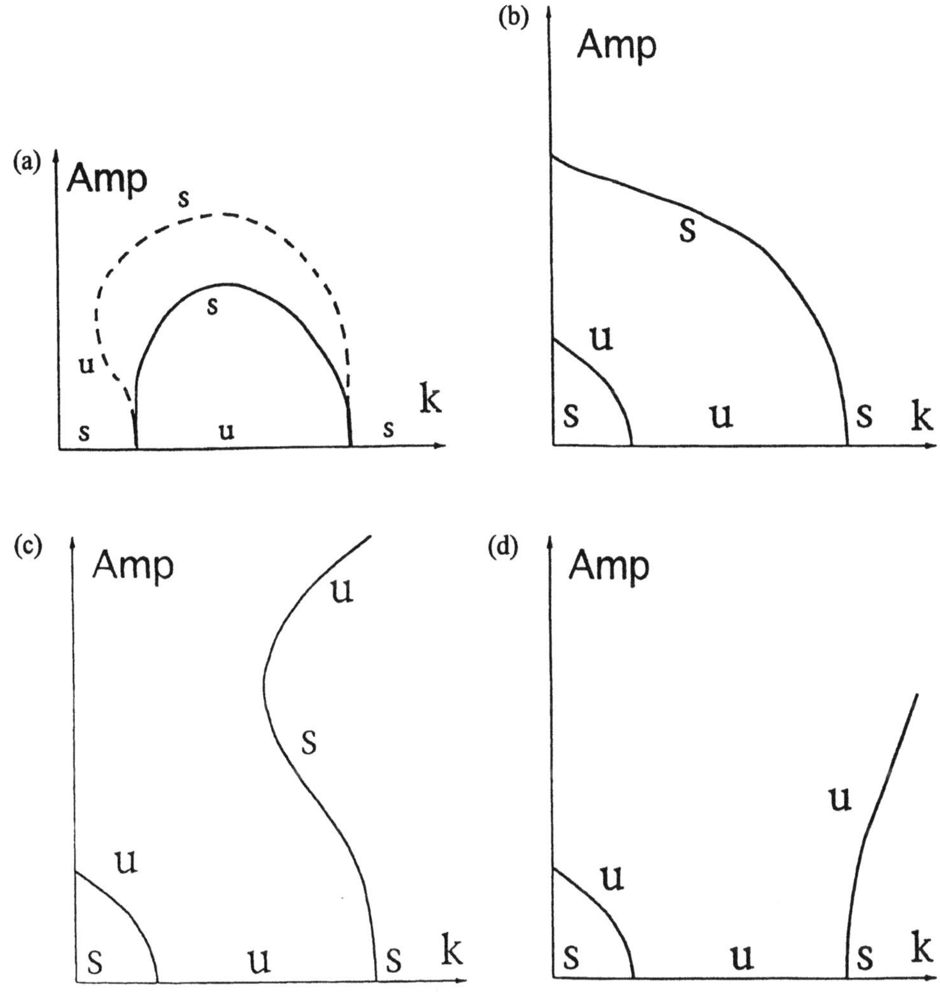

Fig. 1. Cnoidal wave trains: Amplitude versus wavenumber for $0 < \alpha < 13/100$. Transitions from (a) to (d) correspond to increasing values of the nonlinearity parameter D, Equation (1). Marked 's' are stable states and 'u' – unstable ones.

For $0.236 < \alpha < 1/4$, the main difference with the previous case is that one of the solitary wave solutions first disappears at infinity (for $D = 5/12$) and only then (for larger D) the two branches merge.

Figure 3 (a–c) corresponds to $\alpha > 1/4$, when small amplitude sinusoidal waves do not occur and only branches with solitary waves appear. For $D < D_*$ there is no solution other than the rest state. As D is increased over D_* the branch with solitary waves appears (a). Further increasing of D results in (b) and (c). If α is close enough to $1/4$ the dotted curves also occur. The role of the additional nonlinearity ($D \neq 0$) in Equation (1) is crucial. It provides the energy input on the background of a stable motionless state, making possible the excitation of finite amplitude waves.

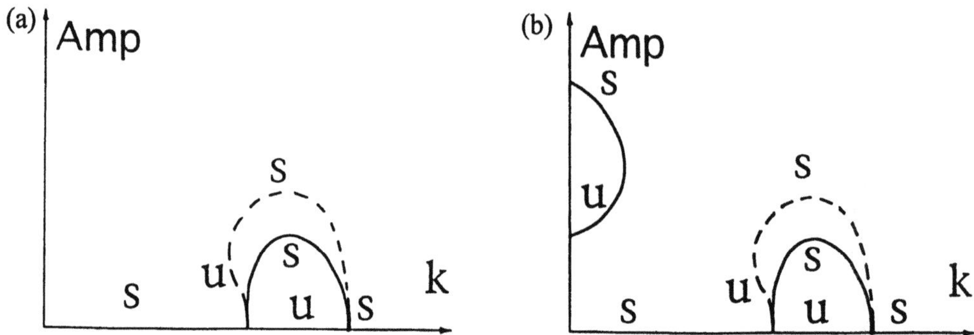

Fig. 2. Cnoidal wave trains: Amplitude versus wavenumber for $13/100 < \alpha < 0.236$. Transitions from (a) to (b) correspond to increasing values of the nonlinearity parameter D, Equation (1). Marked 's' are stable states and 'u' – unstable ones.

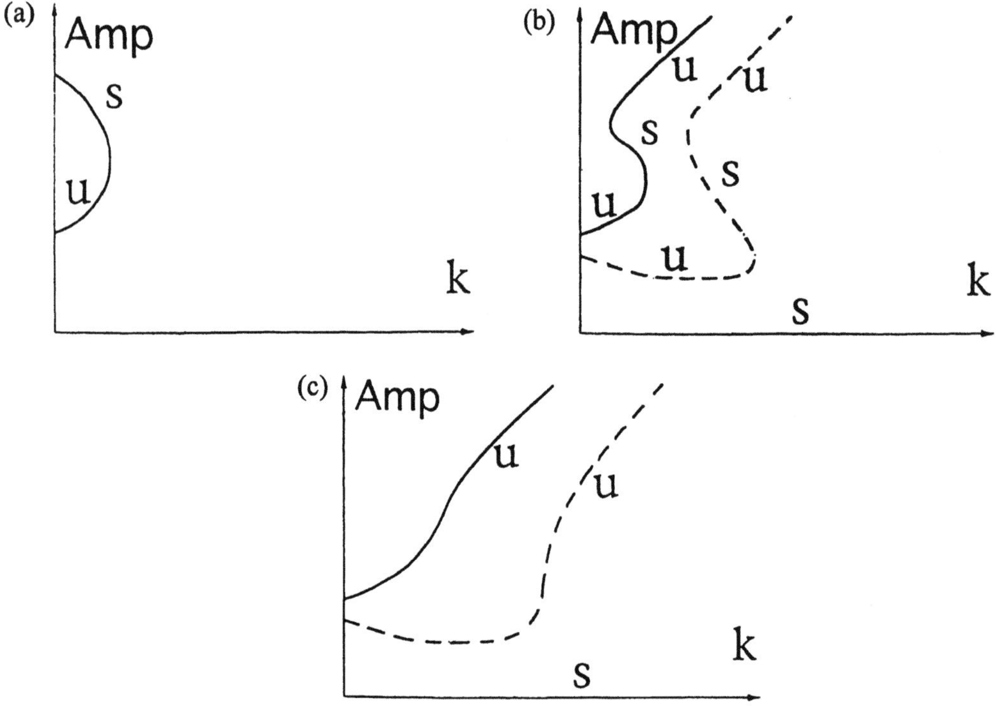

Fig. 3. Cnoidal wave trains: Amplitude versus wavenumber for $\alpha > 1/4$. Transitions from (a) to (c) correspond to increasing values of the nonlinearity parameter D, Equation (1). Marked 's' are stable states and 'u' – unstable ones.

2.2. Stability of cnoidal waves

For any given wavelength, the stability of the solutions can be assessed. To do this let us obtain an evolution equation for the amplitude of a cnoidal wave admitting that the parameters A and s of (2) depend on the slow time scale

$t_1 = \delta t$. Then in the right-hand side (r.h.s.) of (5), a term with time derivative appears.

Repeating the calculations of Subsection 2.1 with the additional term and taking into account that

$$\frac{\mathrm{d}A}{\mathrm{d}t_1} = 2A\frac{E(s) - (1-s^2)K(s)}{s(1-s^2)K(s)}\frac{\mathrm{d}s}{\mathrm{d}t_1}$$

instead of (8) we finally get

$$\frac{\mathrm{d}A}{\mathrm{d}t_1} = -\frac{b_0}{b_3}A - \frac{b_1}{b_3}A^2 - \frac{b_2}{b_3}A^3, \tag{14}$$

where

$$b_3 = \frac{s^2K^2(s) - 2K(s)E(s)s^2 - K^2(s) + 10K(s)E(s) - 9E^2(s)}{8K(s)}.$$

Equation (14) is an evolution equation for A corresponding to a cnoidal wave of given wavelength. In view of (3), Equation (14) can be written in terms of A only. It can be also rewritten in terms of Amp. A limit case corresponding to a solitary wave can be considered. As the r.h.s. of (14) is finite for any $A \geqslant 0$, alternation of the state of stability occurs for steady states which correspond to simple roots of the r.h.s. of (14). Therefore, as we already know the stability of the trivial state, we can easily establish stability-instability of the other regimes, if any, at a given wavelength. According to this, the branches marked 's' are expected to be stable while those marked 'u'– unstable (see Figures 1–3). The points of the curves at which $\mathrm{d}\,Amp/\mathrm{d}k = \infty$ separate stable and unstable branches (these points do not correspond to simple roots of the r.h.s. of (14)). Note that we refer here to 'vertical' state of stability, i.e. stability for a fixed wavelength. Waves, stable in this sense, may be unstable to perturbation of other wavelengths ('horizontal' instability), as we shall see later on.

It appears that in some cases for a certain range of wavelengths there may be no stable solution with the trivial state also being unstable. In other cases a wave with the maximum amplitude is unstable. Hence other solutions for the given wavelength, although linearly stable, may be unstable to finite amplitude disturbances.

2.3. Solitary Waves

Unlike a cnoidal wave of large but finite wavelength, for a solitary wave the solution $h_0 + \delta h_1$ as found in Subsection 2.1 is not uniformly valid in the interval $-\infty < x < +\infty$. For (1) with $\alpha = 0$, this has been shown by Nepomnyashchy and Velarde [3]. For $\alpha \neq 0$, the situation is even more drastic. For a steady solitary wave, (1) yields

$$\int_{-\infty}^{+\infty} h\,\mathrm{d}x = 0 \tag{15}$$

which is not the case if $\alpha = 0$. In the zeroth order, (1) gives a KdV soliton

$$h_0 = \frac{A}{\cosh^2[(A/6)^{1/2}x]} \tag{16}$$

and

$$\int_{-\infty}^{+\infty} h_0 \,\mathrm{d}x = 2\sqrt{6A}. \tag{17}$$

Thus, the uniformly valid solution is irrelevant since it can not satisfy (15). Then a more detailed study is needed to prove the existence of solitary waves. The consideration of Subsection 2.1 can only tell us that if (1) admits a solitary wave solution, its amplitude is given by (13).

First, following the scheme of Subection 2.1, we find the solution in the form $h_0 + \delta h_1$, where h_0 is determined by (16). We get

$$\begin{aligned} g' &= -(6/A)^{1/2} S\left(\frac{2}{15}C^3 + \frac{1}{5}C\right) + (6A)^{1/2} S\left(\frac{4}{63}C^3 + \frac{2}{21}C + \frac{2}{7C}\right) - \\ &\quad - D(6A)^{1/2} S\left(\frac{16}{105}C^3 + \frac{8}{35}C + \frac{2}{7C}\right) + \alpha(6/A^3)^{1/2} SC^3 + \\ &\quad + \frac{3}{4}\frac{c_1}{A}\frac{C^2}{S^2} + \mathrm{const}_1 \frac{C^4}{S^2} + \mathrm{const}_2 \frac{C^6}{S^2} \end{aligned} \tag{18}$$

where $S = \sinh[(A/6)^{1/2}x]$ and $C = \cosh[(A/6)^{1/2}x]$.

To avoid the terms that leave h_1 unbounded for $x \to \infty$ we demand

$$\mathrm{const}_2 = 0 \tag{19}$$

and

$$-\frac{2}{15}\left(\frac{6}{A}\right)^{1/2} + \frac{4}{63}(6A)^{1/2} - \frac{16}{105}D(6A)^{1/2} + \alpha\left(\frac{6}{A^3}\right)^{1/2} = 0 \tag{20}$$

that just gives (13) for the solitary wave amplitude (note that $Amp = A$ in the solitary wave limit $s \to 1$). Integrating (18) taken into account (19) and (20) and recalling that $h_1 = gh_0'$ we get

$$\begin{aligned} h_1 &= -3\alpha\left(\frac{6}{A}\right)^{1/2}\left[\mathrm{const}_1\left(\frac{1}{2} + \frac{3}{2}\left(\frac{A}{6}\right)^{1/2}\frac{S}{C^3}x - \frac{3}{2}\frac{1}{C^2}\right) - \frac{1}{2}\frac{S}{C}\right] - \\ &\quad - \frac{24}{7}(1-D)\left(\frac{A^3}{6}\right)^{1/2}\frac{S}{C^3}\ln C - \\ &\quad - \frac{3}{2}c_1\left(\frac{A}{6}\right)^{1/2}\left[\frac{S}{C^3}x - \left(\frac{6}{A}\right)^{1/2}\frac{1}{C^2}\right] + \mathrm{const}_3\frac{S}{C^3}. \end{aligned} \tag{21}$$

Here const_1 has been redefined for convenience and const_3 denotes another constant of integration.

The expansion $h_0 + \delta h_1$ is not uniformly valid and it fails for large enough $|x| \cong |\ln(\delta)|$. The solution $h_0 + \delta h_1$, correct for $|x| = O(1)$, has to be matched to the solution found for large $|x|$. In the leading approximation, as $h \to 0$ at $x \to \infty$, the latter is just a solution of (1), linearized around $h = 0$:

$$h = \sum_{n=1}^{4} r_n \exp(\lambda_n x), \tag{22}$$

where r_n are some constants and

$$\lambda_{1,2} = \pm(c_0)^{1/2} + \delta\left(-\frac{1 + c_0 + \alpha/c_0}{2} \pm \frac{c_1}{2(c_0)^{1/2}}\right) + O(\delta^2),$$

$$\lambda_3 = -1/\delta + O(1), \qquad \lambda_4 = \delta\alpha/c_0 + O(\delta^2).$$

Consider the asymptotics of $h_0 + \delta h_1$ at large positive x ($x \cong -\ln\delta$):

$$h_0 + \delta h_1 \cong 4A \exp\left[-2\left(\frac{A}{6}\right)^{1/2} x\right] - \frac{3\alpha}{2}\left(\frac{6}{A}\right)^{1/2} (\text{const}_1 - 1).$$

This must be matched to (22), where $r_1 = r_4 = 0$, as $h \to 0$ at $x \to +\infty$. Hence the matching is possible only if

$$\text{const}_1 = 1. \tag{23}$$

In view of (23), the expansion $h_0 + \delta h_1$ is already valid up to $x = O(1/\delta)$ for positive x. Then the asymptotics of $h_0 + \delta h_1$ at large positive x ($x \cong 1/\delta$) reads

$$h_0 + \delta h_1 \cong 4A \exp\left[-2\left(\frac{A}{6}\right)^{1/2} x\right] \times \left[1 - \delta\frac{9\alpha}{2A}x - \delta\frac{4(1-D)A}{7}x - \delta\frac{3c_1}{8}\left(\frac{A}{6}\right)^{1/2}\right]. \tag{24}$$

Then, taking into account (13) and that

$$c_0 = \frac{2A}{3} \tag{25}$$

the demand of (22) matching to (24) yields

$$r_1 = r_3 = r_4 = 0, \qquad r_2 = 4A.$$

The asymptotic form of $h_0 + \delta h_1$ at large negative x ($x \cong \ln\delta$) is

$$h_0 + \delta h_1 \cong 4A \exp\left[2\left(\frac{A}{6}\right)^{1/2} x\right] - 3\alpha\left(\frac{6}{A}\right)^{1/2} \delta.$$

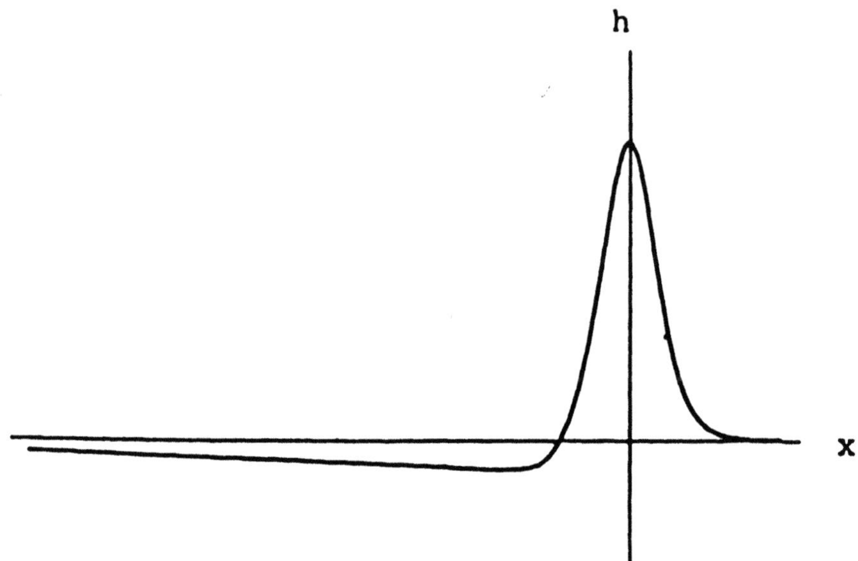

Fig. 4. Schematic profile of a dissipative solitary wave with asymmetry in the mass distribution.

Then matching to (22) gives

$$r_2 = r_3 = 0, \qquad r_1 = 4A, \qquad r_4 = -3\alpha\left(\frac{6}{A}\right)^{1/2}\delta.$$

Note that at this stage we must take $\lambda_1 = (c_0)^{1/2}$, without the term $O(\delta)$.

Summing up, in the region $\ln\delta \leqslant x \leqslant 1/\delta$ the solution is $h_0 + \delta h_1$, where h_0 is defined in (16), h_1 in (21), $A = Amp$ in (13) and const_1 in (23). In the region $x \geqslant 1/\delta$ the solution is

$$h = 4A\exp\left[-\left((c_0)^{1/2} + \delta\frac{1 + c_0 + \alpha/c_0}{2} + \delta\frac{c_1}{2(c_0)^{1/2}}\right)x\right]$$

while in the region $x \leqslant \ln\delta$ we have

$$h = 4A\exp[(c_0)^{1/2}x] - 3\alpha\left(\frac{6}{A}\right)^{1/2}\delta\exp\left[\frac{\delta\alpha}{c_0}x\right]$$

with (25) as the relationship between c_0 and A.

The resulting solitary wave is asymmetric. Its profile is schematically outlined in Figure 4. The head is monotonous, while the tail is not, as a consequence of $\alpha \neq 0$. When $\alpha = 0$, both head and tail are monotonous albeit still the wave is asymmetric as already noted by Nepomnyashchy and Velarde [3].

In the leading order, the solution satisfies (15). Indeed, the leading order is given only by

$$\int_{-\infty}^{+\infty} h \, \mathrm{d}x = \int_{-\infty}^{+\infty} h_0 \, \mathrm{d}x - \int_{-\infty}^{0} 3\alpha\left(\frac{6}{A}\right)^{1/2} \delta \exp\left(\frac{\delta\alpha}{c_0}x\right) \mathrm{d}x + \mathrm{o}(1),$$

and using (17) and (25) we get

$$\int_{-\infty}^{+\infty} h \, \mathrm{d}x = \mathrm{o}(1).$$

Two constants still have been left undetermined in the solution, namely, const_3 (see (21)) and c_1. The first of them accounts for a mere shift along x in (16). They are specified in higher order approximations.

3. Numerical Study with Periodic Boundary Conditions

3.1. Initial Value Problem

To illustrate the evolution and practical stability of wave trains we have integrated numerically Equation (1) with periodic b.c. in an interval of length L. An explicit scheme has been used. Various sets of parameter values for D, α and L have been considered with δ always small, e.g., between 0.01 and 0.1.

Take for instance $\alpha = 0.1, D = 0.2$ and $\delta = 0.1$ which corresponds to Figure 1 (b). For $L = 40$ only cnoidal trains with wavelengths $L, L/2, L/3, L/4, L/5$, and $L/6$ are possible. Trains with wavelengths $L/7, L/8$, etc. have zero amplitude. Although we have used quite a variety of initial conditions, as rather quickly initial conditions decay into the manifold of predicted possible solutions, for all practical purposes it suffices to take $B\sin[(2\pi/L)nx]$, at $t = 0$ with $n = 1, \ldots, 6$ and B large enough. As the computation goes on, the value

$$E = \frac{1}{2L}\int_0^L h^2 \, \mathrm{d}x$$

is examined as a function of time and its stabilization around a constant value within for instance seven digits is considered as a useful criterion of the approach of a stationary state, hence a useful criterion for practical stability.

Figure 5 shows the final profile for the wave $L/2$. The numerical amplitudes agree with the asymptotic results of Subsection 2.1 (within one or two percent). We have also found the waves $L/3$ and $L/4$. The state $L/6$ has not been observed. We shall come back to this point in the Subsection 3.2. The results for $L/5$ are represented in Figure 6 (a) in terms of E versus t. When the time reached 500 the profile with five maxima looks final if we restrict to say three-digit precision (Figure 6 (b)). However within a seven-digit precision no stabilization was observed and upon continuation of the calculation (up to time 1700) the five-hump train yields to a wave train with four humps (Figure 5), that corresponds

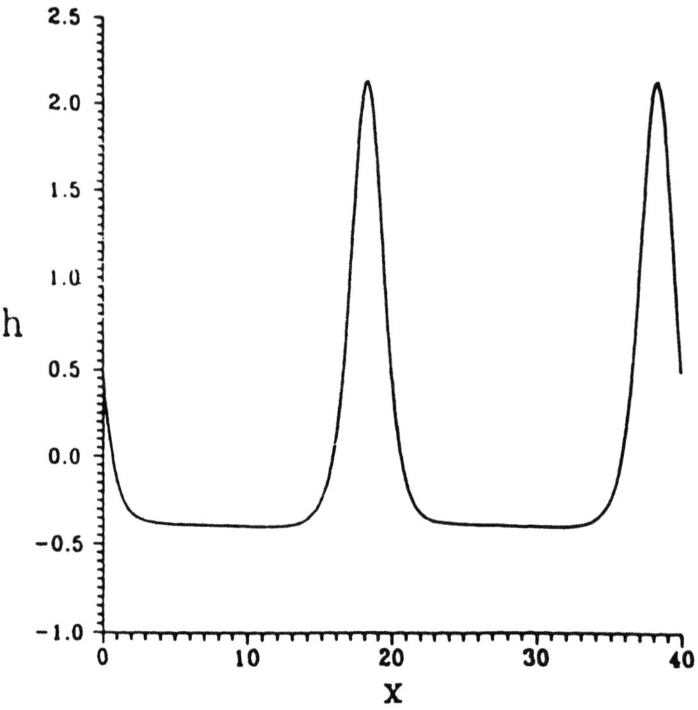

Fig. 5. Cnoidal wave trains for $\alpha = 0.1$, $D = 0.2$ and $\delta = 0.1$ on the interval $L = 40$.

to the second plateau in Figure 6 (a). An example of transition profile is shown in Figure 6 (c).

The wave L has been obtained only using special initial conditions. Indeed as seen in Figure 1 (b) the existence of a unstable wave train (lower branch, left side) demands finite amplitude perturbations to achieve the stable wave train (upper branch). For example, we have taken a KdV solitary wave with suitably large amplitude. The final result is the wave profile shown in Figure 7 for $t \approx 300$ whose amplitude agrees with the analytical results (Subsection 2.1).

In the case $\alpha = 0.24$, $D = 0$, $L = 35$ and $\delta = 0.1$ we expect a wave profile close to sinusoidal. The computation (Figure 8) confirms this and again the number of maxima and the amplitude are in a good agreement with the analytical results (Subsection 2.1).

Take now $\alpha = 0.26$, $D = 0.38$, $L = 35$ and $\delta = 0.05$, which corresponds to Figure 3 (a). In this case the motionless state is linearly stable though unstable to finite amplitude disturbances. Again as initial condition we have used a KdV solitary wave with large enough amplitude. Then at $t = 250$ we obtain a wave with one maximum, similar to Figure 7, but about seven times taller and half thinner.

Finally, it is interesting to note that there are cases when for some wavelengths the motionless state is unstable and yet no cnoidal wave train is predicted (e.g.,

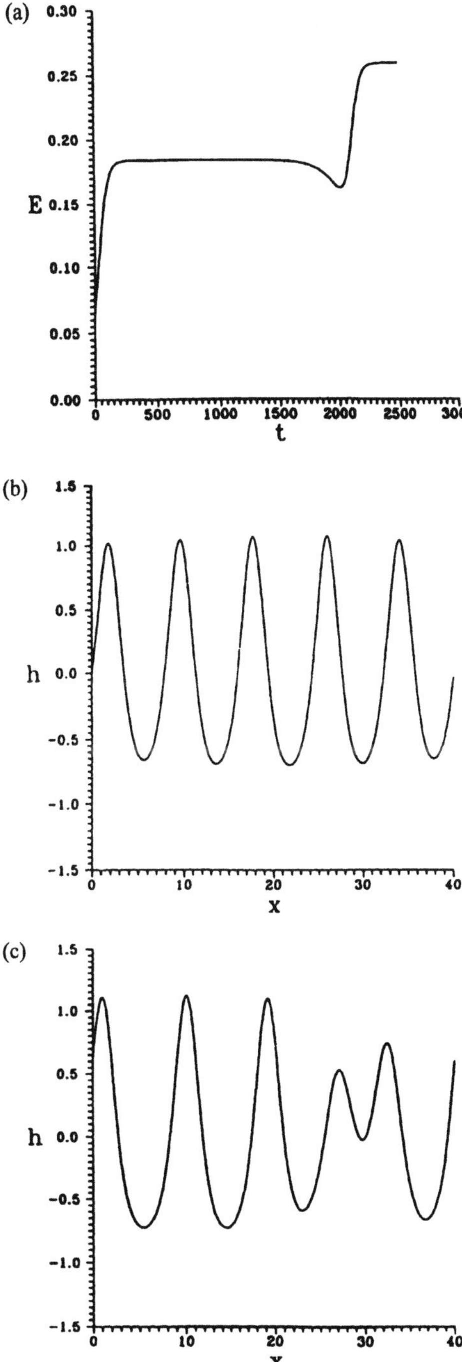

Fig. 6. (a) Energy versus time for an initial wave train with five maxima. (b) A possible, though unstable according to theory, and apparently final state. (c) A profile corresponding to transition from five to four maxima which is actually the stable final state, according to the theory and numerics.

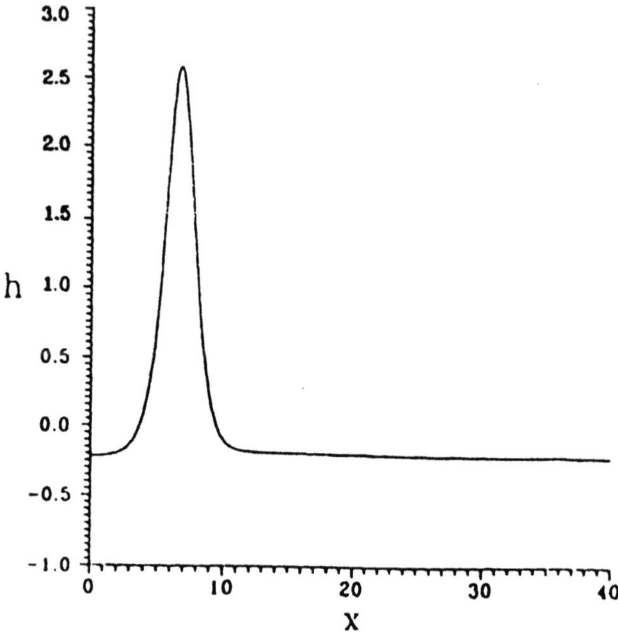

Fig. 7. Solitary wave. Parameter values: $\alpha = 0.1$, $D = 0.2$, $\delta = 0.1$, $L = 40$.

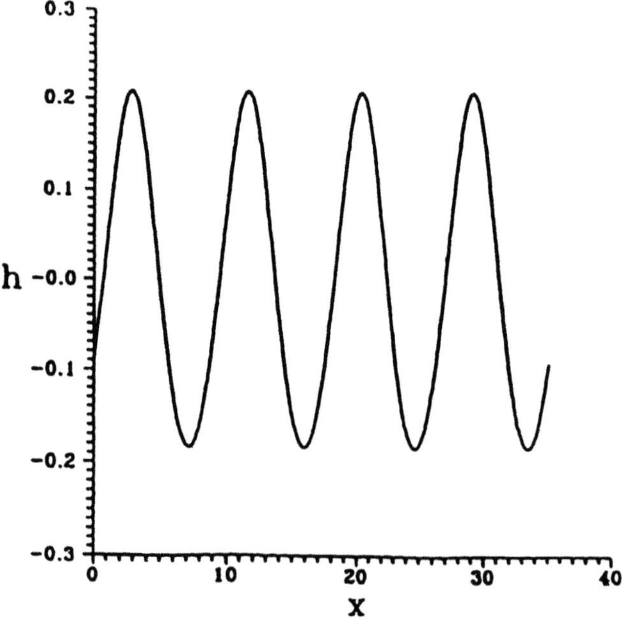

Fig. 8. Cnoidal wave profile for $\alpha = 0.24$, $D = 0$, $\delta = 0.1$ for $L = 35$.

Figure 1 (d)). Take $\alpha = 0.1$, $D = 2$, $L = 20$, $\delta = 0.1$. We do not approach a stable solution but rather we have blow up in finite time. Computation was interrupted. The term responsible for blow-up is the nonlinear term of Equation (1) with coefficient D.

3.2. Stability Analysis

Let us come back to the first case considered in the previous section $\alpha = 0.1$, $D = 0.2$, $L = 40$, $\delta = 0.1$ where wave trains with five and six maxima did not appear as stable cases. With $L = 8$ and the other parameters being the same, upon integration of (1), we easily get the regime with one maximum. For $L = 40$ this state corresponds to five maxima. A similar situation is expected for the case of six maxima. To get an understanding of these and other results we now proceed to a stability analysis of the cnoidal waves.

Let $w(x)\exp(-\lambda t)$ be a perturbation on the periodic wave train $h(x)$ of (1) for some interval L. Linearization yields the equation for $w(x)$

$$\lambda w = cw' + (2hw)' + w''' + \delta\left[w'' + w'''' + D(2hw)'' + \alpha w\right] \tag{26}$$

where the problem is considered in the moving frame. Using a Fourier series description we get

$$w = \sum_{n=1}^{\infty} a_n \phi_n(x),$$

with

$$\phi_n = \begin{cases} \left(\frac{2}{L}\right)^{1/2} \sin(\mu_k x), & n = 2k-1, \\ \left(\frac{2}{L}\right)^{1/2} \cos(\mu_k x), & n = 2k, \end{cases}$$

and

$$\mu_k = \frac{2\pi}{L}k.$$

Note that as

$$\int_0^L w \, \mathrm{d}x = 0$$

no constant term appears in the expansion of w.

Then to obtain the eigenvalues λ of Equation (26) we merely need finding the eigenvalues of an infinite matrix B_{mn}, $m = 1, 2, \ldots, \infty$, $n = 1, 2, \ldots, \infty$, where

$$\begin{aligned} B_{m,n} = {} & (-\mu_m c - \mu_n Q_{n,m+1} - \mu_m^3)\delta_{m+1,n} + \\ & + (\mu_m c + \mu_m Q_{n,m-1} + \mu_m^3)\delta_{m-1,n} + \\ & + \delta\left[(-\mu_m^2 + \mu_m^4 + \alpha)\delta_{m,n} - D\mu_m^2 Q_{m,n}\right] \end{aligned}$$

(a)

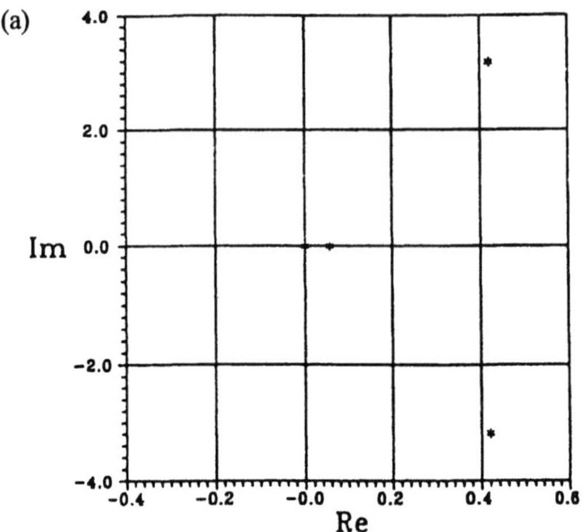

4.0
2.0
Im 0.0
-2.0
-4.0
-0.4
-0.2
-0.0
0.2
0.4
0.6
Re

(b)

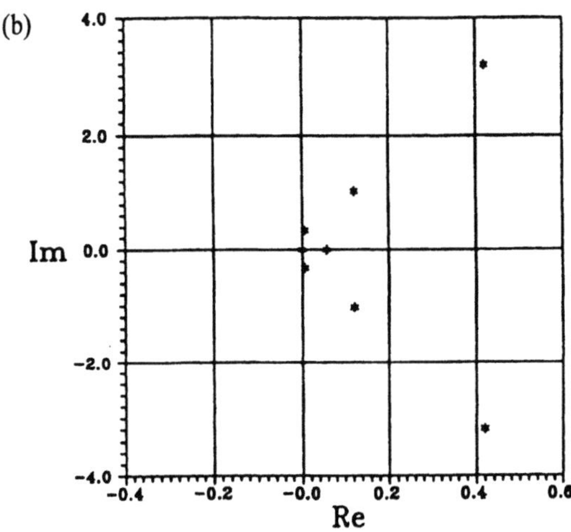

4.0
2.0
Im 0.0
-2.0
-4.0
-0.4
-0.2
-0.0
0.2
0.4
0.6
Re

(c)

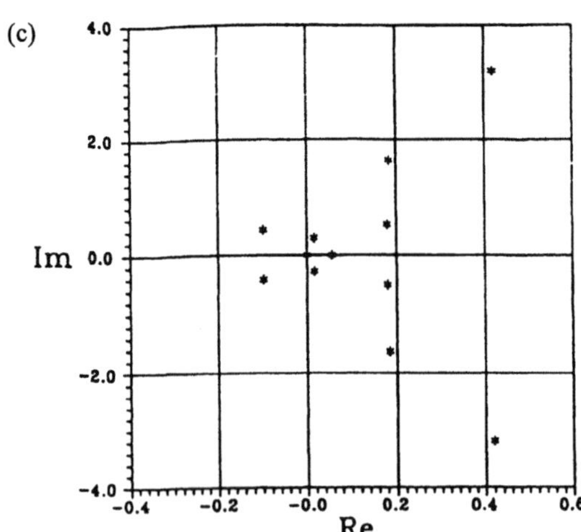

4.0
2.0
Im 0.0
-2.0
-4.0
-0.4
-0.2
-0.0
0.2
0.4
0.6
Re

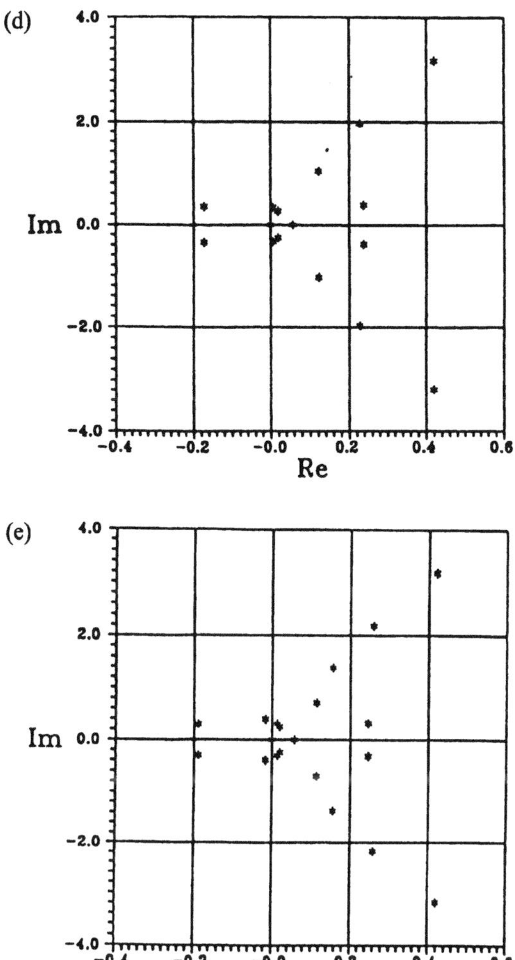

Fig. 9. The most dangerous increments of linear perturbation around (a) one-hump wave for $L = 8$, (b) two-hump solution for $L = 16$, (c) three-hump solution for $L = 24$, (d) four-hump solution for $L = 32$ and (e) five-hump solution for $L = 40$. Parameter values: $\alpha = 0.1$, $D = 0.2$, $\delta = 0.1$.

with

$$Q_{m,n} = \int_0^L 2h\phi_m\phi_n \,\mathrm{d}x,$$

$$\delta_{m,n} = \begin{cases} 1, & \text{if } m = n, \\ 0, & \text{if } m \neq n. \end{cases}$$

To approximately solve the problem we truncate the matrix, with 100×100 elements. Take for instance $h(x)$ for $L = 8$ that according to the numerical criterion indicated in Subsection 3.1 is 'practically' stable. Consider its natural extension to the case $L = 16, 24, 32, 40$, etc. Shown in Figure 9 (a–e) are the lowest eigenvalues for $L = 8, 16, 24, 32$, and 40, respectively. Note that $\lambda = 0$ is always present since it accounts for a mere shift along x. For $L = 8$ and 16 we have no unstable modes, while for $L = 24$ and 32 a conjugate eigenvalues with negative real part appears hence leading to instability. For $L = 40$ we have two pairs of complex conjugate eigenvalues and as their growth rate is rather small the corresponding wave train is unstable yet with a long transient as discussed in Subsection 3.1.

4. Conclusions

Cnoidal wave solutions of a generalized dissipation-modified KdV equation have been considered. Unlike the original KdV equation, where for a given wavelength a continuum of wave amplitudes and velocities exists, here due to the input–output energy balance input–output the dissipative terms select a countable number of amplitudes and velocities. The allowed amplitudes (velocities) have been asymptotically calculated and their dependence on wavelength and the parameters of the problem has been provided (Figures 1–3). Then we have numerically solved the initial value problem with periodic boundary conditions for a container of length L. In accordance with the asymptotic analytical results, the initial conditions select what cnoidal wave is actually realized among the possible ones predicted by the theory for a given period and the boundary conditions. For wave amplitudes, the agreement with the asymptotic results is quite good. We have also provided a stability analysis of the expected waves hence providing an additional selection mechanism and reinforcing the numerical criterion used for practical stability of the final (steady) states. The limit cases of solitary waves and small amplitude sinusoidal waves have been also considered.

Acknowledgements

A. Ye. R. and V. N. K. wish to express their gratitude to the DGICYT of the Ministry of Education and Science (Spain) for a postdoctoral fellowship, and Yu. S. R. for a visiting position at E.T.S.I. Aeronauticos, Universidad Politecnica de Madrid. The authors acknowledge discussions with C. I. Christov, X.-L. Chu, H. Linde, and V. I. Nekorkin. This research has been supported by DGICYT (Spain) under Grant PB 93–81 and by the European Union under Network Grant ERBCHRXCT 93-107.

References

1. Nepomnyashchy, A. A.: Wave motions in a layer of viscous flowing down an inclined plane, *Trans. Permj Univ.* **362** (1976), 114–119, (in Russian).
2. Garazo, A. N. and Velarde, M. G.: Dissipative Korteweg–de Vries description of Marangoni–Bénard oscillatory convection, *Phys. Fluids A* **3** (1991), 2295–2300.

3. Nepomnyashchy, A. A. and Velarde, M. G.: A three-dimensional description of solitary waves and their interaction in Marangoni–Bénard layers, *Phys. Fluids* **6** (1994), 187–198.
4. Rednikov, A. Ye., Velarde, M. G., and Ryazantsev, Yu. S.: Dissipation modified Korteweg–de Vries equation describing Marangoni–Bénard convection: Variations on asymptotics, in M. G. Velarde and C. I. Christov (eds), *Fluid Physics*, World Scientific, Singapore, 1994.
5. Korteweg, D. J. and de Vries, G.: On the change of form of long waves advancing in a rectangular canal, and on a new type of long stationary waves, *Phil. Mag.* **39**(5) (1895), 422–443.
6. Nekorkin, V. I. and Velarde, M. G.: Solitary waves and soliton bound states of a dissipative Korteweg–de Vries equation describing Marangoni–Bénard convection and other thermoconvective instabilities, *Int. J. Bifurcation and Chaos* (1994), to appear.
7. Christov, C. I. and Velarde, M. G.: Dissipative solitons, *Physica D* (1994), to appear.
8. Kawahara, T. and Toh, S.: Nonlinear dispersive periodic waves in the presence of instability and damping, *Phys. Fluids* **28** (1985), 1636–1638.
9. Kawahara, T. and Toh, S.: Pulse interactions in an unstable dissipative-dispersive nonlinear system, *Phys. Fluids* **31** (1988), 2103–2111.
10. Weidman, P. D., Linde, H., and Velarde, M. G.: Evidence for solitary wave behavior in Marangoni–Bénard convection, *Phys. Fluids A* **4** (1992), 921–926.
11. Linde, H., Chu, X.-L., and Velarde, M. G.: Oblique and head-on collisions of solitary waves in Marangoni–Bénard convection, *Phys. Fluids A* **5** (1993), 1068–1070.
12. Linde, H., Chu, X.-L., Velarde, M. G., and Waldhelm, W.: Wall reflections of solitary waves in Marangoni–Bénard convection, *Phys. Fluids A* **5** (1993), 3162–3166.
13. Velarde, M. G., Linde, H., Nepomnyashchy, A. A., and Waldhelm, W.: Further evidence of solitonic behavior in Marangoni–Bénard convection: Periodic wave trains, in M. G. Velarde and C. I. Christov (eds), *Fluid Physics*, World Scientific, Singapore, 1994.
14. Byrd, P. F. and Friedman, M. D.: *Handbook of Elliptic Functions for Engineers and Physicists*, Springer, Berlin, 1954.

Acta Applicandae Mathematicae **39:** 477–487, 1995.

Recent Results on the Generalized Kadomtsev–Petviashvili Equations

JEAN-CLAUDE SAUT
CNRS et Université Paris-Sud, URA 760, Bâtiment 425, Université Paris-Sud, 91405 Orsay, France

(Received: 9 February 1995)

Abstract. We survey some recent results concerning the generalized Kadomtsev–Petviashvili equations, which are natural extensions of KdV type equations to higher dimensions. We will focus on rigorous results of the Cauchy problem and on the existence and properties of localized solitary waves.

Mathematics Subject Classification (1991): 58F07.

Key words: KP equations, Cauchy problem, localized solitary waves.

1. Introduction

Kadomtsev–Petviashvili equations are 'universal' models for dispersive, weakly nonlinear waves, which are essentially unidimensional, when weak transverse effects are taken into account (see, e.g., [14, 16]. In this sense, they are natural extensions of the KdV equations in several spatial dimensions.

They read, for a general nonlinearity $f(u)$,

$$\begin{cases} u_t + f(u)u_x + u_{xxx} + \varepsilon v_y = 0, & \\ u = u(x,y,t), & (x,y) \in \mathbb{R}^2,\ t \geqslant 0, \\ v_x = u_y & \end{cases} \tag{1.1}$$

in dimension 2, and

$$\begin{cases} u_t + f(u)u_x + u_{xxx} + av_y + bw_z = 0, & \\ u = u(x,y,z,t), & (x,y,z) \in \mathbb{R}^3,\ t \geqslant 0, \\ v_x = u_y & \\ w_x = u_z & \end{cases} \tag{1.2}$$

in dimension 3. Equations (1.1) and (1.2) are completed with a prescribed initial value φ.

The constants ε, a, b measure the transverse dispersion effects and are normalized to ± 1. The 'usual' Kadomtsev–Petviashvili equations correspond to $f(u) = u$. We will mainly consider therein power nonlinearities. We recall that (1.1) with $f(u) = u$ is integrable by the inverse scattering method and is classically called KP-I ($\varepsilon = -1$) or KP-II ($\varepsilon = 1$).

Note also that one could consider more general dispersion, that is replace u_{xxx} in (1.1) or (1.2) by $-Lu_x$, where L is defined in Fourier variables by

$$\widehat{Lu}(k) = \alpha(k)\hat{u}(k). \tag{1.3}$$

For instance, the choice $L = H\partial_x$, where H is the Hilbert transform leads to 2 or 3-D extensions of the Benjamin–Ono equation (see [20]).

The main mathematical questions concerning equations such as (1.1) or (1.2) are the well-posedness of the initial-value problem (Cauchy problem), the existence and stability of solitary wave solutions and their role in the dynamics.

Many rigorous results have recently appeared concerning those problems (except the last one) and our aim is to present a review of the more significant of them.

For the sake of simplicity, we will restrict throughout this paper to pure power nonlinearities, $f(u) = u^p$, but more general ones could be easily treated for some of the problems considered therein.

2. The Cauchy Problem

Because of the structure of KP type equations, we will need to use functions which are x derivatives of L^2 functions. Using 'standard' spaces leads to technical complications and does not permit to justify the classical invariants of the equations (see [4] and the references therein). We thus find convenient to introduce the spaces (for $k = 1, 2, \ldots$ and $d = 2, 3$)

$$\dot{H}_x^{-k}(\mathbb{R}^d) = \left\{ f \in \mathcal{S}'(\mathbb{R}^d),\ \frac{1}{\xi_1^k}\hat{f}(\xi) \in L^2(\mathbb{R}^d) \right\}$$

equipped with the norm

$$\|f\|_{-k,x} = \left| \frac{1}{\xi_1^k}\hat{f} \right|_2,$$

and

$$X_s = \left\{ f \in H^s(\mathbb{R}^d);\ F^{-1}\left(\frac{\hat{f}}{\xi_1}\right) \in H^s(\mathbb{R}^d) \right\},$$

$$\|f\|_{X_s} = \|f\|_s + \left\| F^{-1}\left(\frac{\hat{f}}{\xi_1}\right) \right\|_s.$$

We have used the standard notations $|\cdot|_p$ (resp., $\|\cdot\|_s$) for the norm in the Lebesgue space $L^p(\mathbb{R}^d)$ (resp., the Sobolev space $H^s(\mathbb{R}^d)$), and $\hat{}$ or F for the Fourier transform. Finally, $\xi = (\xi_1, \xi_2)$ (resp., (ξ_1, ξ_2, ξ_3)) is the dual variable of (x, y) (resp., (x, y, z)).

The operator ∂_x^{-k} acting on $\dot{H}_x^{-k}(\mathbb{R}^d)$ is defined by $\widehat{\partial_x^{-k} f} = (1/i\xi_1)^k \hat{f}$. We start with a local existence theorem.

THEOREM 2.1. *Let* $\varphi \in X_s$, $s \geqslant 3$, *be such that* $\varphi_{yy} \in \dot{H}_x^{-2}(\mathbb{R}^2)$ (*resp.*, $\varphi_{yy}, \varphi_{zz} \in \dot{H}_x^{-2}(\mathbb{R}^3)$). *There exists* $T > 0$ *such that* (1.1) (*resp.*, (1.2)) *has a unique solution* u *corresponding to the initial data* φ *and such that*

$$u \in C([-T,T]; H^s(\mathbb{R}^d)) \cap C^1([-T,T]; H^{s-3}(\mathbb{R}^d)),$$
$$v \in C([-T,T]; H^{s-1}(\mathbb{R}^2)),$$
$$(\textit{resp., } v, w \in C([-T,T]; H^{s-1}(\mathbb{R}^3))).$$

Moreover

$$\frac{\mathrm{d}}{\mathrm{d}t}\int_{\mathbb{R}^d} |u(\cdot,t)|^2 = 0,$$

and the energy

$$E(t) = \int_{\mathbb{R}} \left[\frac{u^{p+2}}{(p+1)(p+2)} - \frac{1}{2}u_x^2 + \frac{\varepsilon}{2}v^2\right] \mathrm{d}x\,\mathrm{d}y$$

$$\left(\textit{resp., } E(t) = \int_{\mathbb{R}^3} \left[\frac{u^{p+2}}{(p+1)(p+2)} - \frac{1}{2}u_x^2 + \frac{a}{2}v^2 + \frac{b}{2}w^2\right] \mathrm{d}x\,\mathrm{d}y\,\mathrm{d}z\right),$$

is well defined and independent of t *on* $[-T,T]$.

Remark 2.1. The existence part of Theorem 2.1 was essentially established by Ukai [22] (see also [13]) for $d = 2$, except for the assertion concerning v (in [22] v was only shown to belong to $H^{s-1}_{\mathrm{loc}}(\mathbb{R}^2)$). The assertions concerning v and the conserved quantities are proven in [17] for $d = 2$. The extension to $d = 3$ is immediate as is the extension to general dispersions like (1.3).

While Theorem 2.1 is relatively satisfactory as far as local existence is concerned, it is a much tougher question to prove that the local solution is in fact global, or develops a singularity in finite time (blow-up in finite time). Since a rather complete answer is known for the KP-I and KP-II equations, we first consider those cases. Concerning KP-II, Bourgain [6] has proven the following

THEOREM 2.2 ([6]). *The Cauchy problem for KP-II is globally well-posed for data* $\varphi \in H^s(\mathbb{R}^2)$, $s \geqslant 0$.

The proof of Theorem 2.2 is a tour de force. Here are a few ideas of it. Bourgain uses a contraction mapping argument on the integral formulation of the KP-II equation. A new idea is the choice of norm. It is a 'Fourier restriction norm', strongly related to the symbol of the linear KP-II equation:

$$|||u||| = \left(\int_{\mathbb{R}^2\times\mathbb{R}} \left(\frac{1}{\tau} + \left|\lambda - m^3 + \frac{n^2}{m}\right|\right) \times \right.$$
$$\left. \times \left(1 + \frac{|\lambda - m^3 + n^2/m|}{1/\tau + |m|^3}\right)^{1/2} |\hat{u}(m,n,\lambda|^2 \,\mathrm{d}\lambda\,\mathrm{d}m\,\mathrm{d}n\right)^{1/2}$$

where $0 < \tau < 1$ and $\hat{}$ denotes the total Fourier transform with respect to x, y, t.

To regain the loss of an n-factor due to the ∂_x appearing in the nonlinearity $\frac{1}{2}\partial_x(u^2)$, Bourgain states a simple arithmetic identity valid for $m = m_1 + m_2$, $n = n_1 + n_2$, $m, m_1, m_2 \neq 0$:

$$\begin{aligned}&\left(m^3 - \frac{n^2}{m}\right) - \left(m_1^3 - \frac{n_1^2}{m_1}\right) - \left(m_1^3 - \frac{n_2^2}{m_2}\right)\\&\quad = \frac{1}{mm_1m_2}\left[3m^2m_1^2m_2^2 + (n_1m_2 - n_2m_1)^2\right],\end{aligned} \tag{2.1}$$

which implies that for $\lambda = \lambda_1 + \lambda_2$,

$$\begin{aligned}&\operatorname{Max}\left\{\left|\lambda - m^3 + n^2/m\right|, \left|\lambda_1 - m_1^3 + n_1^2/m_1\right|, \left|\lambda_2 - m_2^3 + n_2^2/m_1\right|\right\}\\&\quad \geqslant \tfrac{1}{2}\left|mm_1m_2\right|.\end{aligned}$$

The rest of the proof consists in establishing very delicate L^2 convolution estimates. The globalization is obtained thanks to the conservation of the L^2-norm.

Remark 2.2. (i) It is likely, though not obvious to prove that Theorem 1.2 should be valid when the nonlinear term is replaced by $u^p u_x$, at least for small p's. On the other hand, Bourgain's method does not apply to the KP-I equation because the identity corresponding to (2.1) has a minus sign instead of a plus in the right-hand side.

(ii) In [8], Faminskii obtained the global existence of a weak solution to KP-II in a L^2-weighted space.

The first result concerning global solutions of the KP-I equation is due to Schwarz [19] for the periodic case, $(x, y) \in \mathbb{T}^2$. He proved the global well-posedness of the Cauchy problem for small initial data in

$$V_3 = \left\{v, \int_0^1 v(x,y)\,\mathrm{d}x = 0,\ \|v\|_{V_3} = \left\|\sum_{n=0}^{3}\sum_{\ell=0}^{n}\partial_x^{n-\ell}\left(\partial_x^{-1}\partial_y\right)^{\ell} v\right\|_{L^2} < \infty\right\}.$$

This space is strongly connected to the 7th invariant of the KP-I equation. As was noticed in [6], the smallness assumption can be relaxed by using instead of the Sobolev inequality used in [19]

$$\left|\int_0^1\int_0^1 u^3\,\mathrm{d}x\,\mathrm{d}y\right| \leqslant 2\sqrt{3}|u|_2\,|u_x|_2^{3/2}\left|\partial^{-1}u_y\right|_2^{1/2},$$

the estimate

$$\left|\int u^3\,\mathrm{d}x\,\mathrm{d}y\right| \leqslant 2|u|_2^{3/2}\,|u_x|_2\,|v|_2^{1/2}, \tag{2.2}$$

where $v_x = u_y$, which is valid on $\mathbb{T}^2$ or $\mathbb{R}^2$.

In fact, (2.2) implies that u_x, v are controlled in L^2 by the Hamiltonian

$$H(t) = H(0) = \int \left[\tfrac{1}{2}\, v^2 + \tfrac{1}{2}\, u_x^2 - \tfrac{1}{6}\, u^3\right] \mathrm{d}x\, \mathrm{d}y,$$

thanks to the inequality

$$\frac{1}{2}\int (u_x^2 + v^2)\, \mathrm{d}x\, \mathrm{d}y \leqslant 2H(0) + \tfrac{2}{81}\, |\varphi|_2^6.$$

One can similarly obtain the globall well-posedness for KP-I in $\mathbb{R}^2$ under stronger constraints on the initial data φ (see [18]).

Remark 2.3. It is worth noticing that Zhou [26] and Fokas and Sung [10] have obtained global existence of small solutions to the KP-I equation by inverse scattering techniques. A similar result for KP-II was obtained by Wickerhauser [25]. Purely PDE techniques seem to be more powerful in this context.

The situation for KP type equations distinct from KP-I or KP-II is much less satisfactory as far global existence is concerned and no general result is available.

The only method known so far to prove blow-up of solutions in finite time is still the so-called virial method. It has been introduced in the context of nonlinear Schrödinger equations by the Russian school (Vlasov, Petrishev, and Talanov [23]). A mathematical justification was provided by Glassey [12] and Ginibre and Velo [11]. Note that this method does not work for instance for the generalized Korteweg–de Vries equations

$$u_t + u^p u_x + u_{xxx} = 0, \tag{2.3}$$

though numerical simulations [5] do show that blow-up can occur for $p \geqslant 4$. A rigorous proof of the blow-up for $p \geqslant 4$ is therefore still missing. A similar conclusion can be drawn for the generalized KP-I equation. The estimate

$$\left|\int u^{p+2}\, \mathrm{d}x\, \mathrm{d}y\right| \leqslant c|u|_2^{(4-p)/2}\, |u_x|_2^{p}\, |v|_2^{p/2},$$

valid for $0 \leqslant p \leqslant 4$, shows that $\int (u_x^2 + v^2)\, \mathrm{d}x\, \mathrm{d}y$ is controlled in terms of $|\varphi|_2$ and of the Hamiltonian

$$H(t) = H(0) = \int \left[\tfrac{1}{2}\, v^2 + \tfrac{1}{2}\, u_x^2 - \frac{1}{(p+1)(p+2)} u^{p+2}\right] \mathrm{d}x\, \mathrm{d}y$$

if and only if $p \leqslant 4/3$. It is not unreasonable to conjecture that the solution of the generalized KP-I equation should blow up in finite time for $p \geqslant 4/3$ ($p \geqslant 2$ for integer values). This is supported by some numerical results [25]. The same conjecture for the 2-D Benjamin–Ono KP-I type equations or generalized KP-I equations (i.e. $a = b = -1$ in (1.2)) is that blow up should occur for any $p \geqslant 1$.

Similarly to the generalized KdV equations, the critical value $p = 4/3$ seems to be linked to the critical value for the instability of solitary waves solutions (see Section 3).

This type of blow-up could be said to be due to the interaction of the nonlinearity and the longitudinal dispersion (the u_{xxx} or Lu_x term), and actually the candidate for the critical value of p does depend on this dispersion.

Another type of blow-up does occur for generalized KP-I equations, which is only due to the *transverse* dispersion. To state this result precisely we need to introduce the set for $s \geqslant 3$,

$$\Sigma_s = \left\{\varphi \in X_s,\ \varphi_{yy} \in \dot{H}_x^{-2}(\mathbb{R}^2),\ y\varphi \in L^2(\mathbb{R}^2),\ \varphi \neq 0,\ E(\varphi) \geqslant 0\right\}$$

for $d = 2$, and

$$\Sigma_s = \left\{\varphi \in X_s,\ \varphi_{yy}, \varphi_{zz} \in \dot{H}_x^{-2}(\mathbb{R}^3),\ y\varphi, z\varphi \in L^2(\mathbb{R}^3),\ \varphi \neq 0,\ E(\varphi) \geqslant 0\right\},$$

where $E(t)$ was defined in Theorem 2.1.

THEOREM 2.3. (i) *Let* $\varepsilon = -1$ *in* (1.1). *The set* Σ_s *is not empty for any* $p \geqslant 4$. *Let* $\varphi \in \Sigma_s$. *Then the corresponding solution* u *of* (1.1) *blows-up in finite time. More precisely, there exists* $T^* > 0$ *such that*

$$\lim_{t \to T^*} \left|\frac{\partial u}{\partial y}(\cdot, t)\right|_2 = +\infty.$$

(ii) *Let* $a = b = -1$ *in* (1.2). *The set* Σ_s *is not empty for any* $p \geqslant 2$. *For* $\varphi \in \Sigma_s$, *the corresponding solution* u *of* (1.2) *blows-up in finite time. More precisely, there exists* $T^* > 0$ *such that*

$$\lim_{t \to T^*} \left|\nabla^{\perp} u(\cdot, t)\right|_2 = +\infty$$

where $\nabla^{\perp} = (\partial/\partial y, \partial/\partial z)$.

The proof consists essentially in deriving a virial identity which has been stated formally by Turytsyn and Falkovich [21]. The mathematical justification is given in [17]. It involves in particular the proof of existence results in weighted Sobolev spaces. Let

$$I(t) = \int_{\mathbb{R}^2} y^2 u^2(x, y, t)\, \mathrm{d}x\, \mathrm{d}y \quad \text{for } d = 2,$$

and

$$I(t) = \int_{\mathbb{R}^3} (y^2 + z^2) u^2(x, y, z, t)\, \mathrm{d}x\, \mathrm{d}y\, \mathrm{d}z \quad \text{for } d = 3.$$

Then the virial identity is

$$\frac{d^2}{dt^2}I(t) = 4\varepsilon\left[pE(t) + \frac{p}{2}\int_{\mathbb{R}^2} u_x^2\,dx\,dy + \frac{(4-p)\varepsilon}{2}\int_{\mathbb{R}^2} v^2\,dx\,dy\right]$$

or

$$\frac{d^2}{dt^2}I(t) = 4\varepsilon\left[2pE(t) + p\int_{\mathbb{R}^3} u_x^2\,dx\,dy\,dz + (2-p)\varepsilon\int_{\mathbb{R}^3} (v^2+w^2)\,dx\,dy\,dz\right]$$

(where $\varepsilon = a = b$).

The proof of blow-up then results from the classical inequality

$$|u|_2^2 \leqslant C\left|r^{\perp}u\right|_2\left|\nabla^{\perp}u\right|_2,$$

where $r^{\perp} = y$ (resp., (y, z)), $\nabla^{\perp} = \partial/\partial y$ (resp., $(\partial/\partial y, \partial/\partial z)$), and from the conservation of the L^2-norm.

Remark 2.4. (i) The blow-up result stated in Theorem 2.3 does not depend on the dispersion in x. For instance, in the case of the generalized Benjamin–Ono KP

$$\begin{cases} u_t + u^p u_x - Hu_{xx} + \varepsilon v_y = 0, \\ v_x = u_y, \end{cases}$$

one obtains the virial identity

$$\frac{d^2}{dt^2}I(t) = 4\varepsilon\left[pE(t) + \frac{p}{2}\int_{\mathbb{R}^2} uHu_x\,dx\,dy - \frac{\varepsilon(p-4)}{2}\int_{\mathbb{R}^2} v^2\,dx\,dy\right]$$

where

$$E(t) = E(0) = \int_{\mathbb{R}^2}\left[\frac{u^{p+2}}{(p+1)(p+2)} - \frac{1}{2}uHu_x + \frac{\varepsilon}{2}v^2\right]dx\,dy,$$

which leads to the same blow-up conditions ($\varepsilon = -1,\ p \geqslant 4$) as for the generalized KP-I equations.

(ii) This virial technique is useless for generalized KP-II equations ($\varepsilon = 1$ in (1.1) or $(a, b) \neq (-1, -1)$ in (1.2)). For instance, the virial identity for (1.1) with ε =1 is

$$\frac{d^2}{dt^2}I(t) = 4\int_{\mathbb{R}^2}\left[\frac{u^{p+2}}{(p+1)(p+2)} + 2v^2\right]dx\,dy,$$

and the right-hand side (which is not a conserved quantity) is, for instance, positive for even p's.

The problem of global existence (or blow-up in finite time) remains totally open for the generalized KP-II equations ($p \geqslant 2$).

3. Solitary Waves

The importance of localized solutions of nonlinear dispersive equations cannot be overestimated. On the other hand, until recently, little was known (rigorously) concerning localized solutions of generalized KP equations. Except for the line soliton of the KdV equation, the only known localized solutions were the so-called lump solutions of the KP-I equations (see [1, 2]). It turns out that the problem of the existence of solitary waves of the generalized KP Equations (1.1) (1.2) (with pure power nonlinearity) can be completely classified. Before stating the result we define precisely what we mean by solitary waves.

We shall denote for $d = 2, 3$,

$$X = \left\{\varphi \in H^1(\mathbb{R}^d),\ \partial_x^2\varphi \in L^2(\mathbb{R}^d)\right\},$$

and Y the closure of $\partial_x\ ((C_0^\infty(\mathbb{R}^d))$ for the norm

$$\|\partial_x\varphi\|_Y = \left(\|\nabla\varphi\|_{L^2}^2 + \left\|\partial_x^2\varphi\right\|_{L^2}^2\right)^{1/2},$$

where $\partial_x(C_0^\infty(\mathbb{R}^d))$ denotes the space of functions of the form $\partial_x\varphi$ with $\varphi \in C_0^\infty(\mathbb{R}^d)$ (i.e. the space of functions ψ in $C_0^\infty(\mathbb{R}^d)$ such that $\int_{-\infty}^{\infty}\psi(x, x')\,dx = 0$, for every $x' \in \mathbb{R}^{d-1}$).

DEFINITION 3.1. A solitary wave of (1.1) (resp., (1.2)) is a solution of the type $u(x - ct, y)$ (resp., $u(x - ct, y, z)$), where $u \in Y$ and $c > 0$.

By standard imbedding theorems, if $u \in Y$ and $d = 3$, then $u = \partial_x\varphi$, where $\varphi \in L^6(\mathbb{R}^3)$; if $d = 2$ and $u \in Y$, then $u = \partial_x\varphi$, where $\varphi \in L^q_{\text{loc}}(\mathbb{R}^2)$, for all $q < +\infty$. Note that for $d = 2$, the choice of $\varphi \in L^q_{\text{loc}}$ such that $u = \partial_x\varphi$ is not unique, but two such φ will differ by a function $\psi(y)$ independent of x. Hence, only one of them (up to a constant) satisfies $v = \partial_y\varphi \in L^2(\mathbb{R}^2)$. We assume in all what follows that when $u \in Y$ and when we take $\varphi \in L^q_{\text{loc}}$ with $\partial_x\varphi = u$, we also have $v = \partial_y\varphi \in L^2$. We then denote $v = \partial_y\varphi$ by $\partial_x^{-1}u_y$.

We are thus looking for 'localized' solutions to the systems

$$\begin{cases} -cu_x + u^p u_x + u_{xxx} + \varepsilon v_y = 0, \\ v_x = u_y, \end{cases} \tag{3.1}$$

$$\begin{cases} -cu_x + u^p u_x + u_{xxx} + av_y + bw_z = 0, \\ v_x = u_y, \\ w_x = u_z. \end{cases} \tag{3.2}$$

The following result was established in [7].

THEOREM 3.1 ([7]). (i) *Assume that* $d = 2$. *The Equation* (1.1) *does not admit any nontrivial solitary wave satisfying* $u = \partial_x \varphi$, $\varphi \in X$, $u \in H^1(\mathbb{R}^2) \cap L^\infty_{\mathrm{loc}}(\mathbb{R}^2)$, $\partial_x^2 u$, *and* $\partial_y^2 \varphi \in L^2_{\mathrm{loc}}(\mathbb{R}^2)$, *if*

$$\varepsilon = -1 \quad \textit{and} \quad p \geqslant 4 \tag{3.3}$$

or

$$\varepsilon = +1 \quad \textit{and} \quad p \textit{ is arbitrary.} \tag{3.4}$$

On the other hand, (1.1) *admits a nontrivial solitary wave which belongs to* $H^\infty(\mathbb{R}^2) = \bigcap_{k \geqslant 0} H^k(\mathbb{R}^2)$ *when*

$$\varepsilon = -1 \quad \textit{and} \quad p = 1, 2, 3. \tag{3.5}$$

(ii) *Assume that* $d = 3$. *The Equation* (1.2) *does not admit any nontrivial solitary wave satisfying* $u = \partial_x \varphi$, $\varphi \in X$, $u \in H^1(\mathbb{R}^3) \cap L^{2(p+1)}(\mathbb{R}^3) \cap L^\infty_{\mathrm{loc}}(\mathbb{R}^3)$, $\partial_x^2 u$, $\partial_y^2 \varphi$, *and* $\partial_z^2 \varphi \in L^2_{\mathrm{loc}}(\mathbb{R}^3)$ *if*

$$ab = -1 \quad (\textit{resp., } a = b = 1) \quad \textit{and} \quad p \textit{ is arbitrary} \tag{3.6}$$

or

$$a = b = -1, \quad p \geqslant 2. \tag{3.7}$$

On the other hand, (1.2) *admits a nontrivial solitary wave belonging to* $H^\infty(\mathbb{R}^3)$ *when*

$$\varepsilon = -1, \qquad p = 1. \tag{3.8}$$

Remark 3.1. (1) The existence part of point (i) has been proved independently in [25].

(2) The nonexistence part of Theorem 3.1 is proven via Pohojaev-type identities. The regularity assumptions on u are made to justify them. The existence part is obtained by the concentration-compactness method of P. L. Lions applied to the minimization problem

$$I_\lambda = \mathrm{Inf}\left\{ \|u\|_Y^2,\ u \in Y \text{ with } \int_{\mathbb{R}^d} u^{p+2}\, \mathrm{d}x\, \mathrm{d}x' = \lambda \right\}$$

where $x' = y$ if $d = 2$, $x' = (y, z)$ if $d = 3$ and $\lambda > 0$.

The regularity property of the solitary wave is proven by 'bootstrapping', using the imbedding theorems for anisotropic Sobolev spaces [3] and a variant due to Lizorkin [15] of the Mikhlin–Hörmander multiplier theorem.

(3) The uniqueness of solitary waves to (1.1) and (1.2) (when they exist!) is an open problem.

(4) Contrarily to the solitary waves of Korteweg–de Vries or nonlinear Schrödinger equations, the solitary waves of Kadomtsev–Petviashvili equations are

neither radial, nor positive, and in general do not decay exponentially to 0 at infinity, as shows the example of 'lumps' solutions of the KP-I equation:

$$u(x-ct,y)=\frac{8c\left(1-\frac{c}{3}(x-ct)^2+\frac{c^2}{3}y^2\right)}{\left(1+\frac{c}{3}(x-ct)^2+\frac{c^2}{3}y^2\right)^2}.$$

(5) Theorem 3.1 can be extended to more general nonlinearities and/or dispersions in x (see [7, 8]).

An important question is that of the orbital stability of solitary waves. No rigorous result seems to be known so far (even for the lumps of KP-I). As to instability, it is established in [25] that the solitary waves of (1.1) with $\varepsilon=-1$ and $f(u)=u^p$, $p>4/3$, are unstable. The instability seems to be linked to the blow-up of solutions in finite time (see Section 2). Following the same lines as in [25], one can prove that the solitary wave of (1.2) obtained in Theorem 3.1(ii) is unstable.

References

1. Ablowitz, M. J. and Clarkson, P. A.: *Solitons, Nonlinear Evolution Equations and Inverse Scattering*, Cambridge University Press, Cambridge, 1991.
2. Ablowitz, M. J. and Segur, H.: *Solitons and the Inverse Scattering Transform*, SIAM, Philadelphia, 1981.
3. Besov, O. V., Il'in, V. P., and Nikolskii, S. M.: *Integral Representations of Functions and Imbeddings Theorems*, Vol. 1, J. Wiley, New York, 1978.
4. Boiti, M., Pempinelli, F., and Pogrebkov, A.: Properties of the Kadomtsev–Petviashvili equation, *J. Math. Phys.* **35**(9) (1994), 4683–4718.
5. Bona, J., Dougalis, V. A., and Karakashian, O. A.: Fully discrete Galerkin methods for the Korteweg–de Vries equations, *Comput. Math. Appl.* **7** (1986), 859–884.
6. Bourgain, J.: On the Cauchy problem for the Kadomtsev–Petviashvili equations, *Geom. Funct. Anal.* **3**(4) (1993), 315–341.
7. de Bouard, A. and Saut, J. C.: Solitary waves of generalized Kadomtsev–Petviashvili equations, Preprint, Université Paris-Sud, 1994.
8. de Bouard, A. and Saut, J. C.: in preparation.
9. Faminskii: The Cauchy problem for the Kadomtsev–Petviashvili equation, *Russian Math. Surveys* **5** (1990), 203–204.
10. Fokas, A. S. and Sung, L. Y.: On the solvability of the N-wave, Davey–Stewartson and Kadomtsev–Petviashvili equations, *Inverse Problems* **8** (1992), 673–708.
11. Ginibre, J. and Velo, G.: The global Cauchy problem for the nonlinear Schrödinger equation revisited, *Ann. Inst. Henri Poincaré, Analyse Non Linéaire* **2** (1985), 309–327.
12. Glassey, R.: On the blowing up of solutions to the Cauchy problem for the nonlinear Schrödinger equation, *J. Math. Phys.* **18** (1977), 1794–1797.
13. Isaza, P., Mejia, J., and Stallhohm, V.: Local solution for the Kadomtsev–Petviashvili equation in $\mathbb{R}^2$, Preprint, 1993.
14. Kadomtsev, B. B. and Petviashvili, V. I.: On the stability of solitary waves in weakly dispersive media, *Soviet Phys. Dokl.* **15**(6) (1970), 539–541.
15. Lizorkin, P. I.: Multipliers of Fourier integrals, *Proc. Steklov Math. Inst.* **89** (1967), 269–290.
16. Petviashvili, V. I. and Yan'kov, V. V.: Solitons and turbulence, in B. B. Kadomtsev (ed.), *Rev. Plasma Phys.* XIV, 1989, 1–62.
17. Saut, J. C.: Remarks on the generalized Kadomtsev–Petviashvili equations, *Indiana Math. J.* **42**(3) (1993), 1011–1026.

18. Saut, J. C.: in preparation.
19. Schwarz, M.: Periodic solutions of Kadomtsev–Petviashvili equations, *Adv. in Math.* **66** (1987), 217–233.
20. Spector, M. D. and Miloh, T.: Stability of nonlinear periodic internal waves in a deep stratified fluid, *SIAM J. Appl. Math.* **54**(3) (1994), 688–707.
21. Turitsyn, S. K. and Falkovitch, G. E.: Stability of magnetoelastic solitons and self-focusing of sound in antiferromagnet, *Soviet Phys. JETP* **62** (1985), 146–152.
22. Ukai, S.: Local solutions of the Kadomtsev–Petviashvili equation, *J. Fac. Sci. Univ. Tokyo, Sect. IA Math.* **36** (1989), 193–209.
23. Vlasov, S. B., Petrishev, V. A., and Talanov, V. I.: *Izv. Vyssh. Uchebn. Zaved. Radiofiz.* **14** (1971), 1352; *Radiophys. Quantum Electronics* **14** (1974), 1062.
24. Wang, X. P., Ablowitz, M. J., and Segur, H.: Wave collapse and instability of solitary waves of a generalized Kadomtsev–Petviashvili equation, *Physica* **D78** (1994), 97–113.
25. Wickerhauser, M. V.: Inverse scattering for the heat operators and evolutions in $(2 + 1)$ variables, *Comm. Math. Phys.* **108** (1987), 67–89.
26. Zhou, X.: Inverse scattering transform for the time dependent Schrödinger equation with applications to the KP-I equation, *Comm. Math. Phys.* **128** (1990), 551–564.

Acta Applicandae Mathematicae **39**: 489–505, 1995.

An Explicit Expression for the Korteweg–de Vries Hierarchy

RAINER SCHIMMING
Ernst-Moritz-Arndt Universität, Fachrichtungen Mathematik/Informatik, Friedrich-Ludwig-Jahn-Str. 15a, D-17487 Greifswald, Germany

(Received: 30 September 1994)

Abstract. Several constructions and an explicit expression for the right-hand side of the KdV hierarchy are presented.

Mathematics Subject Classifications (1991): 35Q53, 58F07.

Key words: Korteweg–de Vries equation, solitons.

Introduction

The typical soliton partial differential equations – named Korteweg–de Vries, sine-Gordon, nonlinear Schrödinger, etc. – originally emerged as physical or differential-geometrical models. Mathematical analysis later unveiled several peculiar properties which constitute the solitonic character. As a rule, the given 'seed' equation can be extended to a 'hierarchy'; that means to an infinite sequence of soliton equations of increasing order. In particular the KdV equation

$$\frac{\partial u}{\partial t} = \frac{\partial^3 u}{\partial x^3} + 6u\frac{\partial u}{\partial x} \tag{1}$$

for a function $u = u(x,t)$ of position x and time t extends, according to [8], to a hierarchy which begins as

$$\begin{aligned} u_t &= u_1, \\ u_t &= u_3 + 6uu_1, \\ u_t &= u_5 + 10uu_3 + 20u_1u_2 + 30u^2u_1, \\ u_t &= u_7 + 14uu_5 + \cdots. \end{aligned}$$

Here and in the following $u_t = \partial u/\partial t$, while u_p means the pth derivative of u with respect to x ($p = 1, 2, \ldots$).

The right-hand side of the nth equation ($n = 1, 2, \ldots$) is a differential polynomial, i.e. a polynomial in $u, u_1, u_2, \ldots$ with constant coefficients, of order $2n+1$

and degree n. It is, moreover, the derivative of another differential polynomial $G_n = G_n[u]$; the KdV hierarchy reads

$$u_t = \frac{\partial}{\partial x} G_n[u] \quad (n = 1, 2, \ldots). \tag{2}$$

(Note that the original KdV equation emerges for $n = 2$.) The sequence G_n $(n = 1, 2, \ldots)$ – its constructions, structural properties, explicit expression – is the subject of the present paper.

Each equation (2) admits a Lax representation

$$\frac{\partial L}{\partial t} = [A_n, L] \quad (n = 1, 2, \ldots), \tag{3}$$

where

$$L = D^2 + u \equiv \frac{\partial^2}{\partial x^2} + u(x, t)$$

is the one-dimensional Schrödinger (or Sturm–Liouville) operator with the potential u, and A_n is some linear differential operator with respect to x of order $2n - 1$, the coefficients of which can be expressed as differential polynomials in u. The Lax equation (3) is the starting point for the inverse scattering method, the applicability of which makes (3) solitonic [8, 9]. There are several recursive constructions of the sequences G_n and A_n. The first was given as early as the twenties (!) by J. L. Burchnall and T. W. Chaundy [2, 3]. They searched for semi-commutative pairs, as they call it, of ordinary differential operators and were, of course, not aware of the KdV theory. A second construction is a by-product of the well known 'textbook' relation between the one-dimensional Schrödinger equation and some Riccati equation. This relation can be taken formally, using generating function series (which converge or not), or nonformally, using the Jost solutions of scattering theory.

The differential polynomials G_n are found, by virtue of the so-called Lenard recursion to be proportional to the diagonal values of the heat kernel coefficients H_n to $L = D^2 + u$. This observation is the key to an explicit expression, because we are able to solve the doubly recursive system for the double sequence $[D^p H_n(x, x_0)]_{x=x_0}$. We find for $n \geqslant 2$

$$\frac{2(n!)}{(2n)!} G_n = \sum_{q_2, \ldots, q_n} \prod_{k=1}^{n} \frac{c(q_k, q_{k+1})}{q_k + n - k + 1} u_{q_k - q_{k+1}}, \tag{4}$$

where we use the special conventions

$$u_0 := u, \quad u_{-1} := 0, \quad u_{-2} := 1, \tag{5}$$

$$c(p, q) := \binom{p}{q} + \delta_q^{p+2}, \tag{6}$$

$\binom{p}{q}$: binomial coefficient, δ_q^p: Kronecker symbol

and the sum runs over the integers $q_2, q_3, \dots, q_n$ such that

$$0 \leqslant q_{k+1} \leqslant q_k + 2 \quad \text{for} \quad k = 1, 2, \dots, n-1;$$

$$q_1 := 0, \quad q_{n+1} := 0.$$

Previous attempts in the literature to find a nonrecursive expression for G_n were not completely successful. The formulas in [15] become explicit only after some multiple integrals or some generating functions are evaluated, which is no easy matter. The papers [12, 14] find some terms in G_n but are not able to give all terms.

Sections 3, 4 and 5 of the present paper essentially follow our earlier article [13]. Sections 2 and 6 give a new arrangement of known facts. Section 1 presents new historical facts: the contribution of Burchnall and Chaundy to KdV theory has been overlooked previously, while their contribution to the theory of commuting differential operators has been recognized.

It is a challenge, after we have found the KdV hierarchy explicitly, to construct explicit expressions for other soliton hierarchies too. Moreover, several sequences of differential polynomials associated to the KdV hierarchy wait for an explicit formula.

1. The Burchnall–Chaundy Construction

The papers [2, 3] are devoted to linear differential operators with respect to x

$$P = \alpha_0 D^m + \alpha_1 D^{m-1} + \cdots + \alpha_m, \quad \alpha_0 \not\equiv 0$$

where $D = \mathrm{d}/\mathrm{d}x$, and the coefficients are functions of x: $\alpha_i = \alpha_i(x)$. The special subject of [2, 3] are commuting pairs, i.e. operators P, Q of respective orders m, n such that

$$[P, Q] \equiv PQ - QP = 0. \tag{1.1}$$

(Later we will accompany the commutator $[P, Q]$ by the anti-commutator $\{P, Q\} := PQ + QP$.) The main result reads: P, Q commute if and only if they fulfil an algebraic equation $f(P, Q) = 0$ with constant coefficients and of degree n, m in P, Q, respectively. Further, for numbers p, q such that $f(p, q) = 0$ the vector space of solutions y of the ODE system

$$(P - p)y = 0, \quad (Q - q)y = 0 \tag{1.2}$$

has a positive and finite dimension d. There are then differential operators R, S, T of orders d, $m - d$, $n - d$ respectively such that

$$P - p = S * R, \qquad Q - q = T * R \tag{1.3}$$

and R annihilates the solutions y of (1.2). The pair P, Q generates another commuting pair P_1, Q_1 by the so-called transference relation

$$P_1 - p = R * S, \qquad Q_1 - q = R * T.$$

The transferred operators P_1, Q_1 satisfy the same equation $f(P_1, Q_1) = 0$ as P, Q.

Changing the variable x to $\bar{x} = \bar{x}(x)$ as well as a gauge transformation

$$\overline{P} = \Theta^{-1} P \Theta, \qquad \overline{Q} = \Theta^{-1} Q \Theta$$

with a function $\Theta = \Theta(x) \neq 0$ will leave the commutativity condition invariant. A commutative pair P, Q is called reducible if it can be brought, by means of the two aforesaid transformations, into the form

$$\overline{P} = f(U, V), \qquad \overline{Q} = g(U, V),$$

where U, V is a commutative pair of lower order, i.e.

$$\operatorname{ord} U + \operatorname{ord} V < \operatorname{ord} P + \operatorname{ord} Q,$$

and where f, g denote polynomials with constant coefficients. It is easy to show that the orders m, n of an irreducible (= nonreducible) commutative pair P, Q are relatively prime and $\geqslant 2$.

Burchnall and Chaundy find it convenient to weaken the commutativity to semi-commutativity, defined by

$$\operatorname{ord}[P, Q] \leqslant \min\{m, n\} - 2.$$

They construct then all irreducible semi-commutative pairs P, Q with first member $P = D^2 + u$. (The distinction reducible/irreducible makes sense in the new situation, too.) In order to present the result, let us switch to more modern notation. In particular, $L = D^2 + u$ will be the one-dimensional Schrödinger operator. As usual, a differential operator of order zero, which means multiplication by a function, is notationally identified with this function. A prime means differentation with respect to x.

THEOREM 1.1. *The recursion system for differential operators* A_n *and functions* G_n $(n = 0, 1, 2, \ldots)$.

$$A_{n+1} := 2\{A_n, L\} + \{G_n, D\} \tag{1.4}$$

$$G'_n := [A_n, L], \qquad A_0 := 0, \qquad G_0 := \text{const} \neq 0 \tag{1.5}$$

is solvable. (This means in particular $\operatorname{ord}[A_n, L] = 0$.) *Each solution system satisfies*

$$G'_{n+1} = G'''_n + 4uG'_n + 2u'G_n, \tag{1.6}$$

$$A_{n+1} = \sum_{k=0}^{n} (2G_k D - G'_k)(4L)^{n-k}. \tag{1.7}$$

Proof is by means of mathematical induction with respect to n. The induction step shows $\operatorname{ord}[A_{n+1}, L] = 0$; then G_{n+1} is introduced as the primitive function to the function representing $[A_{n+1}, L]$. In fact, a straightforward calculation shows

$$[A_{n+1}, L] = \cdots = G_n''' + 4uG_n' + 2u'G_n.$$

The recursive definition (1.4) of A_{n+1} is equivalent to

$$A_{n+1} = 4A_nL + B_n,$$

where

$$B_n := 2G_nD - G_n'.$$

This linear first-order recursion can be explicitly solved:

$$A_{n+1} = \sum_{k=0}^{n} B_k(4L)^{n-k}.$$

Equation (1.6) is, unhistorically, attributed to A. Lenard (cf. [4, 9]). It can be written

$$DG_{n+1} = (D^3 + 2uD + 2Du)G_n. \tag{1.8}$$

THEOREM 1.2. *Let* G_n, $\overline{G}_n$ $(n = 0, 1, 2, \ldots)$ *denote two solution systems of* (1.6) *or* (1.8) *with initial conditions* $G_0 = \text{const} \neq 0$, $\overline{G}_0 = \text{const} \neq 0$. *Then there are constants* a_k $(k = 0, 1, 2, \ldots)$ *such that*

$$\overline{G}_n = \sum_{k=0}^{n} a_{n-k}G_k. \tag{1.9}$$

Proof is by mathematical induction. The induction hypothesis implies

$$D\left[\overline{G}_{n+1} - \sum_{k=0}^{n} a_{n-k}G_{k+1}\right]$$
$$= (D^3 + 2uD + 2Du)\left[\overline{G}_n - \sum_{k=0}^{n} a_{n-k}G_k\right] = 0.$$

Hereby the next integration constant a_{n+1} is introduced:

$$\overline{G}_{n+1} - \sum_{k=0}^{n} a_{n-k}G_{k+1} = \text{const} = a_{n+1}G_0.$$

In the terminology of [2], L and A_n form a semi-commutative pair. It is irreducible for $n \geqslant 2$. Moreover, each operator A which completes $L = D^2 + u$

to an irreducible semi-commutative pair is a linear combination, with constant coefficients, of $A_1, A_2, \dots, A_n$ for some n. Burchnall and Chaundy found this latter result with a great portion of skill or intuition.

2. The Riccati Equation Approach

The following relation between the one-dimensional Schrödinger equation

$$(L + \varkappa^2)y \equiv y'' + uy + \varkappa^2 y = 0 \tag{2.1}$$

on the one hand and the Riccati equation

$$Z' + Z^2 + u + \varkappa^2 = 0 \tag{2.2}$$

on the other hand is a well-known textbook matter: if $Z = Z(x)$ solves (2.2) then the following factorisation holds:

$$L + \varkappa^2 = (D + Z)(D - Z). \tag{2.3}$$

As a conclusion, every solution of the first-order equation

$$y' = Zy \tag{2.4}$$

satisfies (2.1). Conversely, if $y = y(x)$ solves (2.1) and does not vanish, then

$$Z := y'y^{-1} \tag{2.5}$$

satisfies (2.2). We assume the constant parameter $\varkappa$ to be real and positive. We admit complex-valued solutions y, Z and set

$$Z = X + iY, \quad X = \Re Z, \quad Y = \Im Z.$$

The Riccati equation (2.2) decomposes by this into

$$X' + X^2 - Y^2 + u + \varkappa^2 = 0, \tag{2.6}$$

$$Y' + 2XY = 0. \tag{2.7}$$

THEOREM 2.1. *If $y \neq 0$ solves the Schrödinger Equation* (2.1) *then*

$$G := \varkappa^{-1}|y|^2 \equiv \varkappa^{-1} y\bar{y} \tag{2.8}$$

solves the nonlinear second-order equation

$$2GG'' - G'^2 + 4(u + \varkappa^2)G^2 = 4c^2 \tag{2.9}$$

as well as the linear third-order equation

$$G''' + 4(u + \varkappa^2)G' + 2u'G = 0, \tag{2.10}$$

where c is a real (integration) constant.

Proof. The quantity

$$\begin{aligned} 2i\varkappa YG &= (Z - \overline{Z})\varkappa G = (y'y^{-1} - \bar{y}'\bar{y}^{-1})y\bar{y} \\ &= y'\bar{y} - \bar{y}'y = \text{const} =: 2i\varkappa c \end{aligned}$$

turns out to be the Wronskian of the two solutions y, $\bar{y}$ of (2.1), and such a Wronskian is known to be constant. We insert

$$Y = cG^{-1}$$

together with

$$2X = -Y'Y^{-1} = -(\log Y)' = (\log G)' = G'G^{-1}$$

into (2.6) and obtain (2.9). Differentiation of (2.9) and cancellation of $G \neq 0$ yields (2.10).

We are not really interested in classical solutions of the differential equations (2.2), (2.9) and (2.10). Rather we want to take $Z = Z(x, \varkappa)$, $G = G(x, \varkappa)$ as generating functions, the expansion of which with respect to $\varkappa$ (convergent or not) produces sequences $Z_n = Z_n(x)$, $G_n = G_n(x)$ which are functionals in the potential $u = u(x)$.

THEOREM 2.2. *There is a unique formal power series of the form*

$$Z = \sum_{n=-1}^{\infty} Z_n(-2i\varkappa)^{-n} \tag{2.11}$$

which solves (2.2). *Its coefficients* $Z_n = Z_n(x)$ *are recursively determined by*

$$Z_{n+1} = Z_n' + \sum_{k=1}^{n-1} Z_k Z_{n-k} \quad \textit{for } n \geqslant 1, \tag{2.12}$$

$$Z_{-1} = i\varkappa, \qquad Z_0 = 0; \quad Z_1 = u. \tag{2.13}$$

As a conclusion,

$$G := Y^{-1} := (\Im Z)^{-1} \tag{2.14}$$

admits a formal expansion as well:

$$G = 4\sum_{n=0}^{\infty} G_n(-1)^n(2\varkappa)^{-2n-1}. \tag{2.15}$$

The coefficients $G_n = G_n(x)$ *are recursively determined by*

$$G_n = 2\sum_{k=1}^{n} Z_{2k-1}G_{n-k} \quad \textit{for } n \geqslant 1, \tag{2.16}$$

$$G_0 = 1/2. \tag{2.17}$$

Moreover, (2.10) *becomes equivalent to the Lenard recursion* (1.6) *or* (1.8) *for our new* G_n.

Proof. Straightforward and is omitted here.

The new G_n are recognized to be special Burchnall–Chaundy quantities G_n. It is plausible to parallel (2.15) by the formal series

$$A := 4\sum_{n=0}^{\infty} A_n(-1)^n(2\varkappa)^{-2n-1}, \tag{2.18}$$

where the A_n are special Burchnall–Chaundy operators. The semi-commutativity result (1.5) then reads

$$G' = [A, L].$$

The operator-valued formal series A admits a remarkably simple expression:

$$4A = (G' - 2GD)(L + \varkappa^2)^{-1}.$$

The proof is straightforward (cf. [3]).

3. Asymptotic Expansions

Let us consider

$$y'' + \varkappa^2 y = 0, \quad \varkappa > 0,$$

that means the case $u \equiv 0$. Obviously,

$$y_+ = \mathrm{e}^{i\varkappa x}, \qquad y_- = \mathrm{e}^{-i\varkappa x}$$

is a fundamental system of solutions. Solutions y_+ or y_- of the full Schrödinger equation

$$(L + \varkappa^2)y \equiv y'' + uy + \varkappa^2 y = 0, \quad \varkappa > 0, \tag{3.1}$$

which behave like $\mathrm{e}^{i\varkappa x}$ as $x \to +\infty$ or like $\mathrm{e}^{-i\varkappa x}$ as $x \to -\infty$, respectively, are called Jost solutions. More precisely, y_+ is assumed to be defined in an interval $[a, +\infty[$ and to satisfy

$$\lim_{x\to+\infty} \mathrm{e}^{-i\varkappa x} y_+ = 1, \qquad \lim_{x\to+\infty} \mathrm{e}^{-i\varkappa x} y_+' = i\varkappa.$$

An analogous definition is taken for y_-. If the potential $u = u(x)$ is defined and continuous for $x \in \mathbb{R}$ and if

$$\int_{-\infty}^{+\infty} (1 + x^2)u(x)\,\mathrm{d}x < \infty$$

then there are unique Jost solutions y_+, y_- of (3.1). Additionally, if u is smooth then so are y_+, y_-. A power series expansion

$$y = y_+ = e^{i\varkappa x} \sum_{n=0}^{\infty} \eta_n (2i\varkappa)^{-n}$$

generally does not exist as a convergent series. But it can be shown to exist as an asymptotic series (which is more than a formal series):

$$y \cong e^{i\varkappa x} \sum_{n=0}^{\infty} \eta_n (2i\varkappa)^{-n} \quad \text{as } \varkappa \to +\infty.$$

As a conclusion, there is an asymptotic expansion of $G = \varkappa^{-1}|y|^2$, namely

$$G \cong 4 \sum_{n=0}^{\infty} G_n (-1)^n (2\varkappa)^{-2n-1} \quad \text{as } \varkappa \to +\infty.$$

4. The Differential Polynomials G_n

Let us, from now on, restrict to the sequence G_n $(n = 0, 1, 2, \ldots)$ from Sections 2 and 3, which is uniquely determined. The more general sequence of Section 1, where the nth member depends on $n+1$ arbitrary constants, is then given by Theorem 2.1. Some notions are needed in order to describe the quantities G_n.

DEFINITION 4.1. A differential polynomial in $u = u(x)$ is a linear combination with constant coefficients of expressions

$$u^{k_0} u_1^{k_1} \cdots u_p^{k_p}, \tag{4.1}$$

with integers $p \geqslant 0$, $k_0 \geqslant 0, \ldots, k_{p-1} \geqslant 0, k_p \geqslant 1$. The monomial (4.1) has the order p, the degree $d := k_0 + k_1 + \cdots + k_p$, and the weight $w := 2k_0 + 3k_1 + \cdots + (p+2)k_p$. A differential polynomial has as its order, degree and weight the maximum of orders, degrees and weights, respectively of its monomial parts. A differential polynomial is called homogeneous if all its monomials have the same weight.

Note that a differential polynomial, as we take it here, has no absolute term.

THEOREM 4.1. *For $n \geqslant 1$, G_n is a homogeneous differential polynomial in u of order $2n-2$, degree n, and weight $2n$. The like, Z_n is a homogeneous differential polynomial of order $n-1$ and weight $n+1$. Both Z_{2k} and Z_{2k-1} have degree k.*

Proof. By mathematical induction, based on (2.12), (2.13), (2.16).

Let us present some of the differential polynomials:

$$\begin{aligned} Z_1 &= u, \\ Z_2 &= u_1, \\ Z_3 &= u_2 + u^2, \\ Z_4 &= u_3 + 4uu_1, \\ Z_5 &= u_4 + 6uu_2 + 5u_1^2 + 2u^3, \\ Z_6 &= u_5 + 8uu_3 + 18u_1u_2 + 16u^2u_1. \end{aligned}$$

$$\begin{aligned} G_1 &= u, \\ G_2 &= u_2 + 3u^2, \\ G_3 &= u_4 + 10uu_2 + 5u_1^2 + 10u^3, \\ G_4 &= u_6 + 14uu_4 + 28u_1u_3 + 21u_2^2 + 70u^2u_2 + 70uu_1^2 + 35u^4. \end{aligned}$$

The coefficients of the differential operators A_n $(n = 1, 2, \ldots)$ belonging to the sequence G_n $(n = 1, 2, \ldots)$ are differential polynomials too. We have, in particular,

$$\begin{aligned} A_1 &= D, \\ A_2 &= 4D^3 + 3Du + 3uD, \\ A_3 &= 16D^5 + 20D^3u + 20uD^3 + D5(3u^2 - u_2) + 5(3u^2 - u_2)D. \end{aligned}$$

THEOREM 4.2. *The dependence of the quantities under consideration – taken as polynomials in $u, u_1, u_2, \ldots$ – on the variables u is given by*

$$Z_n = \sum_{k=0}^{[n/2]} \binom{n/2-1}{k} (4u)^k \left[Z_{n-2k}\right]_{u=0}, \tag{4.2}$$

$$G_n = \sum_{k=0}^{n} \binom{n-1/2}{k} (4u)^k \left[G_{n-k}\right]_{u=0}, \tag{4.3}$$

$$A_n = \sum_{k=0}^{n} \binom{n-1/2}{k} (4u)^k \left[A_{n-k}\right]_{u=0}. \tag{4.4}$$

Proof. All quantities constructed from $L + \varkappa^2 = D^2 + u + \varkappa^2$ depend on u and $\varkappa^2$ only in the combination $u + \varkappa^2$. Therefore our generating functions satisfy

$$\frac{\partial Z}{\partial u} = \frac{\partial Z}{\partial \lambda}, \qquad \frac{\partial G}{\partial u} = \frac{\partial G}{\partial \lambda}, \qquad \frac{\partial A}{\partial u} = \frac{\partial A}{\partial \lambda},$$

where

$$\lambda = \varkappa^2 \quad \text{and} \quad \frac{\partial}{\partial \lambda} = (2\varkappa)^{-1}\frac{\partial}{\partial \varkappa}.$$

As a conclusion, the coefficients satisfy for $n \geqslant 1$

$$\begin{aligned}\frac{\partial Z_n}{\partial u} &= 2(n-2)Z_{n-2},\\ \frac{\partial G_n}{\partial u} &= 2(2n-1)G_{n-1},\\ \frac{\partial A_n}{\partial u} &= 2(2n-1)A_{n-1}.\end{aligned}$$

Iterated differentiation with respect to u gives then

$$\begin{aligned}\frac{\partial^k Z_n}{\partial u^k} &= 2^k(n-2)(n-4)\cdots(n-2k)Z_{n-2k}\\ &= 4^k k!\binom{n/2-1}{k}Z_{n-2k},\end{aligned}$$

and analogous formulas for G_n, A_n. These are inserted into the Taylor expansion of Z_n, G_n, A_n with respect to u and yield (4.2)–(4.4), respectively.

Theorem 4.2, together with the knowledge of Z_k, G_k, A_k for low k, produces the terms of high degree and low order. In particular,

$$60\binom{2n}{n}^{-1}G_n = 30u^n + 10\binom{n}{2}u^{n-2}u_2 + 3\binom{n}{3}u^{n-3}(u_4 + 5u_1^2) + \cdots.$$

This method can be extended as far as one wants, so one knows Z_n, G_n, A_n from this side. The other side, terms of low degree and high order, is far more difficult (except for the very extreme terms).

THEOREM 4.3. *The linear and quadratic terms in Z_n, G_n for $n \geqslant 2$ are exactly given by*

$$Z_n = u_{n-1} + \sum_{k=0}^{n-3}\left[\binom{n-1}{k+1} - 1\right]u_k u_{n-3-k} + \cdots \tag{4.5}$$

$$G_n = u_{2n-2} + \sum_{k=0}^{2n-4}\left[\binom{2n-2}{k+1} + (-1)^k\right]u_k u_{2n-4-k} + \cdots. \tag{4.6}$$

Proof. By mathematical induction with respect to n based on (2.12), (2.13), (2.16) and (2.17). Our method of proof is new, while the quadratic terms in G_n are due to [5].

The right-hand sides of the KdV hierarchy emerge by differentiation with respect to x:

$$60\binom{2n}{n}^{-1} G_n' = 30nu^{n-1}u_1 + 10\binom{n}{2}u^{n-2}u_3 + \\ +3\binom{n}{3}u^{n-3}(u_5 + 20u_1u_2 + 5u_1^3) + \cdots ,$$

where the points indicate terms of lower degree, and

$$G_n' = u_{2n-1} + \sum_{p+q=2n-3} \binom{2n-1}{p+1} u_p u_q + \cdots ,$$

where now the points indicate terms of higher degree.

5. The Heat Equation Approach

The method in the foregoing sections does not suffice to find an explicit expression for the differential polynomials G_n. To that end we have to study the heat equation for L

$$\frac{\partial K}{\partial s} = LK \equiv (D^2 + u)K \tag{5.1}$$

or, more precisely, the heat kernel

$$K = K(x, x_0, s) = \frac{\mathrm{e}^{-(x-x_0)^2/4s}}{\sqrt{4\pi s}} H(x, x_0, s). \tag{5.2}$$

Here $H = H(x, x_0, s)$ is a smooth function of x, x_0 from some interval and of s for small $|s|$, with the initial condition

$$H(x, x_0, 0) = 1. \tag{5.3}$$

By definition, K solves for $s > 0$ the heat equation (5.1), which means for H

$$\left[(x - x_0)D + s\frac{\partial}{\partial s}\right] H = sLH. \tag{5.4}$$

For existence, unicity and further properties of K or H we refer the reader to the literature [1, 6, 11].

THEOREM 5.1. *The diagonal values* $H(x, s) := H(x, x, s)$ *satisfy*

$$2\left(2s\frac{\partial}{\partial s} - 1\right) DH(x, s) = s(D^3 + 2Du + 2uD)H(x, s). \tag{5.5}$$

Proof. H is known to be symmetric in the two points x, x_0: $H(x,x_0,s) = H(x_0,x,s)$. Therefore (5.4) is paralleled by

$$\left[(x_0 - x)D_0 + s\frac{\partial}{\partial s}\right]H = sL_0, \tag{5.6}$$

where now

$$D = \frac{\partial}{\partial x}, \qquad D_0 = \frac{\partial}{\partial x_0}, \qquad L_0 = D_0^2 + u(x_0).$$

We apply $D + 3D_0$ to (5.4), $D_0 + 3D$ to (5.6) and add the results. Considering the identities

$$\begin{aligned}
&(D + 3D_0)(x - x_0)D + (D_0 + 3D)(x_0 - x)D_0 \\
&\quad = -2(D + D_0) + (x - x_0)(D^2 - D_0^2),
\end{aligned}$$

$$(D + 3D_0)D^2 + (D_0 + 3D^2)D_0^2 = (D + D_0)^3,$$

$$R(D + D_0) = DR,$$

$$R(D + D_0)^3 = D^3R,$$

where R denotes the restriction of a two-point quantity to the diagonal $x = x_0$, we arrive at (5.5). This line of proof is based on an idea of [10].

It is well known that the heat kernel admits an asymptotic expansion with respect to small time s (cf. [1, 11]). Namely, there are smooth coefficients $H_n = H_n(x,x_0)$ $(n = 0,1,2,\ldots)$ such that

$$H(x,x_0,s) \cong \sum_{n=0}^{\infty} H_n(x,x_0)s^n \quad \text{as} \quad s \to +0. \tag{5.7}$$

The differential equation (5.4) for H implies the differential-recursion system

$$[(x - x_0)D + n]H_n = LH_{n-1} \quad \text{for} \quad n \geqslant 1, \tag{5.8}$$

$$H_0 \equiv 1. \tag{5.9}$$

This system is known to admit unique smooth solutions $H_n = H_n(x,x_0)$ in some neighbourhood of the diagonal $x = x_0$ [1, 6, 13]. The coefficients H_n are named after J. Hadamard [7], who constructed them for the first time.

THEOREM 5.2. *The diagonal values of Hadamard's coefficients H_n are proportional to the quantities G_n of Sections 2, 3 and 4. More precisely,*

$$(2n)!H_n(x,x) = 2(n!)G_n[u](x). \tag{5.10}$$

Proof. The differential equation (5.5) for the generating functions $H(x,s)$ of the one-point quantities $H_n^0 = H_n(x,x)$ implies

$$2(2n-1)DH_{n-1}^0 = (D^3 + 2uD + 2Du)H_n^0. \tag{5.11}$$

This differs from Lenard's recursion (1.8) only in the numerical factor $2(2n-1)$. The choice of initial values

$$H_0^0 = 1, \qquad G_0 = 1/2 \tag{5.12}$$

makes the solution system of both (1.8) and (5.11) unique in the space of differential polynomials. This uniqueness proves (5.10).

We are now in a position to prove the formula (4) given in the introduction. More generally we derive an explicit expression for the differential polynomials

$$H_n^p = H_n^p(x) := \left[D^p H_n(x,x_0)\right]_{x=x_0}.$$

THEOREM 5.3. *There holds for* $n \geqslant 2$, $p \geqslant 0$

$$H_n^p = \sum_{q_2,\ldots,q_n} \prod_{k=1}^{n} \frac{c(q_k, q_{k+1})}{q_k + n + k - 1} u_{q_k - q_{k+1}}, \tag{5.13}$$

where we use the special conventions (5), (6) *and the sum runs over the integers* $q_2, q_3, \ldots, q_n$ *such that*

$$0 \leqslant q_{k+1} \leqslant q_k + 2 \quad \textit{for} \quad k = 1, 2, \ldots, n-1,$$

$$q_1 := p, \qquad q_{n+1} := 0.$$

Proof. p-fold differentiation of (5.8) and restriction to the diagonal yields

$$\begin{aligned}(p+n)H_n^p &= H_{n-1}^{p+2} + \sum_{q=0}^{p} \binom{p}{q} u_{p-q} H_{n-1}^q \\ &\equiv \sum_{q=0}^{p+2} c(p,q) u_{p-q} H_{n-1}^q.\end{aligned}$$

This recursion system, together with the initial condition $H_0^p = \delta_0^p$, is uniquely solved by (5.13). Our result (4) now follows from (5.13) with $p = 0$ and from (5.10).

6. Conservation Laws for the KdV Equation

Let an evolution equation for $u = u(x,t)$ be given. A pair of functions $X = X(u, u_1, u_2, \ldots)$, $T = T(u, u_1, u_2, \ldots)$ (of finitely many variables) defines a conservation law if

$$\frac{\partial X}{\partial x} + \frac{\partial T}{\partial t} = 0 \tag{6.1}$$

for every solution $u = u(x,t)$ of the evolution equation. Then T is called a conserved density and X is called the current density to T. It is well known how a differential conservation law can be integrated – under certain condition – to an actual conservation law, i.e. the constancy in time of $\int T \, dx$. Let us in the following denote differentiations with respect to x or t by subscripts.

THEOREM 6.1. *If the potential $u = u(x,t)$ is a solution of the KdV equation*

$$u_t = u_{xxx} + 6uu_x$$

then the formal series Z, G of Section 2 satisfy the conservation laws

$$Z_t + (6i\varkappa Z^2 + 2Z^3 - Z_{xx})_x = 0, \tag{6.2}$$

$$G_t + \left(6(|Z|^2 + \varkappa^2)G - G_{xx}\right)_x = 0. \tag{6.3}$$

As a conclusion, every Z_n as well as every G_n $(n = 1, 2, \ldots)$ is a conserved density in the KdV equation.

Proof. Unfortunately, this is tedious, and we omit it. Elements of the proof can be found in the literature [4, 8, 9].

Obviously, a differential conservation law $X_x + T_t = 0$ is invariant under

$$\overline{X} = X + F_t, \qquad \overline{T} = T - F_x, \tag{6.4}$$

where $F = F(u, u_1, u_2, \ldots)$ is a suitable function. The transformations (6.4) can be used to normalize conserved densities. More precisely, 'normalized' here means 'irreducible'.

DEFINITION 6.1. A monomial

$$u^{k_0} u_1^{k_1} \cdots u_p^{k_p}$$

is called irreducible if $k_p \geqslant 2$. A differential polynomial is called irreducible if all its monomials have this property.

DEFINITION 6.2. Two differential polynomials $T, \overline{T}$ are called equivalent if they differ by the derivative (with respect to x) of another differential polynomial F:

$$T - \overline{T} = F_x.$$

The following is well known [16].

THEOREM 6.2. *Every differential polynomial which is a conserved density to some evolution equation has an irreducible equivalent.*

Moreover, the production of the irreducible conserved density in Theorem 6.2 follows a simple recursive algorithm. Let us apply these notions to the KdV equation.

THEOREM 6.3. *The KdV equation admits an infinite sequence*

$$T_n = (-1)^{n-1} u_{n-1}^2 + \cdots \quad (n = 1, 2, \ldots)$$

of irreducible conserved densities, where the points indicate terms of lower order. Moreover, T_n is a homogeneous differential polynomial of degree $n+1$ and weight $2n+2$. The terms of highest degree are given by

$$T_n = c_n u^{n+1} + \cdots ,$$

where the integer

$$c_n := \frac{1}{n+1} \binom{2n}{n}$$

denotes the nth Catalan number.

Proof. By construction: the abovementioned algorithm is applied to the sequence

$$\frac{G_{n+1}}{2n+1} \quad (n = 1, 2, \ldots)$$

of conserved densities. Differential polynomials behave in a controlled manner under the algorithm. We omit the details.

The sequence of Theorem 6.3 begins with

$$\begin{aligned} T_1 &= u^2, \\ T_2 &= -u_1^2 + 2u^3, \\ T_3 &= u_2^2 - 10uu_1^2 + 5u^4, \\ T_4 &= -u_3^2 + 14uu_2^2 - 70u^2u_1^2 + 14u^5. \end{aligned}$$

The discovery of a full explicit expression for the sequence T_n $(n = 1, 2, \ldots)$ seems to be an unsolved problem.

References

1. Berger, M. *et al.*: *Le spectre d'une varieté Riemannienne*, Springer, Berlin, 1971.
2. Burchnall, J. L. and Chaundy, T. W.: Commutative ordinary differential operators, *Proc. London Math. Soc.* **21**(2) (1922), 420–440.

3. Burchnall, J. L. and Chaundy, T. W.: Commutative ordinary differential operators, *Proc. Roy. Soc. London A* **118** (1928), 557–583.
4. Gardner, C. S. *et al.*: Korteweg–de Vries equation and generalizations, VI, *Commun. Pure Appl. Math.* **27** (1974), 97–133.
5. Gilkey, P. B.: Recursion relations and the asymptotic behaviour of the eigenvalues of the Laplacian, *Compos. Math.* **38** (1979), 201–240.
6. Günther, P.: *Huygen's Principle and Hyperbolic Equations*, Academic Press, San Diego, 1988.
7. Hadamard, J.: *Lectures on Cauchy's Problem*, New Haven, Yale Univ. Press, 1923.
8. Lax, P. D.: Integrals of nonlinear equations of evolution and solitary waves, *Commun. Pure Appl. Math.* **21** (1968), 467–490.
9. Lax, P. D.: Periodic solutions of the KdV equation, *Commun. Pure Appl. Math.* **28** (1975), 141–188.
10. McKean, H. P. and Van Moerbeke, P.: The spectrum of Hill's equation, *Invent. Math.* **30** (1975), 217–274.
11. Minakshisundaram, S. and Pleijel, A.: Some properties of the eigenfunctions of the Laplace operator on Riemannian manifolds, *Can. J. Math.* **1** (1949), 242–256.
12. Rosenhouse, A. and Katriel, J.: The Korteweg–de Vries hierarchy od isospectral transformations, *J. Math. Phys.* **28** (1987), 1344–1350.
13. Schimming, R.: An explicit expression for the Korteweg–de Vries hierarchy, *Z. Analysis ihre Anw.* **7** (1988), 203–214.
14. Torriani, H. H.: Conservation laws for the Korteweg–de Vries equation and the theory of partitions, *Phys. Lett. A* **133** (1986), 345–348.
15. Gel'fand, I. M. and Dikii, L. A.: Asymptotics of the resolvent of Sturm–Liouville equations and the algebra of Korteweg–de Vries equations, *Uspechi Mat. Nauk* **30**(5) (1975), 67–100 (in Russian).
16. Kruskal, M. D. *et al.*: Korteweg–de Vries equation and generalizations V, *J. Math. Phys.* **11** (1970), 952–960.

Acta Applicandae Mathematicae **39:** 507–516, 1995.

Evolving Solitons in Bubbly Flows

LEEN VAN WIJNGAARDEN
University of Twente, Enschede, The Netherlands

(Received: 12 September 1994)

Abstract. At the end of the sixties, it was shown that pressure waves in a bubbly liquid obey the KdV equation, the nonlinear term coming from convective acceleration and the dispersive term from volume oscillations of the bubbles.

For a variable u, proportional to $-p$, where p denotes pressure, the appropriate KdV equation can be casted in the form $u_t - 6uu_x + u_{xxx} = 0$. The theory of this equation predicts that, under certain conditions, solitons evolve from an initial profile $u(x, 0)$. In particular, it can be shown that the number N of those solitons can be found from solving the eigenvalue problem $\psi_{xx} - u(x,0)\psi = 0$, with $\psi(0) = 1$ and $\psi'(0) = 0$. N is found from counting the zeros of the solution of this equation between $x = 0$ and $x = Q$, say, Q being determined by the shape of $u(x, 0)$. We took as an initial pressure profile a shockwave, followed by an expansion wave. This can be realised in the laboratory and the problem, formulated above, can be solved exactly.

In this contribution the solution is outlined and it is shown from the experimental results that from the said initial disturbance, indeed solitons evolve in the predicated quantity.

Mathematics Subject Classification (1991): 58F07.

Key words: waves, solitons.

1. Introduction

In the 1960s and 1970s it was recognised that it was not only gravity waves on the free surface of a liquid of finite depth that obeyed the Korteweg–de Vries (KdV) equation: other waves do, too.

Armed with experience, we can today safely predict that waves displaying both nonlinearity and dispersion are governed by a KdV equation, provided dissipation is sufficiently small. Pressure waves in bubbly fluids belong to this class, as was shown by Van Wijngaarden [11] who derived the corresponding KdV equation.

Experiments to verify this were carried out in Russia, as reported in the book by Nakoryakov *et al.* [5] and in our laboratory [6, 7]. While there is no question that these experiments validated the theory qualitatively, it cannot be said that they did this in a completely quantitative way. Various reasons can be given for this. One of them is that many of these experiments concern transitions of the bore or shock-wave type, from one constant level of pressure to another. The Korteweg–de Vries equation has no steady solution for such a transition. In practice, seemingly steady, or quasi-steady pressure profiles are obtained due to an inevitable, though small, dissipation. Such a small dissipation is much less important when the disturbance, from a uniform quiescent state, is of finite spatial extent. Such experiments have, for example, been reported by Nakoryakov *et al.* [5]. However, a mathematical solution of the associated KdV equation was not given these. Also, as is emphasized by Watanabe and Prosperetti [8], thermal effects are extremely important in waves evolving over a long time, obscuring the dynamics.

With a view to these, rather recently gained, insights it seems worthwhile, and especially now that we are celebrating the 100th anniversary of the seminal paper by Korteweg and de Vries [3], to draw attention to a series of experiments which my student P. M. Roelofsen and I did about ten years ago, and which allow a comparison with the exact solution of the pertinent KdV equation.

2. Description of Experiment

The experiments discussed here were carried out in the facility described in detail by Noordzij and Van Wijngaarden [7]. The basic elements are shown in Figure 1.

A long tube of 11 cm diameter contains a mixture of water and small gas bubbles, mainly CO_2 in these experiments. Above the free surface of the mixture, for times $t' < 0$ is a low pressure region with height h_1 and a high pressure region of height h_2.

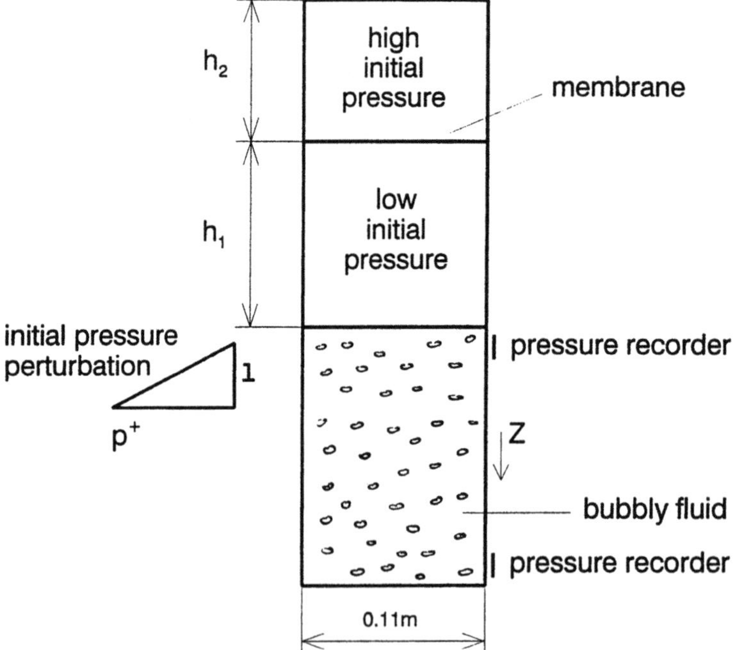

Fig. 1. Schematic representation of shock tube.

Let the pressure for $t' < 0$ be p_0 in the low pressure region and in the mixture. At $t' = 0$ the membrane, which separates the low pressure region from the high pressure region, is punctured. By an adroit choice of h_1, h_2 and the pressure in the high pressure region a pressure perturbation is generated which, at the entrance to the mixture, $z = 0$ in Figure 1, has the shape of a right-angled triangle.

First comes an instantaneous rise of the pressure to a value p^+, followed by an expansion wave in which the pressure linearly returns to p_0. Hence the perturbation is of finite extent, as opposed to a shock wave which produces a semi-infinite perturbation. Since the propagation into the bubbly water is both nonlinear (at sufficiently large $p^+ - p_0$) and dispersive, the initial shape is not preserved but evolves according to the KdV equation.

Before passing on to the way in which this happens, we discuss the most important physical quantities in the experiments:

Bubble radius a, in the undisturbed conditions a_0. In our experiments these are all of the same radius, which varies from one experiment to the other, but is always of the order of 1 mm.

Concentration of gas by volume α. The initial value α_0 can be measured from determining the static pressure distribution along the tube.

Velocity of sound c_0. Denoting the density of the fluid by ρ, the undisturbed pressure as before by p_0, the ratio of specific heats of the gas by γ we have for this the relation [12]:

$$c_0^2 = \frac{\gamma p_0}{\rho \alpha_0 (1 - \alpha_0)}. \tag{2.1}$$

Note that for values of α_0 of a few percent, c_0 is of the order of 10^2 m/s, under atmospheric conditions.

The natural frequency for small amplitude volume oscillations of a bubble ω_B. This has been found by Minnaert [4] to be given by

$$\omega_B^2 = \frac{3\gamma p_0}{\rho a_0^2}. \tag{2.2}$$

At frequencies ω of external disturbances which are far below ω_B, bubbles just follow the external pressure variations, and linear waves pass undeformed through the mixture with wavelength $2\pi c_0/\omega$. The closer the frequencies approach ω_B, the more this changes. The speed of propagation depends on ω for linear waves, long waves travelling faster than short waves. By nonlinear effects, the speed of propagation depends on the wave amplitude. The two phenomena together give rise to interesting behaviour.

3. Some Experimental Results

At various places along the tube (Figure 1), pressure transducers are placed, enabling the recording of the pressure as a function of time. We have done experiments in which the pressure was recorded, generated by the initial signal, in $z = 0$ in the form of a shock wave followed by an expansion wave.

Figures 2 and 3 show some results. In Figure 2 the pressure transducer is located at 47 cm from the free surface; in Figure 3 it is at 97 cm. The particulars of the experiments are given in Table I. In Figure 2 both the profile in $z = 0$ and the following profile in $z = 0.47$ m are shown; in Figure 3 only the profile in $z = 0.97$ m. For this experiment $l = 30$ cm and $p^+ - p_0 = 34$ kPa. Clearly, in both experiments the original disturbance, in the shape of a right-angled triangle, breaks up into a distinct number of wave structures. This is precisely what solution of the KdV equation predicts. The outstanding feature of these experiments is that for this particular initial shape the number of evolving distinct structures, which we shall recognise as solutions, can be exactly calculated and compared with the experimental results.

4. Theory

We denote the dimensionless pressure perturbation by v,

$$v = \frac{p - p_0}{p_0}. \tag{4.1}$$

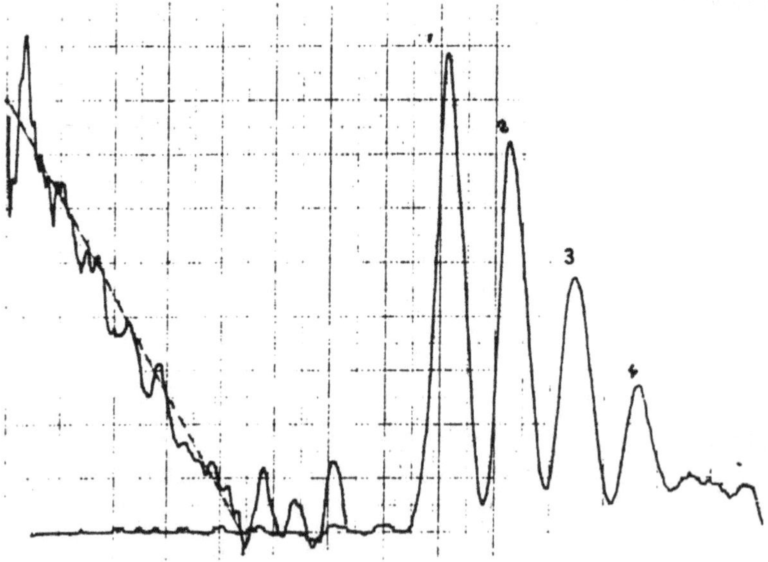

Fig. 2. The pressure profile on entering the mixture (left) and as it appears at the pressure transducer in $z = 0.47$ m below the interface. Data are given in Table I under 'exp 1'.

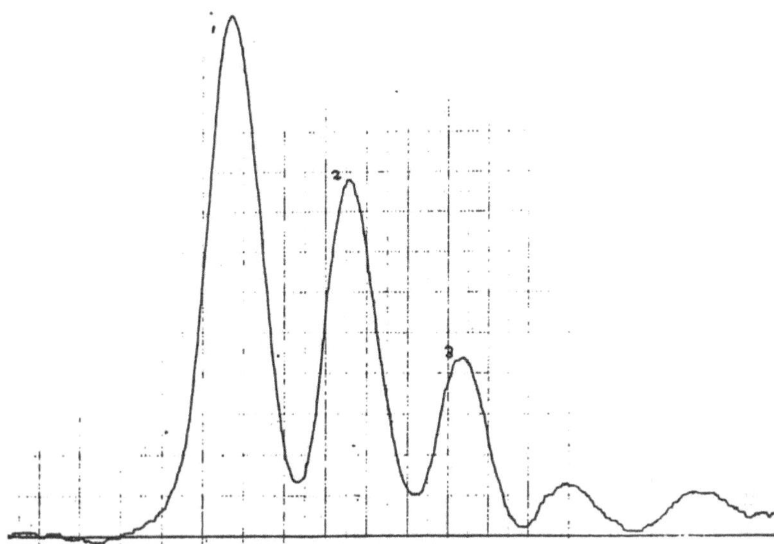

Fig. 3. Pressure profile measured by a transducer in $z = 0.97$ m below the interface following from the original profile with data as given in Table I under 'exp 4'.

Using the quantities defined in Section 2, and in addition the dispersion parameter σ,

$$\sigma = \frac{a_0^2}{\alpha_0(1-\alpha_0)} = \frac{3c_0^2}{\omega_B^2}, \tag{4.2}$$

Van Wijngaarden [11] showed that $v(z,t')$ evolves according to the equation

$$\frac{\partial v}{\partial t'} + c_0\frac{\partial v}{\partial z} + \frac{\gamma+1}{2\gamma}c_0 v\frac{\partial v}{\partial z} + \frac{\sigma c_0}{6}\frac{\partial^3 v}{\partial z^3} = 0. \tag{4.3}$$

It is convenient to choose a new coordinate $x' = z - c_0t'$, moving with the undisturbed speed of sound, and in addition the dimensionless variables x, t and η, defined by

$$x' = \left\{\frac{\sigma l\gamma}{3(\gamma+1)}\right\}^{1/3} x, \tag{4.4}$$

$$t' = \frac{2\gamma l}{(\gamma+1)c_0} t, \tag{4.5}$$

$$v = -\left\{\frac{72\sigma\gamma}{l^2(\gamma+1)}\right\}^{1/3} u. \tag{4.6}$$

We recall (see Figure 1) that l is the length of the initial pressure disturbance.

In terms of the variables x, t and u, (4.3) takes the standard form of the KdV equation employed in the literature (e.g., [9, p. 586])

$$\frac{\partial u}{\partial t} - 6u\frac{\partial u}{\partial x} + \frac{\partial^3 u}{\partial x^3} = 0. \tag{4.7}$$

This equation is valid according to the theory, provided:

(i) dispersion is weak;
(ii) dissipation is negligible; and
(iii) nonlinearity is weak.

The first condition, (i), means that a typical frequency c_0/l must be small with respect to ω_B. The ratio $c_0/\omega_B l$ is about 0.1 in our experiments. The second condition, (ii), is also fulfilled. When we take dissipation (viscous, but mainly thermal) into account, this would in Equation (4.7) lead to an additional term $0.06\,\partial^2 u/\partial x^2$ on the left-hand side. Such a term is negligible in our case, where the scale is determined by (4.4) but becomes important for phenomena of large extent. In other words, the approximation underlying (4.7) is not a uniformly valid one. Regarding (iii) we have chosen pressure disturbances of moderate value, 0.2–0.5, guaranteeing a not-too-strong nonlinearity.

Theory predicts that out of the initial profile $u(x,0)$ evolve

(a) N solitons, N to be found from inverse scattering theory, and travelling in the positive x direction;
(b) a continuous wave train travelling in the negative x direction.

We recall that the x-axis moves with speed c_0 in the z direction. This means that the continuous wave train moves in the laboratory frame rapidly in the negative z direction. These waves can easily be distinguished from the solitons, since they have small amplitude and move quickly. We shall therefore restrict our attention here to the solitons. These emerge from the spectrum of discrete eigenvalues of the Schrödinger equation

$$\frac{\mathrm{d}^2\varphi}{\mathrm{d}x^2} + \{\lambda - u(x,t)\}\varphi = 0, \tag{4.8}$$

in which u serves as potential through Miura's transformation

$$u = \frac{1}{\varphi}\frac{\mathrm{d}^2\varphi}{\mathrm{d}x^2} + \lambda. \tag{4.9}$$

The eigenvalues λ_n of (4.8) are negative, $\lambda_n = -k_n^2$, and, most importantly, independent of time. They are ordered such that $k_1 > k_2 > k_3 > \cdots > k_N > 0$, where N is finite. Because they are independent of time, they can be found from solution of

$$\frac{\mathrm{d}^2\varphi}{\mathrm{d}x^2} + \{\lambda - u(x,0)\}\varphi = 0. \tag{4.10}$$

Asymptotically – that is for large x and t – each eigenvalue corresponds to a soliton. The problem is therefore to find the total number, N, of eigenvalues. In the present problem, this can be found exactly.

First we insert the appropriate form of $u(x,0)$ into (4.10). In physical coordinates x' or z (which are the same at $t'=0$), the pressure profile is

$$\begin{aligned} v = \frac{p-p_0}{p_0} &= \frac{p^+ - p_0}{p_0}\frac{x'}{l}, \quad 0 < x' < l \\ &= 0, \quad x' \geqslant l \quad \text{and} \quad x' \leqslant 0. \end{aligned} \tag{4.11}$$

Using (4.4) and (4.6) this gives for $u(x,0)$

$$\begin{aligned} u(x,0) &= -\frac{p^+ - p_0}{p_0}\frac{x}{6}, \quad x < \left\{\frac{3l^2(\gamma+1)}{\gamma\sigma}\right\}^{2/3} \\ &= 0, \quad x \leqslant 0 \quad \text{and} \quad x \geqslant \left\{\frac{3l^2(\gamma+1)}{\gamma\sigma}\right\}^{2/3}. \end{aligned} \tag{4.12}$$

We consider solutions of (4.10), $u(x,0)$ being given by (4.12), and with boundary conditions

$$x = 0\colon \frac{\partial\varphi}{\partial x} = 0, \quad \varphi = 1. \tag{4.13}$$

Sturm's First Comparison Theorem (see, e.g., [2, Ch. 10]) states the following: suppose a solution, φ_a, corresponding with λ_a, has two zeros, in x_1 and x_2, say, on the x-axis. Then a solution φ_b, for $\lambda_b > \lambda_a$, has an additional zero in between x_1 and x_2. Now we apply this to our problem. For $\lambda \to -\infty$ the solution of (4.10), subjected to (4.13), is $\cosh \lambda x$ which has no zeros for real x. Hence the solution φ_1 corresponding with the largest value of k, k_1, has one zero on the x-axis, the solution φ_2 associated with k_2 has two zeros, etc. Since all eigenvalues are negative, the solution corresponding with $\lambda = 0$ has as many zeros as the one with the nearest eigenvalue which is $\lambda_N = -k_N^2$. Therefore, we have to determine the number of zeros of the solution of (4.10) with $\lambda = 0$ and $u(x,0)$ given by (4.12) and with (4.13) as boundary conditions. (This method has been used before by Hammack and Segur [1] in a problem concerning waterwaves.)

We take $\lambda = 0$ and insert (4.12) into (4.10). It is convenient to use instead of x a new coordinate y, given by

$$x = \left(\frac{6p_0}{p^+ - p_0}\right)^{1/3} y. \tag{4.14}$$

Upon using this in inserting (4.12) into (4.10), the latter equation is transformed into the Airy equation [10],

$$\frac{\mathrm{d}^2\varphi}{\mathrm{d}y^2} + y\varphi = 0, \quad 0 < y < Q, \tag{4.15}$$

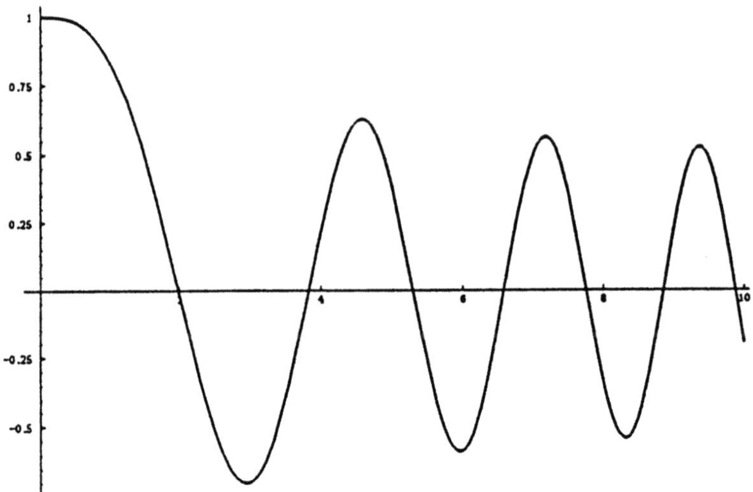

Fig. 4. The function $1.408Ai(-y) + 0.813Bi(-y)$, in the right-hand side of (4.20).

where

$$Q = \left\{ \frac{p^+ - p_0}{p_0} \frac{(\gamma+1)l^2}{2\sigma\gamma} \right\}^{1/3} \tag{4.16}$$

and

$$\frac{\mathrm{d}^2\varphi}{\mathrm{d}y^2} = 0, \qquad y \leqslant 0 \quad \text{and} \quad y \geqslant Q. \tag{4.17}$$

The boundary conditions (4.13) become

$$y = 0\colon \ \frac{\partial \varphi}{\partial y} = 0, \quad \varphi = 1. \tag{4.18}$$

The system (4.15)–(4.18) has the solution

$$y < 0, \qquad \varphi = 1; \tag{4.19}$$

$$0 \leqslant y \leqslant Q, \qquad \varphi = 1.408Ai(-y) + 0.813Bi(-y), \tag{4.20}$$

$$y > Q, \qquad \bar{\varphi} = \varphi = \left(\frac{\mathrm{d}\varphi}{\mathrm{d}y}\right)_{y=Q}(y - Q) + \varphi(Q), \tag{4.21}$$

φ being given by (4.20).

Figure 4 shows a sketch of the function on the right-hand side of (4.20). For a given value of Q we have to count the zeros in $0 < y < Q$. The part in $y < 0$

has no zeros, whereas the part of the solution in $y > Q$ contributes an additional zero when

$$\left\{\frac{d\varphi}{dy}/\varphi\right\}_{y=Q} < 0.$$

5. Comparison with Experiment

In Table I, we collect data of a number of our experiments. In the final columns the number N_t of solitons is given, as following from the theory, expounded above, and the number N_e from the corresponding experiment.

From the table it follows that in all experiments the number of solitons in both theory and experiment agree completely. The experiments shown in Figures 2 and 3 show this for experiments 1 and 4 in the table, respectively. The same agreement holds also for other experiments, not displayed here, verifying the validity of the KdV equation for these waves.

6. Shape of Solitons

We have labelled the evolving structures somewhat loosely as solitons. In fact, the distinct pressure humps shown in Figures 2 and 3 are not yet solitons, because they are still interacting and will only further down the tube become true solitons. If the peak pressure in a soliton is p_s, and accordingly $v_s = (p_s - p_0)/p_0$, a soliton has the shape

$$v = v_s \mathrm{sech}^2\left[\left(\frac{v_s}{2}\frac{\gamma+1}{\sigma\gamma}\right)^{1/2}\left(x' - \frac{1}{3}\frac{\gamma+1}{2\gamma}v_s t'\right)\right]$$

In some experiments, with the pressure transducer at $z = 0.97$ m or further down, this shape could indeed be verified. At closer distance solitons are still undergoing evolution to their final shape. Their number, however, does not change, anymore.

TABLE I.

Exp.	Location press. transd. m	Void fraction α %	l m	C_0 m s^{-1}	$\frac{p^+ - p_0}{p_0}$	Q	N_t	N_e
1	0.47	0.74	0.24	107	0.52	6.6	4	4
2	0.47	0.74	0.19	107	0.30	4.7	3	3
3	0.47	0.74	0.16	107	0.16	3.4	2	2
4	0.97	0.43	0.30	157	0.34	5.8	3	3
5	0.97	1.32	0.13	90	0.13	2.7	1	1
6	0.97	0.74	0.18	118	0.25	4.3	2	2

References

1. Hammack, J. L. and Segur, H.: The Korteweg–de Vries equation and water waves. Part 2, Comparison with experiments, *J. Fluid Mech.* **65**(2), (1974), 289–314.
2. Ince, E. L.: *Ordinary Differential Equations*, Dover Publications, 1956.
3. Korteweg, D. J. and de Vries, G.: On the change of form of long waves advancing in a rectangular canal and on a new type of long stationary waves, *Phil. Mag.* **39**(240) (1895), 422.
4. Minnaert, M.: Musical air bubbles and the sound of running water, *Phil. Mag.* **16** (1933), 235.
5. Nakoryakov, V. E., Pokusaev, B. C., and Shreiber, I. R.: *Wave Propagation in Gas-Liquid Media*, Ch. 3, 1993.
6. Noordzij, L.: Shock waves in bubble-liquid mixtures, in L. I. Sedov (ed.), *Non-Steady Flow of Water at High Speeds*, Proc. IUTAM Symp., Nauka, Moscow, 1973.
7. Noordzij, L. and Wijngaarden, L. van: Relaxation effects, caused by relative motion, on shock waves in gas-bubble/liquid mixtures, *J. Fluid Mech.* **66**(1) (1974), 115–143.
8. Watanabe, M. and Prosperetti, A.: Shock waves in dilute bubbly liquids, *J. Fluid Mech.* **274** (1994), 349–381.
9. Whitham, G. B.: *Linear and Nonlinear Waves*, Wiley, New York, 1974.
10. Whittaker, E. T. and Watson, G. N.: *A Course of Modern Analysis*, Cambridge University Press, Cambridge, 1952.
11. Wijngaarden, L. van: On the equations of motion for mixtures of liquid and gas bubbles, *J. Fluid Mech.* **33**(3) (1968), 465–474.
12. Wijngaarden, L. van: One-dimensional flow of liquids containing small gas bubbles, *Ann. Rev. Fluid Mech.* **4** (1972), 369–396.

MIX
Papier aus verantwortungsvollen Quellen
Paper from responsible sources
FSC® C105338

If you have any concerns about our products,
you can contact us on
ProductSafety@springernature.com

In case Publisher is established outside the EU,
the EU authorized representative is:
Springer Nature Customer Service Center GmbH
Europaplatz 3, 69115 Heidelberg, Germany

Printed by Libri Plureos GmbH
in Hamburg, Germany